# ELECTROCERAMIC-BASED MEMS

## *Fabrication-Technology and Applications*

THE KLUWER INTERNATIONAL SERIES IN

# ELECTRONIC MATERIALS: SCIENCE AND TECHNOLOGY

Series Editor

**HARRY L. TULLER**
***Massachusetts Institute of Technology***

**Other Books in Series**

**DIAMOND: Electronic Properties and Applications**
Lawrence S. Pan and Don R. Kania, Editors

**ION IMPLANTATION: Basics to Device Fabrication**
Emanuel Rimini, Author

**SOLID STATE BATTERIES: Materials Design and Optimization**
C. Julien and G. Nazri, Authors

**SOL-GEL OPTICS: Processing and Applications**
L.C. Klein, Editor

**PHOTOREFRACTIVE EFFECTS AND MATERIALS**
David Nolte, Editor

**MATERIALS FOR OPTOELECTRONICS**
Maurice Quillec, Editor

**PIEZOELECTRIC ACTUATORS AND ULTRASONIC MOTORS**
Ken Uchino, Author

**WIDE-GAP LUMINESCENT MATERIALS: Theory and Applications**
Stanley R. Rotman, Editor

**THIN FILM FERROELECTRIC MATERIALS AND DEVICES**
R. Ramesh, Editor

**MICRO-ACTUATORS: Electrical, Thermal, Optical, Magnetic, Mechanical and Acoustic**
Massood Tabib-Azar, Author

**HIGH-TEMPERATURE SUPERCONDUCTORS: Materials, Properties and Applications**
Rainer Wesche, Author

**NANOCRYSTALLINE METALS AND OXIDES: Selected Properties and Applications**
Philippe Knauth, Joop Schoonman

**NANOSTRUCTURED MATERIALS**
Philippe Knauth, Joop Schoonman

# ELECTROCERAMIC-BASED MEMS

## *Fabrication-Technology and Applications*

*edited by*

**Nava Setter**
*EPFL, Swiss Federal Institute of Technology*

Springer

Library of Congress Cataloging-in-Publication Data

Electroceramic-based MEMS : fabrication-technology and applications/edited by Nava Setter.
p. cm. – (Electronic materials: science and technology)
ISBN 0-387-23310-5 E-ISBN 0-387-23319-9
1. Microelectromechanical systems. I. Setter, N. (Nava), 1949–II. Series

TK7875.E42 2005
621.381—dc22

2004058921



Printed in the United States of America

9 8 7 6 5 4 3 2 1 SPIN 11327813

springeronline.com

# Contents


**Contributors** vii

**Preface** ix

**A. Applications and Devices**

**Chapter 1. MEMS-Based Thin Film and Resonant Chemical Sensors**
H. L. Tuller 3

**Chapter 2. Microactuators Based on Thin Films**
R. Maeda, J.J. Tsaur, S.H. Lee and M. Ichiki 19

**Chapter 3. Micromachined Ultrasonic Transducers and Acoustic Sensors Based on Piezoelectric Thin Films**
P. Muralt 37

**Chapter 4. Thick-Film Piezoelectric and Magnetostrictive Devices**
N.M. White, S.P. Beeby and N.J. Grabham 49

**Chapter 5. Micromachined Infrared Detectors Based on Pyroelectric Thin Films**
P. Muralt 81

**Chapter 6. RF Bulk Acoustic Wave Resonators and Filters**
H.P. Loebl, C. Metzmacher, R.F. Milsom, P. Lok and A. Tuinhout 115

**Chapter 7. High Frequency Tuneable Devices Based on Thin Ferroelectric Films**
S. Gevorgian 133

**Chapter 8. MEMS for Optical Functionality**
S.-G. Kim, G. Barbastathis and H. L. Tuller 157

**B. Materials, Fabrication-Technology, and Functionality**

**Chapter 9. Ceramic Thick Films for MEMS**
R.A. Dorey and R.W. Whatmore 177

**Chapter 10. Thin Film Piezoelectrics for MEMS**
S. Trolier-McKinstry and P. Muralt 199

**Chapter 11. Science and Technology of High Dielectric Constant Thin Films and Materials Integration for Application to High Frequency Devices**
O. Auciello, S. Saha, W. Fan, B. Kabius, S. K. Streiffer D. Y. Kaufman, J. Im and P. Baumann **217**

**Chapter 12. Permittivity, Tunability and Loss in Ferroelectrics for Reconfigurable High Frequency Electronics**
A.K. Tagantsev, V.O. Sherman, K.F. Astafiev, J. Venkatesh and N. Setter **235**

**Chapter 13. Microfabrication of Piezoelectric MEMS**
J. Baborowski **325**

**Chapter 14. Non-Conventional Micro- and Nanopatterning Techniques for Electroceramics**
M. Alexe, C. Harnagea, and D. Hesse **361**

**Chapter 15. Low-Cost Patterning of Ceramic Thin Films**
C. R. Martin and I. A. Aksay **387**

**Index** **411**

# Contributors

**I. Aksay**
University of Princeton, USA

**M. Alexe**
MPI-Halle, Germany

**K. F. Astafiev**
EPFL, Switzerland

**O. Auciello**
ANL, USA

**J. Baborowski**
EPFL, Switzerland

**G. Barbastathis**
MIT, USA

**P. Baumann**
ANL, USA

**S.P. Beeby**
University of Southampton, UK

**R.A. Dorey**
Crannfield University, UK

**S. Gevorgian**
Chalmers University & Ericcson, Sweden

**N.G. Grabham**
University of Southampton, UK

**C. Harnagea**
MPI-Halle, Germany

**D. Hesse**
MPI-Halle, Germany

**M. Ichiki**
NAIST, Japan

**J. Im**
ANL, USA

**B. Kabius**
ANL, USA

**D.Y. Kaufman**
ANL, USA

**S.-G. Kim**
MIT, USA

**S.H. Lee**
NAIST, Japan

**H.P. Loebl**
Philips, Germany

**P. Lok**
Philips, Germany

**R. Maeda**
NAIST, Japan

**C. R. Martin**
University of Princeton, USA

**C. Metzmacher**
Philips, Germany

**R.F. Milsom**
Philips, Germany

**P. Muralt**
EPFL, Switzerland

**S. Saha**
ANL, USA

**N. Setter**
EPFL, Switzerland

**V.O. Sherman**
EPFL, Switzerland

**S.K. Streiffer**
ANL, USA

**A.K. Tagantsev**
EPFL, Switzerland

**S. Trolier-McKinstry**
PSU, USA

**J.J. Tsaur**
NAIST, Japan

**A. Tuinhout**
Philips, Germany

**H.L. Tuller**
MIT, USA

**J. Venkatesh**
EPFL, Switzerland

**R.W. Whatmore**
Crannfield University, UK

**N.M. White**
University of Southampton, UK

# Preface

The concept of micromachining of silicon to form micron-scale structures such as cantilevers, free-standing bridges, membranes, and channels and its combination with microelectronics fabrication methodology and technology has resulted in the emergence of a new category of functional systems called MEMS (microelectomechanical systems). MEMS are miniature systems containing devices or arrays of devices that combine electronics with other components such as sensors, transducers and actuators, and are fabricated by IC (Integrated Circuits) batch processing techniques.

The field of electroceramics (inorganic, non-metallic materials, often polycrystalline, with useful electrical and other functional properties) provides a vast number of active materials for sensors, actuators, and electrical and electronic components. Electroceramic thin films can add therefore many useful functionalities to MEMS. At the same time, because the fabrication of ceramics is commonly a high temperature process that often necessitates an oxygen containing atmosphere, because most ceramics are inert and corrosion resistant, and because the properties of electroceramics are very sensitive to the processing conditions, the integration of electroceramic layers onto silicon or other substrates and their patterning into functional elements need complex technologies that are still under development. This situation is reflected in the current position of electroceramic-based MEMS in the market: While the potential is excellent, first devices are being commercialized only these present days.

A number of books and handbooks have been published recently on state of art MEMS technology and applications, but a book on Electroceramic-based MEMS is not yet available. Only few groups world-wide master already the latter subject to a deep extent. It appears that the MEMS community is short in the know-how of fabrication of electroceramic based devices, and the ceramics community has but a limited understanding of the potential applications and lacks the experience in mainstream MEMS technology. This book may be useful for the two communities and 'bridge the gap' between them. From this perspective, technology and potential applications are equally important and therefore both are equally emphasized in this book: one part of it covers applications and emerging applications and the second part deals with fabrication and functioning.

The book is focused on the use of functional oxide and nitride thin and thick films to extend functionality and application range of MEMS in the large sense, including micro-sensors, micro-actuators, micro-transducers, and electronic components for high frequency communication systems.

The first part of the book consists of reviews on several applications. It presents the diversity and potential of the field and at the same time features some of the recently commercialized and near-to-market applications:

- ***Chemical microsensors*** based on ceramic thin films and resonant structures are reviewed by Tuller emphasizing the advantages that MEMS bring toward improved selectivity of chemical sensors.
- Maeda, Tsaur, Lee and Ichiki introduce the area of ***microactuators***. This area, until now limited practically to electrostatic and capacitive devices may benefit from the high force output with low power dissipation of actuators based on piezoceramic thin films. A related topic, perhaps of a higher complexity is that of ***micromachined ultrasonic transducers*** [MUT] and acoustic sensors. Muralt reviews this topic and compares such transducers based on piezoelectric films [PMUT] to the better known capacitive type miniaturized ultrasonic transducers [CMUT].
- The use of thick films is advantageous in some applications due to functional reasons and in some cases due to lower cost of production. White, Beeby, and Grabham review the area of ***thick-film piezoelectric and magnetostrictive micro-devices***.
- Pyroelectrics constitute a subgroup of piezoelectric materials and are useful for innfrared [IR] detection which is used in a variety of uncooled temperature sensing devices such as fire alarm and IR imaging, and in various chemical sensing devices. Already commercialized pyroelectric thin films based MEMS include linear arrays for gas spectroscopy and two dimensional focal plane arrays for thermal imaging. Muralt, in a comprehensive review on ***pyroelectric microsystems*** covers materials issues, device physics, technology and state of art of applications.
- Another field that benefits already from electroceramic thin films is that of ***rf MEMS***, microsystems for high frequency mobile electronics: One of the first group of products of electroceramic-based microdevices that entered the market was that of thin film bulk acoustic wave devices based on piezoelectric films. They are used as filters and resonators for high frequency mobile communications allowing for the first time the integration of these components with the IC on one substrate. Loebl, Metzmacher, Milsom, Lok, and Tuinhout review this topic addressing both material and electrical performance aspects. Miniaturized tunable capacitors based on tunable ferroelectric thin films open interesting perspectives for new miniaturized and cost effective active rf components such as tunable phase-shifters. Gevorgian describes tunable ferroelectric film performance, compares tunable ferroelectrics technology with other phase agile microwave technologies, and discusses typical applications and system demonstrators.
- Finally, Kim, Barbastathis and Tuller discuss key features of MEMS technology that enable new functionalities of microphotonics, present an overview of various cases, and discuss materials related challenges and future trends for ***optical MEMS*** research and commercialization.

The second part of the book emphasizes materials, fabrication and patterning processes and addresses some functioning issues:

- An overview on standard deposition techniques of electroceramic thin films can be found in a number of recent books and is not included here.
- ***Ceramic thick films*** of 10–100 μm thickness are desired in certain MEMS e.g. for improved sensitivity as shown earlier (chapter 4) by White. Unlike thin films, thick films are usually prepared by powder techniques. Dorey and Whatemore review thick film deposition techniques: screen printing, electrophoretic deposition, hybrid

powder–sol gel technique, and direct writing. An emphasis is given to common issues in thick film fabrication: interfacial reactions, and constrained sintering.

- In contrast to bulk ceramics, ***ceramic thin films*** are always a part of a composite 'substrate-film' system: Trolier-McKinstry and Muralt center on this aspect in piezoelectric thin films. They show how the knowledge accumulated during the past 15 years in the control of growth processes in ferroelectric compositions and in polar materials such as AlN resulted in improved ***piezoelectric properties*** towards various MEMS applications.
- Auciello, Saha, Fan, Kabius, Streiffer, Kaufman, Im, and Baumann present technology and materials ***integration issues*** for tunable ferroelectric applications. In addition to aspects of composition-microstructure-property relationship, they discuss multi-layer structures (composites) which are of particular importance for tunable applications. They discuss also interfacial issues in relation with properties, and the important subject of ***electrodes***. High conductance of the electrodes is essential here and necessitates a special development. The question of the ultimate performance level is an important one particularly in thin films where it is more difficult to separate intrinsic and extrinsic (e.g. processing and integration related) effects: Tagantsev, Sherman, Astafiev, Venkatesh and Setter review the ***dielectric response*** of tunable ferroelectric films, and address bulk materials, thin films, ***nominally pure materials, real (defected) materials and composites***.
- ***Patterning*** is a crucial technology in any MEMS and since electroceramics are new comers, we felt the topic merits a special emphasis. Three reviews are given: Baborowski discusses the 'hard core' micromachining and patterning techniques, namely the adaptation of standard silicon and metalisation patterning methods to patterning of ceramic thin and thick layers on silicon substrates. Methods for stress compensation are of particular importance due to the higher processing temperatures of ceramic films in comparison to standard materials. Processing parameters in dry etching are significantly modified. Silicon-on-Insulator (SOI) or poly-SOI substrates can be used to improve the thickness uniformity on the wafer scale. Alexe, Harnagea, and Hesse focus on nano scale patterning methods. They discuss and compare various top down techniques: focused ion beam patterning, electron beam direct writing, and nano imprint, and comment on potential bottom up techniques which promise smaller lateral sizes but are far from being mature. Finally Martin and Aksay describe, discuss, and compare various low cost, high throughput soft lithography emerging techniques for micron and submicron scale patterning: these are alternative techniques to the IC standard lithography techniques and include techniques such as microtransfer molding, micromolding in capillaries, embossing, and microcontact printing.

In addition to the specific information regarding fabrication and applications of electroceramic MEMS, we hope the reader will find the book informative also in relation to the added value the MEMS approach brings to the area of electroceramic devices: the facilitated fabrication of arrays of sensors, actuators, and other components, the possibilities for direct integration of the functional elements with IC, the cost reduction due to batch processing, the possibility to have a number of different functions on one substrate using various ceramic thin film elements, and of course the improved sensitivity in many cases are some of the advantages that promise to make Electroceramic-based MEMS viable and highly useful.

The publication of this book follows a Special Issue on electroceramic-based MEMS that was published in the Journal of Electroceramics at the beginning of 2004. The ten invited papers of that special issue were adapted by the authors into chapters of the present book and five additional chapters were added. Many thanks to Harry Tuller, the Editor of the Journal, for the suggestion to publish this book. We thank all the authors for their invaluable contributions, in particular we are grateful for the authors who accepted to submit new manuscripts in a very short notice: Paul Muralt, Harry Tuller, Spartak Gevorgian, and Neil White. We thank the European Commission for the financial support of the Network of Excellence MIND—Multifunctional and Integrated Piezoelectric Devices which stimulated the publication of this book. The network MIND is a part of the 6th Framework Programme funded and administrated by European Commission, Directorate-General for Research.

Since the field of Electroceramic-based MEMS is an emerging and fast growing one, the book is certainly incomplete, and the choice of the topics to be included is bound to be unsatisfying for some of the readers. We felt it is better to have the book published quickly for the benefit of those interested in the subject, rather then to wait until the field is mature and the book is more subtle. I shall be grateful therefore for comments and suggestions from the readers, and will take them into account in an eventual second edition of the book.

Lausanne, 8.2004
Nava.setter@epfl.ch

NAVA SETTER
Editor

# Part A

## Applications and Devices

# Chapter 1

# MEMS-Based Thin Film and Resonant Chemical Sensors

H. L. Tuller
*Department of Materials Science and Engineering*
*Massachusetts Institute of Technology*
*Cambridge, MA, 02139 USA*
and
*Boston MicroSystems, Inc.*
*Woburn, MA, 01801 USA*

To satisfy demands for detecting chemicals in the environment and for real time monitoring of chemical processes, ever more sophisticated sensors, often able to operate under harsh conditions, are required to detect a rapidly growing range of chemical species. In this paper, we focus on progress being made to miniaturize and improve the performance of sensors by the integration of electroceramic functional films with MEMS technology offering reduced cost and ability to accommodate large arrays of devices on a single chip. In particular, we discuss recent progress directed towards the growing integration of electroceramics into largely Si-based MEMS structures and more recent efforts to fabricate ceramic MEMS platforms based on refractory materials such as SiC. Towards these ends, the discussion is subdivided with respect to the key types of MEMS sensor platforms used in the production of miniature chemical sensors.

First, micro-hotplates are discussed as a platform for chemoresistive, calorimetric and electrolytic-based sensors. The low thermal mass of these devices leads to markedly reduced power requirements and, in the case of chemoresistive sensors, to further improvements in selectivity by use of programmed heating cycles. A brief discussion is provided regarding the influence of nano-size effects on sensitivity upon use of electroceramic films. A description of recent advances in self-lithographic deposition of films and the microfabrication of SiC-based micro-hotplates is also provided. Microcalorimetric devices, which also benefit from reduced power requirements and increased sensitivity, are also described as are planar electrochemical sensor devices which benefit from solid state thin film reference electrodes.

Second, micro-cantiliver beams and bridges are discussed as platforms with exceptional sensitivity for detecting adsorption of chemical species either by measurement of induced stress or shifts in resonant frequency. These devices are evaluated in terms of ease of readout (e.g. optical vs piezoresistive), geometrical design to achieve a combination of high sensitivity and mechanical robustness and improved selectivity to specific chemical species.

Future work will focus on identifying improved means for integrating electroceramics with Si and/or MEMS fabrication processes, increased use of refractory materials for harsh environment applications and added levels of integration allowing for wireless or remote operation.

## 1. Introduction

Increasing worldwide concerns regarding environmental degradation and health hazards have stimulated a growing interest in means for detecting and monitoring potentially toxic chemicals. The ability to continuously monitor chemical processes and relay this information, via a feedback mechanism, to the process control system can also lead to improved productivity. To satisfy these demands, ever more sophisticated sensors, often able to operate under harsh conditions, are required to detect a rapidly growing range of chemical species. In this chapter, we focus on progress being made to miniaturize and improve the performance of sensors by the integration of electroceramic functional films with MEMS technology.

Chemical sensors must satisfy a number of criteria. Of greatest importance are the criteria represented by the 4 *S*'*s*; *S*ensitivity, *S*electivity, *S*peed and *S*tability. Perhaps the most difficult to satisfy with a single sensor is selectivity. The most effective schemes tend to revolve around the use of arrays of devices with elements or films of differing composition, microstructure and/or catalyst loading. This places additional demands on signal processing, controls and packaging. Furthermore, increasing numbers of applications require portable operation or remote sensing. The trend, in recent years, therefore, has been shifting towards miniaturized sensor systems in which the active sensor elements can be integrated with electronics and micro-electro-mechanical[i] (MEMS) components including membranes, microcantilevers, microheaters, etc.

At first glance, it might not be obvious how MEMS structures relate to chemical specie detection. To begin with, the very thin elastic structures[ii] are extremely sensitive to stress and readily bend or distort in response to tensile or compressive stresses induced by the adsorption of as little as a molecular monolayer as commonly observed in atomic force microscopy (AFM). Likewise, molecular adsorption is readily detected via a measurable shift in resonant frequency of very low mass structures. Furthermore, the ability to under-etch and create membrane-like structures (1–5 microns thick), leads to low thermal mass structures that can readily be heated to hundreds of degrees centigrade with the expenditure of little power. Small feature size (tens to hundreds of microns) also enables the fabrication of large arrays of devices on a single chip, essential for detecting multiple chemical and/or biological species simultaneously.

Ceramic sensors have long been of interest because of their unique transduction capabilities (e.g. electrochemical, semiconductive, piezoelectric, etc.), their low cost and their excellent stability in hostile environments. A prime example has been the zirconia-based potentiometric auto exhaust sensor, utilized in nearly all automobiles. This sensor, as part of a feedback system for controlling the air-to-fuel ratio, enables the operation of the three-way auto exhaust catalyst. This device is designed to operate satisfactorily for over 100,000 kilometers in atmospheres ranging from highly reducing to highly oxidizing conditions while simultaneously exposed to rapidly fluctuating

[i] The acronym MEMS, short for *Micro-Electro-Mechanical Systems*, now has a broader meaning and may include thermal, optical and chemical functions as well. An alternative term, *Microsystems*, is more commonly used in Europe to reflect this broadened meaning of MEMS.

[ii] e.g. micro-cantilever beams, as used in atomic force microscopy (AFM).

temperatures [1]. These sensors, however, are rather large and complex in design. Recent trends have therefore been directed towards integrating electroceramic or other functional ceramic elements into silicon based MEMS structures.

The thrust to integrate sensor materials, traditionally processed at elevated temperatures, with silicon technology has been aided by parallel efforts to integrate ferroelectric and dielectric oxides such as $PbZr_{1-x}Ti_xO_3$ (PZT) (for nonvolatile memories) and $HfO_2$ (high K dielectrics) into silicon-based MOSFET devices. While silicon itself is quite robust, as temperature progressively increases, it first becomes electronically intrinsic (~150–200 °C), begins to soften (~600 °C) and ultimately oxidizes (>800 °C). Wide band gap semiconductors such as SiC and $Al_{1-x}Ga_xN$, on the other hand, retain their extrinsic semiconductor behavior and maintain their elastic mechanical properties and resistance to oxidation to very high temperatures. In the following we discuss recent progress directed towards both of these trends, i.e., the growing integration of ***electroceramics*** into largely Si-based MEMS structures and more recent efforts to fabricate ***ceramic MEMS*** platforms, i.e. based on refractory materials such as SiC. In the following we describe the key types of MEMS sensor platforms used in the production of miniature chemical sensors.

## 2. MEMS-Based Sensor Platforms

Chemical species are most commonly detected by the manner in which they influence the electrical, mechanical, or thermal properties of materials or structures. In the following, we examine a number of MEMS structures utilized in the construction of chemical sensors and examine features of particular importance to their operation.

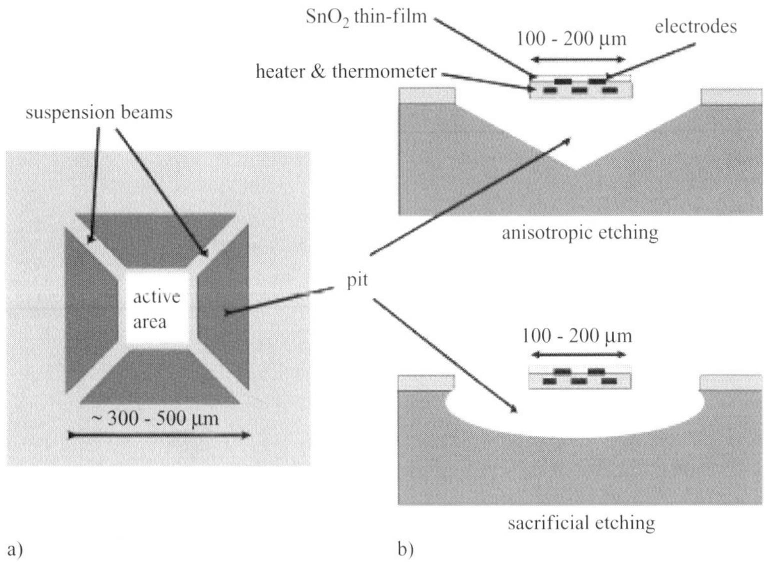


**Fig. 1.** Schematic of suspended membrane microhotplate gas sensor: a) top view; b) side view. From Simon et.al. [9].

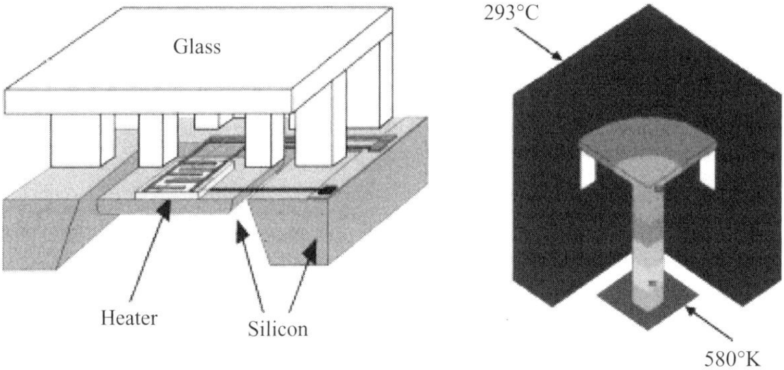


**Fig. 2.** Left: Cross section of hotplate gas sensor in which sensor membrane is mechanically supported by glass pillars while being thermally isolated from the remainder of the silicon chip. Right: Results of finite element analysis used to calculate the temperature distribution in one column in which the hotplate temperature is 580 K and the rim temperature is 293 K (room temperature). From Wöllenstein et. al. [2]

### 2.1. Microhotplate-based sensors: Chemoresistive, Calorimetric, Electrolytic

Nearly all chemical sensors, utilizing chemically active *inorganic* compounds (e.g. $SnO_2$, YSZ), operate at temperatures of approximately 200–800 °C to achieve adequate levels of conductivity and/or insure sufficiently rapid chemical kinetics. This imposes constraints on the use of such sensors consistent with the power requirements, size and cost of compatible heaters. The microhotplate, a self-heated membrane structure, provides a support with exceptionally small thermal mass. The application of microhotplates to gas sensing has received much recent attention [2, 3, 4] particularly by the group of Semancik et al at NIST [5, 6, 7, 8]. An extensive review of the design, fabrication and operation of microhotplate-based gas sensors was published in 2001 by Simon et. al. [9].

A schematic of a typical microhotplate, illustrated in Fig. 1, is commonly 10's–100's of microns on edge, 1–5 microns thick and less than a microgram in mass. This enables well designed microhotplates to operate to temperatures of the order of 400–500 °C on only 50–200 mW of power, rather than approximately 1–2 W required by bulk ceramic devices [9]. Alternative designs use non-perforated membranes (i.e. without tethers) made of thermally insulating materials such as $SiO_2$ or $Si_3N_4$. Generally, power dissipation is minimized by use of low thermal conductivity membranes, reduced heater area, and suspension beams with high length-to-width ratios [9]. Alternatively, the microhotplates may be isolated from the remainder of the chip and supported by thermally insulating (e.g. glass) columns [2], as illustrated in Fig. 2. The low thermal mass of the microhotplate also leads to the ability to rapidly cycle the temperature of the device. Indeed, such devices are reported to exhibit millisecond thermal rise and fall times between room temperature and 400–500 °C [6, 7]. This provides several advantages including, the ability to reduce power consumption further by operating, only periodically, at the designated temperature and by enabling programmed cycling, leading to kinetic differentiation of chemical species by their

**Fig. 3.** Infrared image of a SiC microhotplate (~100 μm on edge) heated to 1070 °C, with molten aluminum (top) and molten gold (bottom) on its surface. (Courtesy: Boston MicroSystems, Inc. Woburn, MA; www.bostonmicrosystems.com).

unique temperature-response signature [6, 7]. In addition to built-in heaters, microhotplates commonly include (1) temperature sensors for monitoring operating temperature and/or reaction induced temperature changes (microcalorimeters) and (2) read-out electrodes for supported chemoresitive or potentiometric sensors as illustrated in Fig 1.

Many gas sensor applications require operation in harsh environments including, for example, in automotive exhausts. Microhotplate sensors fabricated with Si-based IC technology, however, may deteriorate under such harsh environments. Therefore, alternate materials, such as wide band gap semiconductors or insulators, are of interest to serve as the microhotplate platform. Recent progress in the application of photo-electrochemical means for micromachining such refractory materials has enabled such robust MEMS structures to be fabricated [10]. Fig 3. shows recent progress made in the micromachining of single crystal SiC microhotplates capable of operating to temperatures above 1000 °C and in chemically harsh environments.

### 2.1.1. Chemoresistive

Perhaps the most popular means of detecting gas species such as CO, $H_2$ and $NO_2$ are based on chemoresistive principles. Semiconducting oxides, when exposed to different gases, exhibit changes in their resistance or equivalently conductance and are thus referred to as chemoresistive or conductometric sensors. At temperatures below about 400–500 °C, gas molecules adsorb and desorb from the surface of semiconductors without modifying their overall stoichiometry. The adsorbed molecules form surface states which may trap free carriers. This, in turn, induces the formation of depletion layers and space charge barriers at surfaces and grain boundaries [11, 12] as illustrated in Fig. 4. In

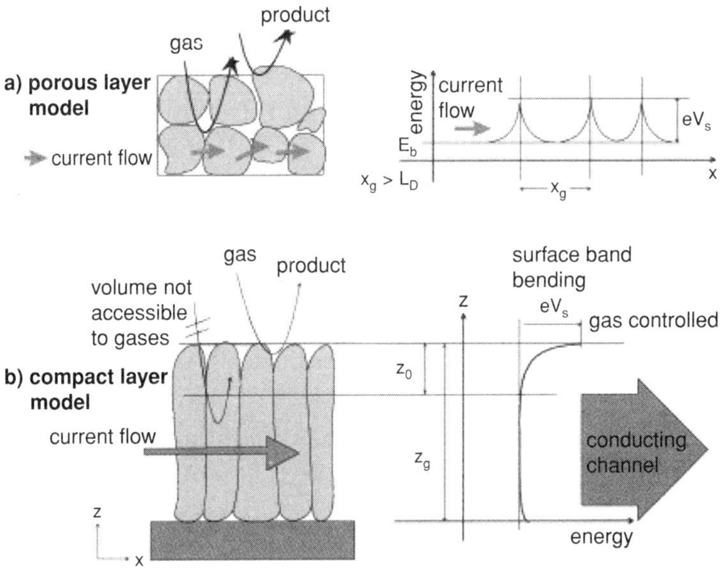


**Fig. 4.** Schematic representation of (a) porous and (b) compact sensing layer with geometry and energy band. $x_g$ grain size, $L_D$ Debye length, $eV_s$ band bending, $z_0$ thickness of the depleted surface layer and $z_g$ layer thickness. From Barsan and Weimar [11].

n-type semiconductors (e.g. $SnO_2$), adsorbed oxygen increases the barrier height while reducing gases decrease it by reacting with the adsorbed oxygen and donating electrons back to the semiconductor. Traditionally, bulk or thick film ceramic sensors have relied on gas modulated potential barriers at the grain boundaries of porous materials for achieving sensor response as illustrated in Fig. 4a. Thin films, commonly prepared by physical or chemical vapor deposition methods such as sputtering or CVD, tend to form as relatively dense films. Consequently, only the top surface is accessible to the gas molecules, as illustrated in Fig. 4b. As is obvious from Fig. 4b, the sensitivity of the sensor will depend on the thickness of the film, since a fraction of the cross section ($z_g - z_0$) of the film, the so-called conducting channel, is unaffected by the gas adsorption. Indeed, as expected, as the films become thinner, they tend to exhibit higher sensitivities [11, 13]. While the majority of work reported on microhotplate sensors utilizes thin film oxides, because porous layers exhibit larger gas active surface areas and reportedly greater immunity to poisoning, there has been growing recent interest in finding means for integrating thick films with micromachined sensor structures [8, 9, 14].

While such sensors exhibit high sensitivity, this detection mechanism, by itself, does not distinguish between different reducing gases. Improved selectivity is achieved by selective addition of catalysts, by morphology differentiation, and/or by temperature modulation. Arrays of different oxides, each with different relative sensitivity to a number of gases of interest, have demonstrated the ability to selectively detect and monitor a number of gases [9]. It should be noted that microhotplates enable *self-lithographic* growth of thin films by MOCVD [15]. Only those hotplates, heated during the flow of organo-metallic precursors over them retain the oxide film due to local pyrolysis of the organic film. This enables one to deposit different oxides on selected microhotplates, within an array, without need for successive levels of masking, thereby substantially simplifying device processing.

### 2.1.2. Calorimetric

Detection of combustible gases can be achieved by detecting the heat generated during the reaction of a combustible gas with oxygen in the atmosphere. Typically, such devices, called pellistors, consist of a small platinum coil embedded in a noble metal catalyst impregnated ceramic bead. The heat generated during combustion is detected as a temperature rise in the ceramic element [16]. This type of sensor suffers from relatively poor sensitivity, slow response time and high power dissipation, given their relatively large thermal mass. Further, ambient temperature fluctuations need to be compensated for with a catalytically inactive reference element, often difficult to match to the active pellistor element. A microhotplate structure, on the other hand, offers the opportunity of lower power consumption, faster response time and well matched reference based on low thermal mass elements fabricated by MEMS fabrication processes.

Zanini et. al. [17] were one of the first to utilize MEMS structures for this application. They fabricated a matched pair of neighboring under-etched $Si_3N_4$ membranes on a silicon wafer, each with a serpentine Pt resistive element to monitor temperature. After the Pt elements were isolated from the atmosphere with another $Si_3N_4$ layer, only one of the membranes was coated with a Pt or Pd catalytic layer. Detection limits on the order of 10 ppm of propylene were measured while thermal response times of 40 ms were recorded.

Cavicchi et al [18] recently reported a micron-scale differential scanning calorimeter (DSC). This consisted of a suspended microhotplate with sample and reference zones at either end, each with a polysilicon microheater for temperature control, and with thermopiles serving to measure the temperature difference between the two zones. A schematic of the device is shown in Fig. 5. A Pd coated device, operating at fixed temperature of 500 °C for varying concentrations of ethanol in air, showed a near linear response from 3–100 ppm. While this mode was useful for level detection, it provided no chemical identification. Using temperature-scanned operation (e.g. periodic ramps from 170 to 570 °C of duration 3.5 s), on other hand, the device responded with periodic waveforms that reflected the kinetics of adsorption, desorption, and reaction of the various gases over the catalyst in air. With different gases exhibiting different response patterns (comparisons were made for methanol, ethanol, hydrogen, benzene and acetone), the use of pattern recognition analytical methods for gas identification became possible.

### 2.1.3. Electrolytic

While a number of potentiometric and amperometric electrochemical principles are utilized in the development of gas sensors, we focus on the more fundamental potentiometric approach. Here, an emf is induced across a solid electrolyte due to the imposition of a chemical activity gradient across the electrolyte. For example, in the automotive exhaust monitor mentioned above, $Zr_{1-x}Y_xO_{2-x/2}$, a solid oxide electrolyte,

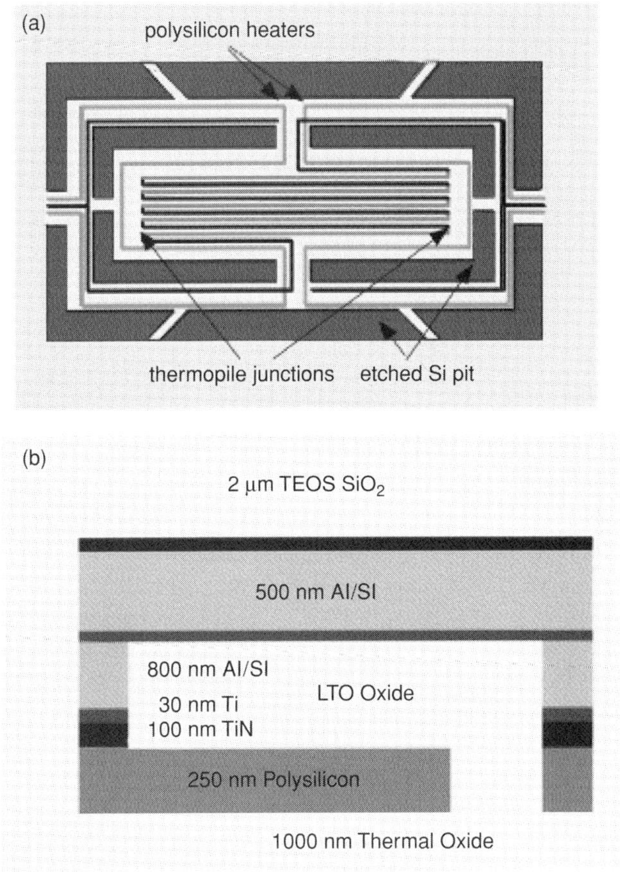


**Fig. 5.** Micro DSC (a) Top view showing sample and reference heaters and five thermopile junctions on each side. b) Side view indicating the layers comprising the thermopile junction. From Cavicchi et. al. [18].

which conducts solely by the motion of oxygen ions (via a vacancy mechanism), is utilized. An emf is induced across the electrolyte tube when the interior of the tube is exposed to an air reference ($P_{O_2} = 0.2$ atm) while the exterior is exposed to the exhaust gas with its $P_{O_2}$ typically fluctuating between $10^{-15}$ and $10^{-2}$ atm [1]. The emf, given by

$$E = (kT/4q)\ln\left(P_{O_2}/P_{O_2}(ref)\right) \tag{2.1}$$

is only weakly temperature dependent, in contrast to semiconductor based sensors.

In a MEMS-based structure, a solid state reference is preferred, thereby negating the need for an isolated gas reference. This is the case for $CO_2$-based electrochemical sensors first proposed by Holzinger et al [19] with cells of the form

$$\text{Au}, O_2, CO_2, Na_2CO_3/\text{Na-}\beta\text{''-}Al_2O_3/Na_2Ti_6O_{13}, Na_2Ti_3O_7, O_2, \text{Au} \tag{2.2}$$

in which the Na ion conductor Na-β"-$Al_2O_3$ is used as the electrolyte to monitor the change in Na activity induced in $Na_2CO_3$ due to changes in $P_{CO_2}$ at the left electrode. The Na activity on the right side of the cell is fixed by the two phase mixture of the sodium titanates. See Ref. 19 for a detailed discussion of the principles of operation.

Microhotplate supported electrochemical-based $CO_2$ sensors have recently been reported which exhibit the expected dependence of emf on $P_{CO_2}$. Bang et. al. utilized a cell of the form [20]

$$\begin{aligned}&\text{Pt, O}_2\text{, CO}_2\text{, carbonate (Na}_2\text{CO}_3\text{:BaCO}_3\\&\quad = 1{:}1.7\text{ mol)/Na ion conductor (NASICON)/Pt}\end{aligned} \tag{2.3}$$

and were able to monitor $CO_2$ concentration from 1000 to 10,000 ppm at an operating temperature of 400 °C while requiring only 52 mW of power. Surprisingly, the cell seemed to operate effectively even without a well defined reference electrode.

## 2.2. Microcantilever beam-based sensors

Scanning probe microscopy has highlighted the use of MEMS technology based on microcantilever beams (MCBs) for achieving exceptionally high sensitivity in the measurement of forces at surfaces on the atomic scale. MCBs with spring constants of 0.01 ⊥ 100 N/m, coupled with vertical displacement resolution of picometers, enables detection of forces down in the pico to nano newton range. Microcantilever beam platforms have demonstrated remarkable sensitivities to thermal, optical, magnetic and chemical stimuli [21], often at sensitivities orders of magnitude greater than those achieved with competing technologies. For example, gravimetric sensitivities determined for silicon nitride and polycrystalline silicon cantilever structures were found to surpass traditional SAW and QCM devices by 2–3 orders of magnitude [22]. Microcantilever structures have been used to measure chemical and biological species at ppb-ppt levels [23, 24] by detecting changes in cantilever static deflection or resonant response. In addition, MCB platforms provide smaller size, lower power consumption, and negligible interference between elements in a multi-element sensor array.

For the most part, detection of MCB deflection has been done optically as in scanning probe microscopes, e.g. atomic force microscopes (AFM). While this approach is appropriate for a laboratory instrument, it is not practical for small and/or portable devices. Optical readout requires precise alignment of the cantilevers with lasers and photodetectors resulting in large, shock sensitive systems requiring high power with associated elevated costs. Alternatively, one can detect deflection piezoresistively. While piezoresistive readout might be an alternative, it suffers from a tradeoff in sensitivity versus manufacturability, since sensitivity requires large area and inherently fragile long, thin (compliant) cantilevers.

Microresonators (cantilevers, bridges, etc.) (see Fig. 6) are highly sensitive structures integrated with piezoelectric drive/sense films and analyte-specific polymer coatings that alter the mass and/or stress of the microresonator when exposed to trace levels

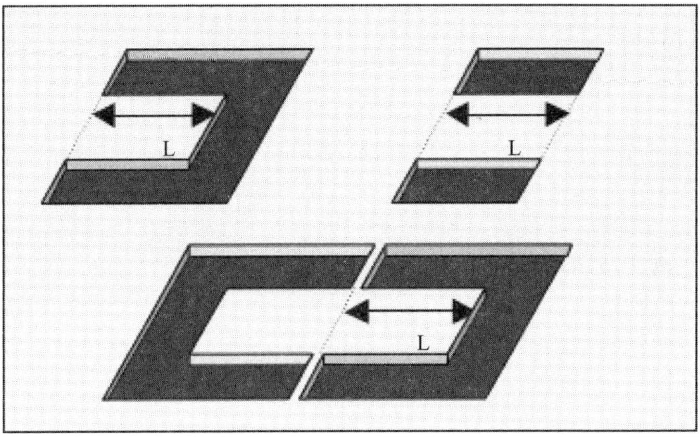


**Fig. 6.** Several resonator geometries: simple cantilever at top left, bridge at top right, and "butterfly" at bottom. From Mlcak et. al. []

of analyte. The resonant frequency of these structures is given by

$$f = (C_1 T/2\pi L^2)\sqrt{E/3\rho_c} \tag{2.4}$$

where the $C_1$ coefficient is 1.758 for cantilevers and "butterflies" or 11.19 for bridges, L is the resonator length, E is Young's modulus, $\rho_c$ is the density of the cantilever, and T is the beam thickness [25]. Piezoelectric microresonators rely on the ability to detect extremely small shifts in resonance frequency. The sensitivity of such devices is given by

$$S = {}^{\alpha f}\!/_{2\rho_c T} \propto {}^{\alpha}\!/_{L^2}\sqrt{E/\rho_c^3} \tag{2.5}$$

where $\alpha$ is the coefficient of adsorption for the chemical onto the absorptive film. One notes that the sensitivity is a function of length, with an inverse squared dependence. Consequently, in contrast to piezoresistive MCBs, microresonator's exhibit enhanced sensitivity with short and thick (stiff) cantilevers [25].

Limitations in current materials processing technologies include, inadequate dimensional, compositional and microstructural control of both the beams and the actuator materials. Piezoelectric films such as PZT, ZnO or AlN, when deposited by sputtering, sol-gel processing, PVD, etc. onto micromachined structures, remain difficult to fabricate reproducibly. This results, in part, from the use of polycrystalline films and etching processes which suffer from inadequate control and reproducibility. Residual stresses have also been found to be difficult to eliminate, leading to additional irreproducibility in resonant characteristics [26]. In contrast, epitaxial films on single crystal micromechanical elements, can be expected to provide more highly reproducible and stable devices.

Much progress has been achieved in recent years in the growth of high quality wide band gap nitrides for blue-green LEDs and lasers. These nitrides, in the AlGaN system, which also happen to be piezoelectric, are unfortunately not compatible with

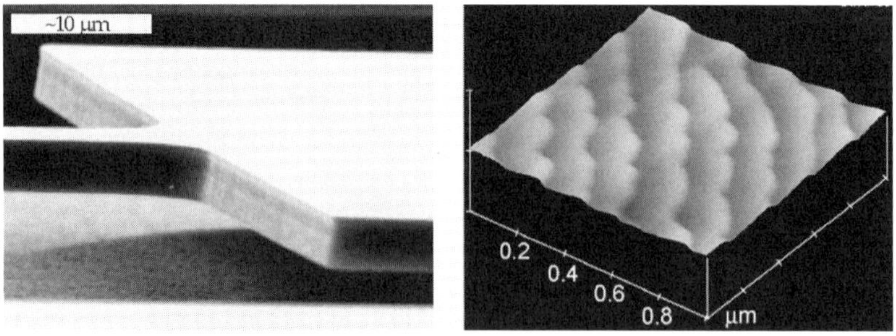


**Fig. 7.** Left: SEM micrographs of single crystal AlN epitaxial film grown by MBE onto a bulk micromachined single crystal SiC structure. Right: AFM micrograph of the AlN surface indicating a step structure replicated from the off-cut SiC surface. From Doppalapudi et. al. [27].

silicon due to high thermal expansion and lattice mismatches. SiC is currently the most suitable substrate (in lattice and thermal expansion match) for epitaxial growth of III- nitrides and, with recent advances in SiC single crystal micromachining technologies, MEMS structures with compatible piezoelectric films have become possible. Figure 7 shows the surface morphology of AlN films grown onto micromachined SiC structures [27]. Such films replicate the steps resulting from the 6-H SiC substrate cut 3,5° off the (0001) plane, and are extremely smooth with RMS roughness less than 1.5 nm (Figure 7b). X-ray diffraction studies showed that the AlN films have excellent crystalline quality and minimal residual stress [27]. Preliminary measurements with SiC-AlN based microresonators, sensitized to a nerve gas simulant, dimethylmethylphosphonate (DMMP), have exhibited sensitivities down to the 100 ppb level [28]. As with all resonant devices, the challenge remains in how to eliminate or minimize cross sensitivity to temperature, humidity and other chemical species. Reference elements, included within an array, provide means for effectively compensating for cross sensitivities.

## 3. Summary and Conclusions

MEMS-based structures provide ideal gas sensor platforms. Their very small size enables the creation of large arrays of devices on a single chip. This allows for the detection of many species in parallel and/or improved selectivity by use of pattern recognition analytical methods. Likewise, the ability to fabricate many devices in parallel within a batch process also insures cost savings.

The corresponding low thermal mass of e.g. micro hot plate structures leads to markedly reduced power requirements, key to portable operation. Low thermal mass also allows for programmed heating cycles, which can be used to achieve further improvements in selectivity and power reduction.

The very thin mechanical structures also exhibit exceptionally small spring constants and gravimetric masses providing for exceptional sensitivity to stress upon bending of microcantiliver beams and to the adsorption/absorption of chemical species onto the elements when excited into resonance. Sensitivities in the ppb levels and below and selectivity to explosives and toxic chemicals are achieved by the application of select polymeric layers to the MCBs.

Electroceramic materials play a key role in this sensor technology as (2.1) the sensitive semiconductive metal oxide films (e.g. $SnO_2$) in micro hotplate sensors, and/or (2) the actuator layers (e.g. PZT, AlN) in microresonators. This requires that alternate fabrication processes, compatible with Si and/or MEMS fabrication processes, be developed and refined. These generally lower temperature processes must be optimized to insure that corresponding degradation in properties of the electroceramic materials does not occur. Much additional work is needed to insure the compatibility of an ever greater number of materials being integrated into MEMS-based systems.

With the growing need for MEMS sensor devices to operate under harsh environments, one is likely to see an increased use of refractory materials serving as structural as well as functional elements (e.g. SiC). Recent breakthroughs in the micromachining of refractory materials will insure continued growth in this direction.

Finally, additional levels of integration can be expected which allow for wireless or remote operation. This includes the integration of micro-power sources such as batteries, fuel cells [29], solar cells, etc. as well as communications capabilities.

## 4. Acknowledgements

The author acknowledges support from the National Science Foundation for work on sensors under grant DMR-0346434 and to the MURI program for support of work on MEMS-based micro fuel cells under grant DAAD19-01-1-0566.

The author wishes further to acknowledge his association with Boston MicroSystems, Inc. as co-founder and consultant and thanks his colleagues at Boston MicroSystems for their ingenuity and camaraderie.

## References

[1] J. Riegel, H.-M. Wiedenmann, and H. Neumann, Exhaust gas sensors for automotive emission control, Solid State Ionics **152–153**, 783 (2002).

[2] J. Wöllenstein, J.A. Plaza, C. Cane', Y. Min, H. Böttner, H.L. Tuller, A novel single chip thin film metal oxide array, Sens. and Actuators B **93,** 350–355 (2003).

[3] S. Möller, J. Lin and E. Obermeier, Material and design considerations for low power microheater modules for gas-sensor applications. Sens. Actuators B **24/25**, 343–346 (1995).

[4] V. Guidi, G. Cardinali, L. Don, G. Faglia, M. Ferroni, G. Martinelli, P. Nelli and G. Sberveglieri, Thin-film gas sensor implemented on a low-power consumption micromachined silicon structure. Sens. Actuators B **49**, 88–92 (1998).

[5] R. Cavicchi, J. Suehle, P. Chaparala, K. Kreider, M. Gaitan, S. Semancik, Micro-hotplate gas sensor, in: Proceedings of the 1994 Solid State Sensor and Actuator Workshop, Hilton Head, SC, USA, 1994, pp. 53–56.

[6] T. Kunt, T. McAvoy, R. Cavicchi and S. Semancik, Optimization of temperature programmed sensing for gas identification using micro-hotplate sensors. Sens. Actuators B **53** (1998), pp. 24–43.

[7] R. Cavicchi, J. Suehle, P. Chaparala, G. Poirier, K. Kreider, M. Gaitan and P. Chaparala, Optimized temperature-pulse sequences for the enhancement of chemically specific response patterns from micro-hotplate gas sensors. Sens. Actuators *B* **33** (1996), pp. 142–146.

[8] R.E. Cavicchi; S. Semancik; F. DiMeo Jr.; C.J. Taylor, Use of microhotplates in the controlled growth and characterization of metal oxides for chemical sensing, J. Electroceram. **9**, 155–164 (2002).

[9] I. Simon, N. Bârsan, M. Bauer and U. Weimar, Micromachined metal oxide gas sensors: opportunities to improve sensor performance, Sens. Actuators B **73**, 1–26 (2001).

[10] H.L. Tuller and R. Mlcak, "Advanced Sensor Technology Based on Oxide Thin Films – MEMS Integration", J. Electroceramics, **4**, 415–425 (2000).

[11] N. Barsan and U. Weimar, Conduction Model of Metal Oxide Gas Sensors, J. Electroceramics,
**7**, 143–167, 2001.

[12] C.C. Wang, S.A. Akbar and M.J. Madou, Ceramic Based Resistive Sensors, J. Electroceram. **2**, 273-282, (1998).

[13] H.L. Tuller, "Solid State Electrochemical Systems: Opportunities for Nanofabricated or Nanostructured Materials", J. Electroceramics. **1**, 211–218 (1997).

[14] M. Heule,U. P. Schönholzer and L. J. Gauckler, Patterning colloidal suspensions by selective wetting of microcontact-printed surfaces, J. European Ceram. Soc. **24**, 2733–2739 (2004).

[15] J. S. Suehle, R.E. Cavicchi, M. Gaitan and S. Semancik, Tin Oxide Gas Sensor fabricated using using CMOS micro-hotplates and in situ processing, IEEE Electron. Device Lett. **14**, 118–120 (1993).

[16] E. Jones, The Pellistor Catalytic Gas Sensor, in P.T. Moseley and B.C. Tofield (eds.), Solid State Gas Sensors, A. Hilger, Bristol, UK, 1987, pp. 17–31.

[17] M. Zanini, J.H. Visser, L. Rimai, R.E. Soltis, A. Kovalchuk, D.W. Hoffman, E.M. Logothetis, U. Bonne, L. Brewer, O.W. Bynum and M.A. Richard, Fabrication and properties of a Si-based high-sensitivity microcalorimetric gas sensor. *Sens. Actuators A:* **48** 3 (1995).

[18] R. E. Cavicchi, G. E. Poirier, N. H. Tea, M. Afridi, D. Berning, A. Hefner, J. Suehle, M. Gaitan, S. Semancik and C. Montgomery, Micro-differential scanning calorimeter for combustible gas sensing, *Sens. Actuators* ***B*** 97, 22–30 (2004).

[19] M. Holzinger, J. Maier, W. Sitte, Fast CO-selective potentiometric sensor with open reference electrode, Solid State Ionics 86–88, 1055–1062 (1996).

[20] Y.-I. Banga, K.-D. Songa, B.-S. Joo, J.-S. Huhb, S.-D. Choi , D.-D. Lee, "Thin film micro carbon dioxide sensor using MEMS process", Sensors and Actuators B, in press.

[21] H.L. Tuller, R. Mlcak, "Inorganic sensors utilizing MEMS and microelectronic technologies", Current opinion in solid state & materials science, **3**, 501 (1998).

[22] P. I. Oden, Gravimetric sensing of metallic deposits using an end-loaded microfabricated beam structure, Sens. Actuators **53**, 191–196 (1998).

[23] T. Thundat, G.Y. Chen, R.J. Warmack, D.P. Allison, E.A. Wachter, Vapor detection using resonating microcantilevers, Anal. Chem. **67**, 519–521 (1995).

[24] L.A. Pinnaduwage, T. Thundat, J.E. Hawk, D.L. Hedden, P.F. Britt, E.J. Houser, S. Stepnowski, R.A. McGill, D. Bubb, Detection of 2,4-dinitrotoluene using microcantilever sensors, Sensors and Actuators **B 99** 223–229 (2004).

[25] R. Mlcak, D. Doppalapudi, J. Chan, A. Sampath, T.D. Moustakas, H.L. Tuller, "Micromachined SiC-AlN Bulk Resonator Array Sensor Platform for Ultra Sensitive Explosives Detection", *The 3rd Intl' Aviation Security Technology Symposium*, Nov. 27–30 (2001).

[26] J. Baborowski, Microfabrication of Piezoelectric MEMS, J. Electroceram. **12**, 33–51 (2004).

[27] D. Doppalapudi, R. Mlcak, J. Chan, H. Tuller, J. Abell, W. Li and T. Moustakas, Sensors based on SiC-AlN MEMS, in Proc. Nitride and Wide Bandgap Semiconductors for Sensors, Photonics, and Electronics V, The Electrochemical Society, Honolulu, Hawaii October 3–8, 2004, to be publ.

[28] R. Mlcak, Boston MicroSystems, Inc., private communication.

[29] C.D. Baertsch, K.F. Jensen, J.L. Hertz, H.L. Tuller, S.T. Vengallatore, S. M.Spearing, and M. Schmidt, "Fabrication and Structural Characterization of Self-Supporting Electrolyte Membranes for a Micro Solid-Oxide Fuel Cell", J. Mat. Sc., in press.

# Chapter 2

## Microactuators Based on Thin Films

R. Maeda, J.J. Tsaur, S.H. Lee and M. Ichiki
*National Institute of Advanced Industrial Science and Technology*
*1-2-1 Namiki, Tsukuba 305-8564, Japan*

MEMS R&D originated from the successes of microactuator device fabrication by Si semiconductor micromachining technology. Although this technology is suitable for fabricating microstructures, the sensing and actuation capability employed is limited only to electrostatic and capacitive devices, which results in the limited functions of the devices. In particular, high force output with low power dissipation cannot be achieved by electrostatic actuation. The integration of piezoelectric materials for MEMS is thus highly encouraged to realize high force output as well as sensing capability using both piezoelectric and inverse piezoelectric effects. This integration then results in simplification of the microstructures. Promising applications of piezoactuators and the difficulties of integrating exotic piezoelectric materials in conventional micromachining processes are discussed in this paper.

# 1. Introduction

Ferroelectric thin film is a target of interest in the fabrication of microelectromechanical systems, memory devices, etc. Piezo-MEMS will be developed with such materials as thin films. In spite of the successful commercialization of FRAM applications, MEMS applications are still being developed. Although there are promising properties, there have been rather few successes in the use of such layers as MEMS devices. Problems in such applications are mainly caused by the very small dimensions required of such devices, in contrast to the limitations and tolerances of the currently used microtechnologies. Although the deposition of piezoelectric PZT-layers using different thin-film technologies has reached an advanced state, layers of more than 1 micron in thickness needed for MEMS applications are still difficult to deposit. When dealing with these problems, researchers must take into account that the structure or function of the devices may be affected by technologically caused constraints, like thickness deviation and deformation by residual stress, and they must also seek to optimize materials processing parameters.

Very much attention has been given to integrating ferroelectric thin films in MEMS for various applications, such as pumps, valves, ink-jet printer heads, ultrasonic fluid drives, and micromirrors, with a focus on changing the length because they have good properties capable of use as driving sources and actuators for microsystems [1–5]. The chief reason for this interest is that we can expect larger force output than that available from electrostatic actuators. However, because of the small strain (usually less than 10%) and high stresses (several MPa) in piezoelectric actuators, their use is suitable for applications with large forces but small displacements [6–7].

This chapter introduces promising applications of piezoactuators. These applications are classified by their shape, cantilever, membrane and bulk types.

# 2. Cantilever Type Actuators

## 2.1. Micro scanning mirrors

Micromirrors derived from microfabrication technology have received a great deal of attention due to their wide applications in optical information systems ranging from barcode readers and laser positioning components to image scanning projection. Additionally, in fiber optical communication systems including DWDM, micromirrors with precisely controlled block-and-pass positioning are used for the shuttering cells in optical switching arrays.

Most micromirror designs consist of one or more cantilevers that support the mirror plate with a hinge/bar connection. Micromachined actuators, derived from electromagnetic, electrostatic, electrothermal or piezoelectric drivers, are integrated with cantilevers to excite bending vibration and twisting rotation of a micromirror with 1D or 2D motion. In recent years, the electrostatically driven mirror has been the mainstream type using electrical comb actuators or a parallel planar electrode scheme. However, high driving voltage of up to a hundred volts is necessary for the large deflection angle requirement. In addition, the electrostatic pull-in on a planar electrode also results in a nonlinearity characteristic relative to the tilting angle. In cases of electrothermal-actuated

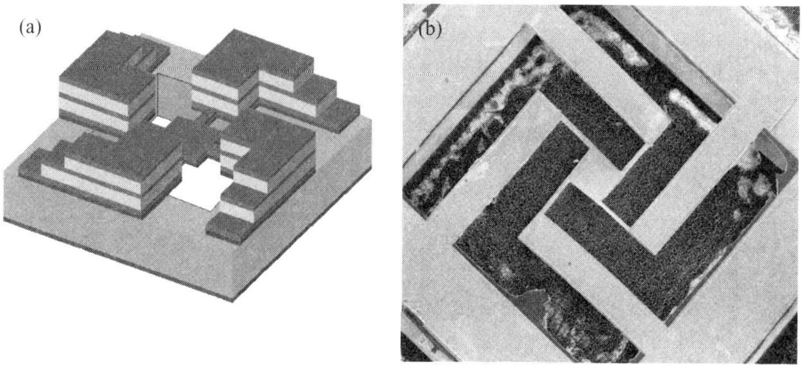


**Fig. 1.** Schematic draw of optical scanning device with double layered PZT layer (a) and the fabricated device (b), Mirror plate: $300 \times 300\ \mu m^2$, DPZT beam: $800 \times 230\ \mu m^2$.

micromirrors, the slow response cannot satisfy the need of wider bandwidth operations. In contrast, the piezoelectric-actuated mirror can provide relatively lower operation voltage, very low power consumption, and better linearity of deflection. The progress of thin film piezoelectric technology made significant contributions toward the reduction of mirror dimensions which resulted in higher resonant frequencies and the improved piezoelectric characteristics for the necessary driving forces.

In 1966, Fowler et al. developed a piezoelectric optical beam scanner using thick lead zirconate titanate (PZT) layers and coated conductive mediums as electrodes [8]. In 1976, Frank et al. reported on a bimorph beam scanner composed of two PZT layers and introduced a pivot design for magnifying angular excursion with one degree of freedom (1D) [9]. To further reduce mirror dimensions and perform motion with two degrees of freedom (2D), in 1991 Goto et al. proposed a silicon micromachined mirror for which a stacked PZT resonator vibrated a torsional silicon spring to perform bending or twisting motions with 2D scanning [10]. With the objective of achieving a large scanning angle and vibration robustness, Ohtsuka et al. developed a four PZT bimorph-cell-actuated optical scanner [11] and Yee et al. proposed a four PZT unimorph-actuated micromirror [11]. Both of them adopted a silicon-based micromirror supported by symmetrically arranged inner and outer frames to perform 2D torsional motion. Regarding PZT materials, since bulk and thick film PZT are not suitable for miniature devices, thin film technology was widely used to prepare the submicron-thick PZT layer between the upper and lower platinum electrodes. In 1999, Schroth and Maeda reported on a sol-gel-deposited PZT film for a 1D and 2D optical scanner [12]. To eliminate the residual strain generated from annealing in the sol-gel process, a bimorph-like double-layered PZT with bi-directional actuation was used for a 2D microscanner as illustrated in Fig. 1 [13].

## 2.2. Piezoelectric AFM

Atomic force microscopes (AFM) are well known as a powerful tool for imaging pictures of atoms or observing surfaces on nanometer scale by scanning a fine ceramic or semiconductor tip over a surface. Typically, the tip is positioned at the end of a silicon-based cantilever beam. As the raster-scan drags the tip over the sample, the cantilever beam deflects. The magnitude of the deflection is captured by a laser that reflects at an oblique angle from the very end of the cantilever, and this indicates the local sample height. An AFM can work in contact mode, which measures the contact repulsive force with the tip touching the sample, or in noncontact/tapping mode, in which the tip can tap across the surface without tip destruction and adhesion problems. In order to increase the sensitivity of force gradients, an AFM cantilever requires a low spring constant which is achieved by scaling down the cantilever size or hollowing it out as a V-shaped structure. To achieve higher scan speed, an AFM is designed with high resonant frequency and integrated with a piezoelectric actuator. Moreover, piezoelectric detectors or piezoresistors are generally built into a silicon cantilever, which is called a self-sensitive probe, in order to improve the complexity of the data readout using an optical system.

The first AFM conceived by Binnig et al. in 1986 [14] stimulated the beginnings of an interest in developing the performance to implement AFM technology in industry such as by integrating it with semiconductor fabrication, refining scanning tip lithography techniques, use in high-density data storage devices and so on.

In 1989, Alexander et al. integrated an irradiation system and a position-sensitive photodetector as an optical lever system to greatly magnify the motion of the tip [15]. In noncontact AFM, the cantilever is vibrated near its resonant frequency and the tip is set at an extremely close distance from the sample. The vibration amplitude changes with the shift of the cantilever, which is modified by force gradients between the tip and the sample. This was used for measuring the vibration amplitude of the lever, and is called a heterodyne detector [16] or optical lever [17]. Due to the complexity of the alignment of the optical lever system, piezoelectric and piezoresistive ceramics were built into the cantilever to simplify the force-detecting system. In 1991, Tortonese et al. developed a piezoresistive cantilever as an integrated sensor by ion implantation and proved its efficient effects in the application of noncontact AFM [18]. Extending this first development, in 1995, Itoh and Suga proposed a single cantilever with two piezoelectric ZnO films for the inherent function of self-excited force sensing and z-axis actuation [19, 20]. Minne et al. then integrated an actuator with ZnO film onto piezoresistive cantilever arrays, which is called parallel AFM [21–23]. This integrated piezoelectric actuator improved the scan speed of the AFM cantilever by eliminating the need for external z-axis actuation and greatly increased the resonant frequency corresponding to the imaging bandwidth. Moreover, since 1995, Fujii and Watanabe have been developing a lead zirconate titanate (PZT)-coated silicon cantilever which can perform either sensing displacement or actuation [24, 25]. Many researchers chose PZT over ZnO due to the higher piezoelectric constant and then improved the performance, for example, by integrating a PZT sensor and actuator in an AFM as illustrated in Figs. 2 [27] and 3 [28] or by eliminating electrical coupling problems at high frequency operation [26–29].

### 2.3. Hard disk drive head

Hard disk drives with conventional single-stage servos cannot satisfy the requirements of precise positioning, high operating speed and frequency due to the development

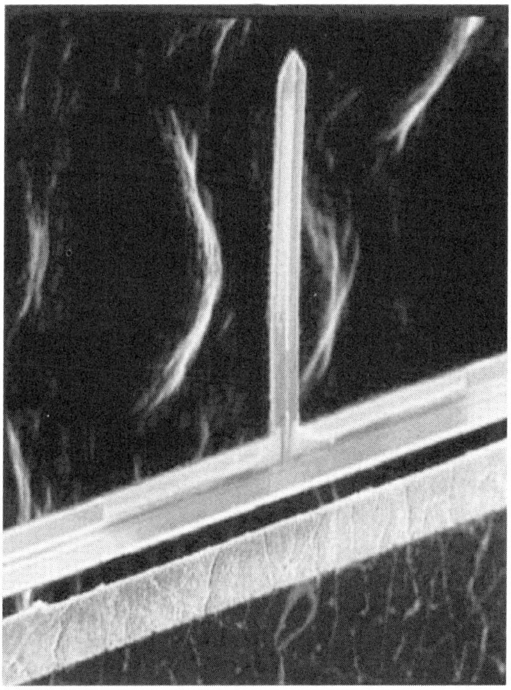

**Fig. 2.** Fabricated AFM cantilever with XYZ actuation by PZT thin film layer.

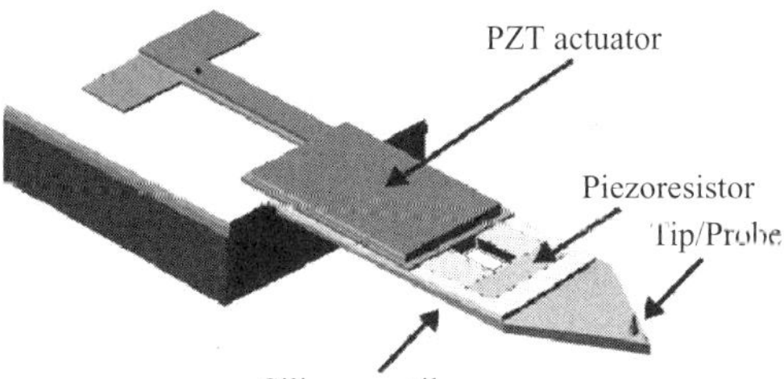


**Fig. 3.** Schematic drawing of self-actuation cantilever with an integrated piezoresistor.

of ever-increasing track densities. A two-stage servo system consisting of a voice coil motor (VCM) and a microactuator was developed. In this two-stage servo scheme, the VCM would move the entire head and arm at low frequency to perform coarse track seeking, and a microactuator would control the slider to perform high-frequency track following. Various microactuators, including piezoelectric, electromagnetic and electrostatic ones, have been developed. However, matching both requirements of necessary displacement stroke and high system resonant frequency is very difficult. In most cases so far, electrostatic and piezoelectric actuators have been applied in two-stage servo systems. Electrostatic actuators mainly using a comb drive scheme can perform linear or rotary actuations with precise positioning control. But the interdigitated finger structure has low stiffness springs, which decrease the resonant frequency of the total system. Additionally, to yield enough driving force, high voltage is necessary, which might possibly result in backlash and stiction problems. In case of piezoelectric actuators, piezoelectric ceramics with high stiffness can achieve high resonant frequency. With relatively lower driving voltage, a large force can be generated with very low power consumption. Their main disadvantage is the relatively low positioning accuracy compared with electrostatic actuators.

# 3. Membrane Type Actuator

## 3.1. What is a membrane type actuator?

Until now, bulk piezoelectric actuators such as piezo-stacks or bimorph piezo-discs, have been used in many piezoelectric actuator applications because of the low actuation force and low displacement in thin film piezoelectric actuators. In the fabrication processes of bulk piezoelectric actuators, epoxy glue is used to bond to a bulk micromachined silicon membrane structure. These processes have many drawbacks in mass production. The use of a membrane (diaphragm) structure can solve the problem of low actuation force and low displacement in thin film piezoelectric actuators. These structures flexures are caused by end clamps. They naturally actuate with an 'S' shape because of increasing hysteresis due to creep in the bonds. This effect can further enhance the displacement and is attractive in piezoelectric applications such as ultrasonic fluid delivery and acoustic wave micromixers.

A membrane type actuator is usually composed of the released silicon structure as shown in Fig. 4. These structures can be patterned by an anisotropic chemical etching process using the enchants like KOH, TMAH [30, 31], or a deep reactive ion etching process [32]. There are various membrane actuator types as shown in Fig. 4(a)–(c). The simple structure has applications in ultrasonic fluid drives and micromirrors, and can be fabricated by silicon anisotropic etching as shown in Fig. 4(a). For pumps, valves, and ink-jet print heads, the membrane type actuator shown in Fig. 4(b) is used with microfluidic channels fabricated with glass or a polymer like PDMS. The structure in Fig. 4(c) is compatible with surface micromachining technology using piezoelectric thin film actuators coated on sacrificial layers.

This process is very simple and the fabricated structure is stable mechanically. For membrane type piezoelectric actuators, thin and thick piezoelectric films are coated on

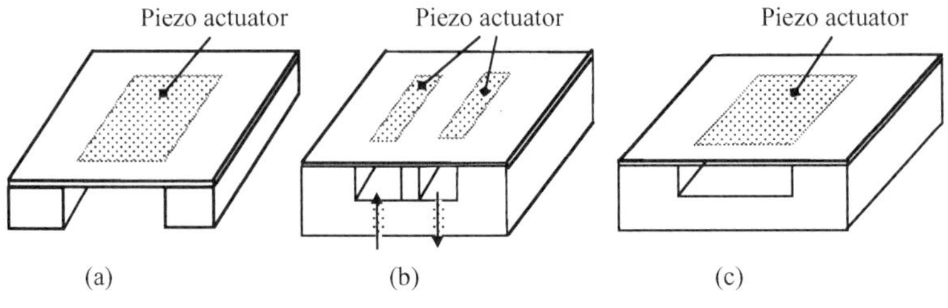


**Fig. 4.** Piezoelectric membrane actuator fabricated using micromachining technology.

silicon membrane structures. There are several methods for coating the piezoelectric layer, such as the sol-gel process, sputtering techniques, chemical vapor deposition for thin piezoelectric film, screen printing methods, and spray coating. For these applications a rather thicker piezoelectric film is desired. However it is not easy to obtain films thicker than 1 mm by conventional deposition methods. Screen-printing is now widely used for such applications, but the machining resolution is not yet satisfactory.

## 3.2. Piezoelectric actuator applications using membrane type

Both thin films and thick films are used for the fabrication of piezoelectric membrane type actuators. Micropumps and valve devices, ink-jet print heads, ultrasonic fluid drives for microfluid applications are the targets of interest for industrial applications.

### *3.2.1. Piezoelectric actuated micro pump or valve in fluid applications*

Microactuators for micropumps or valves are being increasingly applied in total analysis microsystems, labs on a chip, and chemical microreactors. Recently, micropumps driven by piezoelectric actuators have been studied for use in microfluid control systems. The displacement to pump the fluid can be easily controlled by applied voltage and the fabrication is simple. Conventionally bulk piezoelectric materials have been used as actuator materials. In spite of the low force output, piezoelectric thick film actuators are better for miniaturizing these devices than bulk type actuators.

A membrane type actuator using piezoelectric thick film for a micropump was reported by Koch et al. [33]. They utilized screen-printing technology to form a thick piezoelectric film as shown in Fig. 5. The thick film was coated on a fusion bonded silicon membrane whose thickness was controlled by KOH etching. This method can yield higher output force and displacement in the piezoelectric membrane than the sol-gel or physical vapor deposition methods.

A micro-ink-jet application using a screen-printed PZT actuator was reported by a research group at SAIT, Korea [34]. For the piezoelectric actuation method, they used the bending mode for micropumping devices.

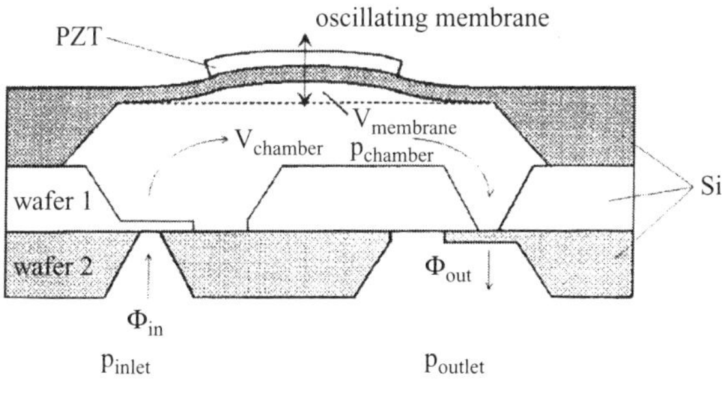


**Fig. 5.** Micropump using screen-printed PZT actuator on silicon membrane. (Courtesy of Neil White, Univ. of Southampton, UK.)

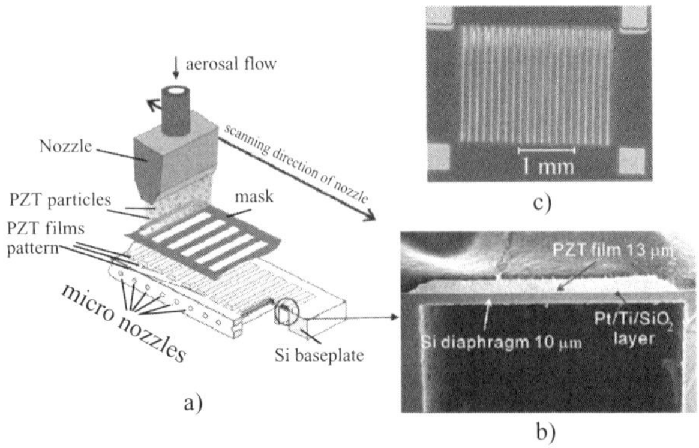


**Fig. 6.** (a) Schematic of aerosol deposition method: fabrication of micropump array, (b) Cross-sectional SEM image of 13-μm-thick Si diaphragm, and (c) Line patterning of PZT (optical image).

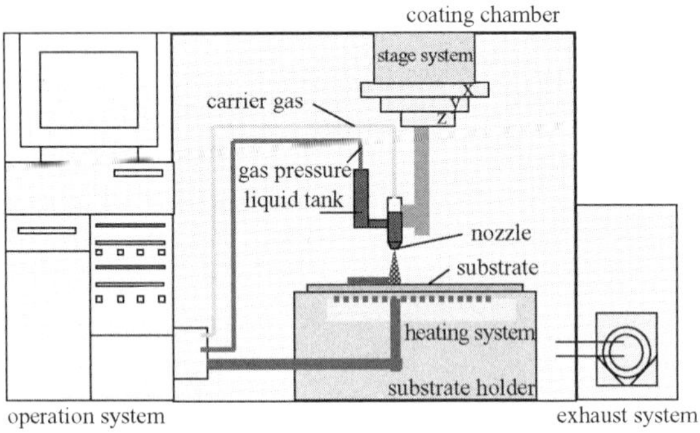


**Fig. 7.** Spray coating system equipment.

### *3.2.2. Piezoelectric actuator application using sprays coating technology*

Recently, various deposition methods for forming thick piezoelectric film such as aerosol deposition, spray coating, and ink-jet printing have been investigated by many researchers. The merits of these deposition methods over the screen-printing method are simplicity without etching and the capability of non-planar surface coating. These methods have been applied to a microparticle delivery system using acoustic waves, micropositioner for hard disk drive, ink-jet printer head, medical applications, and optical microlenses.

A research group has reported fabricating a PZT thick film by the aerosol deposition method (ADM) [35]. This method simplified the fabrication process by eliminating the etching process by using a stencil mask. A high deposition rate of 30 micron/min was attained in a 5 × 5 mm area at temperatures of less than 600 degrees as shown in Fig. 6. To demonstrate applications of PZT thick film by the aerosol deposition method, the group fabricated microfluidic systems such as micropumps, microvalves,

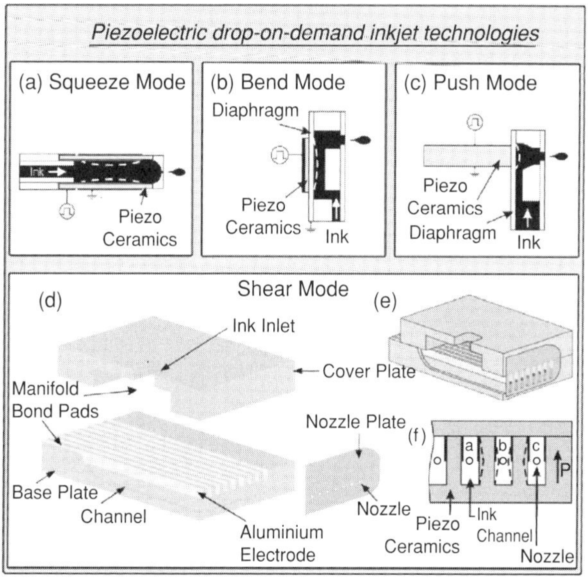


**Fig. 8.** Basic map of piezoelectric drop-on-demand inkjet technologies. (Courtesy of A.M. Grishin, Royal Institute of Technology, Sweden.)

and microparticle delivery systems as illustrated in Fig. 7. They are also interested in fabricating micropositioning actuators for hard disk drives using this deposition method.

Spray coating has been developed at AIST, Japan [36] as shown in Fig. 8. The ejected fine liquid particles were carried by a jet stream of carrier gas onto the silicon substrate. The spray coating method has merits of capability of the dry and wet deposition conditions without damaging the substrate compared with the other thick film deposition methods.

A microactuator for ejecting the ink is necessary for an ink-jet print head system. A piezoelectric actuator for drop-on-demand ink-jet printing was reported by Grishin as shown in Fig. 9 [37]. Piezoelectric drop-on-demand technologies and micromachined droplet ejectors for controlled ink-jet printing can be utilized for piezoelectric thin or thick film formation by generating a transducer wave [38, 39].

Micro actuator for ejecting the ink is necessary for ink jet head system. Piezoelectric actuator for drop-on-demand ink-jet was reported by Grishin as shown in Fig. 8 [37]. Piezoelectric drop-on-demand technologies and micromachined droplet ejector for controlled ink-jet printing can be utilized for piezoelectric thin or thick film formation by generating transducer wave [38–39].

### *3.2.3. Ultrasonic applications*

Thin film lamb waves or thin film flexural-plate wave actuators have advantages over bulk-plate wave actuators for transporting suspensions in fluids and fluid delivery

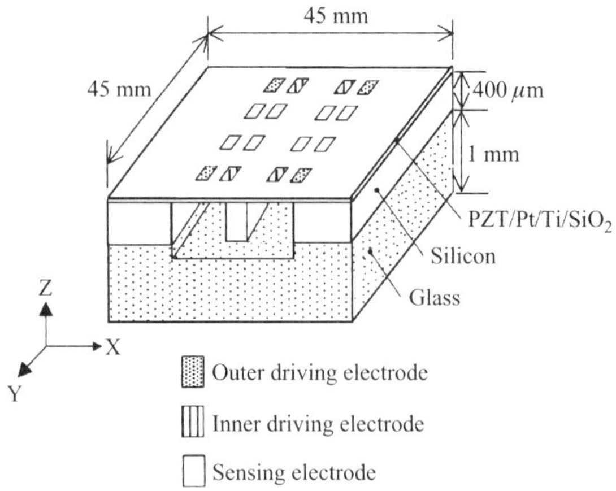


**Fig. 9.** PZT film angular rate sensor using a membrane type actuator

applications [40]. For these applications, piezoelectric materials such as aluminum nitride, zinc oxide, and lead zirconate titanate have been utilized. Controlled membrane thickness less than the acoustic wavelength can be fabricated using the silicon backside etching process. For a piezoelectric thin film interdigitated transducer (IDT), the electrodes are easily fabricated by semiconductor processes. The Lamb wave is suitable for delivery or pumping of minute particles in fluids.

To overcome heat generation during mixing for temperature-sensitive fluidic applications (e.g., DNA chip), an acoustic wave micromixer was used with a piezoelectric thin film membrane actuator. The Fresnel Annular Sector Actuator (FASA) was based on an acoustic-wave transducer and generates strong acoustic waves. A parylene dome-shaped diaphragm using a piezoelectric ZnO thin film acoustic transducer has also been reported [41].

### *3.2.4. Other applications using piezoelectric membrane actuator*

A piezoelectric angular rate sensor with a membrane type actuator was reported by Lee et al as shown in Fig. 9 [42]. This device using a PZT film with 1.5 μm thickness was designed as a structure type utilizing the distribution of driving electrodes for enhancing the displacement. A micromirror for auto-focusing light using a piezoelectric thin film membrane actuator has been demonstrated by Goto et al. [43]. This device was composed of a $Pb(Zr_{0.54}Ti_{0.46})O_3$ piezoelectric diaphragm actuator deposited by radio frequency(RF) magnetron sputtering on a silicon diaphragm fabricated by KOH etching. Power generation applications have been reported [44] that use the inverting electric bias field in PZT to improve the displacement like meanderline actuators. P3 microheat engines based on thin film PZT membrane transducers have been researched by Richards [45]. These devices were composed of two main components, a heat engine with a heat source and a power generator with a piezoelectric transducer. The PZT-based MEMS power generator obtained several mW over 2 volts at frequencies between 300 and 1100 Hz.

### 3.3. Bulk material actuator and high aspect ratio micro structure

The advantages of bulk actuators, compared with cantilever and membrane thin film types, are the high performance of actuation and high reliability of piezoelectric characterization. Established manufacturing methods for bulk materials have better reliability than thin film methods. Bulk actuators have disadvantages in the mass production of MEMS substrates and in reliability because of the existence of the bonding glue layer. It is, therefore, necessary to form direct bonding between the bulk actuator layer and the device structure. In the following section, recent research trends of bulk actuators are introduced. High aspect ratio microstructures (HARMST), which do not fall in the cantilever, membrane or bulk actuator categories, are also introduced.

#### *3.3.1. Bulk actuator*

A parallel-bimorph-type piezoelectric actuator, which consists of two bimorphs arranged parallel to each other, for a high-resolution imager that uses CCD-chip-shift operations for practical applications, has been developed [46]. From an investigation of the basic characteristics, it has been confirmed that the proposed actuator has achieved a sufficiently low voltage drive and high-speed response for the high-resolution imager using CCD-chip-shift operations. As a result, a box-shaped package of an imager, with a 22-pin, dual-in-line format, and dimensions of 30(L) $\times$ 20(W) $\times$ 11(H) mm was obtained as a usable size. A demand displacement of $8.5 \times 10^{-3}$ mm was obtained at a driving voltage of 13 Vp-p (30 Hz pulses).

A valveless micropump based on a novel flow rectification principle using the temperature dependence of the liquid viscosity has been proposed [47]. A prototype device was fabricated by micromachining technology, as shown in Fig. 10. There is a linear relationship between the flow rate and pulse length of the heating signals. The

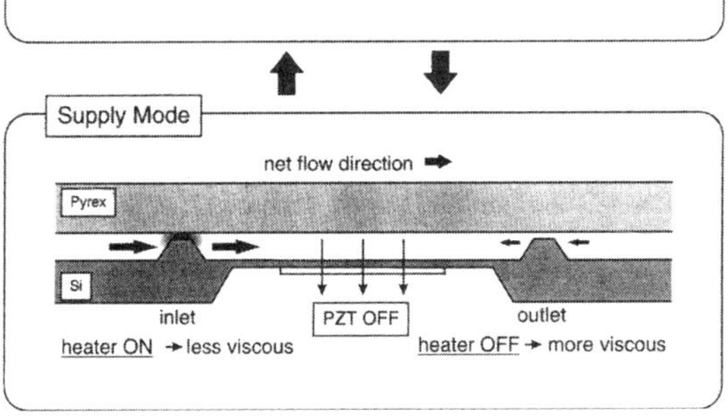


**Fig. 10.** Structure and operating sequence of a viscosity-based dynamic valve pump.

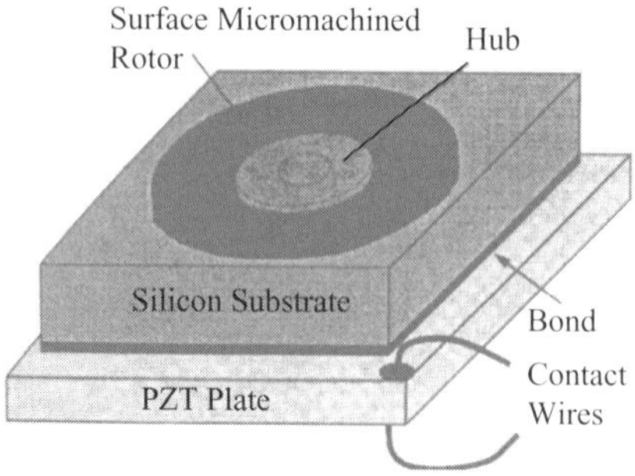


**Fig. 11.** Surface micromachined rotor actuated by a PZT plate.

contribution of local preheating of the liquid was not detected. These findings imply that a strong damping mode was realized. The maximum flow rate was limited by the decrease of the temperature difference and the generation of the vapor bubbles. The feasibility of this new pump principle has been confirmed, including the capability of bi-directional pumping.

An active micromixer for continuous flow has been proposed [48]. Mixing occurs directly by ultrasonic vibration. The patterns of inlets, outlets and mixing chamber were formed in glass. The entire flow path was encapsulated by anodic bonding of a Si wafer to the glass. A diaphragm ($6 \times 6 \times 0.15$ mm) was etched on the Si side to prevent ultrasonic radiation from escaping to the other parts of the device. The ultrasonic vibration originated from a bulk piezoelectric lead-zirconate-titanate (PZT) ceramic ($5 \times 4 \times 0.15$ mm). The PZT was adhered to the diaphragm and was excited by a 60 kHz square wave at 50 V (peak-to peak). Liquids were mixed in a chamber ($6 \times 6 \times 0.06$ mm) with the Si oscillating diaphragm driven by the PZT. A solution of uranine and water was used to evaluate the effectiveness of the mixing. The entire process was recorded using a fluorescent microscope equipped with a digital camera. The laminar flows of the uranine solution (5 ml/min) and water (5 ml/min) were mixed continuously and effectively when the PZT was excited. The temperature rise of the device was 15 °C due to the ultrasonic irradiation.

The first-ever all-surface micromachined ultrasonic microrotor has been proposed [49]. The construction of the device is shown in Fig. 11. The rotor is actuated by electrically driving a piezoelectric PZT plate mounted at the back of a silicon die, eliminating the need for interconnects and space-consuming surface actuators. The rotor operates with single-phase sub-5V peak-to-peak excitation in atmospheric pressure. The piezoelectric plate is adhesively mounted making the method suitable for actuating micromachines from any surface micromachine process. Two different modes of operation have been demonstrated: pulsed and resonant. The pulse actuation results in a low rotation rate (0.5 to 3 RPM) while resonant actuation results in fast rotation (10 to 100 RPM). The ability to drive a geared-down rotor (50:7), much smaller than the driving rotor, indicates high torque output capability.

The miniaturization of low power (typically 10 $\mu$W) and low rotation speed (about 100 rpm) motors for wristwatch applications has been proposed [50]. These motors are based on the conversion of the standing wave of the piezoelectric activated silicon stator membrane into rotation. This conversion is realized by a microfabricated rotor. To achieve the miniaturization of actuators, their construction requires simple-structured components and easy configuration. The new micromotor stably rotates at any posture, and the starting torque was measured to be about 2.6 μNm.

This new type of piezoelectric micromotor consists of components that structure function flexibly such as the Smart structure, which is different from conventional actuators. This research has achieved a micromotor for practical use in micromechanism fields [51].

A centimeter-sized robot for 5 DOF micromanipulation has been proposed [52]. Each leg ($1.5 \times 1.5 \times 4.0$ mm) can move independently and in three orthogonal directions. The main component in the robot is a six-legged monolithic piezoceramic unit suitable for large-scale production. The mechanical walking mechanism uses solid multilayer piezoactuators that make it possible to achieve both functionality and high precision.

Optical micromachines are easily miniaturized in fabrication sizes and are controllable by optical energy supplied wireless from a remote point. An optically driven moving machine that moves like a caterpillar using optoelectronic principles has been proposed [53]. It would consist of two parts, a body made of a spring and temperature sensitive ferrite and feet made of piezoelectric devices (PZT).

Electrostatic optical motor was developed, which is used with the photovoltaic effect of ferroelectric lead lanthanum zirconate titanate(PLZT) [54]. The caused voltage is transferred into the stator of the optical motor. The rotor is cited between the stator and could move in accordance with the attractive electrostatic force.

### *3.3.2. High aspect ratio structure*

An array of lead zirconate titanate (PZT) rods whose cross section is 25 μm square has been fabricated using synchrotron radiation (SR) lithography, and the effects of the aspect ratio (height/width) of the PZT rods on 1–3 piezoelectric composite properties were studied [55]. The cross section is shown in Fig. 12. The aspect ratio was between 3 and 9. The mechanical quality factor and acoustic impedance were found to depend weakly on the aspect ratio.

**Fig. 12.** PZT rod array (cross section: 25 μm).

**Table 1.** Representative character of fabrication technologies for piezoelectric actuator

| | Thin film on Si | Thick film on Si | Bulk actuator on Si (Bonding) |
|---|---|---|---|
| Actuator thickness | Less than 1 μm | 1 to 50 μm | 50 μm to 1 mm |
| Process temperature | 500 to 700 °C | 800 to 900 °C | R.T. to 400 °C |
| Crystallinity and orientation | Polycrystal, orientated or epitaxially grown crystal | Polycrystal, random orientation | Polycrystal, random orientation |
| Elecrical properties | Inferior to bulk | Inferior to bulk | Good |
| Process | PVD, CVD, sputtering, laser ablation | Screen printing, jet printing system | Solid-state reaction method |
| Compatibility with IC process | Good | Intermediate | Bad |

A novel Silicon (Si) mold process has been proposed to fabricate stacked PZT/Si microactuators which consist of parallel plates of piezoelectric PZT intercalated by Si electrodes [56]. The batch-fabrication process is capable of creating a multi-layer PZT structure with plates as thin as 18 mm, meaning that an actuator driving voltage of as low as 18V would be sufficient for operation.

### 3.3.3. *Future problem*

The direct bonding method has continued to be developed in recent years. Direct bonding between Si-wafer and Si-wafer, and between Si-wafer and single crystal have already been established [57, 58]. This technology should become the key technology for mass production on bulk materials attached to MEMS. On the other hand, development of high-k ferroelectric single crystals has also been progressing recently [59]. The main application fields are limited to ultrasonic diagnosis at present. But some of these applications have the high performance of ferroelectrics and piezoelectric characteristics. Therefore these are promising for bulk actuator applications.

### 3.3.4. *Summary of fabrication technologies*

A comparison of fabrication technologies for piezoelectric actuators is given in Table 1.

The thickness of the film actuator is typically less than 1 μm. The film has a flat surface and is suitable for micromachining, e.g., photolithography, ion-milling, which could achieve a microactuator that is less than 1 μm thick. It is also easy to control the crystal orientation and to grow the crystal structure epitaxially in thin film. The improvement of material properties is, therefore, easier than with other technologies. On the other hand, it is also possible to form thick films from 1 to 50 μm in thickness. The firing temperature is relatively lower than bulk, but the film density and electrical properties are inferior to those of bulk. The flatness of the surface is not yet good and not still problematic for microfabrication. On the other hand, the output power of thick film is much better than that of thin film and is suitable for realizing relatively large-size devices. Therefore, if its problems are overcome, thick film is a good choice for microactuators. Bonding technology enables us to use the material properties of bulk directly. The actuator properties are, without a doubt, superior to those of other technologies. It is possible obtain the plate that thickness is less than 100 μm after the bonding process. On the other hand, microfabrication on bulk material has already been established. The bulk fabricated device size is over 100 μm and is larger than thin or thick film actuators. Room temperature bonding technology has been developed recently and could become a useful technology for MEMS applications.

## 4. Conclusion

Piezoelectric microactuators for MEMS were introduced classified by their shapes, i.e. cantilever, membrane and bulk types. Various applications were presented, but thicker films with high and reliable electrical properties have yet to be developed, and etching technology must be established for further successful applications. The alternative method of bonding bulk ceramics is being employed for fabrication, but the bonding technology is still under development.

## References

[1] P. Muralt, *J. Micromech. Microeng.*, **10**, 136 (2000).
[2] E. Quandt and H. Holleck, *Microsystem Technologies*, **1**, 178 (1995).
[3] D.L. Polla, *Proceedings of SPIE* (Austin, 1997), p. 24.
[4] M. Okuyama, *Sensors Update*, **6**, 79 (2000).
[5] D. Hauden, *Piezoelectric Materials: Advances in Science, Technology and Applications*, edited by C. Galassi (Kluwer Academic Publishers, Netherland, 2000), p. 335.
[6] T. Ikeda, *Fundamentals of piezoelectricity* (Oxford University Press, Oxford, 1990).
[7] K. Uchino, *Piezoelectric Actuators and Ultrasonic Motors* (Kluwer Academic Publisher, 1997).
[8] V. J. Fowler and J. Schlafer, *Proc. IEEE*, **54**, 1437 (1966).
[9] L. F. Frank and J. K. Lee, U.S. Patent 3,981,566 (21 Sep. 1976).
[10] H. Goto and K. Imanaka, *Proc. SPIE, Miniature and Micro-optics/Fabrication and System Applications*, **V1544**, 272 (1991).
[11] Y. Ohtuka, H. Nishikawa, T. Koumura, and T. Hattori, *Proc. IEEE MEMS*, 306 (1995).
[12] A. Schroth, C. Lee, S. Matsumoto, and R. Maeda, *Sensors and Actuators*, **A73**, 144 (1999).
[13] J. Tsaur, L. Zhang, R. Maeda, and S. Matsumoto, *Jpn. J. Appl. Phys.*, **V41**, 4321 (2002).
[14] G. Binnig, C.F. Quate, and Ch. Gerber, *Phys. Rev. Lett.*, **56**, 930 (1986).
[15] S. Alexander, L. Hellemans, O. Marti, J. Schneir, V. Elings, P.K. Hansma, M. Longmire, and J. Gurley, *J. Appl. Phys.*, **65**, 164 (1989).
[16] Y. Martin, C.C. Williams, and H.K. Wickramasinghe, *J. Appl. Phys.*, **61**, 4723 (1987).
[17] G. Meyer and N.M. Amer, *Appl. Phys. Lett.*, **53**, 2400 (1988).
[18] M. Tortonese, H. Yamada, R.C. Barrett, and C.F. Quate, *Proc. 6th Int. Conf. Solid-State Sensors and Actuators,Tranducers*, 448 (1991).
[19] T. Itoh and T. Suga, *Sensors and Actuators*, **A54**, 477 (1996).
[20] T. Itoh, T. Ohashi, and T. Suga, *Proc. IEEE MEMS*, San Deigo, USA 451 (1996).

[21] S.C. Minne, S.R. Manalis, and C.F. Quate, *Appl. Phys. Lett.*, **67**, 3918 (1995).
[22] S.R. Manalis, S.C. Minne, and C.F. Quate, *Appl. Phys. Lett.*, **68**, 871 (1996).
[23] S.C. Minne, G. Yaralioglu, S.R. Manalis, J.D. Adams, J. Zesch, A. Atalar, and C.F. Quate, *Appl. Phys. Lett.*, 72, 2340 (1998).
[24] T. Fujii, S. Watanabe, M. Suzuki, and T. Fujiu, *J. Vac. Sci. Technol. B*, **13**, 1119 (1995).
[25] T. Fujii and S. Watanabe, *Appl. Phys. Lett.*, **68**, 467 (1996).
[26] T. Itoh, T. Suga, and C. Lee, *Appl. Phys. Lett.*, **69**, 2036 (1996).
[27] J. Chu, R. Maeda, T. Itoh, K. Kataoka, and T. Suga, *Jpn. J. Appl. Phys.*, **V38**, 7180 (1999).
[28] Y. Kim, H. Nam, S. Cho, J. Hong, D. Kim, and J. Bu, *Sensors and Actuators*, A103, 122 (2003).
[29] C. Lee, T. Itoh, and T. Suga, *Sensors and Actuators*, **A72**, 179 (1999).
[30] D.B. Lee, *J. Appl. Phys.*, **40**, 4569 (1969).
[31] H. Seidel, L. Csepregi, A. Heuberger, and H. Baumgärtel, *J. Electrochem. Soc.*, **137**(11), 3612 (1990).
[32] A.A. Ayon, X. Zhang, and R. Khanna, *Solid-State Sensor and Actuator Workshop* (Hilton Head Island, South Carolina, June 4–8, 2000), p. 339.
[33] M. Koch, N. Harris, R. Maas, A.G.R. Evans, N.M. White, and A. Brunnschweiler, *Measurement Science and Technology*, **8**, 49 (1997).
[34] S.H. Lee, J. Chung, S. Lim, and C.S. Lee, *Proceedings of the 2001 International Conference on Robotics and Automation* (Seoul, Korea, May 21–26, 2001), p. 616.
[35] J. Akodo and M. Lebedev, *Nano- and Micortechnology: Materials, Processes, Packaging, and System, SPIE*, vol. 4936 (2002) p. 234.
[36] M. Ichiki, L. Zhang, Z. Yang, T. Ikehara, and R. Maeda, Microsystem Technologies, vol. 10 (2004) p. 360.
[37] J. Brünahl and A.M. Grishin, *Sensors and Actuators*, A **101**, 371 (2002).
[38] S. Okamura, R. Takeuchi and T. Shiosaki, *Japanese Journal of Applied Physics*, **41**, 6714 (2002).
[39] G. Percin and B.T. Khuri-Yakub, *Review of Scientific Instruments*, **73**(5), 2193 (2002).
[40] P. Luginbuhl, S.D. Collins, G.-A. Racine, M.-A. Grétillat, N.F. de Rooij, K.G. Brooks, and N. Setter, *Journal of Microelectromechanical systems*, **6**(4), 337 (1997).
[41] V. Vivek, Y. Zeng, and E.S. Kim, *The 13th Annual International Conference on Micro Electro Mechanical Systems* (2000) p. 668.
[42] S.H. Lee and M. Esashi, *Sensors and Actuators*, **A114**, 88 (2004).
[43] M. Sakata, S. Wakabayashi, H. Totani, M. Ikeda, H. Goto, M. Takeuchi, and T. Yada, *8th International Conference on Solid-State Sensors and Actuators, and Eurosensors 95*, p. 422.
[44] W.P. Robbins, D.L. Polla, T. Tamagawa, D.E. Glumac, and W. Tjhen, *J. Micromech. Microeng.*, **1**, 247 (1991).
[45] C. Xu, J. Hall, C. Richards, D. Bahr, and R. Richards, *ASME IMECE MEMS Symposium MEMS 2* (2000), p. 261.
[46] C. Tanuma, *Jpn. J. Appl. Phys.*, **38**, 5603 (1999).
[47] S. Matsumoto, R. Maeda, and A. Klein, *Micro. Thermo. Eng.* **3**, 31 (1999).
[48] Z. Yang, S. Matsumoto, H. Goto, M. Matsumoto, and R. Maeda, *Sensors and Actuators*, **A93**, 266 (2001).
[49] V. Kaajakari, S. Rodgers, and A. Lal, *Proc. IEEE MEMS*, 40 (2000).
[50] L. Dellmann, S. Gautsch, G.A. Racine, and N.F. de Rooij, *Proc. Transducers '99*, 1752 (1999).
[51] Y. Suzuki, K. Tani, and T. Sakuhara, *Proc. Transducers '99*, 1748 (1999).
[52] U. Simu and S. Johnsson, *Proc. Transducers01*, 690 (2001).
[53] K. Fukushima, Y. Otani, and T. Yoshizawa, *J. Pre. Mech. Eng.*, **64**, 1512 (1998).

[54] Y. Morikawa, M. Ichiki and T. Nakada, *Jpn. Soc. Mech. Eng.*, **C69**, 161, 2003.
[55] Y. Hirata, T. Numazawa, and H. Takada, *Jpn. J. Appl. Phys.*, **36**, 6062 (1997).
[56] S.N. Wang, K. Wakabayashi, J.F. Li, and M. Esashi, *Proc. of Transducers '99*, 1762 (1999).
[57] H. Takagi, R. Maeda and T. Suga, *Sensors and Actuators*, **A105**, 98(2003).
[58] H. Takagi, R. Maeda and T. Suga, *J.Micromech. Microeng.*, **11**, 348(2001).
[59] Y. Hosono, K. Harada, T. Kobayashi, K. Itsumi, M. Izumi, Y. Yamashita and N. Ichinose, *Jpn. J. Appl. Phys.*, **41**, 7084(2002).

# Chapter 3

## Micromachined Ultrasonic Transducers and Acoustic Sensors Based on Piezoelectric Thin Films

P. Muralt

*Ceramics Laboratory, Swiss Federal Institute of Technology EPFL, CH-1015 Lausanne, Switzerland*

A review is given on the current state of the art in piezoelectric micromachined ultrasonic transducers (pMUT). It is attempted to quantify the limits of pMUT's with respect to the electromechanical coupling, and to relate current achievements. Best reported results so far is a coupling coefficient $k^2$ of 6 %. This was achieved with 2 μm thick PZT thin films on a 5 μm thick silicon layer obtained by micromachining from a SOI wafer. The material properties are the most important part for a good performance, however, design contributes as well. Electrode size, layer thickness relation, and border conditions must be dimensioned correctly. Main needs for future research are identified in design, micromachining and further improvements of PZT films. Applications are shortly reviewed.

## 1. Introduction

Ultrasonic transducers are best known for their applications in medical ultrasonic imaging (see, e.g., [1]). For this purpose, an ultrasound wave is emitted into the body by a transmitter. The intensities of waves reflected from the various parts of the body are measured as a function of time by a receiver. Transmitter and receiver function can be delivered by the same device if the excitation and reception are separated in different time windows. The frequency depends on the depth of the investigated organs, and amounts to 3 to 5 MHz for deep organs, 15 MHz for eyes, and up to 50 MHz for skin, artery walls, and other thin tissues. The excitation signal is either a pulse (pulse-echo method) or a continuous wave. In the first case, the pulse contains a broad frequency range, and the imaging is based on the intensities and delay times with which reflected wave packets are captured by the receiver. In the second case the frequency band is narrow. Frequency shifts due to the Doppler effect at moving parts can be observed. In order to get efficient excitation and sensitivity, the devices are operated at resonance. However, short pulses, or equivalently broad emission bands, are only achieved if the quality factor $Q$ is small (The acoustical $Q$ of an echographic transducer should be less than one [2]). The receiver should of course be very sensitive and thus efficiently transform the received mechanical power into an electrical signal. For this purpose a large coupling coefficient ($k^2$) is required. Doppler shift measurements require narrower bands, and thus allow for larger $Q$ factors. However, in the practice, the same probe should cover both operating methods. At present, ultrasound transducers are based on piezoelectric $\lambda/2$ bulk wave resonators. The piezoelectric material is an optimized PZT ceramics. Modern transducers are linear arrays in order to perform beam steering and shaping, and to increase resolution. The optimal element density is two per sound wavelength of the investigated medium. In medical applications this corresponds to a pitch of about 300 μm for deep organs, 100 μm for eyes, and 30 μm for thin tissue investigations.

There are several other applications of ultrasound in sensors and materials testing. Closely related to medical imaging are acoustic microscopy and non-destructive testing. The main differences are related to the fact that the acoustic impedances, sound velocities, and acoustic damping of the investigated materials are different from the ones of the human body, requiring adaptions of geometries and frequencies. The pulse-echo method is also applied in sensors to measure distances or to detect the presence of an object, similar to the ultrasonic "vision" system of bats. Doppler shift measurements are used in flow meters measuring liquid flows. Gas detectors can be made using the principle of resonating cavities, whose resonance frequency depends on the sound velocity of the gas in the cavity [3]. Microphones fall also into the category of sensors for acoustic waves. Finally, there are actuators based on ultrasonic excitation, such as ultrasonic motors and droplet ejectors as used in ink jets. Both have been realized as micromachined versions [4, 5].

In the recent years, micromachined versions of transducers attracted very much interest [6–10]. In some cases it is motivated by integrating a MUT directly onto a CMOS chip [11]. The main reason for imaging is the prospect to dispose of 2-dimensional arrays offering excellent beam steering (phased array operation), high-resolution and real-time imaging capabilities. It is clear that micromachining techniques allow for a much higher

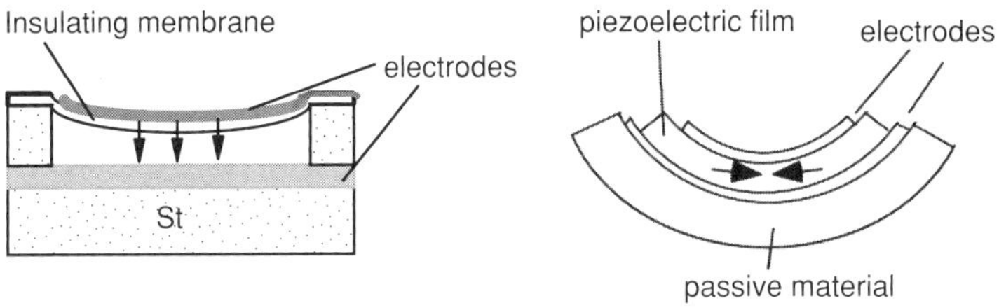


**Fig. 1.** Schematic cross section through (a) electrostatic and (b) piezoelectric MUT structure. The arrows indicate the direction of the force developed by the driving principle.

precision than the reticulation techniques used for the fabrication of bulk ceramics elements. This is especially true for transducers operating above 10 MHz. Although progress has been made in reticulation techniques [12], the wiring of 2-d transducers is difficult to solve in case of reticulated ceramics. Silicon micromachining is quite advanced in the fabrication of deep through holes [13, 14] and buried conductor lines needed to realize wiring of 2-d arrays. So it seems to be quite clear that the interconnect part of a 2-d array will be fabricated by means of silicon micro machining (realized in a "connectic" waver). The question is whether the acoustic part is made as well in silicon technology—in this case the joining of the two parts is a mastered problem—or whether the acoustic part looks more like a reticulated ceramics (hybrid approach). The latter approach has indeed been successfully realized in case of pyroelectric focal plane arrays for infrared vision systems [15] based on a ceramic (Ba,Sr)$TiO_3$ ceramics.

For the pure silicon micromachining approach, electrostatic as well as piezoelectric thin film structures are investigated (see fig. 1). Both are based on membrane structures. A priori, these have an advantage for medical applications: Membranes are much softer than bulk PZT and match better the acoustic impedance of soft matter. Most important advantage of capacitor type structures (capacitive micromachined ultrasonic transducer cMUT) is the high coupling coefficient. Close to the collapse, 100 % conversion efficiency is theoretically calculated (see, e.g. [16]). Although impossible to operate too close to the collapse, the practically achievable values are quite large. Coupling coefficients of more than 70 % have been demonstrated [17, 18]. One of the reasons for the large coupling is the fact that the electrostatic force acts vertical to the membrane, i.e. parallel to the membrane deflection and force output. In piezoelectric materials or structures, the coupling coefficient $k^2$ is defined as the ratio of the input mechanical energy to the stored electrical energy. $k^2$ depends on the stiffness of the material or the spring constant of the structure. This is also the case for piezoelectric MUT's (see below). In cMUT's, however, the effective spring constant softens with increasing voltage and becomes zero at the collapse voltage [19]. The coupling coefficient in this case must be rather understood as the ratio of the available mechanical work to the electrical energy delivered to the device. In piezoelectric devices the available work depends very much on the matching. In pMUT's, the exploited effect is based on the transverse coupling $k_p$. The piezoelectric force is in fact in the membrane plane, thus perpendicular to the force to be delivered. The mechanism to turn the output force by 90° is the bending of the membrane by the transverse piezoelectric stress or strain (see fig. 1). Bimorph structure and hetero-morph structures can be applied. The latter contains a passive layer (Si, $Si_3N_4$, etc.). In order that this mechanism works well, the neutral plane of the membrane must not lie within the piezoelectric film. Otherwise the lower part of the film would work against the upper part of the film. It should be in the passive material. Using hand waving arguments, we can say that in the best case the coupling coefficient of a piezoelectric heteromorphous membrane amounts to $k^2 = k_p^2/2$ since half of the vibrating structure is passive. In bulk PZT, $k_p$ amounts to 60 % typically, which is 10 to 20 % less than $k_{33}$ used in bulk transducers. The best possible coupling coefficient amounts thus to about $k = 42$ % ($k^2 = 18$ %). So even if one argues that the better acoustic matching would correspond to a $k^2 = 36$ % for a bulk transducer, we still miss by a factor 1.5 the effective coupling of today's probes. This figure possibly might be acceptable for high resolution, high frequency probes. The question remains, whether the above optimistic numbers are reached or not, and whether there are other applications that do not need such a high coupling. This paper will give a short review on pMUTs and tries to answer some of the questions.

## 2. Clamped Plate Structure

The above optimistic estimation is based on a piezoelectric laminated plate that is freely suspended. This means that there is no force acting on the border of this plate. A rough estimation of the coupling factor for this case is given in ref. [20]:

$$k^2 \approx 6\frac{1-\nu}{Y}\frac{e_{31,f}^2}{\varepsilon_0\varepsilon_{33,f}}\frac{t_p}{h}, \quad t_p < h/2 \tag{2.1}$$

where the thickness of the piezoelectric material $t_p$ must be smaller than half the total thickness $h$. $Y$ and $\nu$ are Youngs modulus and Poisson ratio of the passive material, $e_{31,f}$ is the effective transverse piezoelectric coefficient $e_{31,f} = d_{31}/(s_{11}^E + s_{12}^E)$ that is easily determined from thin film structures [21, 22]. Taking silicon as passive material, considering an $e_{31,f}$ of 12 C/m$^2$ that applies for dense PZT{100} films [23], we arrive at $k^2 = 16$ % for $t_p = 0.4h$. This value is in fact close to the first estimation.

The essential point of micro machining and design is to solve this problem of turning the output force by 90°. The simplest possibility, which is also best understood in the frame of analytical mathematics, is the clamped disk or plate (The first derivative of the deflection function is zero at the border of the plate). A flexural resonance is excited as schematically shown in figure 2. The whole structure consists of a few films with a minimal amount of patterning steps. The membrane is either liberated by surface micromachining (as shown in figure 2) or by bulk micromachining. The first such

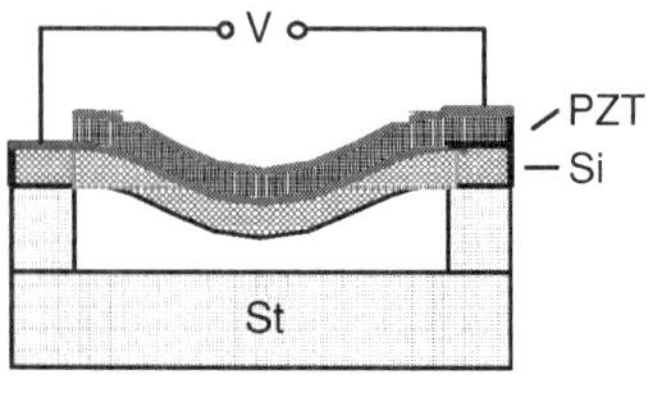


**Fig. 2.** Schematic cross section through clamped plate.

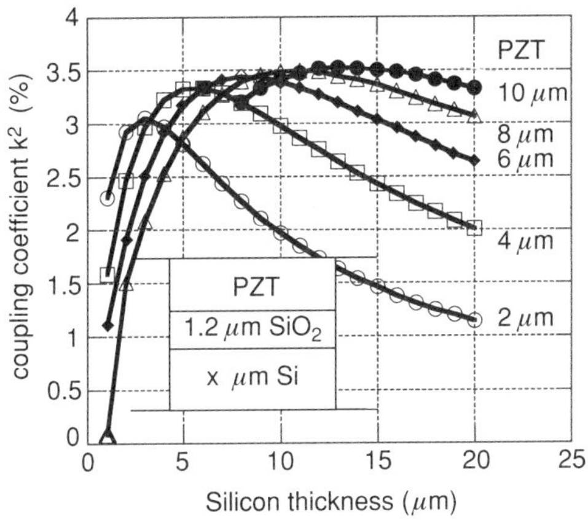


**Fig. 3.** Simulation results of $k^2$ for the basic deflection mode in case of a stress free diaphragm of 300 μm diameter with a centered electrode having the optimal diameter of 65 % of the plate diameter. PZT film parameters: $e_{31,f} = 12$ C/m$^2$, $\varepsilon_{33,f} = 1200$. The calculations are explained in ref. [28].

structures containing PZT films have been fabricated for micromachined ultrasonic stators for micromotors [4, 24]. They also have been applied for the first demonstrations of ultrasonic imaging with pMUTs [9]. Such plate structures can be handled by analytical mathematics if it is assumed that the deflection is small as compared to the diameter of the plate. The problem is usually solved for radial symmetrical structures. Solutions have been calculated following Kirchhoff's plate theory by implementing the equation of state of the piezoelectric material into a multilayered plate obeying the plate equation, the boundary conditions for clamping, and the electrode geometries. Differential equations or complete plate equations have been derived [25, 26], or in a more approximate way using the energy method [27, 28], which in fact goes also back to Kirchhoff [29]. In this case, the elastic energy is numerically calculated based on the deflection functions of the fundamental resonance (Bessel functions). Stiffness and position of the neutral plane are calculated according known algorithms [30]. Using the latter method, $k^2$ has been calculated for various thickness values of PZT as a function of the thickness of the passive part as shown in figure 3.

The maximal coupling is achieved when the Si layer is slightly thicker than the PZT film. The curves show that the coupling is drastically reduced when the neutral plane is situated inside the PZT material. The maximal value does not increase much with PZT film thickness, and saturates at about $k^2 = 3.5$ %. However, the maximum becomes broader with increasing thickness. In addition, the peak values are less reduced by film stresses that may stretch the membrane and may reduce very much the peak in coupling coefficient. Stretching forces add a correction term to $k^2$ proportionally to $a^2/h^3$ ($a$: plate radius, [27]). With increasing thickness $h$, the influence of stretching is thus dramatically reduced. The resonance frequency increases roughly as $f_0 \propto h/a^2$. If the application defines the size of the plate (as e.g. the pitch of an array) and the frequency, the thickness of PZT and silicon are fixed by the condition of maximal coupling. The PZT film thickness has—of course—a direct impact on the amplitude and the output force in the emitter mode, and on signal voltage and signal-to-noise in the receiver mode.

Above calculations show that the clamped plate is not an ideal realization of a flexural transducer. The clamping of the plate reduces the coupling factor considerably. The question is whether one can approximate the feely suspended plate to increase coupling.

## 3. Approximating Freely Suspended Plates

The free plate is best approximated by a structure keeping the plate in the center. The whole border would be free to move. However, it is not evident how to mount the electrical connections to such a structure. It is much easier to keep the membrane by a number of bridges around the border and leave the center free to move. The realization of such a bridge is shown in figure 4. (see ref. [31] for fabrication details).

An improved control of the membrane thickness is achieved by means of SOI wafers. Such a wafer contains a silicon layer of defined thickness separated from the bulk wafer by a buried oxide layer. The latter serves as etch stop layer and allows the fabrication of very homogeneous silicon membranes of well-defined thickness. Figure 5 shows the admittance curve of a 300 μm plate laminated with a 2 μm thick PZT film, obtained from a SOI substrate [33]. The coupling coefficient was derived from an equivalent circuit model (as depicted in figure 5): $k^2 = CN^2/C_0$. The parasitic capacities due to wiring should be minimized, otherwise these parasitics add to $C_0$. $k^2$ was obtained as 5.3 %, which is considerably larger than the expected result of a clamped plate (2.9 % according to fig. 3). For the interpretation we have to include the fact that a bias voltage was applied. This increases $e_{31,f}$ and decreases the dielectric constant. (At equivalent films, an increase of $d_{33,f}$ by 35 % has been measured at 100 kV/cm [23]). The coupling increases accordingly. Figure 6 shows the increase of $k^2$ as a function of bias field (different device). The calculated 2.9 % could well increase by 80 % to 5.2 % considering the bias of 100 kV/cm as applied for the measurement of fig. 5. The

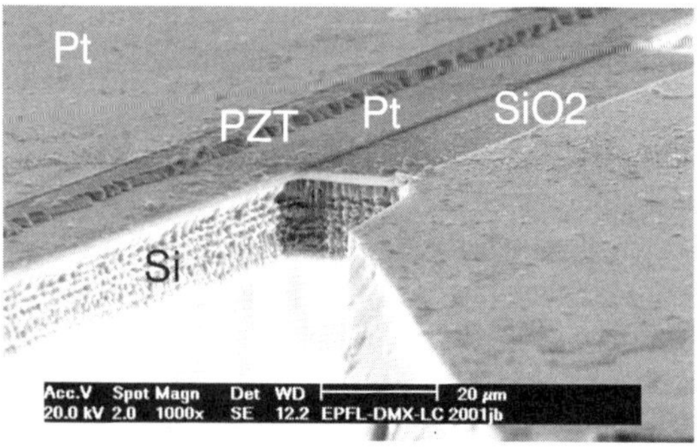


**Fig. 4.** SEM viewgraph showing the micromachined bridge of a suspended membrane with the etched Pt/PZT/Pt sandwich. The membrane was obtained by thinning the silicon down from the backside by deep silicon etching. The final thickness was defined by time control. (from ref. [32]).

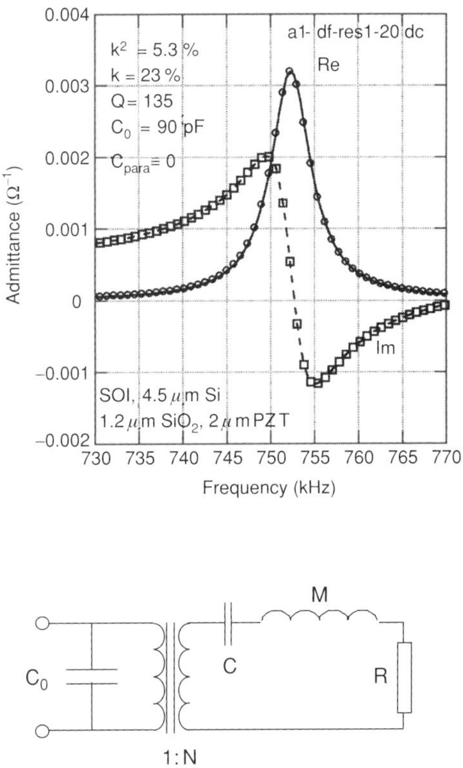


**Fig. 5.** Admittance curves of 300 μm diameter, round, suspended, single transducer at 100 kV/cm dc bias field (from ref. [33]), together with the equivalent circuit model used for the derivation of parameters.

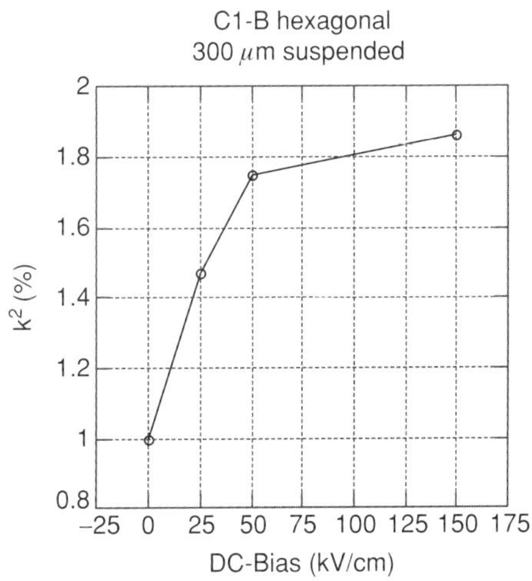


**Fig. 6.** Increase of small signal coupling by the application of a bias voltage. (from ref. [32]).

increase by partial liberation of the plate appears to be rather marginal. More designs and measurements are needed to clarify this point.

## 4. Emission Power

The emission power of an ultrasonic device is an additional important parameter. In case of good acoustic matching the emitted power should approach the mechanical power $P_{mech}$ in the vibrating structure:

$$P_{mech} = k^2 P_{el} = k^2 \omega C_0 U_{rms}^2 \tag{4.1}$$

where $\omega$ is angular frequency and $U_{rms}$ the root mean square voltage applied to the capacitor. Thin films usually dispose of quite large break-down fields. So in principle one can operate the device at much larger fields $E_{rms}$ than what is applied in case of bulk transducers (about 10 times larger). The energy density $\varepsilon_0 \varepsilon_r E_{rms}^2$ may be increased substantially and compensates to a certain extent the smaller thickness of the piezoelectric materials and lower coupling coefficient. In this way, relatively large output powers are achievable and the power specifications are met in theory [32]. However, there is a price to pay: an increased loss due to dielectric loss (small signal) or hysteretic behavior (large signal). This issue requires special attention in order to avoid overheating of the device.

## 5. Quality Factor—Immersed Operation

The investigated pMUT's exhibited high quality factors when operated in air. The example shown in figure 5 (wavelength in air / diameter = 1.3) yielded a Q-factor of 135, the $k^2{*}Q$ product amounted to 7.2. When immersed into a liquid, resonance frequency and Q-factor decrease considerably. The frequency decrease is compatible with Lamb's theory using double side loading by a liquid [35]. Figure 7 shows the admittance of a 1 mm diameter transducer in air and in Fluorinert (an electrically insulating liquid) [33]. The Q factor decreased by a factor 5. Given that the sound velocity in fluorinert is 600 m/s, the 19 kHz resonance corresponds to a wavelength of 32 mm, i.e. a value much larger than the device diameter of 1 mm. In this regime, the impedance of the liquid is low and mostly reactive [34], explaining the large Q observed. Figure 8 shows a 450 μm element of a 16 element array. In this case the resonance in fluorinert is at 190 kHz, corresponding to a wavelength of 3.2 mm. The reduction in Q amounts to a factor 6. Most interesting to achieve larger bandwidths is the modal behavior in immersion. Simulation calculations show that it might be possible to enlarge the bandwidth due to higher modes that are getting close to the basic mode in the liquid due to the bridge geometry [32, 36].

## 6. Discussion and Outlook

The processes available today allow for the fabrication of planar pMUT structures with few micrometer thick PZT films. While such pMUT's may achieve theoretically

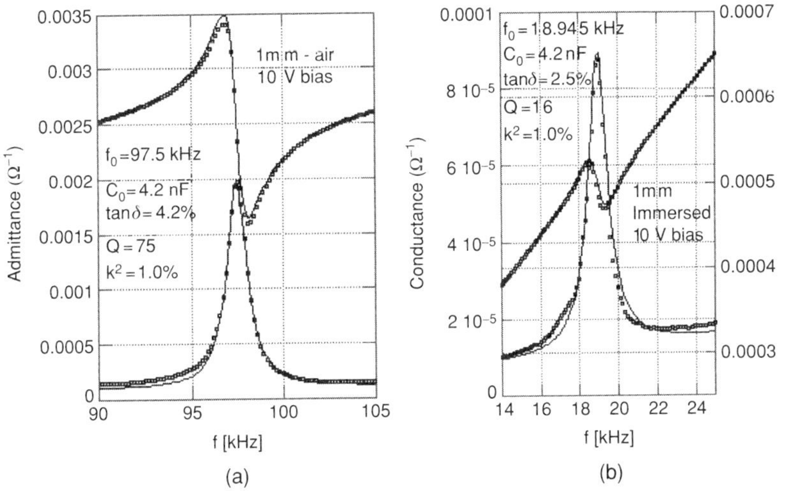


**Fig. 7.** Admittance curves of 1000 μm diameter, round, suspended, single transducer at 50kV/cm dc bias field: (a) in air; (b) in Fluorinert. The full line is a fit to the equivalent circuit model (from ref. [33]).

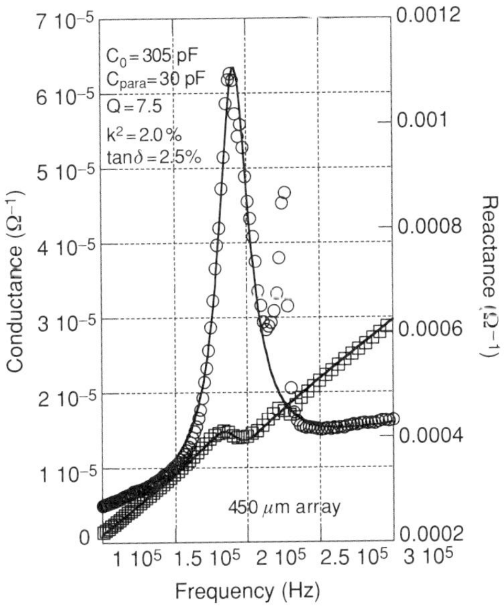


**Fig. 8.** Admittance curves of round, suspended pMUT immersed in Fluorinert. (Diameter: 450 μm, silicon thickness: 5 μm, PZT thickness: 2 μm). It is part of a linear array of 16 elements.

coupling factors of about $k^2 = 10 - 20$ %, as argued earlier in this paper, the best coupling factors achieved to date amounts to 6 %. Improvements of the material seem to be still possible. Comparing the $e_{31,f}$ of PZT films with best ceramics data [23], and prospects with other solid solution systems of $PbTiO_3$ (see [37]), an improvement

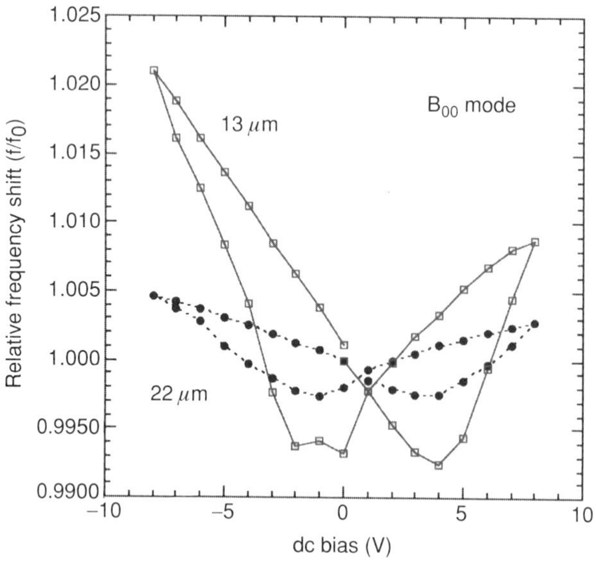


**Fig. 9.** Frequency shift of the fundamental resonance by piezoelectric stresses. Plate diameter: 2 mm; PZT thickness: 0.5 μm; silicon thickness: 13 and 22 μm, electrode: concentric with diameter of 1 mm. (from ref. [27]).

of this coefficient by 50 % looks realistic. This means a factor two increase of $k^2$. Bimorph structures have to be considered as well. They could be suitable to increase the sensitivity, or the bandwidth in reception, due to the fact that bimorphs exhibit a larger piezoelectric part in the vibrating structure. For the emission they do not improve much, since one cannot apply much field opposite to the polarization without switching. So the compressive piezoelectric stresses are limited in value. This can be seen from the frequency shift induced by the piezoelectric stress ($e_{31,f}$) as shown in figure 9. The up-shift corresponds to tensile stress when the electric field is parallel to the polarization, the down-shift to compressive stress when the electric field is anti-parallel to the polarization. The frequency shift is proportional to the piezoelectric stress [27]. Further improvements can be expected with optimized designs to approach an ideal suspended structure. For immersed operation, a further issue needs to be solved. In the present design, we opened the membranes to increase the vibration amplitudes. For immersed operation, it is necessary to seal them again. Ideally, sealing should be done by polymer film. A further idea to increase bandwidth in liquids consists in overlapping resonance modes [36]. Important are simulation calculations to predict the modal behavior in liquids [38] and with sealing layers.

It is clear that pMUT's are more suited for applications with no need of large bandwidths. Possible applications are presence sensors working on short distances as we have demonstrated in transmit-receive experiments [33], as well as by other authors [39]. The actuator area is another promising field. Demonstrations of micropumps and droplet ejectors [40] look very promising.

## 7. Acknowledgements

This work was supported by the European Program "Growth/Innovative Products" and the Swiss Office for Education and Research OFES in the frame of the project "PARMENIDE". I would like to thank all the partners of this project, particularly to J. Baborowski and N. Ledermann from our laboratory, J.-F. Gelly from Thales-Microsonics, and D. Schmitt from the Frauenhofergesellschaft IBMT.

## References

[1] J.C. Bamber and M. Tristam, *Diagnostic Ultrasound*, in *The physics of medical imaging*, S. Webb, Editor. 1988, Adam Hilger: Bristol. p. 319–388.

[2] J.L. Vernet, W. Steichen, R. Lardat, O. Garcia, and J.F. Gelly. *PMUTs design optimization for medical probes applications*. in *IEEEE Ultrasonic Symposium*. 2001. Atlanta (USA): IEEE.

[3] V. Mecea, Tunable gas sensors, *Sensors and Actuators B* **15–16**, 265–269 (1993)

[4] P. Muralt, M. Kohli, T. Maeder, A. Kholkin, K.G. Brooks, N. Setter, et al., Fabrication and characterization of PZT thin-film vibrators for micromotors, *Sensors and Actuators A* **48**, 157–165 (1995)

[5] G. Percin, T.S. Lundgren, and B.T. Khuri-Yakub, Controlled ink-jet printing and deposition of organic polymers and solid particles, *Appl.Phys.Lett.* **73**, 2375–2377 (1998)

[6] M.I. Haller and B.T. Khuri-Jakub. *A surface micromachined electrostatic ultrasonic air transducer*. in *IEEE Ultrasonic Symposium*. 1994.

[7] M.I. Haller and B.T. Khuri-Jakub, A surface micromachined ultrasonic air transducer, *IEEE Trans. UFFC* **43**, 1–6 (1996)

[8] H.T. Soh, I. Ladabaum, A. Atalar, C.F. Quate, and B.T. Khuri-Jakub, Silicon micromachined ultrasonic immersion transducer, *Appl.Phys.Lett.* **69**, 3674–3676 (1996)

[9] J.J. Bernstein, S.L. Finberg, K. Houston, L.C. Niles, H.D. Chen, L.E. Cross, et al., Micromachined high frequency ferroelectric sonar transducers, *IEEE Trans.UFFC* **44**, 960–969 (1997)

[10] O. Oralkan, X. Jin, L. Degertekin, and P. Khuri-Yakub, Simulation and experimental characterization of a 2D capacitive micromachined ultrasonic transducer element, *IEEE Trans. UFFC* **46**, (1999)

[11] P.C. Eccardt and K. Niederer, Micromachined Ultrasonic Transducers with improved coupling factors from a CMOS compatible process, *Ultrasonics* **38**, 774–780 (2000)

[12] R. Farlow, W. Galbraith, M. Knowless, and G. Heyward, Micromachining of piezocomposite transducer using copper vapor laser, *IEEE Trans. UFFC* **48**, 639–640 (2001)

[13] S.J. Ok, C.H. Kim, and D.F. Baldwin, High density, high aspect ratio through-wafer electrical interconnect vias for MEMS packaging, *IEEE Trans. Adv. Packaging* **26**, 302–309 (2003)

[14] X.G. Li, T. Abe, X.Y. Liu, and M. Esashi, Fabrication of high density electrical feed-throughs by deep reactive ion etching pf Pyrex glass, *J.Microelectromech.Systems* **11**, 625–630 (2002)

[15] C.M. Hanson, Hybrid Pyroelectric-Ferroelectric Bolometer Arrays, *Semiconductors and Semimetals* **47**, 123–174 (1997)

[16] A.S. Ergun, G.G. Yaralioglu, and B.T. Khuri-Yakub, Capacitive Micromachined Ultrasonic Transducers: Theory and Technology, *J. Aerospace Eng.* **April 2003**, 76–84 (2003)

[17] J. Johnson, O. Oralkan, U. Demirci, S. Ergun, M. Karaman, and P. Khuri-Yakub, Medical imaging using capacitive micromachined ultrasonic trandsducer arrays, *Ultrasonics* **40**, 471–476 (2002)

[18] G.G. Yaralioglu, A.S. Ergun, B. Bayram, E. Haeggstrom, and B.T. Khuri-Jakub, Calculation and measurements of electromechanical coupling coefficients of capacitive micromachined ultrasonic transducers, *IEEE Trans. UFFC* **50**, 449–456 (2003)

[19] F.V. Hunt, *Electroacoustics; the analysis of transduction and its historical background*. 2nd ed. ed. 1982, Cambridge MA: Harvard University Press.

[20] P. Muralt, Piezoelectric thin films for MEMS, *Integrated Ferroelectrics* **17**, 297–307 (1997)

[21] M.-A. Dubois and P. Muralt, Measurement of the effective transverse piezoelectric coefficient e31,f of AlN and PZT thin films, *Sensors and Actuators A* **77**, 106–112 (1999)

[22] J.F. Shepard, P.J. Moses, and S. Trolier-McKinstry, The wafer flexure technique for the determination of the transverse piezoelectric coefficient (d31) of PZT thin films, *Sensors and Actuators A* **71**, 133–138 (1998)

[23] N. Ledermann, P. Muralt, J. Baborowski, S. Gentil, K. Mukati, M. Cantoni, et al., {100}-textured, piezoelectric Pb(Zrx, Ti1-x)O3 thin films for MEMS: integration, deposition and properties, *Sensors and Actuators A* **105**, 162–170 (2003)

[24] A.M. Flynn, L.S. Tavrow, S.F. Bart, R.A. Brooks, D.J. Ehrlich, K.R. Udajakumar, et al., Piezoelectric micromotors for microrobots, *J.Microelectromechanical Systems* **1**, 44–51 (1992)

[25] G. Percin and B.T. Khuri-Yakub, Piezoelectrically actuated flextensional micromachined ultrasound transducers—I. Theory, *IEEE Trans. UFFC* **49**, 573–583 (2002)

[26] G. Percin, Plate equations for piezoelectrically actuated flexural mode ultrasound transducers, *IEEE Trans. UFFC* **50**, 81–88 (2003)

[27] P. Muralt, A. Kholkin, M. Kohli, and T. Maeder, Piezoelectric actuation of PZT thin film diaphragms at static and resonant conditions, *Sensors and Actuators A* **53**, 397–403 (1996)

[28] M.-A. Dubois and P. Muralt, PZT thin film actuated elastic fin micromotor, *IEEE Trans. Ultrasonics, Ferroelectrics, and Frequency Control* **45**, 1169–1177 (1998)

[29] I. Szabo, *Höhere Technische Mechanilk*. 1956, Berlin, Göttingen, Heidelberg: Springer.

[30] J. Söderquist, Similarities between piezoelectric, thermal and other internal means of excitation vibrations, *J.Micromech.Microeng.* **3**, 24–31 (1993)

[31] J. Baborowski, Microfabrication of piezoelectric MEMS, *J.Electroceramics* (2003)

[32] P. Muralt, D. Schmitt, N. Ledermann, J. Baborowski, P.K. Weber, W. Steichen, et al. *Study of PZT Coated Membrane Structures for Micromachined Ultrasonic Transducers*. in *IEEE Ultrasonic Symposium*. 2001. Atlanta (USA).

[33] J. Baborowski, N. Ledermann, and P. Muralt. *Piezoelectric Micromachined Ultrasonic Transducers based on PZT thin films*. in *IEEE Ultrasonic Symposium*. 2002. Munich (Germany).

[34] D. Royer and E. Dieulesaint, *Ondes elastiques dans les solides*. Vol. 1. 1996, Paris: Masson.

[35] H. Lamb, On the vibration of an elastic plate in contact with water, *Proc.Royal Soc London* **A 98**, 205–216 (1920)

[36] J. Baborowski, N. Ledermann, P. Muralt, and D. Schmitt, Simulation and characterization of piezoelectric micromachined ultrasonic transducers (pMUT's) based on PZT/SOI membranes, *Int. J. Comp. Eng. Sci.* **4**, 471–475 (2003)

[37] S. Troyler-McKinstrey and P. Muralt, see Chapter 10, this book, (2005)

[38] J.F. Gelly, F. Lanteri, and S. Ballandras. *MUT: Benchmarking with State of the Art Transducers for Medical Imaging Purpose*. in *3rd MUT workshop*. 2003. Lausanne.

[39] K. Yamshita, H. Katata, M. Okuyama, H. Miyoshi, G. Kato, S. Aoyagi, et al., Arrayed ultrasonic microsensors with high directivity for in-air use using PZT thin film on silicon diaphragms, *Sensors and Actuators A* **97–98**, 302–307 (2002)

[40] G. Percin and B.T. Khuri-Yakub, Micromachined droplet ejector arrays for controlled ink-jet printing and deposition, *Rev. Sci. Instr.* **73**, 2193–2196 (2002)

# Chapter 4

# Thick-Film Piezoelectric and Magnetostrictive Devices

N.M. White, S.P. Beeby and N.J. Grabham
*School of Electronics and Computer Science University of Southampton SO17 1BJ UK*

## Synopsis

Thick-film technology exhibits a variety of desirable characteristics, many of which are particularly relevant to the realisation of micro-sensors and actuators. This chapter describes how screen printed thick-films can be used as the basis for a variety of piezoelectric and magnetostrictive transducers. The technology has also been combined with MicroElectroMechanical Systems (MEMS) to generate new types of micro-engineered structure. The evolution of the technology to a successful enabling mechanism for solid state sensors is described.

The chapter begins with a brief overview of piezoelectric thick-films including a discussion of the main factors relating to paste formulation, characterisation and techniques for fabricating devices. There is also a description of methods for fabricating thick-films on silicon, which opens up the possibility of using thick-film technology in the field of MEMS. A number of specific sensors and actuators are described, including accelerometers, micropumps, ultrasonic motors, slip sensors for prosthetic hands, resonators, elastic wave sensors and ultrasonic separators.

The latter stages of the chapter deal with the evolution of thick-film magnetostrictive devices. Coverage is given to the fundamental concept of magnetostriction, with reference to the giant magnetostrictive material Terfenol-D. The formulation and characterisation of magnetostrictive thick-films is discussed, together with a review of some applications of films in a variety of application areas including micropumps, pressure sensors, microvalves, optical switching and displacement sensors.

# 1. Introduction

Transducer development continues to expand, and is largely driven by the advances in the field of microelectronics. Thick-film technology was introduced about forty years ago as a means of producing *hybrid* circuits. These comprise semiconductor devices, monolithic ICs and other discrete devices in addition to the thick-films themselves. Thick-film hybrid circuits are generally regarded as being compact, robust and relatively inexpensive and have found application in areas such as televisions, calculators, telephones, automotive electronics and consumer products. Thick-film sensors are a relatively new application of the technology, with a majority of the published working appearing in the past 15 years or so. This paper will demonstrate how some of the essential aspects and materials used in thick-film technology have been exploited to provide an alternative method of sensor fabrication. The integrability of the technology also allows combination with other enabling technologies, such as silicon, to provide powerful and economically-viable solid state sensors [1, 2].

Thick-film technology therefore allows for the provision of electronic circuits within the transducer housing, enabling self-contained modules capable of delivering sensor data that may have been processed and conditioned in some manner (i.e. filtered, calibrated etc.). Additionally, the technology provides for the creation of support structures upon which other sensor materials can be deposited, perhaps using other fabrication techniques (thin film, spraying etc.). Furthermore, the thick-film material itself can be exploited as the main transduction element. This latter point is evidenced by the development of piezoelectric and magnetostrictive thick-films, which will now be described in more detail.

# 2. Thick-Film Piezoelectric Devices

The word 'piezo' is derived from the Greek *piezein* meaning to press, and hence piezoelectricity is literally 'pressure electricity'. Lead Zirconate Titanate (PZT) is a piezoelectric ceramic material that was discovered almost half a century ago. It has a relatively high value of charge sensitivity ($d_{33}$ in the range 150–600 pC/N) and Curie temperature in the range 150 to 400 °C. These properties make PZT a good choice for sensor and actuator applications.

Screen printed piezoelectric films are based on a combination of PZT powder, together with a glass frit and suitable vehicle. The films are deposited as pastes using a screen printing process. After drying at a temperature of around 150 °C in an infra red belt drier, the films are then exposed to a high temperature firing cycle. This takes place in a belt furnace at temperatures up to 1000 °C and during this stage, the powders sinter to form a fired, composite structure.

Some of the fundamental, linear relationships between the electrical and mechanical quantities of a piezoelectric material are given below:

$$\mathbf{D} = d\mathbf{T} + \varepsilon^{\mathrm{T}}\mathbf{E} \tag{2.1}$$

$$\mathbf{S} = s^{\mathrm{E}}\mathbf{T} + d\mathbf{E} \tag{2.2}$$

where **D** is the dielectric displacement, **T** the stress, $\varepsilon$ the dielectric constant, **S** the strain, **E** the applied electric field (V/m), $s$ the compliance (inverse of modulus, $m^2$/N) and $d$ (C/N) is a piezoelectric coefficient. The superscripts denote the quantity being kept constant at the boundary conditions. For example $s^E$ is the compliance at a constant electric field (electrodes short circuited).

Equation (2.1) describes the *direct* piezoelectric effect and equation (2.2) describes the *reverse* piezoelectric effect, relating the strain produced in response to an applied stress and electric field. For vibration or motion applications, such as sounders, high $d$ coefficients are desirable. A further quantity is the voltage coefficient $g$, given by

$$g = \frac{d}{\varepsilon_o \varepsilon_r} \tag{2.3}$$

This represents the open circuit voltage generated by an applied stress and is expressed in Vm/N.

In general, the $d$ and $g$ coefficients have different values depending upon the orientation of the sample. Numerical subscripts are used to specify the directional properties. Direction 3 is considered to be the direction along which the sample has been polarised and directions 1 and 2 are the other perpendicular dimensions. The $d$ and $g$ coefficients generally have two subscripts, the first of which refers to the electrical quantity, and the second represents the mechanical quantity. As an example, $d_{33}$ is a coefficient of the material measured when the applied force is along the same direction as the poling axis.

## 2.1. History

Piezoelectric thick-films were first described by Baudry [3] in 1987, who fabricated an acoustic coupler to demonstrate an application of the technology. The thick-film material was based on a combination of PZT powder, a small amount of glass frit and an organic vehicle. Finite element analysis was performed on various geometries of alumina substrate in order to find the optimum dimensions for thick-film piezoelectric buzzers. A further application addressed the design of an acoustic coupler. Such devices can be used as a means of communicating across an electrically-insulating region.

Morten *et al* [4, 5] describe the preparation and characterisation of piezoelectric thick-films based on PZT. The quantity of glass binder in the formulation was varied from 3 to 30% and it was found that the lower the glass content, the higher the value of $d_{33}$. The authors also suggest that the $d_{33}$ coefficients of pastes prepared with a lead oxide binder were higher than those measured on similar pastes using lead-alumina-silicate glass.

Work undertaken by Moilanen *et al* studied ways to reduce the number of print cycles required for multilayer actuators. A novel screen-printing technique, known as double-paste printing, was developed for this purpose [6]. The technique maximises the efficiency of the printing process as it only requires a single screen to be aligned. The method will not work for all designs, but is suggested to be particularly beneficial for reducing costs in the mass production of multilayer stacked actuators.

Investigations concerning the use of dispersants in PZT films as a means of optimising the particle distribution have also been conducted. In order to improve the quality of the printed films it is essential that the paste spreads evenly on the substrate. Dispersants act to create repulsive forces between the particles, thus reducing clumping and increasing the homogeneity of the film. Thiele and Setter [7] have studied various dispersants and their effects on the rheology of thick-film pastes. Their experiments looked at eight commercially available dispersants with a Sensor Technology BM500 PZT powder. The results showed that a phosphate ester oligomer was the optimum dispersant. Rheological and optical microscopy measurements of the fired film surfaces highlighted the benefits of using a dispersant in the thick-film paste formulations. A decrease of 40% in mean surface roughness was observed for the film containing the dispersant compared with the sample containing none. The addition of only 0.1 wt.% of the dispersant result of in a dramatic decrease in viscosity (2.42 Pa.s to 0.22 Pa.s), which allows for a higher concentration of solids. They also showed that the dispersant has no adverse effects upon the dielectric and piezoelectric properties.

Studies by Papakostas *et al* [8] have looked into the use of polymer binders in thick-film pastes. This replaces the usual glass frit in an effort to lower the processing temperatures. The experiments used PZT5H powder and Ronascreen Green OPSR 500 polymer. Three pastes were analysed, 90, 85 and 75 wt.% PZT powder with their respective amounts of polymer paste. These pastes were printed on to an alumina substrate with a capacitor structure of two electrode layers with the test paste in between. The bottom electrode was a standard cermet silver/palladium ESL 9635-A paste used for convenience in the experiment. The top electrode was a silver polymer (ESL 1107-S) paste.

## 2.2. Fabrication

Thick-films are deposited by a screen-printing process. This is an effective and inexpensive method for producing the desired film geometry on the surface of the substrate (typically alumina, insulated stainless steel or silicon). The process is suitable for mass production and offers the ability to produce films of thickness 10–100 $\mu$m in a single print stroke. Limitations of conventional screen-printing are the minimum feature size and geometry achievable; typical minimum line width and space being 100–150 microns. Research by Robertson *et al* has demonstrated the production of a thin stainless steel foil into which the required pattern is etched. This has been termed the $\mu$-Screen [9]. This custom-etched mesh is mounted in a conventional screen frame and is therefore compatible with industrial screen printers. The $\mu$-Screen allows the production of screen-printed features of 50 μm line width and spacing.

For the production of glass-ceramic (cermet) piezoelectric films, it is necessary to sinter them at a suitable temperature (typically around 900 °C) to cause the binding material (a lead-based glass) to melt with the PZT powder and form a bonding matrix. The films form a fired, composite material often having different electro-mechanical properties from those of the bulk material. Studies by Glynne-Jones *et al* have been conducted into the firing profile for thick-film layers on silicon [10]. The films were based on PZT type 5H supplied by Morgan Electro Ceramics Ltd., and were fired over a range of temperatures from 750 °C to 1000 °C. The results showed that films at the lower firing temperature of 750 °C exhibited poor sintering. At temperatures above 800 °C, samples showed acceptable adhesion and sintering. The samples fired at the lower temperature exhibited $d_{33}$ values of between 101 and 109 pC/N. A maximum value of 169 pC/N was obtained from the samples fired at 1000 °C. This is due to the improved sintering of the film at these higher temperatures. One drawback, however, is that the increased firing temperature causes lead to migrate from the film and pollute the surface of the silicon.

## 2.3. Silicon processes (MEMS)

The combination of the screen printing process with silicon micromachining is, perhaps, not an obvious choice. The screen printing process is a physical process that exerts pressure on the substrate. Micromachined silicon devices typically have sub-micron feature sizes and are too fragile to withstand the printing process. In addition, the resolution and accuracy of alignment registration is generally poorer than that achievable with photolithographic processes. There is, however, certainly a place for micromachined silicon devices incorporating thick-film piezoelectric films. Micropumps are an example of a larger device that requires more substantial actuating forces than those achievable with typical thin-film piezoelectric layers. Other advantages include the fact the screen printing process deposits the material in the desired pattern, thus negating the need for subsequent photolithographic and etching steps. The screen printing process is essentially a simple, batch compatible low-cost process, which is accessible to a wide range of MEMS developers. Combining screen printable piezoelectric layers with micromachined structures opens up a large number of opportunities for various sensing and actuating applications.

The first published research into printing piezoelectric layers on silicon was published in 1995 by Chen *et al* [11]. This work successfully demonstrated that films up to 12 μm thickness could be printed onto oxide-coated wafers with 15O nm thick sputtered platinum film on a 20 nm thick titanium adhesion layer. The PZT film was dried at 110 °C and fired at a peak temperature of 850 °C for 15 minutes. This firing profile produced blistering of the titanium/platinum films and resulted in a porous PZT film that has not completely densified. A top platinum electrode was also sputtered onto the PZT thereby forming the standard capacitor structure. The electrical properties of the film depend upon the PZT powder used and the processing parameters. This work used a hard piezoelectric material that exhibited a relative permittivity of 200 and a $d_{33}$ coefficient of 50 pC/N.

The effects of high-temperature firing were also highlighted by Maas *et al* [12]. PZT-based films where fired onto silicon wafers at a peak temperature of 950 °C. Lead was found to diffuse out of the PZT and into the silicon and could not be prevented by using oxide or nitride barrier layers. The contaminated areas around the PZT become conductive and the only material that was able to prevent the lead diffusion was a printed layer of commercial dielectric (IP211 from Heraus). Incorporating such a barrier layer, however, substantially increases the thickness of the overall structure and reduces the mechanical coupling efficiency of the piezoelectric material.

The combination of screen printed piezoelectric films and micromachined silicon structures was explored further by Beeby *et al* [13, 14]. This work investigated the spread of the reaction caused by the lead evaporation versus firing temperature. The investigation confirmed that the reaction begins at a temperature of 850 °C and continues

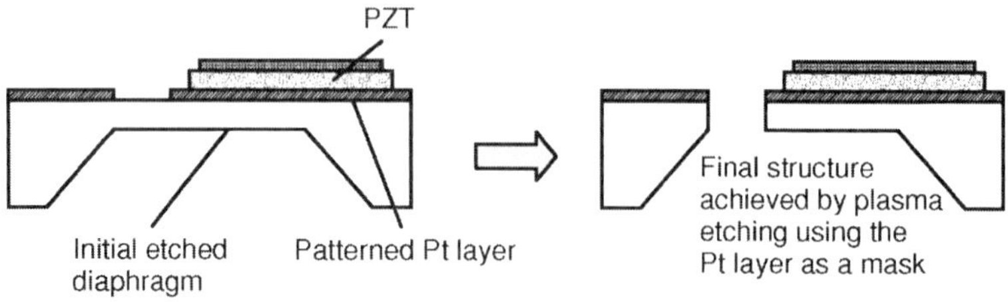


**Fig. 1.** Combined silicon micromachining and thick-film piezoelectric fabrication process.

as the temperature increases. The reaction process has also been studied by Thiele *et al* who have postulated that the reaction product is a lead silicate [15]. In contrast to the findings of Mass, Thiele found the reaction material to be non-conducting. Thiele also reported evidence of the diffusion of silicon ions into the film at high firing temperatures. This diffusion results in the formation of zircon crystals ($ZrSiO_4$) in PZT films fired at 950 °C for one hour, and reduces the dielectric properties of the PZT film.

Beeby's work [14] also explored the possibilities of performing standard micromachining processes on wafers with thick-film piezoelectric films already deposited. The wafers were exposed to a variety of photolithographic and wet and dry etching processes. In summary, the plasma based dry etch processes were found not to affect the PZT or electrode films. The lower platinum electrode could therefore be used as a masking layer for subsequent silicon etches. This development led to the definition of a process whereby structures could be defined and released after printing of the PZT and top electrode films (see Figure 1). The lead migration reaction product, however, was found to inhibit these etches and could potentially cause difficulties in defining structures accurately. Some of the devices discussed in the following sections have been fabricated using this process.

## 2.4. Characterisation

The piezoelectric constant $d_{33}$ is an important parameter for sensor applications as it relates the electrical charge produced for a given applied mechanical stress. Hence, this gives an indication of the sensitivity of the film. This property is particularly significant because the majority of applications using planar technology have electrodes on the upper and lower surfaces of the film and thus the poling field and measurement direction are both in the '3' axis. The most common technique for measuring $d_{33}$ is the so-called *direct* or *Berlincourt* method, applying a force to the piezoceramic and measuring the charge generated. Care must be taken when using the direct method with thick-film samples, as the substrate can bend during the measurement and lead to inaccurate readings. The Take-Control PM35 piezometer [16] is an example of a commercial instrument designed to measure $d_{33}$ values in bulk piezoelectrics. This has been shown to produce consistent values of $d_{33}$ provided that allowance is made for the effect of bending stresses within the substrate and the sample is held firmly in the rig [17]. By altering the substrate supporting structure it has been shown that the Berlincourt piezometer can also be used to measure $e_{31}$(and therefore $d_{31}$) [18]. The sample is supported by a ring structure whilst the upper force is applied on the sample surface at the centre of the ring. The analysis requires some basic mathematical treatment which is presented in reference 18.

Torah *et al* [18] suggest that the effect of clamping reduces the measured piezoelectric coefficient of the film. The reduction is due to the effect of the $d_{31}$ coefficient. Theoretical analysis predicts a reduction in the measured $d_{33}$ of 62%. Typical thick-films have a measured, clamped $d_{33}$ value of 130 pC/N, which is therefore the equivalent of an unconstrained value of around 350 pC/N. Work carried out by Dorey and Whatmore on sol-gel thick films also explored the effect of substrate clamping on the measured $d_{33}$[19]. The substrate clamping effect introduces stresses in the PZT film which distort the unit cell and reduce the effectiveness of the polarisation process.

## 2.5. Devices

Thick-film piezoelectric devices have been used in a variety of applications. In the first paper on thick-film piezoelectrics, published in 1987, Baudry [3] describes the design of two devices. The first is a thick-film piezoelectric buzzer. These have the advantages of light weight, low power consumption and robustness. The second application was that of an acoustic coupler. Such devices can be used as a means of communicating across an electrically insulating region. Two piezoelectric layers were fabricated onto both sides of an alumina substrate so that they were mechanically bound but electrically isolated. The frequency of operation was designed to be 100 kHz, well beyond the audible range.

### *2.5.1. Micropumps*

An early example of combining screen printed thick-film PZT layers with silicon micromachined structures is described by Koch *et al* [20, 21]. A micropump was fabricated comprising three silicon wafers that were fusion bonded together. The device is depicted in Figure 2. The passive cantilever valves were produced by a boron etch stop technique and fusion bonding. Tests of the valves showed good performance, as no flow could be detected in the reverse direction. Initial experiments on a thick-film screen printed piezoelectric membrane actuator were undertaken. A study of suitable pastes for electrodes on different insulation layers on silicon revealed silicon dioxide and cermet gold paste as the most satisfactory combination. Deflection measurements of a 7 mm $\times$ 3 mm PZT and 8 mm $\times$ 4 mm silicon bimorph membrane gave 1 μm movement at an applied voltage of 100 V. A quasi-static simulation package of the flow through the micropump was also undertaken. The valve action was simulated using the ANSYS finite element package coupled with FLOW3D. Pump rates of up to 400 micro litres per minute with a maximum backpressure of up to 35 kPa at 100 V driving voltage across the PZT were measured. The existence of lead pollution around the area of the PZT was noted and this had the effect of causing the $SiO_2$ barrier layer to become conductive as mentioned previously.

A similar diffuser micropump has been developed by Kim *et al* [22] and uses a TMAH-etched diaphragm with a screen printed PZT-PCW actuator on rectangular or square diaphragms of approximately 20 microns thickness (see Figure 3). A lead-based complex oxide (PCW) additive has been infiltrated into the screen printed PZT film to

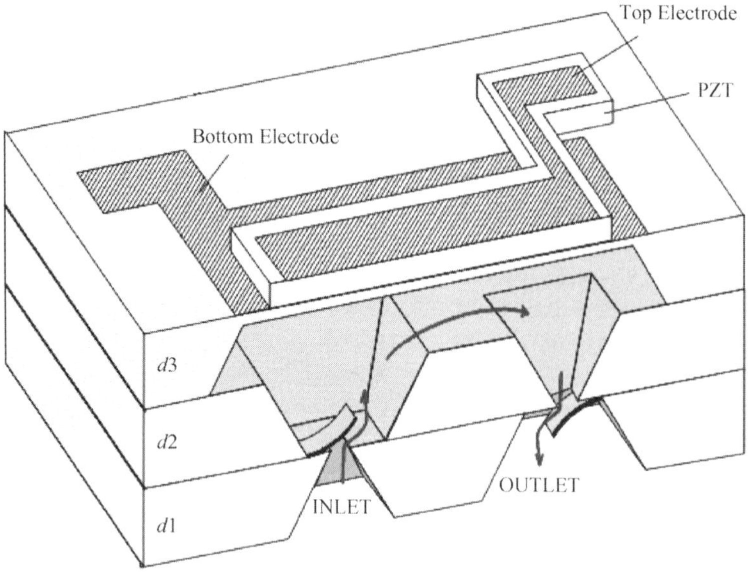


**Fig. 2.** Combined silicon/thick-film micropump (after Koch *et al* [21]).

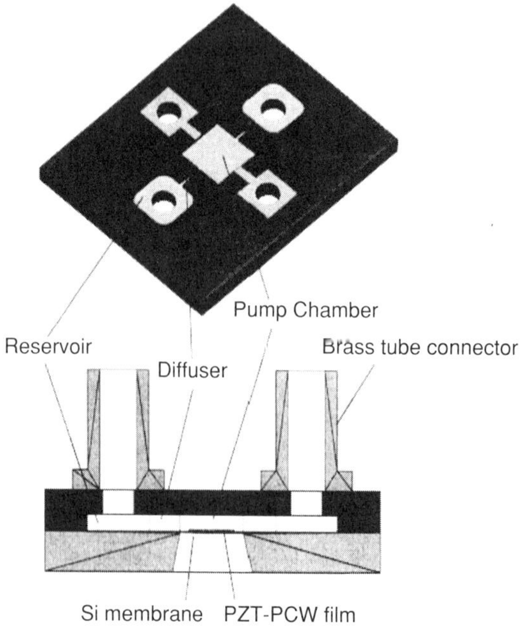


**Fig. 3.** Schematic of diffuser type micropump (after Kim *et al* [22]).

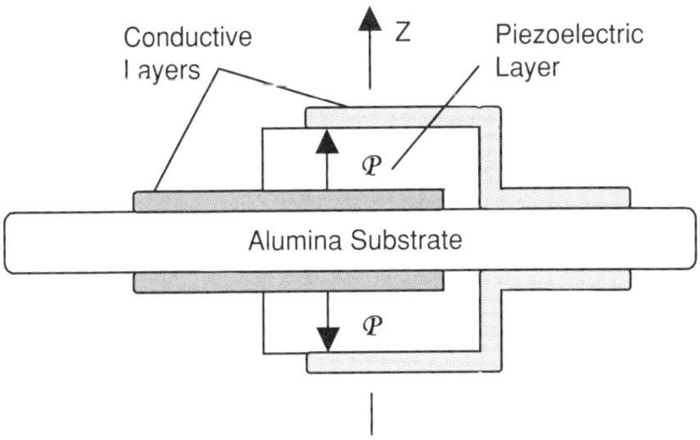


**Fig. 4.** Cross-sectional view of a thick-film accelerometer (after Crescini *et al* [23]).

improve both density and the electrical properties. The pump chamber was etched into either a Pyrex or silicon wafer, which is bonded to the actuator wafer to form the sealed cavity. The resonant frequency of a typical pump design is 53.3 kHz in air and this falls to 32.4 kHz when filled with a fluid. The overall size of the pump is 9 × 11 mm in size and the optimum geometry of PZT (1.5 × 1.5 mm) produces a mean flow rate of 5.3 μl/min.

### 2.5.2. *Accelerometers*

Several different configurations of piezoelectric thick-film accelerometer have been documented. Crescini *et al* have developed a thermally compensated thick-film accelerometer [23]. The device uses two sensing elements, printed on either side of the substrate in a planar capacitor structure, which allows for the reduction in thermal drift. By connecting the two piezoelectric elements in series and packaging the sensing elements and electronics close together, it is possible to reduce the effects of the extra charge generated with increase in temperature. This, however, requires that the manufacturing limits of the devices are extremely close together. The two devices are polarised in the same axis, but have opposite polarity. This means that thermal drift in one device will be the opposite polarity in the other and the effect is therefore nulled. Figure 4 shows a schematic of the device. The high frequency cut-off is determined only by the resonance properties of the alumina substrate, the elastic properties of the case and the bonding techniques used. The printed films were 40-50 μm thick, with a $d_{33}$ coefficient of around 180 pC/N, when poled at 3 MV/m.

Another example of a combined thick-film and silicon micromachined structure is the accelerometer described by Beeby *et al* [24]. The device is a deflection-based dynamic mechanical accelerometer having a 17 mg inertial mass suspended by four silicon beams, each located at a corner of the seismic mass. The deflection of the inertial mass due to an applied acceleration is sensed using thick-film piezoelectric elements located on the supporting beams. Each piezoelectric element is printed as a planar capacitor structure with the active layer being sandwiched between an upper and lower electrode. As the mass moves relative to the chip frame, the beams deflect causing the piezoelectric layer to deform and hence a charge is induced. The amount of

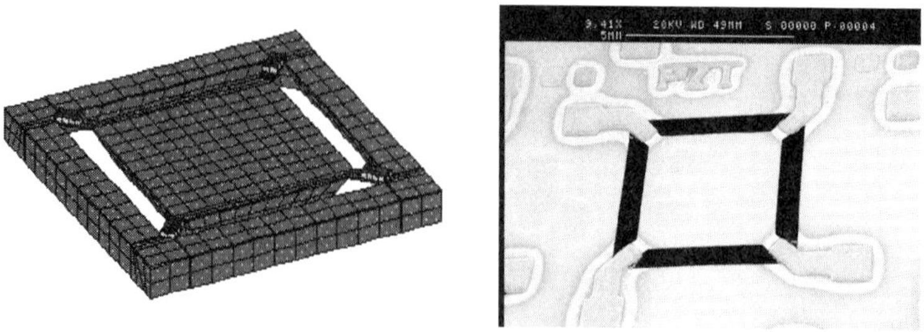


(a) (b)

**Fig. 5.** (a) Finite element model and (b) electron micrograph of a combined thick-film/silicon accelerometer (after Beeby *et al* [24]).

charge generated depends upon the piezoelectric properties of the printed layer and is proportional to the deflection generated by applied accelerations. The device is depicted in Figure 5, which shows a finite element model of the device, together with a scanning electron micrograph of the actual device. The total capacitance of the elements in this arrangement was measured at 360pF. The accelerometer was tested on a Goodmans vibration generator type V50 Mk1. This was driven via a Goodmans PA50 VA power amplifier by an HP 35660A dynamic signal analyser. The system was calibrated using a Bruel & Kjaer accelerometer type 4369 and the charge was measured using a Kistler 5001 charge amplifier in order to determine the fundamental sensitivity in terms of pC/g. The sensitivity of the device was found to be 16 pC/g in the $z$ (vertical) direction. The $x$- and $y$-axis cross-sensitivity was measured at 0.64 pC/g for the accelerometer, which is equivalent to 4% of the full scale output. The overall fabrication process was relatively straightforward, providing a yield in excess of 95%, which is quite impressive for a micromachined device. Evidence of the lead pollution around the PZT, as described in section 2.2, can be clearly seen in the electron micrograph of Figure 5.

### *2.5.3. Ultrasonic motors*

Ultrasonic motors are based or the principle of converting vibrations into rotary motion. They offer the advantage over electromagnetic motors of removing the need for wire-wound components, such as the stator and rotor, and hence are amenable to integration using micro-fabrication techniques [23, 26, 27]. Aoyagi *et al* [28] described an ultrasonic motor that uses thick-film PZT actuators to induce specific resonant modes into a substrate. These modes ($B_{22}$, $B_{30}$ and $B_{03}$) are induced in different directions according to the electrode design. They can then be combined in order to vibrate the substrate. On the upper side of the substrate there is a ceramic column, which is moved as a result of the induced vibrations. A ball bearing located on top of the column, has freedom of movement in all directions, and hence this type of motor is sometimes referred to as a multi-degree-of-freedom (MDOF) device. The principle of operation is illustrated in Figure 6.

The stator of the ultrasonic motor consists of a 20 mm square alumina plate, 250 μm thick with an alumina cylinder, 3 mm outside diameter, 0.5 mm thick and 3 mm long, attached to the centre of the plate. This amplifies the lateral displacements of the plate

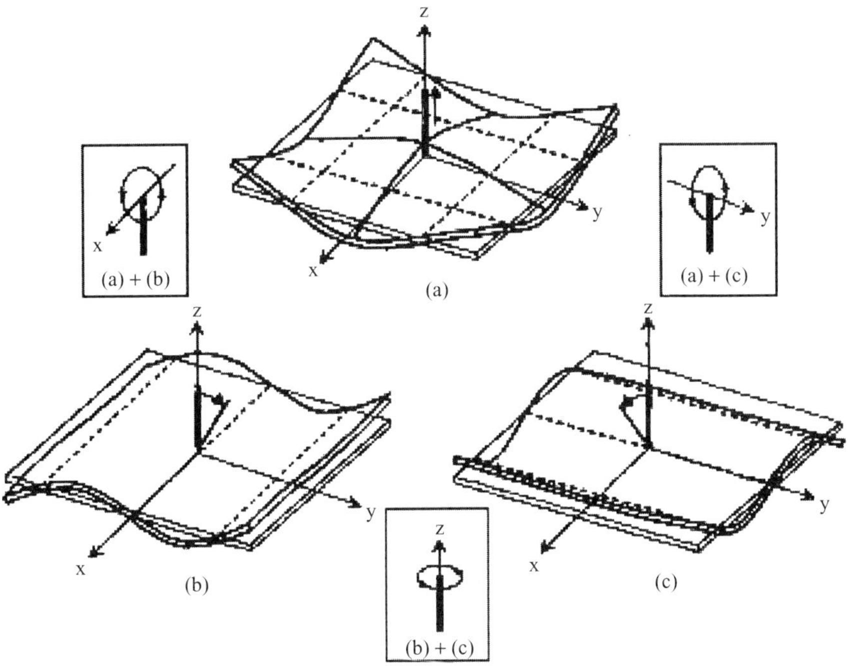


**Fig. 6.** Resonant modes in a multi-degree-of-freedom ultrasonic motor, $B_{22}$, $B_{30}$ and $B_{03}$ respectively, and their associated displacements, (after Aoyagi *et al* [28]).

associated with the bending modes, which on the surface of the plate, are relatively small. The alumina cylinder also holds a small ball bearing, which serves as the rotor at the centre of the vibrator. The thick-film PZT electrodes were printed on the underside of the alumina plate.

The arrangement of the thick-film electrodes and piezoelectric elements is shown in Figure 7. The structure was analysed on a purpose-built optical fibre interferometer. The applied excitation voltage for these measurements was 200 V peak-to-peak. The resonant frequencies of the $B_{30}$ and $B_{22}$-modes were found to be to 15.45 kHz and 16.2 kHz respectively. Finite element analysis had predicted that the resonance frequencies of the $B_{30}$ ($B_{03}$) and $B_{22}$ modes were 14.47 kHz and 16.49 kHz, respectively. These frequencies should, ideally, be the same in order to reduce both the mechanical and, as a result, electrical impedance. The resonance responses were, however, rather small and this is because the thick-film printed PZT layer has a relatively low piezoelectric coefficient, $d_{31}$, of approximately $-33$ pC/N compared to the bulk value of $-274$ pC/N. Rotation angles of the rotor were changed by applying short bursts containing a number of sinusoidal pulses with a frequency of 16.23 kHz and an amplitude of 200V peak-to-peak. Precision positioning of the rotor was found to be possible with this device and a resolution of rotor angle of approximately 0.08 degrees per unit pulse after 5 ms burst of drive pulses is achievable.

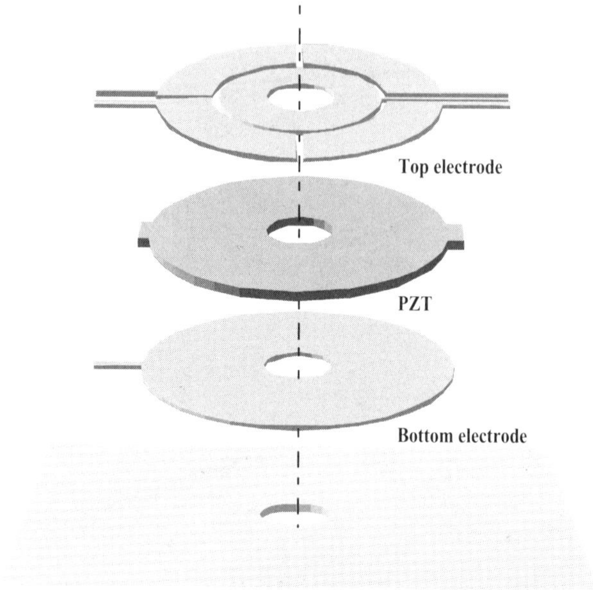


**Fig. 7.** A schematic of the arrangement of the PZT and electrode layers of the thick-film ultrasonic motor (after Aoyagi *et al* [28]).

### 2.5.4. Prosthetic hands—slip sensors

One of the problems with a majority of prosthetic devices is the lack of feedback control. In the case of a prosthetic hand, for example, the operator cannot feel an item within the grasp of the hand—the operator has no 'sense' of what it is that they are holding, beyond that being visually assessed. The inability to monitor the grip force imposed on a grasped item means that the user cannot be totally sure of the security of the grip and is unaware when the object begins to slip from their grasp. An example of a prosthetic hand is shown in Figure 8 [29]. This is a prototype myoelectrically-driven device, which was designed at the University of Southampton. In use, the hand is controlled by the electrical signals produced by any convenient flexor-extensor muscle pair; one signal opens the hand and the other closes it. Each finger on the hand is individually controlled by its own dedicated motor, allowing independent flexion and extension of each mechanical digit. Thick-film piezoelectric sensors have been fabricated onto the stainless steel finger tip as described by Cranny *et al* [30], and shown in Figure 9. These are primarily intended to act as slip sensors. If an object is being held within the hand, the PZT sensors can detect the onset of slip by detecting the vibrations within the finger tip. A signal can then be sent to the motors to cause the hand to grip tighter.

### 2.5.5. Resonators

Resonant sensors confer a number of benefits over those with traditional analogue outputs. They can be very sensitive and versatile, allowing the measurement of a variety of variables (pressure, acceleration, mass, viscosity etc.). Also frequency can be a very

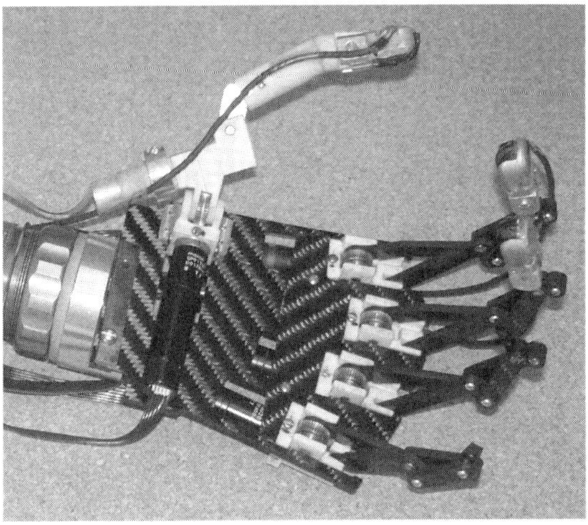

**Fig. 8.** The Southampton-Remedi prosthetic hand (after Light and Chappell [29]).

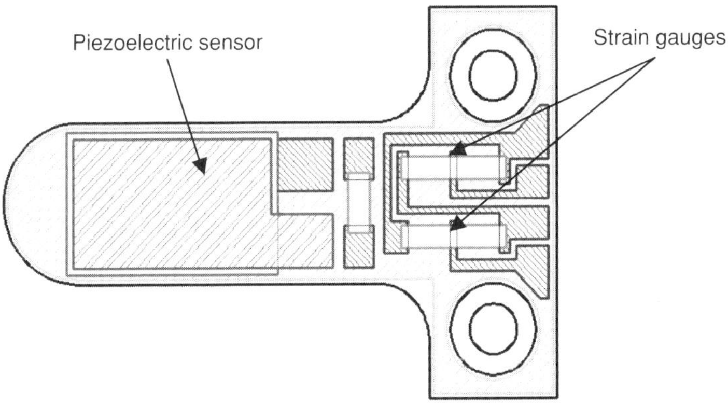


**Fig. 9.** Prosthetic hand fingertip comprising piezoelectric slip sensor and thick-film strain gauges (after Cranny *et al* [30]).

convenient form of information encoding, as it permits the acquisition of data via simple timer-counters such as those provided with modern microcontrollers. A major advantage is that the resonant frequency is determined mainly by the physical structure and the measurand, and is relatively insensitive to the tolerances of added films [31]. Yan *et al* describe an application for thick-film piezoelectric materials as the excitation and detection mechanism of a triple beam resonator [32].The device consists of three beams (or tines) aligned alongside each other and joined at a de-coupled region at each end to the surrounding material. The central beam has twice the width of the two outer beams. The resonating element has an overall length of 15.5 mm, a thickness of 0.25 mm and beam widths of 2 mm and 1 mm, respectively. The distance between the beams is 0.5 mm. A photograph of the actual device is shown in Figure 10.

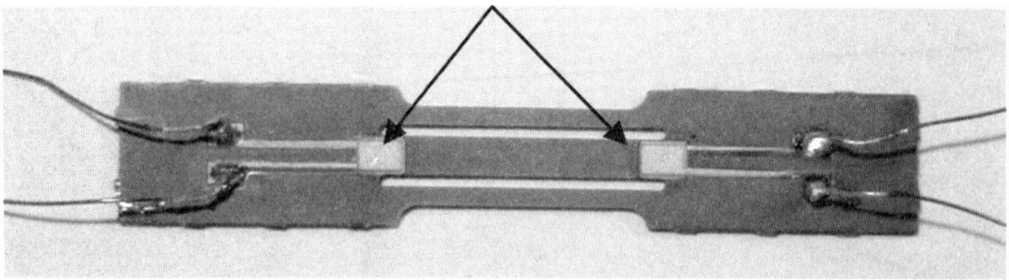


**Fig. 10.** A triple-beam tuning fork resonator (after Yan *et al* [32]).

Thick-film PZT elements were screen printed on separate regions at each end of the central beam, where maximum surface strains exist. The PZT element at one end acts as an actuator, driving the vibrations, whilst the other element detects them. Positioning the PZT driving and sensing elements on the regions of maximum stress, optimises the driving forces and sensing signals. The substrate of the resonator was etched from a 0.5 mm thick 430S17 stainless steel thin plate. A dielectric layer was screen printed, dried and fired at the defined driving and sensing regions on the top surface of the resonator using a standard screen-printing process. Consecutive layers of lower gold electrode, piezoelectric paste and upper gold electrode were then deposited and fired sequentially. The dielectric layer was required to isolate the lower electrode from the resonator substrate. The device is operated in its third mode of vibration, as this provides out-of-plane motion of the tines, whose moments of inertia cancel out thereby providing a dynamically-balanced structure with a quality factor, Q, in excess of 4000.

Micromachined silicon beam resonators with thick-film piezoelectric drive and detection elements have also been demonstrated by Beeby *et al* [33, 34]. The devices consisted of an encastre beam etched into a silicon membrane as shown in the SEM photo in Figure 11. The fabrication process followed that described in section 2.3 and depicted in Figure 1. The screen printed PZT film of 40 mm thickness is sandwiched between a common evaporated Pt bottom electrode and a Au printed top electrode. FEA was used to simulate the device and this predicted a fundamental resonant frequency of 68.4 kHz which compared to the experimentally observed 56.5 kHz. The difference was predominantly due to the slight over-etching of the membrane producing a thinner than expected resonator. The Q-factor of this device was found to be 70 in air which highlights the increasing importance of gas damping effects on resonator structures as they reduce in size. It is also indicative of an unbalanced resonator structure, as opposed to the triple beam tuning fork discussed above.

The screen printed piezoelectric layer itself has also been used as a resonant structure. A planar capacitor structure, as shown in Figure 12, printed onto an alumina substrate can be operated as a thickness mode resonant structure. When operated in such a mode, the structure is sensitive to mass loads applied on the free surface. Sensors have been fabricated by Ferrari *et al* [35, 36] that demonstrate a mass sensitivity of 114 kHz $mg^{-1}$ on devices with a thickness of 140–150 μm. Applications of such a structure include chemical sensing, whereby the thick-film structure is coated by a chemically sensitive material that adsorbs certain chemical molecules. The oscillator circuit that controls the resonance has been fabricated as a hybrid circuit around the

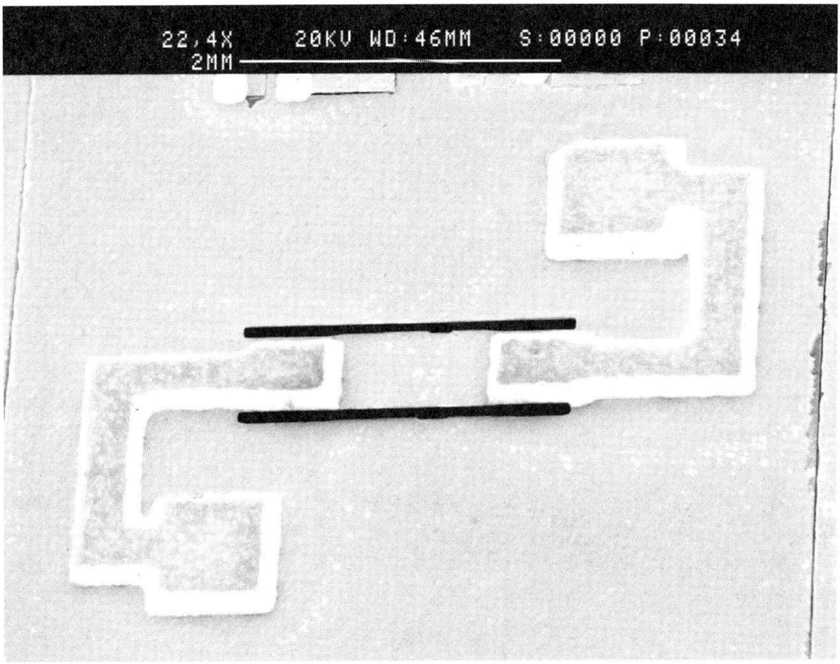


**Fig. 11.** Silicon encastre beam resonator with thick-film PZT drive and detection elements.

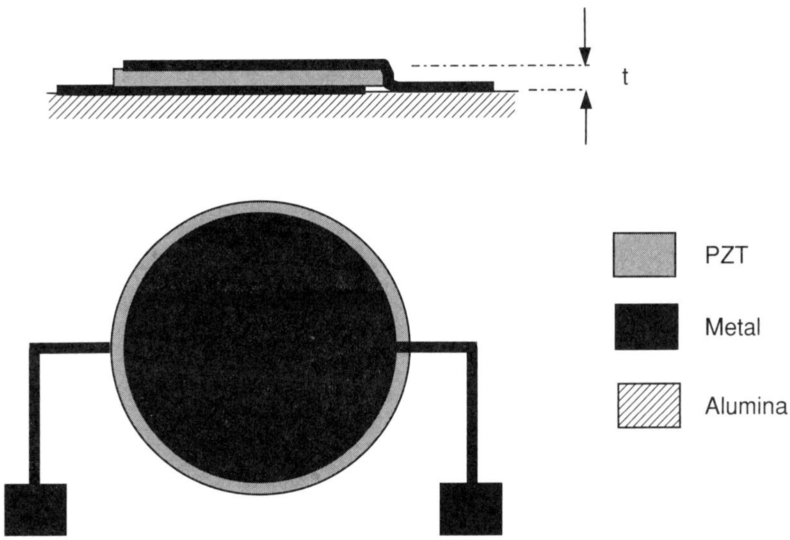


**Fig. 12.** Thickness mode piezoelectric thick-film gravimetric sensor (after Ferrari *et al* [36]).

sensor, demonstrating a compact solution readily achievable by exploiting thick-film sensor technology.

The PZT-PCW film used in the diffuser micropump discussed previously has also been employed in the fabrication of a resonant cantilever also designed as a mass sensitive resonator [37]. The cantilever is fabricated form silicon using a TMAH wet etch after which the PZT film is printed and the sol-gel solution infiltrated. Cantilevers from 850 μm to 2450 μm in length and 500 μm to 1100 μm in width were fabricated. The sensitivity of these cantilevers was determined by sputtering a gold film on the top surface. The short cantilevers proved to be the most sensitive with an additional 1 ng mass producing a change of a few 10s of Hertz with a base resonance of 25 kHz. The intended application of this device is in biosensors where the cantilever would be used to detect minute changes in mass due to the adhesion of protein cells. The method of coating the cantilever with the protein sensitive film is, however, not discussed.

### 2.5.6. Elastic wave sensors

Piezoelectric thick-films have also been used as elastic wave sensors, in particular surface acoustic wave (SAW) devices [38, 39, 40]. An example, not drawn to scale, is shown in Figure 13. The sensor comprises two interdigitated electrode patterns, screen printed and fired onto an alumina substrate of thickness 630 μm. The piezoelectric thick-film (thickness 40 μm) is deposited over the electrodes. The width and gap between the electrodes is 200 μm. The piezoceramic material is polarised between the 'fingers' of the electrode pattern.

One of the interdigitated transducers (IDTs) acts as an input device. A sinusoidal voltage is applied to the electrodes and, by virtue of the reverse piezoelectric effect, an elastic wave is generated within the film and travels towards the other IDT. At the receiver, the acoustic wave is translated back into an electrical signal as a result of the direct piezoelectric effect. The wave velocity, $v$, is determined by the acoustic properties of the medium. The wavelength, $\lambda$, is equal to the centre-to-centre distance between adjacent pairs of fingers. The frequency of the SAW device is therefore $F = v/\lambda$, and is typically around 5 MHz. Commercial SAWs have frequencies in the region 30 MHz—1 GHz. The device is usually operated as a delay-line oscillator by amplifying the signal from the output transducer and feeding back to the input. A phase shifter may

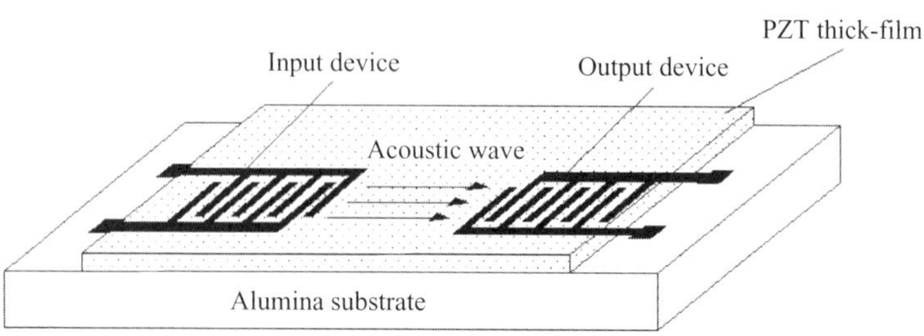


**Fig. 13.** A thick-film elastic wave sensor (after White and Ko [40]).

also be required to ensure a total phase shift of a multiple of $2\pi$ around the loop. A number of mechanisms may be exploited in order to use the device as a sensor. The distance between the two IDTs is often referred to as the 'acoustic gap' and as the wave travels across this, an interfering medium such as a chemical, gas, biological or humidity sensitive coating will change a particular characteristic (amplitude, phase or frequency) of the wave. SAW sensors are therefore extremely versatile and find use in a wide variety of applications.

### 2.5.7. *Multi-modal sensors*

There is often a requirement to take multiple measurements in a given application. As an example, when crude oil is extracted from the ground, it is in a mixture containing water, gas and particulates. In order for the oil to be in a suitable state for refining, it needs to be separated from the mixture. This process is performed inside a separator vessel, which allows the mixture to settle into discrete phases. Typical operating conditions of such installations are characterised by high temperatures reaching in excess of 150 °C, pressures in excess of 150 bar, and the presence of aggressive chemical compounds. There are currently no commercial instruments that identify where the phase interfaces lie. This is partly because of the harsh environment, but also because the interfaces themselves are not clear. There is a gradual separation between the phases between the emulsion of oil and water and the foam of oil and gas. Hale *et al* [41] and Dyakowski *et al* [42] have considered the use of thick-film technology in solving this problem.

Figure 14 shows a schematic of the sensor probe, which contains thick-film capacitive and ultrasonic sensors together with a commercial, high power ultrasonic cleaning transducer. The capacitance sensors are used as part of a tomographic imaging system that distinguishes the various phases within the separator. The thick-film ultrasonic sensors are used to take a time-of-flight measurement across the fluid mixture and hence determine the velocity. The high power ultrasonic transducer, used for cleaning the sensor faces, is mounted on the back face of one of the stainless steel support plates. The

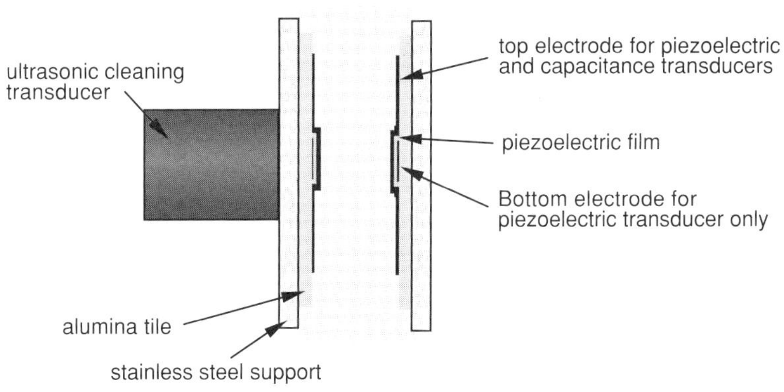


**Fig. 14.** A dual modality, thick-film sensor probe (after Hale *et al* [41]).

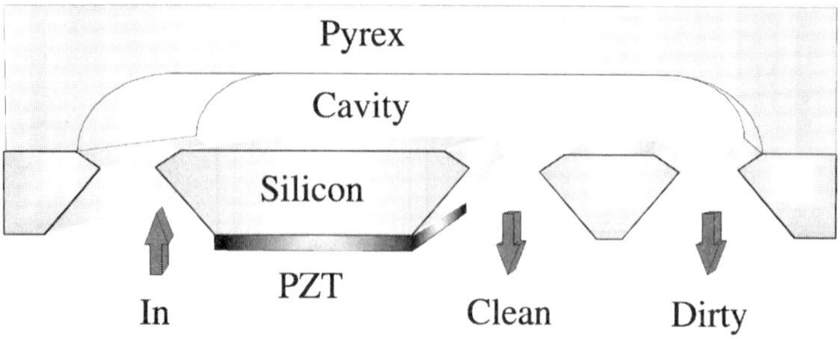


**Fig. 15.** Schematic cross-section of a microfluidic separator (after Harris *et al* [43]).

ultrasound excites the adjacent sensor tile and couples into the process fluid, through which it propagates to the other sensor tile. This configuration allows both tiles to be cleaned using a single ultrasonic transducer. Owing to the relatively high operating temperature, PZT type 5A was used to fabricate the acoustic sensors. This material has a higher Curie temperature (350 °C) than type 5H (180 °C).

### *2.5.8. Microfluidic separators*

Ultrasonic standing waves can be used to generate small forces on particles within an acoustic field. This has the effect of moving the particles to a particular point within a fluid where the pressure forms a node or anti-node. The effect can be used to: aggregate particles in a specific area; manipulate and fractionate particular particles; or to separate groups of particles from the carrying fluid. Figure 15 shows a cross-sectional view of a micromachined device as described by Harris *et al* [43, 44, 45].

The device uses a dual layer of co-fired PZT thick-film as the acoustic transducer, which is used to generate a standing wave field within the chamber. A fluid/particle mixture is drawn into the device through the port on the left of the figure. An acoustic standing wave, of a half wavelength, is maintained within the cavity and, as the particles move through the field, they migrate to the pressure node at the centre of the cavity. A fraction of the particle-free fluid can then be drawn from the 'clean' outlet and the remainder of the fluid/particle mixture can be drawn from the final ('dirty') outlet. Such a scheme allows for a relatively small quantity of clarified fluid to be produced, but different geometries have the potential to increase the proportion of recovered clarified fluid, or alternatively to increase the concentration of the particles recovered from the 'dirty' outlet. Tests have shown that this type of transducer is more efficient at converting electrical energy into acoustic energy than the traditional solution of bonding bulk PZT to the substrate. Efficiencies of 43% have been achieved, compared with, typically, 18% for bulk PZT having the same configuration.

The device has been manufactured in a combination of silicon and Pyrex. The latter, being transparent, allows visual inspection of the separation process. Figure 16 shows the device separating a continuous flow of water and yeast particles. Flow is from right to left and the view is looking down on the two outlets. The outlet at the far left of each picture is the outlet for the 'clean' flow and the outlet in the centre of each picture is the outlet for the 'dirty' flow.

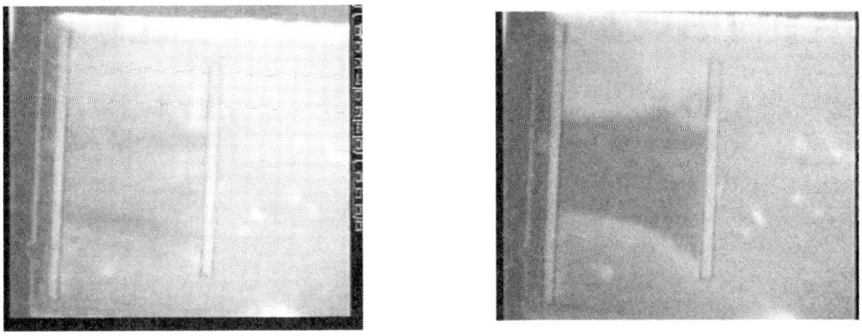

**Fig. 16.** View of the outlets of the barrier-less filter with power off and power on, respectively (after Harris *et al* [45]).

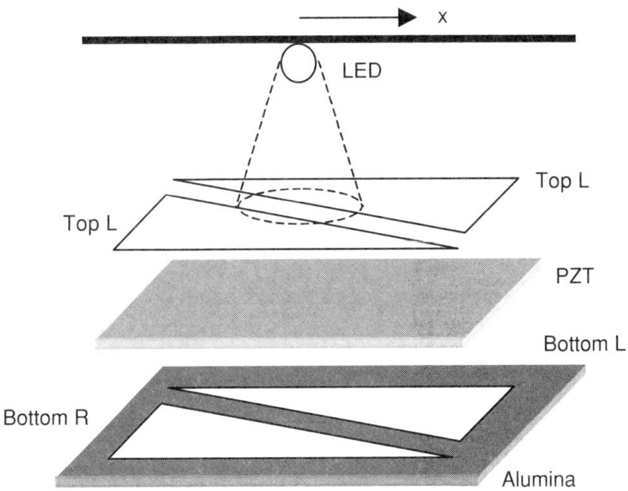


**Fig. 17.** Expanded view of pyroelectric displacement sensor (after Ferrari *et al* [46]).

### *2.5.9. Pyroelectric thick film sensors*

Sensors have also been developed that exploit the pyroelectric effect of PZT thick-films. A displacement sensor based upon a moving LED and a fixed PZT pyroelectric sensor has been demonstrated by Ferrari *et al* [46]. The pyroelectric response of the PZT layer is intrinsically AC and therefore the source is modulated at 80 Hz. The sensor design uses a standard planar capacitor structure with four silver/palladium electrodes of triangular shape as shown in Figure 17. The two pairs of electrodes are differentially poled and the output from one pair is subtracted from the other. At the central position ($x = 0$) the net output is zero and this goes positive for movement to the right, negative for movement to the left. The differential poling arrangement also serves to cancel out the piezoelectric response of the sensor to mechanical vibrations. The pyroelectric coefficient of the PZT film was measured to be $1.4 \times 10^{-4}$ $Cm^{-2}K^{-1}$. The sensor achieved a displacement resolution of better than $1.5 \times 10^{-3}$ of the measured span.

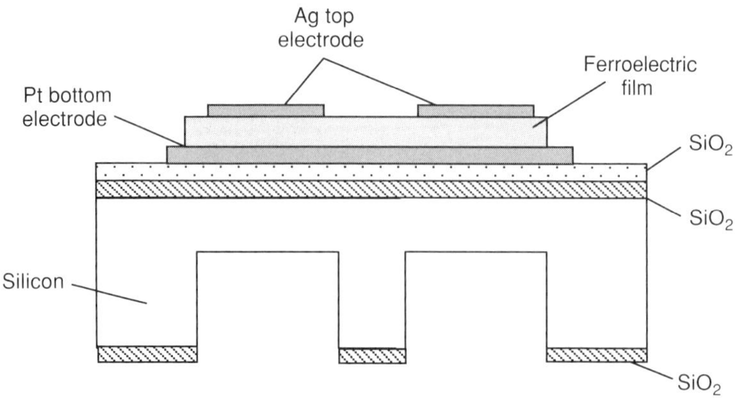


**Fig. 18.** Thick-film ferroelectric silicon actuator.

### *2.5.10. Thick-film actuators*

Silicon micromachining techniques have also been applied to the fabrication of an array of actuators described by Futakuchi *et al* [47]. The actuator structure, shown in Figure 18, was first wet etched in KOH before processing the screen printed lower platinum electrode, ferroelectric film and silver top electrode. The ferroelectric film is composed of $Pb[Zr_{0.7}Ti_{0.3}(Mg_{1/3}Nb_{2/3})_{0.3}(Zn_{1/3}Nb_{2/3})_{0.1}(Mg_{1/2}W_{1/2})_{0.1}]O_3$. The actuator membrane has an area of $3 \times 10$ mm$^2$ and the maximum displacement equalled approximately 0.5 μm for an applied voltage of 30 V and a membrane thickness of 100 μm.

## 3. Thick-Film Magnetostrictive Devices

Magnetostriction is defined as the dimensional change of a magnetic material caused by a change in its magnetic state, and also the reverse effect by which an applied stress causes a change in the material's magnetic state. All magnetic materials exhibit varying degrees of magnetostriction [48]. The magnetostrictive effect was discovered in 1847 by J.P. Joule, who noted that the length of a bar of iron changed when it was magnetised. The dimensional change form of magnetostriction is often called *Joule magnetostriction*, and the change in magnetic state by an applied stress is termed inverse Joule magnetostriction. In the case of a material exhibiting positive magnetostriction there is a corresponding increase in length of the material in the direction of the applied magnetic field. Negative magnetostriction occurs in some materials, in which case an applied field will reduce the length of the material in the direction of the applied magnetic field. The magnetostrictive effect is independent of the polarity of the applied magnetic field.

Joule magnetostriction can be demonstrated using a rod of a magnetostrictive material placed inside a solenoid coil as illustrated in Figure 19.

In the figure, (a) represents the rod within the solenoid when no current is passed through the coil and (b) represents the effect of passing a current, $I$, through the coil.

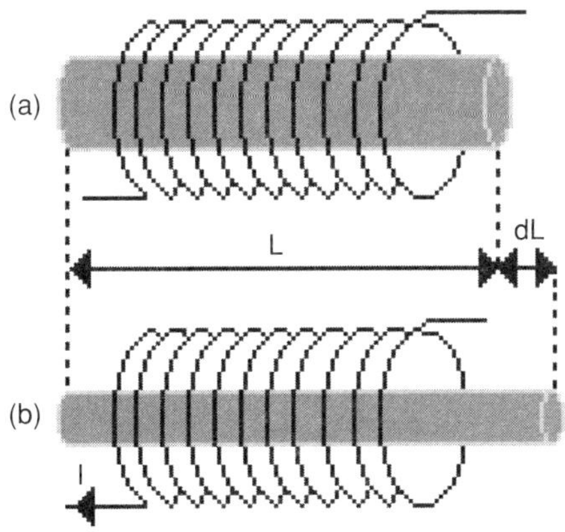


**Fig. 19.** Dimensional change due to magnetostriction.

The major effect of passing current through the coil is the increase in length, $dL$, of the rod. As the volume of the rod stays the same during the extension, the diameter of the rod will reduce proportionally.

The magnetostriction, $\lambda$, of the material forming the rod shown in Figure 19 is given by,

$$\lambda = \frac{dL}{L} \tag{3.1}$$

The magnitude of the magnetostriction is a function of the magnetic field strength, which is in turn a function of the current in the coil. There is also the inverse magnetostrictive effect, whereby the application of an external force causes a change in the magnetic properties of the material. This effect allows magnetostrictive materials to be used for sensing applications in addition to as an actuator material. The sensing properties of the material can enable simultaneous measurement and application of force through one device, as described by Clephas *et al* [49].

Many of the current applications of magnetostrictive materials can be found in the field of actuators. One of the reasons for this is that large strains are achievable and also that there is the potential for remote excitation of the material. This removes the need for direct connection to the active material, enabling it to be embedded within sealed enclosures, for example in applications involving hazardous substances.

### 3.1. History

All magnetic materials exhibit some degree of magnetostriction. There is, however, only a limited range of materials that exhibit sufficient levels of magnetostriction for practical use. Early magnetostrictive devices were based on nickel and found use in sonar applications [50]. In the 1960s, the magnetostrictive properties of the rare earth metals terbium and dysprosium were identified, but due to the low Curie temperatures of these two elements, they were not suitable for use at room temperature. To produce a material that would exhibit a large magnetostriction at room temperature, terbium and dysprosium were alloyed with transition metals and, in 1971, this led to giant magnetostriction at room temperature using the alloys $TbFe_2$ and $DyFe_2$. A problem with these materials is the relatively large magnetic anisotropy, which means that very high magnetic fields, typically in excess of 2MA/m, are required to achieve the giant magnetostrictive effect [51].The first anisotropy constant for $TbFe_2$ is negative and for $DyFe_2$ it is positive but both materials have a positive magnetostriction. Alloys of the two materials have been produced, with the aim of combining the magnetostriction of both materials whilst reducing the overall anisotropy, by virtue of the opposing polarities of the first anisotropy constants. The alloying of terbium, dysprosium and iron lead to the discovery in the early 1970s of Terfenol-D which has the composition $Tb_{0.3}Dy_{0.7}Fe_2$; this provides a magnetostriction at room temperature of over $1000{*}10^{-6}$ and exhibits low anisotropy. The combination of giant magnetostriction and low anisotropy have made Terfenol-D a popular magnetostrictive material for applications that are required to operate at room temperature.

Magnetostrictive materials can be fabricated in several forms. The main forms are:

- **Bulk**—Where the material is used in its as grown, crystallographic form, such as that studied by Wang *et al* [52];
- **Melt spinning**—Where bulk material is melted and ejected onto a wheel to form a ribbon, which is much thinner than that machined from the bulk material (typically 50 to 450 μm thick). Such material is described by Lindgren *et al* [53];
- **Thin-film**—Where a thin layer of magnetostrictive material is deposited onto a substrate using a sputtering operation, this process may use a target made of bulk material, or a composite target to give the desired composition. Quandt provides an overview of this material form [54];
- **Composite**—Where the magnetostrictive bulk material is ground into a powder and a matrix material is used to bind the magnetostrictive particles together, such as the epoxy bonded material reported by Duenas *et al* [55].

The thick-film version of the material shares commonality with both the thin-film and composite forms. Because the thick-film material must be deposited onto a substrate, the displacement of the material is constrained, in a similar manner to that associated with thin-film materials. The combination of the active material and binding matrix in the thick-film is similar in nature to composite materials.

### 3.2. Fabrication process

Bulk magnetostrictive materials such as Terfenol-D are produced from their component elements using the process described by Cook *et al* [56]. To produce Terfenol-D rods, the three constituent elements (terbium, dysprosium and iron) are alloyed together using an arc melting method. This yields an alloy having an irregular crystallographic alignment. The alloy is then fed into a zone-refining, directional solidification process where the alloy is melted and grown onto a suitable seed crystal with the desired crystallographic structure. As the alloy melts and subsequently re-solidifies along its length, the structure will alter to that of the seed crystal. The crystallographic structure of the seed crystal, and thus the finished rod, is chosen to given optimum operation with the magnetic field applied along the length of the rod. Once the alloy has been directionally solidified, the material can be ground into shape. For flat-sided samples, the rods are cut using diamond-coated abrasive saws. As the material is brittle and difficult to machine, intricately shaped samples are very expensive to make. The dimensions of the finished samples are also limited by the maximum size of the rods that can be produced using the zone-refining process. The magnetostriction of bulk materials at higher frequencies can be improved by laminating the material. This is accomplished by cutting the bulk material into thin strips and then laminating these together. This improves the magnetostriction by reducing the eddy currents in the material, which oppose the externally applied actuation field, therefore reducing the operating field to which the magnetostrictive material is subjected. This laminating process can be expensive to carry out due to the problems encountered when machining the brittle material.

The detrimental effects of the eddy currents at higher operation frequencies can be greatly reduced by combining the magnetostrictive material in powdered form with a non-conductive matrix material and forming it into a composite material. Composite fabrication also offers the advantage over the brittle, bulk magnetostrictives in that more complex shapes can be formed through moulding processes.

The fabrication process used to prepare a thick-film magnetostrictive, based on Terfenol-D as the active material, is similar to that of the piezoelectric thick-films described earlier. The active particles are mixed with a glass frit, which acts as a binder, prior to printing and firing to form the desired composite. The process for fabricating magnetostrictive thick-film differs from that of the piezoelectric films in that ground alumina is added as an inert filler material to the mix of active and binder material. The purpose of this filler material is to improve the structure of the resulting thick-film because the irregular dimensions of the magnetostrictive particles can lead to the presence of voids within the film. These lead to a reduction in the net magnetostriction of the film because the active material displaces into the voids, rather than coupling the force into the rigid matrix.

Owing to the detrimental effects of oxidisation on the magnetostrictive material used, including the risk of ignition due to the pyrophoric nature of Terfenol-D, the firing process is carried out in an inert nitrogen atmosphere at a temperature in the region of 500 °C. In order to enable successful firing of the thick-film under these conditions, it is necessary to select a suitable glass frit [57]. A full description of the process for fabricating magnetostrictive thick-films, based on Terfenol-D, is described by Grabham *et al* [58].

## 3.3. Silicon integration

To integrate thick-film materials with MEMS devices, it is necessary to identify a suitable means of bonding the thick-film material to the silicon wafer. This is a problem because of the smooth surface of the silicon wafer, compared to the somewhat rougher surface of the alumina substrates that are used with thick-films. Additionally, the different thermal expansion coefficients of silicon ($2.33*10^{-6}/^{\circ}C$) and alumina ($6.8*10^{-6}/^{\circ}C$) result in thermally-induced stresses between the substrate and the thick-film. The thick-film must therefore be securely bonded to the silicon substrate in order to prevent delamination. Depositing the films directly onto the surface of the silicon wafer does not provide sufficient adhesion. A means of improving the bonding between the

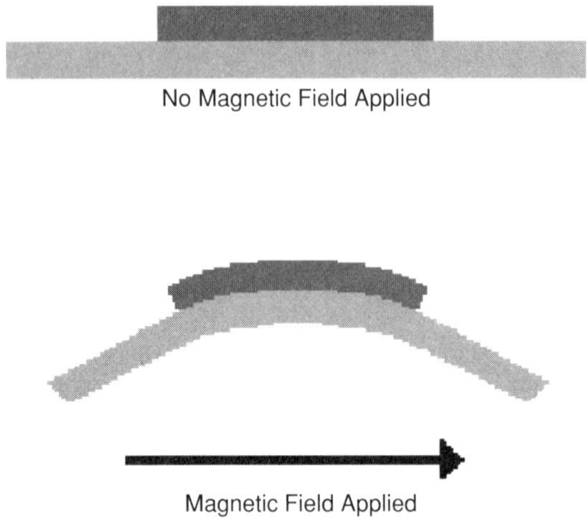


**Fig. 20.** Bending of thick-film device with applied magnetic field.

thick-film and substrate is therefore required. One method of improving the chemical bonding between the silicon and the thick-film is to grow an interface layer, such as silicon dioxide, onto the silicon. This has been found to produce improved bonding. The $SiO_2$ layer is grown by heating the silicon wafer in an air atmosphere, which then causes $SiO_2$ to grow across the entire surface of the wafer. The duration of the heating process controls the thickness of the layer. In order to leave regions of the wafer free of $SiO_2$ for other processes, it is possible to mask the areas where the $SiO_2$ is desired and remove the oxide from the unmasked areas using standard silicon processing steps. The magnetostrictive thick-film can then be deposited and processed in the usual manner onto the $SiO_2$ layer as desired.

### 3.4. Characterisation

To quantify the properties of a deposited magnetostrictive material it is necessary to calculate the magnetostriction of the thick-film as a function of the applied magnetic field. In the case of bulk samples, this can be readily achieved by recording the elongation of the sample in the direction of the applied magnetic field, and using the relationship between change in length and initial length to give the magnetostriction of the sample. In the case of a film on a substrate, however, the constraints produced by the substrate need to be considered. An extension in the thick-film along the direction of the applied field will be result in bending of the sample under test. This is illustrated in Figure 20, where the darker grey represents the magnetostrictive thick-film and the lighter grey is the substrate. (The upper diagram shows the device with no magnetic field applied and the lower with a field applied.)

As the magnetostriction of the thick-film results in bending of the substrate, it is possible to calculate the extension of the thick-film, and thus the magnetostriction, by measuring either the deflection of the substrate, as used by Farber *et al* [59], or the strain produced in the substrate, as has been used with thin-film materials by Arai *et al* [60].

To relate the deformation to the magnetostriction, the technique described by van de Riet [61] can be used. This requires a suitable test sample with constraints applied to the relative dimensions of film and substrate. In cases where such an analytical approach can not be applied, for example due to complex sample geometry, it is possible to apply finite element modelling to the problem to obtain a computed value. In both cases, it is first necessary to determine, experimentally, the various mechanical properties for both the magnetostrictive film and the substrate.

## 3.5. Devices

A variety of sensors and actuators, utilising magnetostrictive films, have been reported in the literature. One area that is receiving significant attention is that of microfluidics. Pumps and valves have been designed that can be incorporated into "Lab-on-a-chip" type applications to control the movement of the substances of interest through the device(s) on the chip. Magnetostrictive materials have a significant advantage over piezoelectric devices in this area because the actuation is achieved through the application of a magnetic field and does not require the high voltages associated with piezoelectric materials. The magnetic field can also be applied remotely, thus allowing the design of fully-sealed devices. For generation of the actuating magnetic fields, either permanent magnets or electromagnets can be used. In the case of electromagnets, these can be located either externally or integrated into the device.

### *3.5.1. Micropumps*

A thick-film magnetostrictively-driven micropump, with integrated actuation coils, is being developed at the University of Southampton [62]. This is similar in nature to the device described in section 2.5.1. A diaphragm is chemically etched into the silicon to form the actuation membrane in a pump. On the non-chamber side of the diaphragm, a multi-layer coil is micromachined and the magnetostrictive thick-film subsequently deposited onto the surface of the coils. The location of the coils, in close proximity to the magnetostrictive material, permits selective operation of devices in the case where a number of individual pumps are required in close proximity, for example in lab-on-a-chip applications.

The micromachined diaphragm with integrated coils is shown in Figure 21. The rectangular region on the surface of the coils is the magnetostrictive thick-film.

**Fig. 21.** Micromachined diaphragm with integrated coils and thick-film magnetostrictive.

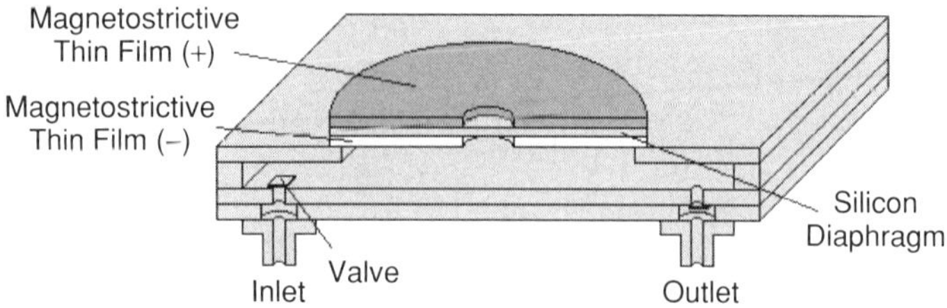


**Fig. 22.** Thin film magnetostrictive micropump (after Quandt *et al* [63]).

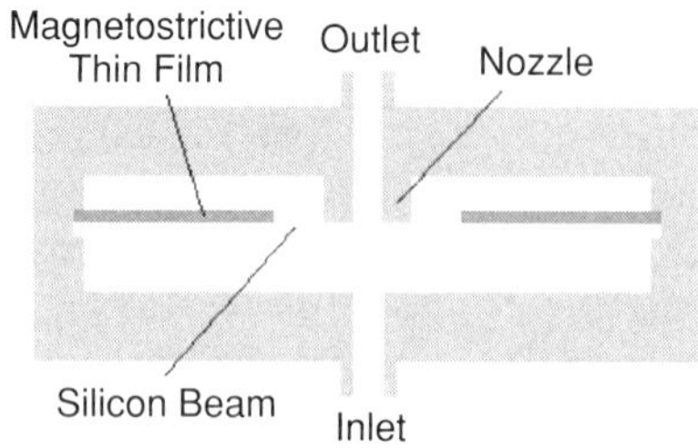


**Fig. 23.** Thin film magnetostrictive micro valve (after Quandt *et al* [64]).

An alternative fabrication process is used by Quandt *et al* [63], who fabricated a magnetostrictive-driven micropump using laser machining of silicon wafers to form the pump chamber and diaphragm prior to deposition of magnetostrictive thin films onto the diaphragm surfaces. The pump elements are then subsequently assembled using epoxy adhesives to produce the assembled device. Both positive and negative magnetostrictive materials have been used on opposing faces of the diaphragm to increase the operational stroke. To produce the actuation field, an external pot wound coil is used, which is positioned above the diaphragm. A cross section of the structure is illustrated in Figure 22.

### 3.5.2. Micro valve

The fabrication of a magnetostrictively excited micromachined valve has been reported by Quandt *et al* [64]. Their device consists of a silicon beam supported at both ends, with magnetostrictive thin films applied to the top surface at both ends of the beam. The plain centre portion of the beam is held against the output nozzle of the valve until the magnetostrictive thin film is actuated. When this occurs, the beam moves away from the nozzle and opens the valve. A cross-section of the micro valve in the 'off' state is shown in Figure 23.

### 3.5.3. Optical switching

An application of magnetostrictive thin films as actuators is the optical switch described by Moon *et al* [65]. This device switches an optical signal from one input optical fibre into one of three output fibres. The switching is achieved using a cantilever beam

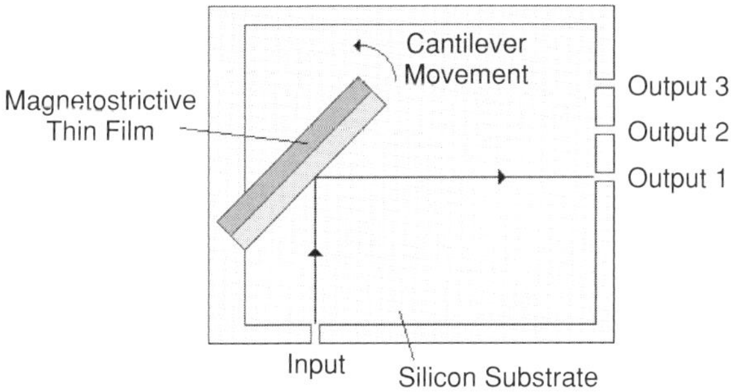


**Fig. 24.** Thin film magnetostrictive optical switch (after Moon *et al* [65]).

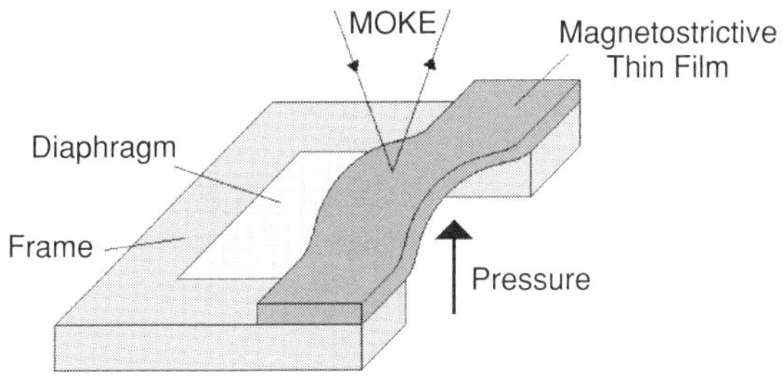


**Fig. 25.** Thin film magnetostrictive pressure sensor (after Karl *et al* [66]).

with the optical signal reflected off the beam using the arrangement shown in Figure 24. The deflection of the beam occurs because of contraction of the magnetostrictive thin film layer (due to the use of a material with negative magnetostriction) and this changes the incident angle of the light source to the beam and therefore the angle of reflection. By varying the magnetic field it is possible to select which of the output fibres the input signal is reflected into.

### *3.5.4. Pressure sensing*

A thin film pressure sensor is reported by Karl *et al* [66]. This type of sensor consists of a diaphragm with a Metglas-based magnetostrictive thin film deposited onto it. An illustration of a cross section of the device is shown in Figure 25. As pressure is applied to the diaphragm, the permeability of the magnetostrictive thin film changes, and is detected using the magneto-optical Kerr effect (MOKE). The Kerr effect is the rotation of polarisation of light reflected by a magnetised material.

### *3.5.5. Displacement sensor*

A bimorph structure for the measurement of displacement is reported by Kaniusas *et al* [67]. The device is illustrated in Figure 26 and consists of a magnetostrictive ribbon glued onto a non-magnetic metal support strip. Whilst the magnetostrictive material used in this example device is formed by melt spinning, the same device could be

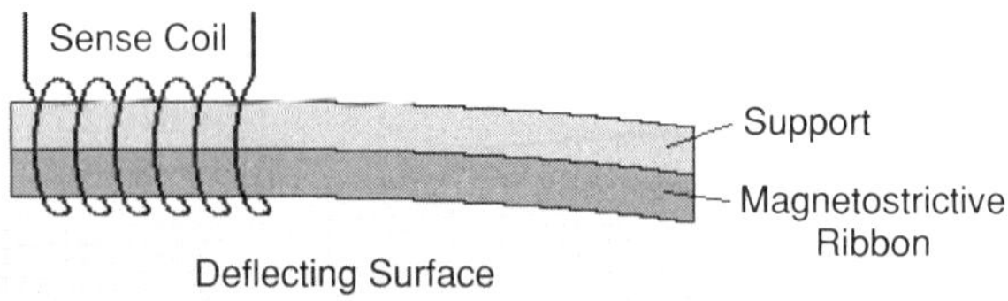


**Fig. 26.** Melt spun magnetostrictive deflection sensor (after Kaniusas *et al* [67]).

fabricated by depositing a magnetostrictive thin or thick-film onto the support in place of the ribbon and its associated adhesive. A coil is wound at one end round a section of the strip and ribbon, this coil is used to measure the permeability of the magnetostrictive ribbon, the change in which is used to calculate the deformation of the device.

## References

[1] Brignell JE, White NM and Cranny AWJ, Sensor applications of thick-film technology, IEE Proceedings Part I, Vol. 135, No. 4, 1988, pp. 77–84
[2] White NM and Turner JD, Thick-film sensors: Past, present and future, Measurement Science and Technology, Vol. 8, 1997, pp. 1–20
[3] Baudry H, Screen printing piezoelectric devices, Proceedings of the sixth European Microelectronics Conference, 1987 pp. 456–63
[4] Morten B, De Cicco G and Prudenziati M, Piezoelectric properties of PZT thick-films Proc. 7th European Hybrid Microelectronics Conference paper 8.4, 1989
[5] Morten B, De Cicco G, Gandolfi, A and Tonelli C, Advances in ferroelectric thick-film materials and sensors Proc. 8th European Hybrid Microelectronics Conference, 1991 392–9
[6] Moilanen H, Leppavuori S and Uusimaki A, Fabrication of piezoelectric thick-film low-voltage multilayer actuators using a new double-paste printing technique, Sensors and Actuators A, Vol. 37–38, pp. 106–111, 1993.
[7] Thiele ES and Setter N, Lead zirconate titanate particle dispersion in thick-film ink formulations, J.Am.Ceram.Soc., Vol. 83, 2000, pp. 1407–12
[8] Papakostas TP, Harris, NR, Beeby, SP and White NM, Piezoelectric thick-film polymer pastes, Eurosensors XII, 1998, pp. 461–464
[9] Robertson C, Shipton RD and Gray DR, Miniature sensors using high density screen printing, Sensor Review, Vol. 19, No. 1, 1999, pp. 33–36
[10] Glynne-Jones P, Beeby SP, Dargie PG, Papakostas TP and White NM, An investigation into the effect of modified firing profiles on the piezoelectric properties of thick-film PZT layers on silicon, Measurement Science and Technology, 2000, Vol. 11, pp. 526–531
[11] Chen HD, Udayakumar KR, Cross LE, Bernstein JJ and Niles LC, Dielectric, ferroelectric, and piezoelectric properties of lead zirconate titanate thick films on silicon, J. Appl. Phys., Vol. 77 (7), 1 April 1995, pp. 3349–3353.
[12] Maas R, Koch M, Harris NR, White NM and Evans AGR, Thick-film printing of PZT onto silicon, Materials Letters 31, 1997, pp. 109–112.
[13] Beeby SP, Blackburn A, White NM, Silicon micromachining processes combined with thick-film printed lead zirconate titanate actuators for microelectromechanical systems, Materials Letters, 40, 1999, pp. 187–191.

[14] Beeby SP, Blackburn A, White NM, Processing of PZT piezoelectric thick films on silicon for micromechanical systems, J. Micromech. Microeng., 9, 1999, pp. 218–229.

[15] Thiele ES, Damjanovic D and Setter N, Processing and properties of screen-printed lead zirconate titanate piezoelectric thick films on electroded silicon, J. Am. Ceram. Soc., 84 (12), 2001, pp. 2863–2868.

[16] Take-Control Piezometer: http://www.take-control.demon.co.uk

[17] Southin JE, Wilson SA, Schmitt D and Whatmore RW, e31,f determination for PZT films using a conventional '$d_{33}$' meter, J. Phys. D: Appl. Phys, 2001, Vol. 34, pp. 1456–1460,

[18] Torah RN, Beeby SP and White NM, Experimental investigation into the effect of substrate clamping on the piezoelectric behaviour of thick-film PZT elements, Journal of Applied Physics, 2004, Vol. 37, pp. 1–5

[19] Dorey R A and Whatmore RW, Apparent reduction in the value of $d_{33}$ piezoelectric coefficient in PZT thick films, Integrated Ferroelectrics, Vol. 50, 2002, pp. 111–119.

[20] Koch M, Harris NR, Maas R, Evans AGR, White NM and Brunnschweiler A, A novel micro pump design with thick-film piezoelectric actuation, *Meas. Sci. Technol.*, (1997), Vol. 8, pp. 49–57

[21] Koch M, Harris N, Evans AGR, White NM and Brunnschweiler A, A novel micromachined pump based on thick-film piezoelectric actuation, Sensors and Actuators A—Physical, Vol. A70, No. 1–2, pp. 98–103, 1998.

[22] Kim Y-B, Kim H-J, Cheon CI, Choi D-J and Kim TS, Preparation of diffuser type micropump using screen-printed PZT-PCW thick films, Integrated Ferroelectrics, Vol. 50, 2001, pp. 61–70.

[23] Crescini D, Marioli D, Sardini E and Taroni A, Large bandwidth and thermal compensated piezoelectric thick-film acceleration transducer, Sensors and Actuators A, Vol. 87, pp. 131–138, 2001

[24] Beeby, SP, Ross JN and White NM, Thick-film PZT/micromachined silicon accelerometer, IEE Electronics Letters, 1999, Vol. 35, No 23, pp. 2060–2

[25] Kurosawa M, Inagaki H, and Higuchi T, Investigation of stator transducer for a column-shaped ultrasonic motor, Ultrasonics, Vol. 34, pp. 271–274, 1996

[26] Morita T, Kurosawa M and Higuchi T, A cylindrical micro ultrasonic motor using PZT thin film deposited by single process hydrothermal method ($\phi$ 2.4 mm, L = 10 mm stator transducer), IEEE Transactions on Ultrasonics, Ferroelectrics, and Frequency Control, Vol. 45, No. 5, pp. 1178–1186, 1998.

[27] Hu M, Du H, S Ling -F, Teo J -K, A piezoelectric spherical motor with two degree-of-freedom, Sensors and Actuators A, Vol. 94, pp. 113–116, 2001

[28] Aoyagi M, Beeby S.P and White NM, A novel multi-degree-of-freedom thick-film ultrasonic motor, IEEE Transactions on Ultrasonics, Ferroelectrics, and Frequency Control, Vol. 49, No.2, pp. 151–158, 2002.

[29] Light CM and Chappell PH, Development of a lightweight and adaptable multiple-axis hand prosthesis, Medical Engineering and Physics, Vol. 22 pp. 679–684, 2000

[30] Cranny AWJ, Chappell PH, Beeby SP and White NM, Improving the functionality of a prosthetic hand through the use of thick-film force sensors, *Proc. Eurosensors XVII*, Portugal, 2003

[31] White N M and Brignell J E, Excitation of thick-film resonant structures IEE Proceedings Sci.ence, Measurement and Technology, vol. 42, pp. 244–48, 1995

[32] Yan T, Jones BE, Rakowski, RT, Tudor MJ, Beeby, SP and White NM, Metallic Triple Beam Resonator with Thick-film Printed Drive and Pickup, IEE Electronics Letters, Vol. 39, No. 13, pp. 982–983, 2003

[33] Beeby SP and White NM, Thick-film PZT-silicon micromechanical resonator, Electronics Letters, Vol. 36, No. 19, 14th September 2000, pp. 1661–1662.

[34] Beeby SP and White NM, Silicon micromechanical resonator with thick-film printed vibration excitation and detection mechanisms, Sensors and Actuators A88, 2001, pp. 189–197.

[35] Ferrari V, Marioli D and Taroni A, Self-resonant piezoelectric thick-films for sensing applications, Electronics Letters, Vol. 32, No. 9, April 1996, pp. 855–856.

[36] Ferrari V, Marioli D and Taroni A, Thick-film resonant piezo-layers as new gravimetric sensors, Meas. Sci. Technol., Vol. 8, 1997, pp. 42–48.

[37] Kim HJ, Kim Y-B, Kang J-Y and Kim TS, Fabrication and resonant behaviour of PZT thick film cantilever for biochip, Integrated Ferroelectrics, Vol. 50, 2002, pp. 11–20.

[38] Morten B, De Cicco G and Prudenziati M, A novel torque sensor based on elastic waves generated and detected by piezoelectric thick-film, Sensors and Actuators, Vol. 41–42, pp. 33–38, 1994

[39] White N M and Leach G R, Fabrication of a thick-film sensor employing an ultrasonic oscillator, IEE Proceedings Science Measurement and Technology 42 249–54, 1995

[40] White N M and Ko V T K, Thick-film acoustic wave sensor structure Electronics Letters, Vol. 29, pp. 1807–8, 1993

[41] Hale J, Dyakowski T, Jaworski A, White NM and Harris, N, A Sensor System for Oil-Water Separators: Materials Considerations. Proc. of Sensors and their Applications XII, Limerick, pp. 71–76, 2003.

[42] Dyakowski T, Hale J, Harris N, Jaworski A, Nowakowski A, White NM and Zhang Y, Characterisation of heterogeneous mixtures by using thick-film ultrasonic transducers, 3rd World Congress in Industrial Process Tomography, Alberta, Canada, 2–5 September 2003

[43] Harris N, Hill M, Shen Y, Townsend R, Beeby SP and White NM. A dual frequency, ultrasonic, microengineered particle manipulator, Ultrasonics, Vol. 42, pp. 139–144, 2004.

[44] Hill M, Harris N, Townsend R, White NM and Beeby SP. A micro fabricated ultrasonic particle manipulator with frequency selectable nodal planes, World Congress on Ultrasonics 2003 Paris, pp. 1647–1650, 2003.

[45] Harris NR, Hill M, Beeby S, Shen Y, White NM, Hawkes JJ, Coakley WT, A Silicon Microfluidic Ultrasonic Separator. Sensors and Actuators B 95, pp. 425–434, 2003

[46] Ferrari V, Marioli D and Taroni A, Displacment sensor based on pyroelectric thick films and contactless light-spot cursor, IEEE Trans. On Instrumentation and Measurement, Vol. 51, No. 4, August 2002, pp. 819–823.

[47] Futakuchi T, Yamano H and Adachi M, Preparation of ferroelectric thick-film actuator on silicon substrate by screen printing, Jpn. J. Appl. Phy., Vol. 40, 2001, pp. 5687–5689.

[48] Cullen JR, Materials with large magnetostrains. Scripta Metallurgica *et* Materialia, Vol. 33, No. 10/11, pp. 1849–1867, 1995.

[49] Clephas B, Janocha H, Simultaneous sensing and actuation of a magnetostrictive transducer, SPIE, Vol. 3329, pp. 174–184, March 1998.

[50] Jenner AG, Smith RJE, Wilkinson AJ, Greenough RD, Actuation and transduction by giant magnetostrictive alloys, Mechatronics, Vol. 10, pp. 457–466, 2000.

[51] Jiles DC, The development of highly magnetostrictive rare earth-iron alloys, Journal of Physics—D: Applied Physics, Vol. 27, No. 1, pp. 1–11, 14th January 1994.

[52] Wang BW, Busbridge SC, Li YX, Wu GH and Piercy AR, Magnetostriction and magnetization process of $Tb_{0.27}Dy_{0.73}Fe_2$ single crystal, Journal of Magnetism and Magnetic Materials, Vol. 218, pp. 198–202, 2000.

[53] Lindgren EA, Haroush S, Port JC, Mazzatesta AD, Rosen M, Wun-Fogle M, Restorff JB, Clark AE and Lindberg JF, Development of Terfenol-D transducer material, Journal of Applied Physics, Vol. 83, No. 11, pp. 7282–7284, 1st June 1998.

[54] Quandt E, Giant magnetostrictive thin film materials and applications, Journal of Alloys and Compounds, Vol. 258, pp. 126–132, 1997.

[55] Duenas T.A. and Carman G.P., Experimental results for magnetostrictive composites, Adaptive Structures and Material Systems, pp. 63–71, 1998.

[56] Cook B.A., Harringa J.L and Hansen T., Electrical and thermal properties of $Tb_{0.3}Dy_{0.7}Fe_{2-x}$, Journal of Applied Physics, Vol. 87, No. 2, pp. 776–780, 15th January 2000.

[57] Grabham N.J., Beeby S.P. and White N.M., Effects of the binder material on the mechanical properties of thick-film magnetostrictive materials, Sensors and Actuators A—Physical, Vol. 110, pp. 365–370, 2004.

[58] Grabham N.J., Beeby S.P. and White N.M., The Formulation and processing of a thick-film magnetostrictive material, Measurement Science and Technology, Vol. 13, No. 1, pp. 59–64, January 2002.

[59] Farber P and Kronmuller H., Crystallization behaviour and magnetic properties of highly magnetostrictive Fe-Tb-Dy thin films, Journal of Magnetism and Magnetic Materials, Vol. 214, pp. 159–166, 2002.

[60] Arai K.I., Yamaguchi M., Muranaka C.S., Measurement of thin film's magnetostriction with piezoelectric substrates, IEEE Transactions on Magnetics, Vol. 25, No. 5, pp. 4201–4203, September 1989.

[61] van de Riet E., Deflection of a substrate induced by an isotropic thin-film stress, Journal of Applied Physics, Vol. 76, No. 1, pp. 584–586, 1st July 1994.

[62] Grabham N.J., Beeby S.P., White N.M., Micromachined Diaphragms with Integrated Actuation Coils, Proceedings of Eurosensors XVII, September 2003.

[63] Quandt E and Seemann K., Magnetostrictive thin film microflow devices, Micro Systems Technologies 96, pp. 451–456, 1996.

[64] Quandt E and Seemann K., Fabrication and simulation of magnetostrictive thin-film actuators, Sensors and Actuators A—Physical, Vol. 50, pp. 105–109, 1995.

[65] Moon S., Lim S.H., Lee S.B., Kang H.K., Kim M.C., Han S.H and Oh M.H., Optical switch driven by giant magnetostrictive thin films, Proceedings of SPIE, pp. 854–862, March 1999.

[66] Karl W.J., Powell A.L., Watts R., Gibbs M.R.J and Whitehouse C.R., A micromachined magnetostrictive pressure sensor using magneto-optical interrogation, Sensors and Actuators A—Physical, Vol. 81, pp. 137–141, 2000.

[67] Kaniusas E., Mehnen L., Krell C and Pfutzner H., A magnetostrictive acceleration sensor for registration of chest wall displacements, Journal of Magnetism and Magnetic Materials, Vol. 215–216, pp. 776–778, 2000.

# Chapter 5

## Micromachined Infrared Detectors Based on Pyroelectric Thin Films

P. Muralt
*Ceramics Laboratory, Swiss Federal Institute of Technology EPFL, Lausanne, Switzerland*

An overview is given on infrared detector technology based on pyroelectric thin films. Pyroelectricity allows the realization of uncooled, thermal infrared detectors. The pyroelectric effect is introduced as well as the relevant materials family, namely the ferroelectrics. Properties are discussed in view of the infrared detector application. The product of dielectric constant and loss determine the intrinsic noise power of the pyroelectric element and must be kept under control while optimizing pyroelectric coefficients by materials choice and operating conditions. Two possible exploitation modes exist. In a first one, the true pyroelectric effect of a stable and poled ferroelectric element is employed. There is no need for temperature stabilization and the detector is operated at ambient conditions. In a second mode, ferroelectric materials are stabilized by an electric field closely above the para-to-ferroelectric phase transition in order to profit from the large and strongly temperature dependent anomaly related to the polarization fluctuations. This requires a precise temperature control and the phase transition must be of slight first order. An overview is given on pyroelectric thin film materials. Microstructural, process, substrate, and clamping effects are discussed. The physics of IR detectors for pyroelectrics is introduced highlighting the situations relevant for thin film detectors in theory and experiment. As thin film elements exhibit small thermal capacities, they are especially apt to be used at higher frequencies measuring the current response, provided that the thermal insulation of the element is well done. Micromachining is briefly introduced and some examples of realized detectors are given. The state of the art of applications is presented including point detectors for human presence sensors, linear arrays for gas spectroscopy, and two-dimensional focal plane arrays for thermal imaging. Finally, some comments are made on the potential of thin film pyroelectric devices.

## 1. Introduction

Pyroelectricity provides one of the best performing principles for the detection of temperature changes. It is a phenomenon of polar matter and results in the appearance of surface charges proportional to a temperature change $\Delta T$. The basic structure is a planar capacitor whose charge $Q$ varies according to $\Delta Q = Ap\Delta T$, where $A$ is the area of the capacitor and $p$ the pyroelectric coefficient. A pyroelectric sensor thus directly generates an electric signal, facilitating the integration into electronic circuits. Pyroelectric crystals, ceramics of ferroelectric materials, as well as polymers have therefore been used since the 1960's in *thermal* infrared (IR) detectors (see [1–3] for reviews), joining the earlier thermal IR detection techniques of bolometers and thermopiles (see [4] for a historical review on thermal infrared detectors). Since the late 1980's, pyroelectrics are also investigated in the form of thin films and integrated onto micromachined structures (see [5] for a review). Pyroelectricity thus made its way from a scientific curiosity known since 2500 years [6] to a very useful physical effect that for instance enables firemen to see through smoke.

## 2. Fundamentals

Pyroelectricity is the electrical response of a polar, dielectric material to a change in temperature. The origin of this effect lies in polar features that are lined up with the same orientation along at least one direction in the material. In polymers, the polar features are polar molecules or groups. One can thus identify microscopic dipoles (for a review, see [3]). In inorganic materials it is often not possible to identify dipoles. It is rather the anisotropy and anharmonicity of chemical bonds that causes the centre of gravity of the positives charges moving differently than the centre of gravity of the negative charges upon changing the temperature. An equivalent effect results if negative charges (electrons) from the negative ions move towards the positive ions by a charge transfer process. More insight into such mechanisms of covalent nature in ionic crystals have recently been gained by first-principle calculations [7]. In crystalline matter, *pyroelectricity* occurs in all materials with symmetries that allow the existence of a polar direction. No symmetry element may exist in the point group that inverts or rotates this direction by 180° (see, e.g, [8]). Such materials are also *piezoelectric*, i.e. they respond with an electric field to a change of the shape. From the 20 piezoelectric point groups, 10 are polar and allow for pyroelectricity (see Figure 1). The pyroelectric effect is anisotropic and depends on the directions of the electroded faces with respect to the crystal axes: $\boldsymbol{p} = (p_x, p_y, p_z)$. The highest pyroelectric coefficients are measured in *ferroelectrics*. In such substances there is a spontaneous electric polarization $\boldsymbol{P}_s$ that can be inverted by the application of an external electric field, and consequently is a measurable physical quantity. Upon heating up, $\boldsymbol{P}_s$ diminishes and becomes zero at the critical temperature of the ferroelectric—para-electric phase transition as depicted in Figure 2 for a second order phase transition. The symmetry group of the ferroelectric phase is a sub-group of the symmetry group of the para-electric phase, i.e. a number of symmetry elements are lost (certainly the inversion element, if present in the paraelectric phase). The direction of $\boldsymbol{P}_s$ is not uniquely defined. Turning an existing solution by

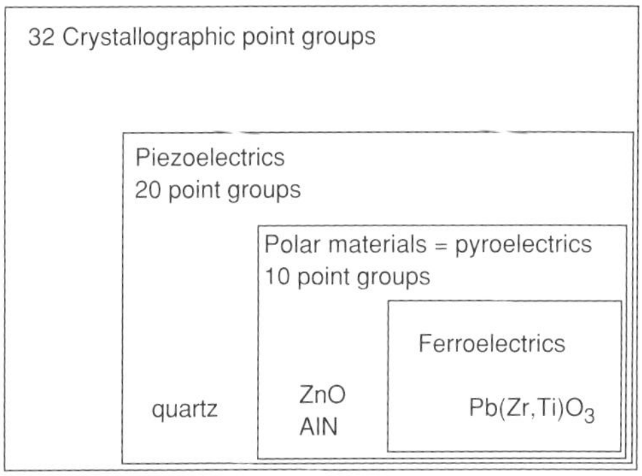


**Fig. 1.** Classification of space groups according to phenomena of polar symmetry including typical examples.

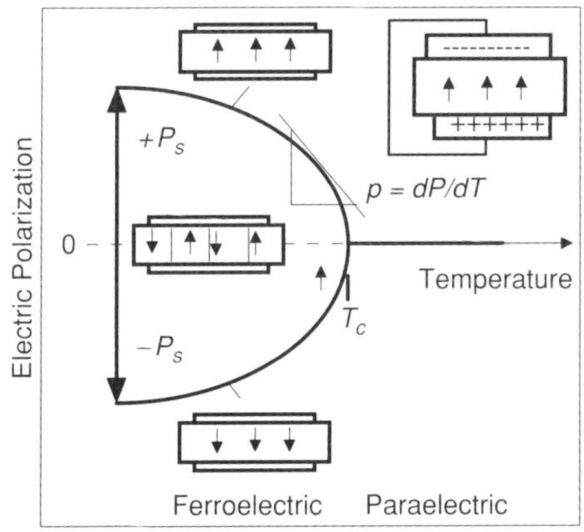


**Fig. 2.** Schematic graph showing the bifurcation (second order phase transition) of the electrical polarization at the critical temperature from $P = O$ (paraelectric phase) to the two equally probable solutions $+/-P_s$ for the spontaneous polarization in the ferroelectric phase. The inserted capacitors show extreme cases of 180° domain configurations.

180° (applying the lost inversion, mirror or rotation symmetry) yields always another, equally probable solution. Regions with homogeneous polarization are called domains. For achieving a macroscopic polarization, a poling process in an external electric field—often combined with heating above room temperature—must be applied to eliminate the 180° domains and to switch possibly also ferroelastic domains (as e.g. 90° domains). The macroscopic orientation obtained after removal of the external field is called remanent polarization $\boldsymbol{P}_r$.

The charges on the electrodes are given by the electric displacement field $\boldsymbol{D}$ perpendicular to the electrode faces. Giving the index 3 to this direction, the measured pyroelectric coefficient is equal to $\partial D_3 / \partial T$. The electric displacement field is be written as $\boldsymbol{D}(\boldsymbol{E}) = \varepsilon_0 \boldsymbol{E} + \boldsymbol{P}(\boldsymbol{E})$, where $\varepsilon_0$ means the permittivity of the vacuum. The in ferroelectrics polarization part is much larger and the only relevant temperature dependent

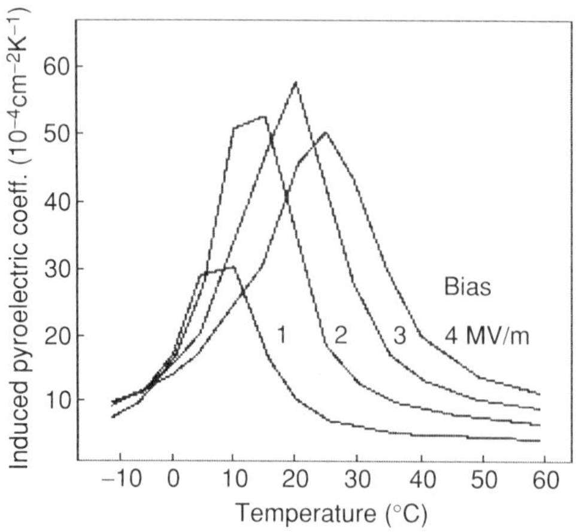


**Fig. 3.** Experimental curves for the induced pyro-electric coefficient in lead scandium tantalate (from [9]);

term. For the present discussion one can set $\boldsymbol{D}$ equal to $\boldsymbol{P}$.

$$\Delta D_3 = \frac{\partial P_3(E)}{\partial T}\Delta T \simeq \frac{\partial P_{r,3}}{\partial T}\Delta T + \frac{\partial \varepsilon_{33}}{\partial T}E_3 \cdot \Delta T \tag{2.1}$$

The first term is called the true pyroelectric effect and $p_3 = \partial P_{r,3}/\partial T$ is the true pyroelectric coefficient. The second term arises from the first order correction for $P_r(E)$. Pyroelectricity arising due to this second term is called induced pyroelectricity. Accordingly there are two modes of operation for pyroelectric IR detectors. The first one works without external electric field at a temperature much below the para-to-ferroelectric phase transition. The polarization $P_r$ (omitting the index 3 in the following) has previously been maximized by hot poling at an elevated electric field, assuring a good time stability of $p$ at the application temperature around room temperature. The second mode exploits the peaking of $p$ near the phase transition. A dc electric field needs to be applied to obtain a non-zero average polarization. One speaks of induced pyroelectric currents or sometimes of dielectric bolometer mode, since the field dependence of $P$ can be considered as a dielectric property. The applied field has not only the effect of eliminating 180° domains, it broadens and shifts (first order transition only) the phase transition to higher temperatures. This leads to a reduction of the signal. There is thus an optimal dc field (a few MV/m) and temperature (see Figure 3) at which the response is peaking. The temperature of such a device needs to be stabilized. The advantage of the first method is the simplicity of operation conditions: no dc field and no temperature stabilization are needed. The second method allows higher sensitivities.

In the following the phenomenological theory of Landau, Ginzburg and Devonshire (LDG, see MS 435) is used to gain some better insight into the difference of true and induced pyroelectricity (using the notation in [10]) and to prepare the ground for discussing materials figure of merits. The free energy density $G$ of the ferroelectric phase is written as (monodomain, $\boldsymbol{P}$ perpendicular to electroded faces):

$$G(T, P) = \frac{1}{2}\beta(T - T_0)P^2 + \frac{1}{4}\gamma P^4 + \frac{1}{6}\delta P^6$$
$$E = \beta(T - T_0)P + \gamma P^3 + \delta P^5 \tag{2.2}$$

The derivation with respect to $D$ yields the electric field $E$. The lowest energy at $E = 0$ yields the spontaneous polarization $P_s$. The inverse permittivity due to the ferroelectric effect is equal the second derivative of $G$ with respect to $D$. A general relation is found for the *true pyroelectric coefficient* as follows: $p = -\beta\varepsilon P_s$. High pyroelectric coefficients are thus found in materials with high polarization and high coefficient $\beta$. The true pyroelectric *voltage response*, which is proportional to $p/\varepsilon$, follows the spontaneous polarization and is thus optimal at low temperatures, far below the phase transition. In case of a *second order phase transition* ($\gamma > 0$), and neglecting the sixth order term ($\delta = 0$), the spontaneous polarization is readily obtained as $P_s = (\frac{\beta}{\gamma}(T_0 - T))^{1/2}$. The true pyroelectric coefficient is derived as $p = \frac{-1}{2}(\frac{\beta}{\gamma(T_0 - T)})^{1/2}$. The pyroelectric coefficient diverges at the phase transition temperature $T_0$. So does the permittivity, for which below the critical temperature $T_0$ the relation $\varepsilon^{-1} = \frac{\partial^2 G}{\partial P^2} = 2\beta(T_0 - T)$ is obtained. The pyroelectric coefficient obviously behaves as the square root of the permittivity, i.e. $\frac{p}{\sqrt{\varepsilon}} = -\beta\sqrt{\frac{2}{\gamma}} = const.$ In *ideal first order transitions* ($\gamma < 0, \delta > 0$) the critical temperature $T_c = T_0 + \frac{3\gamma^2}{16\beta\delta}$ is above $T_0$ and the polarization jumps form 0 to $\sqrt{-3\gamma/4\delta}$. The permittivity remains finite but jumps as well at $T_c$. The jump of $P_s$ at $T_c$ gives rise to large pyroelectric currents. Induced pyroelectricity is thus very large in materials with first order phase transitions. In addition, the field needed for induced pyroelectricity does not simply broaden the transition as in case of second order phase transitions, but the discontinuity of $P$ is preserved up to a certain critical field [9]. The critical temperature is shifted to higher temperatures up to a maximal possible value of $T_{max} = T_0 + \frac{9\gamma^2}{20\beta\delta}$. For illustration, calculated curves are depicted in Figure 4 for a model substance.

A further type of phase transitions that attracts attention for being used in the induced pyroelectric mode is the relaxor-ferroelectric transition. It occurs typically in specifically disordered ferroelectrics, where for instance the B-site of the perovskite $ABO_3$ is a mixture of randomly distributed ions of different valency. In the above mentioned $PbSc_{0.5}Ta_{0.5}O_3$, $Sc^{+3}$ and $Ta^{+5}$ ions occupy in a 1:1 mixture a site that should have a 4+ ion. If the two different ions are ordered in a periodic manner, the above discussed, ideal first order phase transition takes place, if they are disordered, the ferroelectric phase is shifted to lower temperatures [11]. Usually, very high low-frequency permittivities are obtained around the phase transition. It is thought that random internal fields prevent a homogeneous, ferroelectric polarization. Instead, local polar regions appear that easily grow in an external field thus providing for the huge dielectric response. There are chances that the absence of domain or phase fluctuations in the relaxor phase allow for working points with a larger *dP/dT* and a lower *tanδ*. A modelling of this issue is missing to date.

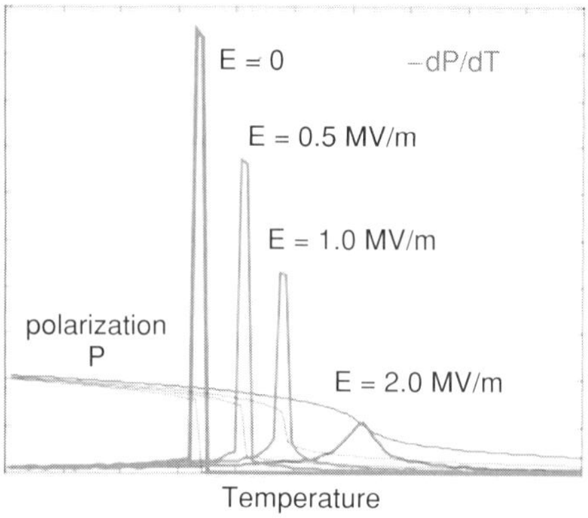


**Fig. 4.** Theoretical behavior of the pyroelectric coefficient as a function of temperature and for various electric fields, as derived from the Landau-Devonshire model. The behavior near the phase transition in absence of an external field is purely theoretical, since the average polarization decays to zero by thermal fluctuations.

The pyroelectric charge is usually not directly measured as a charge but as a current $I$ trough a capacitor of surface A. Taking the time derivative of eqn 2.1:

$$I = A \cdot \frac{dD}{dt} = A \cdot p \cdot \frac{dT}{dt} \tag{2.3}$$

The current measurement is much less prone to drifts and also faster than charge measurements. In addition, charges are compensated with time because of leakage currents through the pyroelectric capacitor, or by conduction through air or along the device surface. For imaging static pictures or temperatures a chopper is applied that modulates the infrared radiation falling onto the detector with a constant angular frequency $\omega = 2\pi f$. The frequency $f$ amounts typically to 30 to 60 Hz for imaging applications.

Several methods have been proposed for the measurement of the pyroelectric effect of bulk (see e.g. [6]) or thin film samples. The most frequently applied techniques are based on the pyroelectric current generated by a temperature ramp of constant rate (i.e. $dT/dt$ = const), yielding a constant pyroelectric current according to eqn 2.3. One speaks of the Byer-Roundy technique when a single ramp is applied. It is important to check heating and cooling curves, because the release of trapped charges (thermally induced currents) might falsify the heating curve. Such trapped charges are injected into the sample during the poling procedure. An improvement of the method is obtained by cycling the temperature periodically. The temperature is swept up and down with small amplitudes (e.g. 1 K) and with constant rates (e.g. 0.2 K/s) around a given average temperature. The pyroelectric current, being proportional to the slope of the temperature, will switch between two constant values after some stabilization time. The temperature is best controlled on a small chuck equipped with Joule heating and Peltier cooling elements, placed in a chamber protecting from perturbations [9, 12]. For higher modulation frequencies, temperature changes are induced by chopped light absorption (Chynoweth). However, this technique is more qualitative than quantitative because it requires the knowledge of thermal properties for deriving the actual temperature (see the next section). Sometimes the polarization is measured as a function of temperature. This is rather dangerous, since the polarization must be derived from a hysteresis loop obtained by switching the polarization. Firstly, it is not sure whether the complete polarization is switching, secondly fatigue could reduce polarization more efficiently than the temperature increase.

The quality of a sensor does not only depend on sensitivity, but also on the signal-to-noise ratio (S/N ratio), or in other words, the noise level. The intrinsic noise type of pyroelectrics is the Johnson noise of the capacitor element (see, e.g., [1, 2]). For a capacitor of area $A$ and thickness $t$, the noise current is obtained as:

$$I_n = \sqrt{\frac{A}{t} 4kT\,(\sigma + \omega\varepsilon \tan\delta)\,\Delta f} \tag{2.4}$$

where $\sigma$ is the leakage conductivity of the pyroelectric thin film (in bulk detectors: the parallel resistor), $\omega$ the angular frequency, $tan\sigma$ the loss tangent, $\Delta f$ the frequency bandwidth, $A$ the surface, and $t$ the thickness of the element. The cross over frequency $\omega_{cn} = \sigma/(\varepsilon \cdot tan\sigma)$ between resistor type noise and dielectric noise amounts to typically 0.1–10 Hz. At usual modulation frequencies of 10 to 60 Hz the dielectric noise is thus dominating. The intrinsic figure of merit of the dielectric and pyroelectric materials parameters for an optimal S/N ratio can thus formulated as:

$$FM = \frac{p}{\sqrt{\varepsilon \tan\delta}} \tag{2.5}$$

Various phenomena can contribute to the dielectric loss. At high frequencies (GHz and higher) the dielectric loss increases due to interactions of phonons and ac electric fields (see [13] and [14]). In non-centro-symmetric materials the dominating phonon mechanism is the quasi-Debye loss mechanism. Thermal fluctuations and relaxation of phonons lead to charge fluctuations in a similar way as dipoles give rise to loss in the Debye theory. In ferroelectrics, domain wall relaxations play a role as well. In the 1 to 10 GHz range there is usually a relaxation of 180° domain wall motions. All these mechanisms yield a $tan\delta$ that increases at least linearly with the dielectric constant, i.e. $\tan\delta \propto \varepsilon^{\alpha}$, where a is between 1 and 4 [13]. It is difficult to make a general theory on loss mechanisms at the very low frequencies of a few tens of Hertz. In this region dielectric loss is governed by defects and ferroelectric domains. Charged or dipolar point defects relax with respect to position or orientation, structural defects may interfere with electric fields by means of the piezoelectric coupling, and defects may delay domain wall motions or give rise to a creeping of domain walls. Many of the low frequency relaxation mechanisms show again a $tan\delta$ proportional to $\varepsilon$. Near the critical temperature, polarization fluctuation and phonon losses (soft mode) were found to follow a Curie-Weiss law like the dielectric constant, meaning that $tan\delta$ diverges as well at the critical temperature. This has for instance been found in ref [9] where the loss tangent and the dielectric constant increase by abut the same factors upon approaching the critical temperature ($PbSc_{0.5}Ta_{0.5}O_3$). The materials figure of merit *FM* behaves thus rather like $p\varepsilon^{-1}$ than $p\varepsilon^{-1/2}$ especially near critical temperatures.

Evaluating the *FM* by means of the LDG theory leads to the conclusion that in case of second order phase transitions there is no improvement of the intrinsic signal to noise ratio when going closer to the phase transition. $p\varepsilon^{-1/2}$ is in fact independent of temperature and taking into account the increasing dielectric losses, the intrinsic signal to noise is expected to decrease with temperature. The case of first order transitions is more difficult to judge. The permittivity does not diverge at the phase transition, there is only a discontinuity. Lead scandium tantalate showed an increase of *FM* of nearly a factor 3 [9], hence a clear advantage for induced pyroelectricity in this case. The optimal dc field is, however, not the one at which the pyroelectric current is peaking, i.e. inside the hysteretic first order transition with mixed phase noise, but just somewhat above the critical field at which the jump in polarization is lost.

## 3. Operation Principles of Pyroelectric Infrared Detectors

The main application of pyroelectric materials is detection of infrared (IR) radiation. The latter is measured indirectly by means of a temperature change $\Delta T$ of an absorbing structure as a result of the absorbed radiation power. The thermal properties of the absorbing structure have to be suitably designed for obtaining a maximal temperature increase $\Delta T$ at the required response frequency. $\Delta T$ is obtained from the balance of the heat flows, as sketched in Figure 5. First of all, some fraction $\eta$ of the infra-red radiation (power W) falling onto the element is absorbed, i.e., transformed into heat. The temperature change at a given heat input $\eta \cdot W \cdot \Delta t$ depends on the heat conductivity $G$ to the surrounding heat reservoir (heat sink) at temperature $T_O$ and the heat capacity $H$ of the element. It is supposed that the element exhibits a uniform temperature. Thermal wavelength effects are thus neglected. This is justified in thin structures with uniform illumination by the IR radiation. (In $PbTiO_3$ with specific heat capacity $c_p = 3.2\ \mathrm{MJ/m^3/K}$ and specific heat conductivity $\kappa = 3.8\ \mathrm{W/m/K}$, the thermal wavelength amounts to 600 μm at 1 Hz and 20 μm at 1 kHz.) Considering stationary solutions in the frequency domain for an IR power of $W = W_\omega \cdot e^{i\omega t}$, the temperature modulation is readily obtained as:

$$\Delta T_\omega = \frac{\eta \cdot W_\omega}{G + i\omega \cdot H}, \quad |\Delta T_\omega| = \frac{\eta \cdot W_\omega}{G\sqrt{1 + \omega^2 \cdot \tau_{th}^2}} \tag{3.1}$$

The thermal time constant $\tau_{th} = H/G$ has been introduced. There are thus two frequency domains separated by the inverse thermal time constant. The *current responsivity*, i.e. the current response per Watt of radiation power falling onto the detector

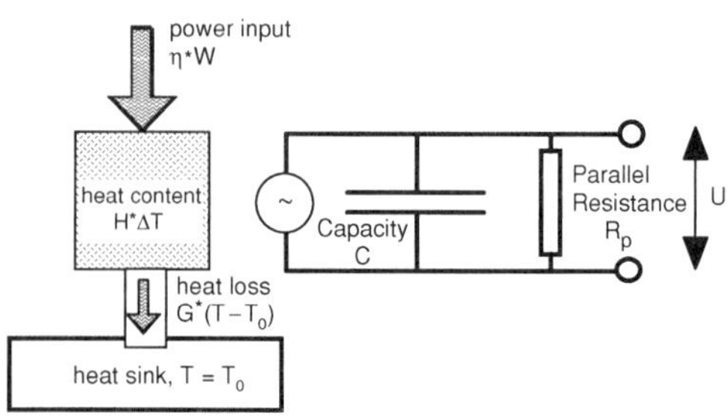


**Fig. 5.** Schematic of thermal and electrical circuit of a pyroelectric element.

element is obtained as:

$$\Re_J(\omega) = \frac{J_\omega}{W_\omega} = \frac{p \cdot \eta \cdot \omega \cdot A}{G\sqrt{1 + \omega^2 \cdot \tau_{th}^2}} \tag{3.2}$$

Alternatively, the voltage across a parallel resistor $R_p$ can be measured. In bulk devices, a value of 10 G W is typically applied ($R_p$ should not exceed the gate impedance of the amplifier.) In thin film devices, leakage through the film may yield an $R_p$ in a good value range. At higher frequencies the dielectric loss tangent *tan d* comes also into play. The conductance $Y$ of the complete element (thus including the parallel resistance) is obtained as:

$$|Y(\omega)| = \frac{1}{R_p}\sqrt{(1 + \omega \cdot \tau_{el} \cdot \tan\delta)^2 + \omega^2 \cdot \tau_{el}^2} \tag{3.3}$$

The electrical time constant $\tau_{el} = R_p \cdot C$ has been introduced. The voltage response is proportional to the impedance $Z = Y^{-1}$. Above the angular frequency $\tau_{el}^{-1}$, the impedance decreases as $\omega^{-1}$. The pyroelectric element thus works like a low pass filter, cutting off the high frequency voltage response. Omitting the small term in *tanδ,* the complete voltage responsivity is written as:

$$\Re_V(\omega) = |Z(\omega)| \cdot \Re_J(\omega) = \frac{p \cdot \eta \cdot \omega \cdot A \cdot R_p}{G \cdot \sqrt{1 + \omega^2 \cdot \tau_{th}^2} \cdot \sqrt{1 + \omega^2 \cdot \tau_{el}^2}} \tag{3.4}$$

The general trends of the two responsivity functions depend on the relative size of the two time constants. In *bulk devices* one generally encounters $\tau_{el} < \tau_{th}$ [1, 15]. The inverse situation, i.e. $\tau_{el} > \tau_{th}$ is typically thin film structures [16]. The resultant frequency behavior for thin films is schematically shown in Figure 6. In thin film devices the parallel resistance is not directly involved in the *voltage response*. This is very advantageous, as film leakage is not easy to control precisely. The mounting of parallel resistors can be avoided if the leakage is sufficiently large to avoid discharging through the input gate of the voltage amplifier. Observing that $C = \varepsilon A/t_p$ we can further develop the response in the intermediate region as $\Re_v \cong \frac{\eta \cdot p \cdot t_p}{\varepsilon \cdot G}$ in case of thin films, and $\Re_v \cong \frac{\eta \cdot p \cdot A \cdot R_p}{H}$ in the bulk case. The figure of merit given in general, $F_V s = \frac{p}{\varepsilon c_p}$, corresponds to a mix of those two situations, and is relevant in any case for the high frequency case. It includes both, dielectric constant (not that $\varepsilon = \varepsilon_0 \varepsilon_r$) and specific heat capacity $c_p$. For the thin film situation $F_V^* = p/\varepsilon$ is correct at the peaking of the voltage response. A small thickness is favorable to drive the current response up. The impact of the heat conductivity $G$ at intermediate and low frequencies, and its unimportance at high frequencies, is demonstrated in Figure 7, which depicts the response of a pyroelectric detector in air and in vacuum, i.e. once with air convection (large $G$), and once without air convection (small $G$).

The *ultimate detection limit* of a sensor is given by the intrinsic noise of the detector element. In order to reach this limit with the complete sensor set-up, the noise contribution from the amplifier and from external sources of the environment have to be smaller than the intrinsic noise. Using low-noise amplifiers like J-FET transistors, the

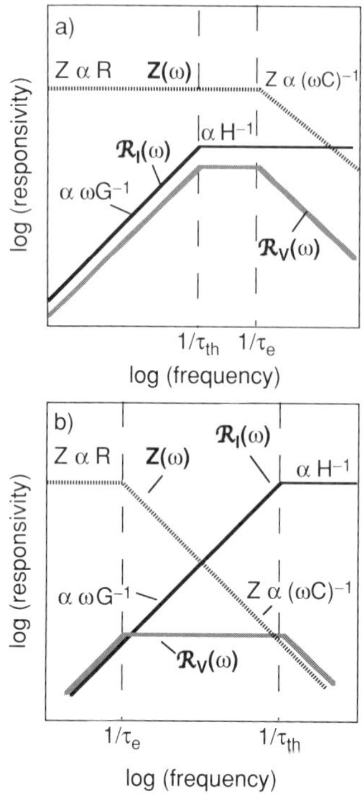


**Fig. 6.** Log-log scheme of current and voltage response, together with the impedance for the typical bulk (top) and thin film situation (bottom).

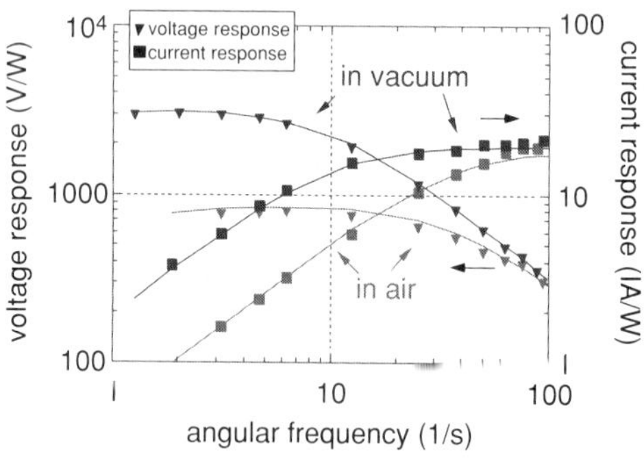


**Fig. 7.** Voltage and current responsivity of a linear array element (1 × 12) measured in air and in vacuum. One element was 0.4 × 0.9 mm large. The curves are calculated according to eqns 3.4 and 3.6, the points are measured ([16]).

theoretical limit of the intrinsic noise can indeed be reached. The thermal noise power $P_{thn} = \sqrt{4kT^2GB}$ is proportional to the square root of the heat conduction $G$, and is very small for optimized thin film devices. The Johnson noise is normally dominating. The contributing noise sources are given by the real part $g$ of the admittance $Y$. The

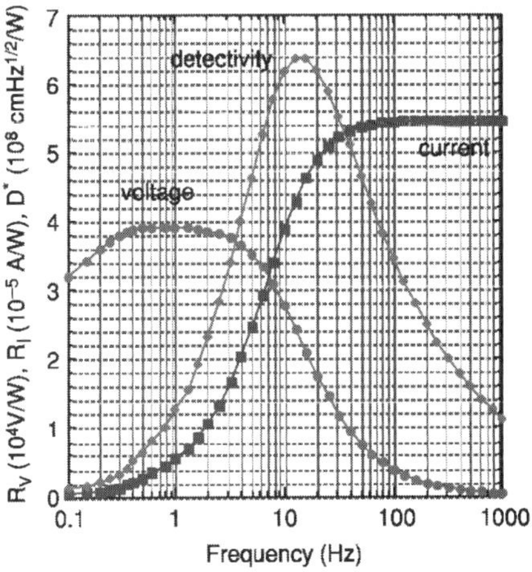


**Fig. 8.** Calculated responsivities and detectivity for a $0.01\,\text{mm}^2$ large PLT element, taking typical published parameters (from [5]).

noise current $J_n$ for a bandwidth $B$ is obtained as:

$$J_n = \sqrt{4kTgB} \quad with \quad g = R_p^{-1} + \omega C \tan\delta \tag{3.5}$$

The minimal detectable power is the noise equivalent power defined by $NEP = J_n/\Re_J$. Very often, the detectivity $D^*$ is given instead of the NEP:

$$D^* = \frac{A^{1/2}B^{1/2}}{NEP} = \frac{\Re_J(\omega)\cdot A^{1/2}B^{1/2}}{J_n} \tag{3.6}$$

The multiplication with $A^{1/2}B^{1/2}$ cancels surface and bandwidth contribution from the noise. At high frequencies, $D^*$ is independent of detection area and bandwidth. For this reason $D^*$ is frequently used as a measure of performance to compare different detectors. The best operating conditions are found near the inverse thermal time constant, where the signal to noise ratio is peaking (see Figure 8). Ideally, $\tau_{th}$ should be matched to $\tau_{el}\cdot\tan\delta$. The materials figure of merit for the detectivity is generally given as:

$$F_D = \frac{p}{c_p\sqrt{\varepsilon\tan\delta}}. \tag{3.7}$$

The detectivity $D^*$ is a useful quantity to characterize an IR detector independently of size and detection bandwidth. However, for thermal imaging applications, $D^*$ alone does not sufficiently characterize the complete device, which also contains an optical part. The quantity of interest is the minimal temperature difference of a black body target object with respect to the background temperature one still can detect. It is called the "noise equivalent temperature difference", NETD. The procedure to evaluate this quantity is as follows: The noise voltage $V_n$ of the detector is measured when the target temperature is in equilibrium with the background temperature. The target temperature is increased by $\Delta T$ and the signal voltage $V_s(\Delta T)$ is measured. The NETD is derived as [17]:

$$NETD = \frac{\Delta T}{V_s / V_n} \tag{3.8}$$

$\Delta T$ is equal to the NETD if the signal is just equal to the noise voltage. Of course, the NETD can be related to the NEP and $D^*$ of the detector knowing the optical properties of the imaging system. The temperature increase $\Delta T$ has to be such that the increase of the power falling onto one pixel of the focal plane array (FPA) equals the NEP. The power increase is proportional to $\frac{dW}{dT}\Delta T$ in the wavelength interval $\Delta\lambda$ detected, where $W$ is the power density emitted by the target surface. It is further proportional to the FPA pixel area $A_D$, and the transmissivity $\tau$ of the optics. The reader is referred to special literature on this topic [17] for the derivation of general formulas. The formula given below finds frequent use [18]:

$$NETD = \frac{(4F^2 + 1)}{A_D \cdot \tau \cdot (dW/dT)_{\Delta\lambda}} \cdot NEP = \frac{(4F^2 + 1)}{A_D^{1/2} \cdot \tau \cdot (dW/dT)_{\Delta\lambda}} \cdot \frac{B^{1/2}}{D^*} \tag{3.9}$$

F is the f-Number of the optics. Typically, a NETD of less than 100 mK is required for uncooled IR-imaging systems [19].

# 4. Thin Film Materials

## 4.1. Properties and figure of merits

Properties of thin film materials differ from those of bulk materials in as much as microstructure and substrate influence are of importance. Table 1 contains a list of representative materials and their characteristics for pyroelectric applications. It can be noted that thin film materials show as good characteristics as bulk materials, especially when they are grown epitaxially. In contrast to bulk ceramics, thin films can be grown textured or even completely oriented in the case of epitaxy. The optimal texture is achieved when the polar axis (polarization) stays perpendicular to the electrodes everywhere in the film. In this case, a performance similar to the one of single crystal materials is obtained. This leads to a considerable improvement of properties in the case of substances that only exist as polycrystalline ceramics in bulk form (as e.g., PZT, PLT). A good demonstration of this case is epitaxial $PbTiO_3$, whose figure of merit $F_V^*$ was measured as 291 $kVm^{-1}K^{-1}$ [20] at thin films, whereas only 107 $kVm^{-1}K^{-1}$ is reached in bulk ceramics.

The discussion of bulk material properties made clear that there is a trade off between temperature stability and size of the pyroelectric effect. If a cheap, simple and reliable device is the goal, materials with high critical temperatures ($T_c$) such as $LiTaO_3$ and $PbTiO_3$ are more adequate. If the operating temperature can be held precisely, close to or exactly at the critical temperature, materials with near room temperature $T_c$, with ideal first order or relaxor-type behavior, yield much larger responses. The materials choice taken in the published thin film work indicates that $PbTiO_3$ derived compounds are the clear favorites until now, and much less work has been spent on the second category

**Table 1.** Bulk and thin film pyroelectric materials (sources: [1–3, 5, 21]. The figure of merits $F_V$, and $F_D$ are given in the text

| Material | p $[\mu Cm^{-2}K^{-1}]$ | $e_r$ | tand | $c_p$ $[10^6 Jm^{-3}K^{-1}]$ | $F_V$ $[m^2C^{-1}]$ | $F_D$ $[10^{-5}Pa^{-1/2}]$ | $T_c$ $[^\circ C]$ |
|---|---|---|---|---|---|---|---|
| $LiTaO_3$, single crystal | 230 | 47 | < 0.01 | 3.2 | 0.17 | 5–35 | 620 |
| $PbZrTiO_3$ (PZFNTU) bulk ceramic | 380 | 290 | 0.003 | 2.5 | 0.06 | 5.5 | 230 |
| $PbTiO_3$, bulk ceramics | 180 | 190 | 0.01 | 3.0 | 0.04 | 1.5 | 490 |
| $PbTiO_3$ (PCWT4-24) bulk ceramic | 380 | 220 | 0.01 | 2.5 | 0.08 | 3.4 | 255 |
| P(VDF/TrFE) 80/20 co-polymer film | 31 | 7 | 0.015 | 2.3 | 0.22 | 1.4 | 135 |
| PZT 15/85 (111) polycrystalline thin film on silicon (MEMS) | 160–220 | 200–230 | 0.01–0.015 | 2.7 | 0.03–0.04 | 1.3–1.5 | ~420 |
| $PbLaTiO_3$ 10–20% La polycrystalline thin film on silicon | 200–576 | 153–550 | 0.01–0.024 | 2.7 | 0.02–0.15 | 0.7–4.1 | |
| $PbTiO_3$ (001) thin film epitaxy on Pt/MgO | 250 | 97 | 0.006 | 3.2 | 0.09 | 3.4 | ~490 |

of materials (see citations in [5]). Pure $PbTiO_3$ has been mostly abandoned because of too high dielectric losses and difficulties to pole. The most advanced integration work on silicon has been achieved with PZT (15 to 30 %) Zr films. This is a kind of spin-off from the work done in ferroelectric memories and actuators. The results collected in table 3 suggests that PZT is going to be replaced by PLT or PCT. The table also reveals that the research in $LiTaO_3$ thin films for pyroelectric applications is far away from being as advanced as its applications in bulk detectors. In general, films grown in epitaxial quality on single crystal substrates such as MgO, $SrTiO_3$ or sapphire exhibit figures of merit that are roughly two times higher than the ones of films deposited onto silicon. The reason for that will be discussed below. The earlier work on relaxor-type materials and lead scandium tantalate (PST) dealt with depositions on inert substrates only. Only recent efforts have been successful using ordinary platinized silicon substrates. The properties are, however, by a factor 3 to 10 lower. Main problem are the higher nucleation and growth temperatures as compared to lead titanate family, resulting in a degradation of the platinum bottom electrode and its interfaces. Suitable seeding layers [22–24] are applied to lower the nucleation temperature. Improvements in sol-gel chemistry still seem to be possible. Nevertheless it looks difficult to synthesize PST without second phases and with B-site ordering below 700 °C [25]. Low deposition temperatures have been achieved by MOCVD, however, no material properties were reported [26]. Comparing results obtained on equivalent substrates, one observes that induced pyroelectricity did not prove to be superior to the best PLT or PCT thin films showing the true pyroelectric effect so far (with the possible exception of PMZN).

For low frequency applications, thicker pyroelectric films might offer an advantage, since the voltage response increases with increasing thickness. Using thin film deposition techniques, ferroelectric thin films on silicon crack above 2 to 3 μm thickness, typically. The range of 5 to 20 μm is more apt for so-called thick film techniques. A slurry containing pyroelectric particles is deposited by screen printing on a substrate before drying and annealing. The published work (see table 6) indicates, however, a loss in performance by a factor 2 to 3. In addition, it is quite difficult to combine silicon micromachining techniques with thick film techniques. Most interesting looks the composite approach. The dilution of the pyroelectric material seems to reduce the dielectric constant much more than the pyroelectric coefficient, yielding a large $\Re_V$ [27].

Finally, it has to be mentioned that in some of the works the pyroelectric current has been measured, without determining the pyroelectric coefficient, so for $NaNO_2$ [28], and $LiTaO_3$ [29].

The figure of merit $F_D$ has been determined very rarely by experiment. In most cases, it is just calculated using a literature bulk value for the volume heat capacity $c_p$. The values reported in table 1 are copied from the cited literature or else, calculated using reasonable values for $c_p$.

## 4.2. Substrate effects

Thin film materials cannot be considered independently from the substrate on which they are grown. The substrate may well influence microstructure and properties of the ferroelectric thin films. Three types of substrate influence can be identified:

(1) Growth phenomena related to the surface on which the film is grown (usually the bottom electrode).
(2) Thermal strains imposed by the bulk of the substrate during cool down from the growth temperature (the substrate during growth is typically is 100 to 1000 times thicker than the film).
(3) Thermal expansion mismatch with thin film support during operation of the pyroelectric film yielding a piezoelectric contribution to the pyroelectric effect.

The chemical nature, lattice and microstructure of the growth-substrate have important impacts on the nucleation of the films in general. The ferroelectric materials considered here often exhibit a pronounced nucleation controlled growth. This means that considerable activation energies are needed for nucleation of the correct phase, so for instance in case of PZT [30]. The growth-substrate may thus have a seeding function. If it is possible to reduce the interface energy for a given orientation, the nucleation energy is reduced for this given orientation and textured films can be obtained. This can be exploited to obtain a higher pyroelectric coefficient. The capability to grow one or the other orientation of $PbTiO_3$ derived compounds on platinized silicon substrates is an important issue of ferroelectric film integration. It is generally found that the (100)/(111) texture ratio is controlled by the Pb/Ti ratio during nucleation. This ratio is either influenced through growth parameters (e.g., high temperature favors Ti, thus (111)-orientation) or by suitable seed layers, such as a nanometer thick Ti or $TiO_2$ layer, leading to a (111)-orientation [31, 32]. Thin $PbTiO_3$ {001} layers of some 10 nm may be used to seed perovskite (100) orientation [33, 34].

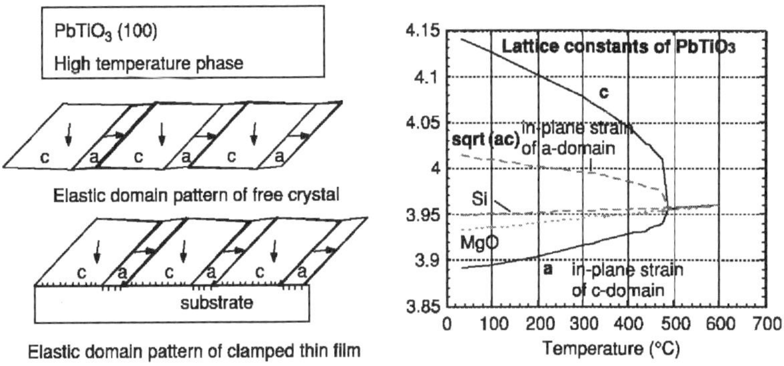


**Fig. 9.** Schematics showing domain formation in tetragonal {100} thin film together with spontaneous strains of $PbTiO_3$. Clamping to substrate induces angular distortions and dislocations.

Pyroelectric thin films of the perovskite family are typically grown at 500 to 700 °C in the paraelectric phase. The para-to-ferroelectric phase transition occurs when cooling down from the growth temperature. The best-suited true pyroelectrics known to date are all tetragonal materials derived from $PbTiO_3$, a material with a large polarization and large spontaneous strains (see Figure 9). The tetragonal and polar c-axis grows with decreasing temperature, the $a$-axis shrinks. If the film would be a homogeneous and free crystal, the c-axis would point with equal probability of $^1/_3$ along one of the former cubic axis resulting in a pattern of 90° domains. If during poling, 180° switching occurred exclusively, a maximal polarization of $P_s/3$ could be expected at best along one of the former cubic axis. The thin film case is much more complex. Very close to the substrate (bottom electrode) epitaxial relationships are established between the perovskite and the substrate lattice. The epitaxial misfit might be higher for one axis than for the other. For instance the $a$-axis of $PbTiO_3$ ($a$ = 3.96 Å) is closer to the lattice constant of platinum (3.92 Å) than the c-axis of $PbTiO_3$ (4.14 Å). The interface $PbTiO_3$/Pt thus has lower energy with (001) planes than with (100) planes of $PbTiO_3$. For the bulk of the film, it is rather the average elastic energy due to misfit strains with the bulk of the substrate (as e.g., silicon) that needs to be minimized. In the ordinary case this happens by introducing misfit dislocations. However, in materials with ferroelastic domains, elastic energy can also be released by appropriately changing the volume fraction of the different types of domains [35]. The creation of domain walls is in competition with the creation of dislocations. The phenomenon has been studied mostly at the epitaxial system $PbTiO_3${100}/$SrTiO_3$(100) [36–43] and $PbTiO_3${100}/MgO [44].

In order to achieve more than $P_s/3$ polarization of a {100}-oriented $PbTiO_3$ film, it is necessary that the misfit strains with the substrate are negative (the film is compressed), which results in an increase of c-domains. This mechanism is especially effective in a temperature interval just below the critical temperature, as nicely demonstrated in Figure 10. Ideal for a good c-axis orientation is a compressive stress of the film at the phase transition which may happen in the system $PbTiO_3${100}/MgO. Well $c$-axis oriented films are thus obtained on $SrTiO_3$ [45] and MgO substrates [20] (see also table 4). Such a stress is built up during cool down from the growth temperature to the phase transition if the thermal expansion coefficient of the substrate is larger than

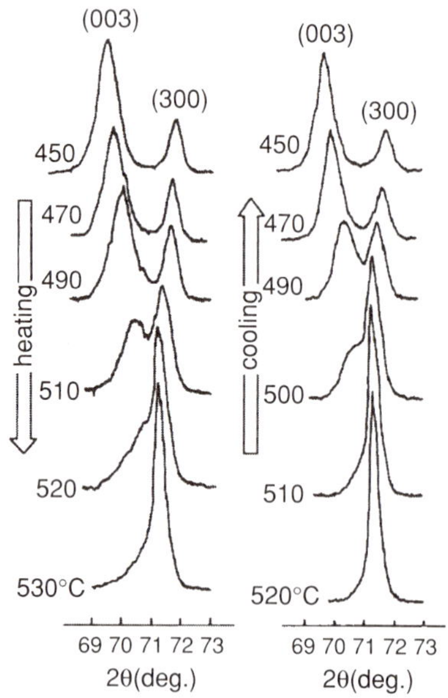


**Fig. 10.** X-ray diffraction intensities of (003) and (300) reflections as a function of temperature for epitaxial $PbTiO_3$ thin films grown on Pt(100)/MgO(100) ([20]).

the one of the paraelectric $PbTiO_3$. Residual growth stresses such as tensile stresses after crystallization of sol-gel processed films play a role as well. The domain issue is an important technological problem. The final goal is mostly the integration onto silicon substrates, which is a stringent condition for combining sensor and electronics on the same chip. However, for optimal polarization in tetragonal ferroelectrics such as $PbTiO_3$, silicon has a too low thermal expansion coefficient. Other substrates like MgO yield superior results. For small arrays only, a hybrid approach might be acceptable. MgO is a suitable substrate in these case, for it is possible to micromachine MgO [46, 47]. In some device concepts the pyroelectric thin film is first grown on a more ideal substrate than silicon, and later assembled on silicon [48, 49]. The initial substrate is dissolved or grinded down.

On silicon substrates, $PbTiO_3\{100\}$ films exhibit usually less than 50 % c-axis orientation. This phenomenon is accentuated if the deposition process yields a tensile stress, as can be caused by the crystallization shrinkage of the sol-gel preparation technique. Sputter processes allow incorporation of a compressive stress during in-situ deposition, and a hence stronger $c$-axis orientation [50]. The degree of orientation does not only influence the pyroelectric ceofficient, but also increases the dielectric constant, which renders the figures of merit even lower. The permittivity amounts to 80 to 120 in single-crystalline $PbTiO_3$ [51]. Thin films grown on silicon substrates, exhibit much larger values, typically 250, especially those processed by sol-gel techniques. In epitaxial films on MgO, this increase is well correlated with the degree of $c$-axis orientation (see Figure 11). According to these results, the dielectric constant of 250 corresponds to a degree of orientation of $^1/_3$, which is indeed observed in sol-gel films [52]. A possible origin of this behavior is the contribution of 90°-domain walls

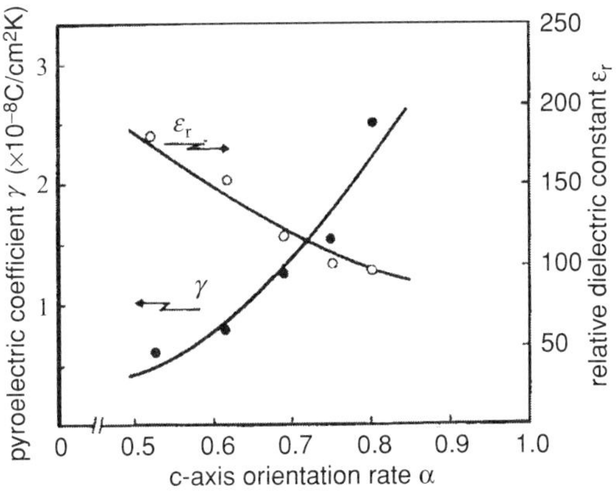


**Fig. 11.** Relation between pyroelectric coefficient $\gamma$, permittivity $\varepsilon_r$ and the degree of a c-axis orientation of epitaxial $PbTiO_3$ thin films grown on Pt(100)/MgO(100), as obtained from $\theta$-$2\theta$ x-ray diffraction experiments and defined as $\alpha = \frac{I(001)}{I(001)+I(100)}$ ([20]).

to the dielectric constant. Their density necessarily increases with lowering of the degree of $c$-axis orientation. The not so much convincing results with {100}-oriented films on silicon has motivated to chose the second best orientation, namely the (111)-orientation [53, 54]. However, it was found that optimized poling leads to about the same figures of merit for both orientations [55–57] since a portion of the a-domains can be flipped.

### 4.3. Piezoelectric contributions to pyroelectricity

In order to consider the third effect lest us briefly recall the definition of the pyroelectric coefficient. Thc "true" or primary pyroelectric effect is measured when volume and shape are kept constant (coefficient $p^{(strain)}$). This means that the thermal expansion is not included. The latter is causing the so-called secondary pyroelectric effect. When the pyroelectric effect is measured at constant stress ($p^{(stress)}$), what is usually done, it is composed of the primary and secondary effects:

$$p_3^{(stress)} = p_3^{(strain)} + e_{3i} \cdot \alpha_i \tag{4.1}$$

The thermal expansion coefficients $\alpha_i$ contribute to the measured charge through the piezoelectric effect, which in this case is described by the $e_{ij}$-coefficients. A thin film is free perpendicular to the plane (direction 3), but is forced to dilatate together with the substrate. With respect to free expansion, the film suffers a strain of $\alpha_s - \alpha_f$ (indexes s for substrate and f for film), given that film and substrate have a common temperature. This strain yields a stress which results in an additional secondary term that needs to be added to $p^{(stress)}$ [58]:

$$p_3^{(film)} = p_3^{(stress)} + \frac{2d_{31}}{s_{12}^E + s_{12}^E}(\alpha_s - \alpha_f). \tag{4.2}$$

$e_{31,f} = \frac{d_{31}}{s_{12}^E + s_{12}^E}$ is the effective piezoelectric coefficient of the thin film, which describes the charge density $D_3$ as a function of in plane strains [59]. Equation 3.9 can thus be derived from a thin film constituent equation that can be written as:

$$dD_3 = \varepsilon_{33}^{(x_1,x_2,\sigma_3,T)} \varepsilon_0 \cdot E_3 + p_3^{(stress)} \cdot dT + 2e_{31,f}^{(T)} \cdot (\alpha_s - \alpha_f) \cdot dT + e_{31,f}^{(T)} \cdot (x_1 + x_2) + d_{33,f}^{(T)} \cdot \sigma_3 \quad (4.3)$$

In perovskite based ferroelectric materials the contribution due to thermal mismatch is smaller than $p^{(stress)}$ [58]. Taking typical values for PZT films ($e_{31,f} = -2$ to $-8$ Cm$^{-2}$, $\alpha_s - \alpha_f = -4$ to 4 ppm/K) a piezoelectric contribution in the range of $+/-50$ μCm$^{-2}$K$^{-1}$ is estimated. This value depends very much on the orientation of a tetragonal material. In some cases it might be negligible, in other cases it might be large.

## 4.4. Poling

There are two ways to deal with the poling issue: to apply a poling procedure on the finished device, or to grow self-poled films. The second possibility avoids a time consuming, and possibly also yield reducing production step. However, the growth of highly poled films works only with plasma or vacuum processes [61]. A charge gradient due to different diffusivities of electrons and oxygen vacancies after oxygen release in vacuum was given as possible origin [62]. The self-poling may also occur in sol-gel processed films, but is a too small effect. An optimal poling requires temperatures in the 100 to 200 °C range. The good pyroelectric materials often have too large coercive fields at room temperature for switching the polarization of the 180 ° domains. Large internal bias-fields indicate the existence of defect charges that pin domain walls and cause backswitching at zero field. At increased temperature, where the conductivity in the film is enhanced, these charges can be redistributed for pinning aligned domains. In the PZT 15/85 system the pyroelectric coefficient was observed to increase until 130 to 200 °C poling temperature (see Figure 12), above which some irreversible deterioration mechanisms take place. The dielectric constant decreases by about 30% during poling. In case of tetragonal PZT films it was observed that the c-axis domain volume fraction is enhanced at the cost of the a-axis domain part [56, 63].

## 4.5. Porosity and composites

Finally it has to be pointed out that not only perfect films in terms of microstructure, but also defective films may exhibit a good pyroelectric performance. The composite material of $BaTiO_3$ particles in a polymer matrix was an example encountered above. A further technique to dilute the ferroelectric material in order to decrease the dielectric constant was found in sol-gel processed PCT [64–66]. Porosity was installed due to a slow heating rate in the annealing step. In a randomly porous film one would also expect a decrease of the pyroelectric constant. This was, however, not necessarily observed. The dielectric constant decreases by almost a factor 3, whereas the pyroelectric coefficient stays about the same (see Figure 13), resulting in an increase of $F_V$"=p/ε to 3.4 μCm$^2$K$^{-1}$ (using relative dielectric constant). A further improvement to 4.5 μCm$^2$ K$^{-1}$ was accomplished by introducing polymer particles into the solution [67]. These

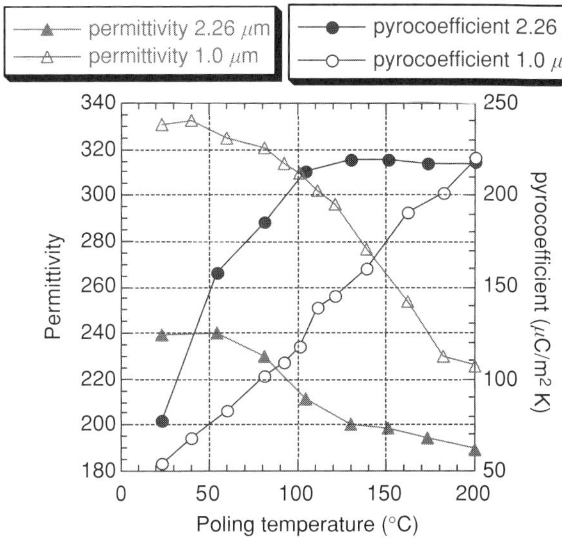


**Fig. 12.** Hot poling of sol-gel prepared PZT15/85 thin films with (111) orientation grown on Pt(111)/Ta/$Si_3N_4$/$SiO_2$/Si. Pyroelectric coefficient $p$ and relative permittivity $\varepsilon_r$ are shown as a function of poling temperature for two different films. A poling field of $2 \times 10^7$ V/m was applied. [55, 60]). The figure of merit $F_V$ increases from 17 and 35 to 110 and 130 kV/m/K upon increasing the poling temperature from 25 to 200 °C.

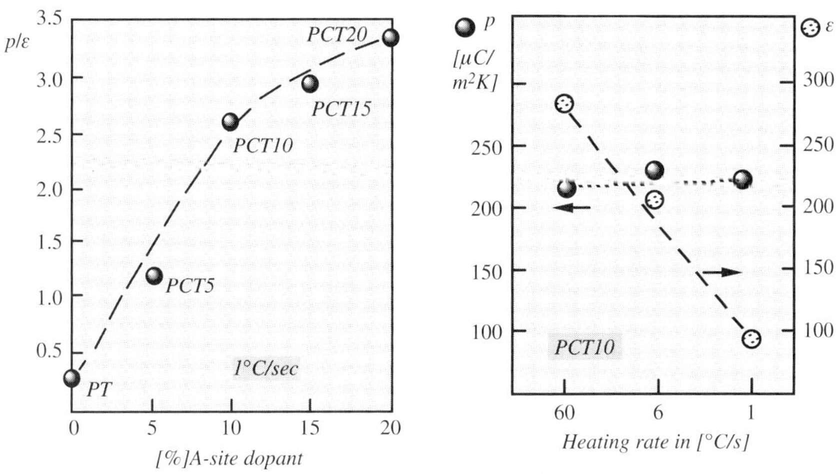


**Fig. 13.** Porous Ca modified $PbTiO_3$ obtained by sol-gel processing. The porosity °C is increased by slowing down the heating rate of the high-temperature anneal. Left: p/ε ration vs. Ca concentration. (The value 1 corresponds to a figure of merit $F_V$ of 113 kV/m/K). Left: Pyroelectric coefficient p and permittivity ε as a function of heating rate (porosity). ([64, 66])

were burned away during annealing. An additional advantage of porosity is the lowering of thermal conductivity.

# 5. Micromachining

## 5.1. Fabrication technology

There are various fabrication concepts for pyroelectric devices. As discussed in the previous section, silicon is not the most suited substrate for pyroelectric thin film growth. On the other hand, silicon is the most suited (and cheapest) substrate for all other fabrication steps, such as semiconductor processing and micromachining. For this reason some concepts have been developed to grow the pyroelectric film on some other substrate and later glue it onto silicon structures, instead of directly depositing onto silicon wafers. Another variation is introduced by different methods of micromachining, which are needed for tailoring thermal properties. Small or linear arrays are rather fabricated by bulk micromachining, whereas large arrays have to be fabricated by surface micromachining techniques. An important question is finally whether the electronics should be integrated on the same chip, or whether a suitable packaging for hybrid assembly can be applied.

### 5.1.1. *Silicon micromachining*

The theory in section 4 showed the importance of the thermal insulation of the pyroelectric element. Silicon, which is best with respect to manufacturing and on-chip integration, is a very good thermal conductor. The pyroelectric element thus needs to be placed on some insulating supporting structure that is well isolated from the silicon, preferentially by a vacuum gap. There are various possibilities: membrane and bridge structures obtained by bulk micromachining, or bridge structures obtained by surface micromachining techniques. The schematic drawings of fig. 14 shows the three possibilities using wet etching. For all cases there are nowadays also dry etching (deep silicon dry etching) techniques allowing for fabrication of deep recessions with vertical walls and independently of the orientation of the feature with respect to the crystal planes. All of the possibilities have their convenient and their inconvenient aspects. Bulk micromachining works fine for small arrays and linear arrays, but the brittleness

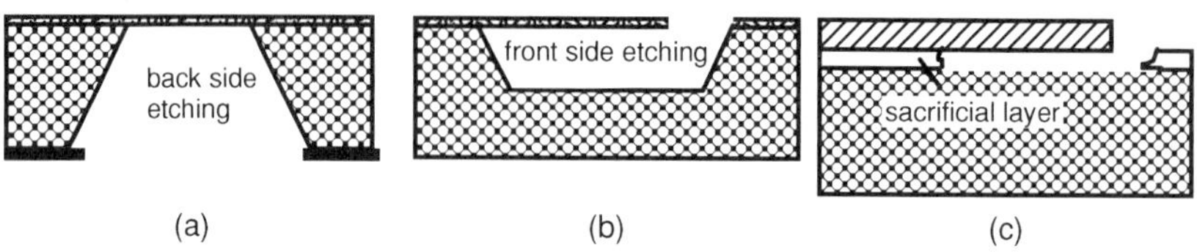


**Fig. 14.** Different micromachining techniques for thermal insulation of suspended structures: (a) bulk micromachining, (b) surface micromachining by front-side etching of the silicon substrate, (c) surface micromachining by sacrificial layers. The shown geometries in (a) and (b) correspond to anisotropic wet etching. (For a textbook, see e.g. [71].)

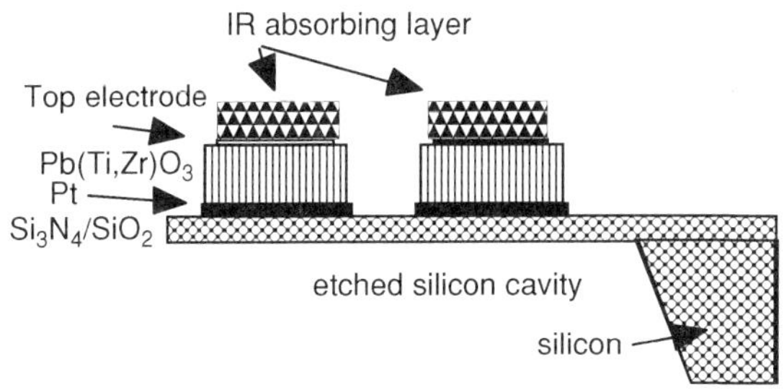


**Fig. 15.** Typical structure of pyroelectric elements of a linear array on a thin membrane fabricated by means of micromachining [75]. The elements are contacted to pads in the other direction than the one seen in this cut.

of the membrane poses problems for large two-dimensional arrays. Additionally, no transistors can be integrated in the pixel cell. Bridge structures exhibit a better thermal insulation than full membranes, but they need more advanced fabrication tools. Front side etching does not need double side aligning of masks, however, the masking of the devices during silicon etching can be a big obstacle when applying wet etching. This problem can be avoided by the use of an easily removable sacrificial layer. The latter is typically a layer of boron-phosphor-silicon glass (BPSG). Recently also porous silicon has been demonstrated for that purpose [68, 69]. A further possibility is the use of more or less isotropic dry etching of silicon in $SF_6$. With surface micromachining and sacrificial layer techniques, only small distances to the silicon substrate are achieved, making the use of vacuum imperative. Two-dimensional modelling of thermal properties of small gap structures is treated in [70]. It is not the goal of this section to go into details of silicon micromachining. The interested reader is referred the book of Madou [71] for an introduction into this topic.

*Example 1: Membrane structure obtained by bulk micromachining for 1-d arrays*

A stress compensated membrane layer of $Si_3N_4/SiO_2$ [72], or low stress nitride is coated on both sides of a double side polished wafer (see fig. 15). This coating fulfils the following functions: It serves as a mask for back-side etching in KOH or in an equivalent base (see, e.g. [73]). Second it serves as a support of the pyroelectric elements (membrane) exhibiting a low thermal conductivity. Bottom electrode and pyroelectric film (PZT15/85) are deposited by sputtering and sol-gel respectively. The top electrode is deposited and patterned by a lift-off technique before a quartz layer is sputter deposited for reduction of parasitic capacity below the contact pads [74]. Windows down to the top electrodes are opened by a $CF_4$ reactive ion etching. The PZT elements on the membrane part are etched free in a HCl:F solution, leaving only narrow bridges between the elements and the bulk silicon part, as needed for separation of bottom and top conductor. The platinum bottom electrode is removed between the elements by electrochemical etching. This etching technique does not attack the membrane material. After deposition and patterning of the conductor lines, pads (Au/Cr) and absorbing layer, the silicon is removed below the elements by back-side etching, as defined by a window in the back-side nitride layer, in order to obtain the result shown in Fig. 16. The 0.9 μm thick membrane with a specific conductivity of 2 W/m/K gives a fairly good thermal

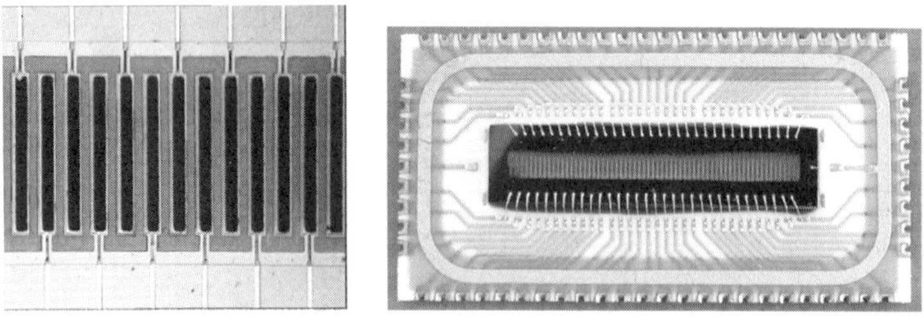

(a) (b)

**Fig. 16.** (a) Top view on 50 element array with 200 μm period obtained with bulk micromachining showing the black platinum absorbers, the Cr-Au contact lines, the membrane layers between the elements, and the $SiO_2$ layer for reduction of parasitic capacitance (Membrane size: 2 × 11 mm) (from [76, 77]) (b) packaged 64 element array using same technology (Courtesy IR-Microsystems).

**Fig. 17.** Example of surface micromachining for focal plane arrays in monolithic technology, as designed for pyroelectric thin film focal plane array by GEC-Marconi. (from [53])

insulation, which allowed for rather high voltage responses at 1 Hz of 800V/W in air [75]. The membrane roughly doubles the heat capacity of the pyroelectric element. Rather long thermal time constants of 28 ms in air, and 104 ms in vacuum have been obtained with such devices [16].

*Example 2: Surface micromachining for 2-d arrays*

For 2 dimensional arrays bulk micromachining is not practicable. Suitable surface micromachining processes are thus required. The example presented here [53] is based on 0.5 μm silicon nitride layer on a 1 to 2 μm thick sacrificial layer. The silicon nitride is the mechanical supporting layer for the pyroelectric capacitor stack. After removal of the sacrificial layer, the suspended structure is hold by two bridges (see Figure 17). The geometrical arrangement combines in an optimal way the need for long bridges on one hand and the need for an as large as possible detector density (filling factor). The extremities of the bridges rest on previously prepared via hole studs, which also serve

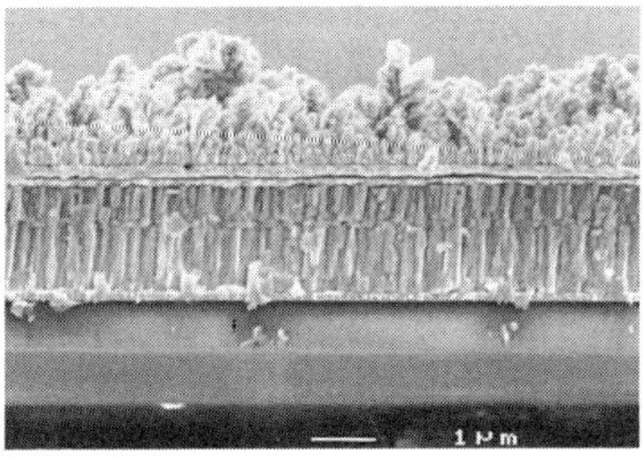


**Fig. 18.** Electro-chemically deposited black platinum grown on a Cr-Au top electrode on a PZT/Pt/Ta/$Si_3N_4$/$SiO_2$ layer stack [16].

to contact bottom and top electrode. The CMOS processing for the readout electronics, including the conductor lines to the transistors, must be performed before ferroelectric film processing and micromachining is carried out. This poses one of the major problems since the high processing temperatures for pyroelectric films usually exceed the stability limit of the standard aluminum conductor lines used in CMOS technology.

### 5.2. IR absorption techniques

There are several possibilities for increasing the absorption of the IR light. One is the addition of a black absorbing layer. Very high absorption coefficients are obtained by this method. Electrochemically deposited black platinum [78] yields 90 % absorption up to 10 µm wavelength. The blackness results from the dendritic and porous microstructure (see Figure 18). There is also the possibility to produce black gold by evaporation in a nitrogen atmosphere. For good absorption, a thickness of about 1 to 2 µm is needed. This is quite a lot compared to the pyroelectric film thickness and leads to an enhancement of the heat capacity, and consequently to a reduction of the current response. The black absorbing layer is thus ideal for low frequency, voltage detection applications.

A second method applies very thin, semi-transparent Nickel or Cr-Ni films as absorbing top electrodes. The absorption depends on the refractive indices on the two sides of the metal film, and of the absorption coefficient of the latter [79, 80]. On PVDF absorptivities of 0.6 (@ 8–14 µm) have been achieved. The formulas given by Bauer and coworkers indicate that the high refractive index of ferroelectric materials might reduce this value to 0.4 to 0.5.

The third method profits from the fact that a transparent layer of $\lambda/4$ thickness on a totally reflecting substrate (e.g. a metal) exhibits a low reflection coefficient for the wavelength 1. The wave is entrapped in the $\lambda/4$ cavity and absorbed during multiple reflections. Taking the pyroelectric film as this transparent material, about 1.0 µm (n = 2.6) thickness would be needed for 10 µm wavelength. This thickness is ideal for growing high quality thin films at reasonable costs by various deposition techniques. This method requires a transparent top electrode, for which usually a very thin, semitransparent metal film is applied. It is clear that this method can be combined with the previous one [81]. A further modification of this method is the combination with the micromachined surface gap [82]. The bottom of the gap is formed by a mirror (e.g. a metal film) and the cavity consists of the vacuum gap and the pyroelectric thin film, which now

**Fig. 19.** Top view on point detector, mounted on TO-39 header. The dice is 4.5 × 4.5 mm large. Two rectangular pyroelectric elements are sitting on a transparent membrane. A JFET is glued onto the dice at the upper left corner. (from [16]).

must be thinner than a micrometer. In addition, a transparent bottom electrode must be provided.

### 5.3. Connection to read-out chip

In order to exploit the noise limit of the sensor element, the electrical connections to the first amplification stage of the electronics have to be as short as possible. The ideal solution is certainly the monolithic integration, i.e., a direct integration of sensors and read-out electronics. However, this increases very much the complexity of fabrication. As a consequence, the yield goes down, and the costs go up. For point detectors and linear arrays, the hybrid approach is certainly the more efficient and cheaper method. In the case of point detectors, the necessary preamplifier, a simple JFET, finds place in one corner of the sensor dice (see Figure 19).

In case of (large) two-dimensional arrays the vertical stacking of detector elements and read-out circuit is inevitable. In order to achieve a good filling factor, the detector elements have to cover most of the surface. There is hardly any place for horizontal electrical interconnects on the same level. Two concepts are conceivable: Either the detector structure is obtained by surface micromachining on top of the read-out wafer, or the detector array is fabricated on a different wafer and joined later to the read-out wafer by a suitable bonding technique assuring vertical electrical contacts. The second possibility is a stacked wafer configuration. Deep silicon etching techniques also contributed to this field. It is possible to etch vertical holes through a complete waver (300 to 400 μm) that exhibit smaller diameters than the pitch of the array (40 to 50 μm). The holes can be filled with a conductor (doped poly-Si deposited by CVD, electrochemically deposited Ni or Cu) providing for a contact between surface micromachined thermal structures on one side, and a ROIC wafer joined on the other side by flip chip techniques. No such 2-d stacked thin film detector exists yet, but some inspiration can be gained from reticulated ceramic detectors. Quite complex hybrid focal plane arrays with

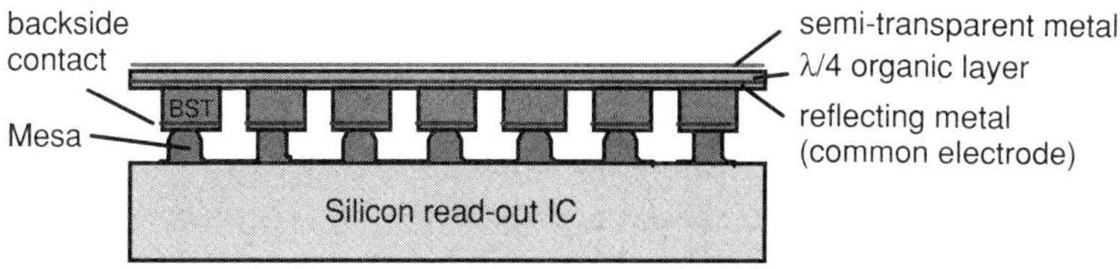


**Fig. 20.** Schematic cross section through focal plane array obtained by reticulation of ceramic BST wafer that is grinded down and bonded to the read-out chip by a mesa structure (TI uncooled FR hybrid technology [84].) The pixel thickness (BST) amounts to 10 to 18 μm, the pitch is 48.5 μm.

384 × 288 and 328 × 245 pixels were recently presented by GEC-Marconi and Texas Instruments [83, 84]. They work either with pyroelectric PZT ceramic or with field-induced pyroelectricity (dielectric bolometer) of $BaSrTiO_3$ (BST) or $PbSc_{1/2}Ta_{1/2}O_3$. The version of Texas Instruments operates with BST at a stabilized temperature near the ferroelectric critical temperature around 20 °C. It is still called an uncooled imaging device, because the necessary heater/Peltier element combination consumes much less power than is required for the standard cooled devices. The starting point of fabrication is a ceramic BST wafer. After reticulation by laser or ion milling processes, polishing down, electrode and absorber layer deposition, the so obtained BST array is bonded to a read-out IC via a mesa structure (see Figure 20) (GEC-Marconi utilizes solder bumps). The absorber layer is a transparent organic λ/4 layer, sandwiched between a semitransparent metal layer and the common electrode. The mesa structure is grown on the read-out IC. The latter is either a charge-coupled device (CCD) or exhibits amplifiers and sample-and-holds for each pixel, combined with a multiplexed reading for the output. Such devices achieve a NETD of 0.1 K or less.

The performance of the above hybrid FPA's is very sufficient at present. Today they are inside many commercial products. However, thermal properties are not optimal yet: the heat capacity of the ceramic pixel is about ten times larger than with a thin film pixel. Also the heat conduction through the mesa studs is certainly larger than what could be achieved with silicon micromachining. So it is expected that thin film versions still could improve the performance and reduce costs.

## 6. Applications

There are, in fact, only few reported works, where a complete device has been manufactured and characterized. Most of the industrial thin film array projects are currently in an intermediate stage of development or in the stage to be introduced into the market. So quite some time elapsed after the first work of Schopf, Ruppel and Würfel with $NaNO_2$ thin films in 1989 [85].

*Point detectors*

The point detector shown in Figure 19 is the simplest of all devices. No special electronics was needed. The sensor dice simply replaced the $LiTaO_3$ single crystal detector. The thin film version was able to detect a moving person on a distance of maximal 30 m [16, 86]. The relatively large flakes resulted in a large thermal coupling by convection, reducing the D* to $1.0 \times 10^8$ cmHz$^{1/2}$/W. Point detectors are in general too simple devices to justify micromachined thin film versions.

A linear array working with MgO substrates has been developed by Matsushita [87]. PLT thin films were deposited onto MgO(100) substrates. These were later etched from the backside to obtain the desired thermal properties, similar to silicon bulk micromachining. A detectivity of $3.5 \times 10^8$ cmHz$^{1/2}$/W was measured.

*Linear arrays*

The linear array shown in Figure 16 has been developed for an infrared spectroscopy gas sensor. IR-light source, optics, focal plane array and read-out electronics are integrated in one housing. The light source is a filament or a micromachined thermal infrared source. The heater current is modulated to obtain a modulation of the signal. Each element of the array receives a certain wavelength section, as defined by a grating. In this way a complete absorption spectrum can be sampled in parallel operation [89]. Interesting wavelength intervals are the range between 8 and 12 μm for hydrocarbons and between 4 to 5 μm for carbon oxides (see Figure 21). The pyroelectric material was a sol-gel deposited, (111)-oriented PZT15/85 thin film. With a chopping frequency of 10 Hz (measurement bandwidth of 1.25 Hz) a NEP of 0.2 nW and a detectivity of $2 \times 10^8$ cmHz$^{1/2}$/W was measured after the CMOS readout chip [88]. The read-out chip increased the noise by about 30 %.

*Pyroelectric 2-d focal plane arrays*

Development of pyroelectric focal plane arrays started in the 1980's. A device consisting of a reticulated, 60 μm thick pyroelectric ceramic wafer that was bonded to a CCD readout wafer by means of contact solder bumps, exhibiting a pixel diameter of 70 μm showed a NETD of 400 mK at 25 Hz [2, 90]. A presently sold commercial product consists of a 14 μm thick reticulated ceramic wafer with a 50 μm pitch [91], as shown in Figure 20. The NETD could be reduced to 75 mK, and the spatial resolution could be

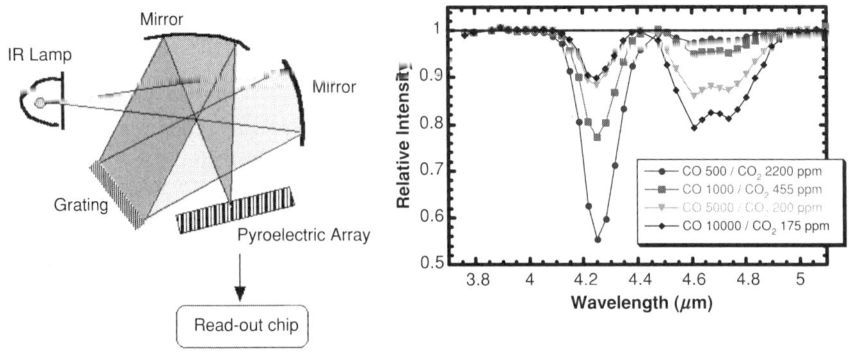


**Fig. 21.** Absorption spectra of $CO_2$ and CO mixtures measured by means of a thin film pyroelectric array (from [89]).

**Fig. 22.** View on 11 × 6 array of Siemens [92].

increased. The fundamental limits are not yet reached with this camera. However, in the frame of this ceramic wafer technology, a further shrinkage of wafer thickness and pitch will become increasingly difficult. Much more promising are silicon micromachining techniques enabling still smaller and thinner pixels. Thermal cross talk between pixels is reduced very much as well.

Siemens has developed an 11 × 6 array for thermal imaging (see Figure 22) potentially useful for small security cameras. Sputter deposited PZT 20/80 films were applied and D* of $3 \times 10^8$ cmHz$^{1/2}$/W is reported for 10 Hz. The suspended structures have been obtained by bulk micromachining of (110) silicon wafers. This silicon cut allows fabrication of narrower silicon bridges, which are needed in this design for carrying the electrical connection lines to the pads to which the read-out electronics is connected [92].

Raytheon-TI Systems reported the successful development of a monolithically integrated focal plane array based on a pyroelectric PLZT thin film, and containing 320 × 240 pixels of 50 × 50 μm size [82, 91, 93]. Crucial to this achievement was the balancing of the thermal budget of the sol-gel deposition process to allow perovskite PLZT crystallization on the one hand, and to avoid degradation of the aluminum metallization on the other hand. This allowed the fabrication of the sensor structure directly on top of the interconnecting metal level of the CMOS read-out circuit. The critical temperature for the CMOS metallization was evaluated to be 550 °C during short anneals. The sensor structure consisted of suspended pyroelectric thin film plates obtained by a sacrificial layer technique (see Figs. 23 and 24) similar to the structure shown in fig. 18), with the difference that the gap is used to trap the IR light as discussed above. Thermal imaging with a NETD of around 200 mK has been demonstrated. This value is higher than expected from basic pyroelectric and thermal properties. Major problems in transferring processes from the laboratory stage to a complex device fabrication have

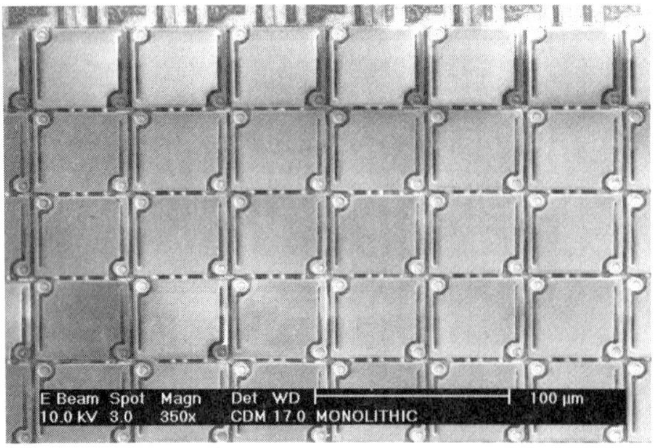


**Fig. 23.** Partial view on 320 × 240 pixel array of Raytheon Systems (from [94]) (Courtesy Raytheon Systems).

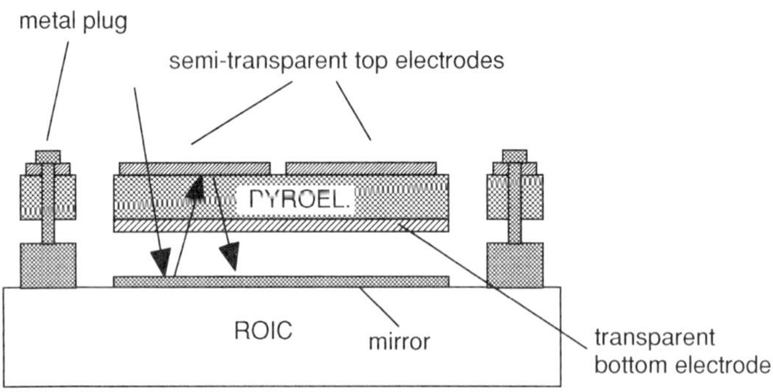


**Fig. 24.** Schematic cross section through pyroelectric IR detector pixel of Raytheon systems (after [82]). One pixel consists of two serial capacitors facilitating the fabrication.

been encountered. Film stresses, cracking, warping, metal oxidation and interdiffusion reactions resulted in a lowering of pyroelectric properties and absorption coefficient.

It has also been tried to avoid micromachining completely by applying silica aerogels as thermal insulation layer [95]. A NETD of 700 mK was reported. The problem is essentially that at 30 Hz the thermal wavelength in such an aerogel amounts still to about 20 μm (see section 3). With a thickness of 1 μm, the heat conductivity is about a factor of 10 higher than with bulk micromachined membrane structures in air (as estimated from aerogel data given in [96]). The increase of the aerogel layer thickness is problematic since the surface becomes more and more rough. A further problem is the limitation of the processing temperature to less than 600 °C at which the aerogel starts to collapse.

In the previous discussion, the NETD was taken as measure of performance. However, this is not the only one. Crosstalk, dynamic range, fix pattern noise, minimum resolvable temperature, and smallest resolvable feature are other important parameters. The human eye reacts also differently on dc-coupled (as e.g. thermoresistive bolometers) devices and ac-coupled devices (as e.g. pyroelectrics). The picture quality in the latter case appears to be better than in the first case, even though the NETD is equal [82].

## 7. Conclusions

This article reviewed the status of research in pyroelectric thin films, of pyroelectric thin film devices and applications. The sensitivity limit of pyroelectric IR detectors was derived and several ways of improving the present performance for approaching the theoretical limits were pointed out. It must be pointed out that fundamental limits of thermal detectors, as achievable by micromachining techniques in general, could also be reached by pyroelectric micromachined versions. In principle, micromachined thin film versions could decrease the NETD down to 10–20 mK[5]. Unfortunately there are still too many processing difficulties to surmount. Quite some materials engineering combined with thin film processing skills are still required to improve the relevant figures of merits. This includes not only an increase of the pyroelectric coefficient, but also a decrease of dielectric constant and dielectric loss. The material to be used, or whether the true or the induced pyroelectric effect should be applied are still open questions. On the device side, integration technology is of paramount importance for achieving the optimal combination with the required thermal properties and the needs for electrical read-out. Many things will depend on the practicability of fabrication. Pyroelectricity has good prerequisites for use in high-end thermal detectors. However, the mastery of fabrication is the key for obtaining competitive pyroelectric devices.

## References


[1] S.G. Porter, A brief guide to pyroelectric detectors, *Ferroelectrics* 33, 193–206 (1981)

[2] R.W. Whatmore, Pyroelectric devices and materials, *Reports on progress in physics* 49, 1335–1386 (1986)

[3] S. Bauer and S.B. Lang, Pyroelectric polymer electrets, *IEEE Trans. Dielectrics Electr. Insulation* 3, 647–676 (1996)

[4] R.G. Buser and M.F. Tompsett, *Historical overview*, in *Uncooled infrared imaging arrays and systems*, P.W. Kruse and D.D. Skatrud, Editors. 1997, Academic Press. p. 1–16.

[5] P. Muralt, Micromachined infrared detectors based on pyroelectric thin films, *Rep. Progr. Phys.* 64, 1339–1388 (2001)

[6] S.B. Lang, *Sourcebook of pyroelectricity*. 1974, Lonon: Gordon and Breach Science Publishers.

[7] S. Tinte, J. Iniguez, K.M. Rabe, and D. Vanderbilt, Quantitative analysis of the first principles effective Hamiltonian approach to ferroelectric perovskites, *Phys. Rev. B* 67, 064106 (2003)

[8] Y.F. Nye, *Physical properties of crystals*. 1985: Oxford University Press.

[9] R. Watton and M.A. Todd, Induced pyroelectricity in sputtered lead scandium tantalate films and their merit for IR detector arrays, *Ferroelectrics* 118, 279–295 (1991)

[10] M.E. Lines and A.M. Glass, *Principles and applications of ferroelectrics and related materials*. 1977, Oxford: Oxford University Press.

[11] F. Chu, N. Setter, and A.K. Tagantsev, The spontaneous relaxor-ferroelectric transition of PST, *J. Appl. Phys.* 74, 5129–5134 (1993)

[12] M. Daglish, A dynamic method for determining the pyroelectric response of thin films, *Integrated Ferroelectrics* 22, 473–488 (1998)

[13] A.K. Tagantsev, Mechanisms of dielectric loss in microwave materials, *MRS Symp. Proc.* 603, 221–232 (2000)

[14] V.L. Gurevich and A.K. Tagantsev, Intrinsic dielectric loss in crystals, *Adv. Physics* 40, 719–767 (1991)

[15] E.H. Putley, The pyroelectric detector, *Semiconductors and Semimetals* 5, 259–285 (1970)

[16] M. Kohli, C. Wüthrich, K.G. Brooks, B. Willing, M. Forster, P. Muralt, et al., Pyroelectric thin film sensor array, *Sensors and Actuators A* 60, 147–153 (1997)

[17] J.M. Lloyd, *Thermal imaging systems*. Optical physics and Engineering, ed. W.L. Wolfe. 1975: Plenum Press.

[18] P.W. Kruse, Uncooled IR focal plane arrays, *SPIE* 2552, 556–563 (1995)

[19] P.W. Kruse, A comparison of the limits to the performance of thermal and photon detector imaging arrays, *Infrared Physics & Technology* 36, 869–882 (1995)

[20] K. Iijima, Y. Tomita, R. Takayama, and I. Ueda, Preparation of c-axis oriented PbTiO3 thin films and their crystallographic, dielectric, and pyroelectric porperties, *J. Appl. Phys.* 60, 361–367 (1986)

[21] A.J. Moulson and J.M. Herbert, *Electroceramics*. 1990, London: Chapman & Hall.

[22] V. Fuflyigin, E. Salley, P. Vakhutinsky, A. Osinsky, and J. Zhao, Free-standing films of PST for uncooled infrared detectors, *Appl. Phys. Lett.* 78, 365–367 (2001)

[23] A.P.D. Kroon, S.C. Dunn, and R.W. Whatmore, Piezo and pyroelectric properties of lead scandium tantalate thin films, *Integrated Ferroelectrics* 35, 209–218 (2001)

[24] Z. Kighelman, D. Damjanovic, A. Seifert, L. Sagalowicz, and N. Setter, Relaxor behavior and electromechanical properties of Pb(Mg,Nb)O3 thin films, *Appl. Phys. Lett.* 73, 2281–83 (1998)

[25] D. Liu and D. Payne, Lower temperature crystallization and ordering in sol-gel derived PST powerders and thin layers, *J. Appl. Phys.* 77, 3361–64 (1995)

[26] D. Liu and H. Chen, Low-temperature preparation of perovskite PST thin films using MOCVD, *Materials Letters* 28, 17–20 (1996)

[27] C.E. Murphy and P.J. Dobson, Pyroelectric properties of fine BaTiO3 PVDFT thin film composites, *Ferroelectrics* 152, 127–132 (1994)

[28] H. Vogt, P. Würfel, U. Hetzler, and W. Ruppel, NaNO2 layers as pyroelectric radiation detectors, *Ferroelectrics* 33, 243–248 (1981)

[29] C.-H. Kohli, P.E. Schmid, and F. Lévy. *Ferro- and pyroelectric properties of LiTaO3 thin films*. in *IMF-9*. 1997. Séoul.

[30] C.K. Kwok and S.B. Desu, Formation kinetics of PZT thin films, *J. Mater. Res.* 9, 1728–1733 (1994)

[31] K. Aoki, Y. Fukuda, K. Numata, and A. Nishimura, Effects of titanmium buffer layer on PZT crystallization porecess in sol-gel deposition technique, *Jpn. J. Appl. Phys.* 34, 192–195 (1995)

[32] P. Muralt, T. Maeder, L. Sagalowicz, S. Hiboux, S. Scalese, D. Naumovic, et al., Texture control of PbTiO3 and PZT thin films with TiO2 seeding, *J. Appl. Phys.* 83, 3835–3841 (1998)

[33] T. Maeder, P. Muralt, M. Kohli, A. Kholkin, and N. Setter, $Pb(Zr,Ti)O_3$Thin Films by In-situ Reactive Sputtering on Micromachined Membranes for Micromechanical Applications, *British Ceram. Proc.* 54, 206–218 (1995)

[34] A. Seifert, N. Ledermann, S. Hiboux, and P. Muralt, Study of the transverse piezoelectric coefficient of PZT thin films as a function of texture and composition, *Mat. Res. Soc. Symp. Proc.* 596, 535–540 (2000)

[35] A.L. Roitburd, Equilibrium structure of epitaxial layers, *phys. stat. sol.(a)* 37, 329–339 (1976)

[36] C.M. Foster, Z. Li, M. Buckett, D. Miller, P.M. Baldo, L.E. Rehn, et al., Substrate effects on the structure of epitaxial PbTiO3 thin films prepared on MgO, LaAlO3, and SrTiO3 by metallorganic chemical vapor deposition, *J. Appl. Phys.* 78, 2607–2622 (1995)

[37] C.M. Foster, W. Pompe, A.C. Daykin, and J.S. Speck, Relative coherency strain and phase transformation history in epitaxial ferroelectric thin films, *J. Appl. Phys.* 79, 1405–1415 (1996)

[38] W. Pompe, X. Gong, Z. Suo, and J.S. Speck, Elastic energy release due to domain formation in the strained epitaxy of ferroelectric and ferroelastic thin films, *J. Appl. Phys.* 74, 6012–6019 (1993)

[39] A. Seifert, A. Vojta, J.S. Speck, and F.F. Lange, Microstructural instability of single crystal thin films, *J. Mater. Res.* 11, 1470–1582 (1996)

[40] J.S. Speck and W. Pompe, Domain configurations due to multiple misfit relaxation mechanisms in epitaxial ferroelectric films I. Theory, *J. Appl. Phys.* 76, 466–476 (1994)

[41] J.S. Speck, A. Seifert, W. Pompe, and R. Ramesh, Domain configurations due to multiple misfit relaxation mechanisms in epitaxial ferroelectric films II: Experimental verification and implications, *J. Appl. Phys.* 76, 477–483 (1994)

[42] J.S. Speck, A.C. Daykin, A. Seifert, E.A. Romanov, and W. Pompe, Domain configurations due to multiple misfit relaxation mechanisms in epitaxial ferroelectric films III: Intefacial defects and domain misorientations, *J. Appl .Phys.* 78, 1696–1706 (1994)

[43] N.A. Pertsev and A.G. Zembilgotov, Domain population in epitaxial ferroelectric thin films: Theoretical calculations and comparison with experiment, *J. Appl. Phys.* 80, 6401–6406 (1996)

[44] E. Sviridov, V. Alyoshin, Y. Golovko, I. Zakharchenko, V. Mukhortov, and V. Dudkevich, Phase transition and domain structure of heteroepitaxial PbTiO3/(001) MgO films, *Ferroelectrics* 128, 1–6 (1992)

[45] A. Seifert, F.F. Lange, and J.S. Speck, Epitaxial growth of PbTiO3 thin films on (001) SrTiO3 from solution precursors, *J. Mater. Res.* 10, 680–691 (1995)

[46] K. Iijima, N. Nagao, T. Takeuchi, I. Ueda, Y. Tomita, and R. Takayama, Sputtering of lead-based ferroelectrics, *MRS Symp. Proc* 310, 455–465 (1993)

[47] T. Kotani, T. Nakanishi, and K. Nomura, Fabrication of a new pyroelectric infrared sensor using MgO surface micromachining, *Jpn. J. Appl. Phys.* 32, 6297–6300 (1993)

[48] Z. Sitar, R. Gutmann, H. Pierhöfer, and P. Günter. *Liquid phase epitaxy of Na-K-Ta-Nb-O for pyroelectric applications*. in *MRS Symposium*. 1995. Boston: MRS.

[49] R. Takayama and etal. in *ISIF*. 1995.

[50] P. Muralt. *Piezoelectric and pyroelectric microsystems based on ferroelectric thin films*. in *10th IEEE Symp.Appl.Ferroelectrics*. 1996. East Brunswick (USA): IEEE.

[51] Z. Li, M. Grimsditch, X. Xu, and S.K. Chan, The elastic, piezoelectric and dielectric constants of tetragonal PbTiO3 single crystals, *Ferroelectrics* 141, 313–325 (1993)

[52] Y.H. Huang, *Sol-gel PbTiO3 thin films for pyroelectric applications*. 1995, Swiss Fed.Inst.Technology: Lausanne.

[53] N.M. Shorrocks, A. Patel, M.J. Walker, and A.D. Parsons, Integrated thin film PZT pyroelectric detector arrays, *Microelectronic Eng.* 29, 59–66 (1995)

[54] M. Kohli, C. Wüthrich, K. Brooks, M. Forster, P. Muralt, N. Setter, et al. *Pyroelectric thin film sensor array*. in *Eurosensors*. 1996. Leuven (Belgium).

[55] M. Kohli, A. Seifert, and P. Muralt, Poling of pyroelectric thin films, *J. Integrated Ferroelectrics* 22, 453–463 (1998)

[56] M. Kohli, P. Muralt, and N. Setter, Removal of 90°-domain pinning in (100) PZT 15/85 thin films by pulsed operation, *Appl. Phys. Lett.* 72, 3217–19 (1998)

[57] M. Kohli and P. Muralt, Poling of ferroelectric films, *Ferroelectrics* 225, 155–162 (1998)

[58] J.D. Zook and S.T. Liu, Pyroelectric effects in thin film, *J. Appl. Phys.* 49, 4604–4606 (1978)

[59] P. Muralt, Piezoelectric thin films for MEMS, *Integrated Ferroelectrics* 17, 297–307 (1997)

[60] M. Kohli, A. Seifert, B. Willing, K. Brooks, and P. Muralt, Thin film pyroelectric arrays as infra-red detectors, *J. Int. Ferroelectrics* 17, 359–367 (1997)

[61] R. Bruchhaus, D. Pitzer, R. Primig, W. Wersing, and Y. Xu, Deposition of self-polarized PZT films by planar multi-target sputtering, *Integrated Ferroelectrics* 14, 141–149 (1997)

[62] G.E. Pike, W.L. Warren, D. Dimos, B.A. Tuttle, R. Ramesh, J. Lee, et al., Voltage offsets in (Pb,La)(Zr,Ti)O3 thin films, *Appl. Phys. Lett* 66, 484–486 (1995)

[63] H. Hector, N. Floquet, J.C. Niepce, P. Gaucher, and J.P. Ganne, Texture, structure and domain microstructure of ferroelectric PZT thin films, *Microelectronic Eng.* 29, 285–288 (1995)

[64] A. Seifert, P. Muralt, and N. Setter. *Ca-modified PbTiO3 thin films for pyroelectric devices.* in *Electroceramics V.* 1996. Aveiro (Portugal): University of Aveiro.

[65] A. Seifert, L. Sagalowicz, P. Muralt, and N. Setter, Microstructural evolution of dense and porous (PbCa)TiO3 thin films for pyroelectric applications, *J. Mater. Res.* 14, 2012–2022 (1999)

[66] A. Seifert, P. Muralt, and N. Setter, High figure of merit porous (Pb,Ca)TiO3 thin films for pyroelectric applications, *Appl. Phys. Lett.* 72, 2409–11 (1998)

[67] G. Suyal, A. Seifert, and N. Setter, Pyroelectric nanoporous films: Synthesis and properties, *Appl. Phys. Lett.* 81, 1059–1061 (2002)

[68] T. Bischoff, G. Müller, W. Welser, and F. Koch, Frontside micromachining using porous silicon sacrificial layer technologies, *Sensors and Actuators A* 60, 228–234 (1997)

[69] C. Dücsö, E. Vazsonyi, M. Adam, I. Szabo, I. Barsony, J.G.E. Gardeniers, et al., Porous silicon bulk micromachining for thermally isolated membrane formation, *Sensors and Actuators A* 60, 235–239 (1997)

[70] B. Ploss, D. Lienhard, and F. Sieber, Thermal simulation of micromachined bridges for integrated pyroelectric sensor arrays, *Microelectronic engineering* 29, 75–78 (1995)

[71] M. Madou, *Fundamentals of microfabrication.* 1997, Boca Raton: CRC Press.

[72] A. Bell, Y. Huang, O. Paul, Y. Nemirovsky, and N. Setter, A thin film pyroelectric detector, *Int. Ferroelectrics* 6, 231–240 (1995)

[73] H. Seidel, A. Heuberer, and H. Baumgärtel, *J. Electrochem. Soc.* 137, 3612 (1990)

[74] M.-A. Dubois and P. Muralt, PZT thin film actuated elastic fin micromotor, *IEEE Trans. Ultrasonics, Ferroelectrics, and Frequency Control* 45, 1169–1177 (1998)

[75] P. Muralt, K. Brooks, M. Kohli, T. Maeder, and C. Wüthrich. *Ferroelectric thin films for microsystems.* in *ECASIA.* 1995. Montreux (Switzerland): John Wiley & Sons.

[76] B. Willing, M. Kohli, P. Muralt, N. Setter, and O. Oehler. *Gas spectrometry based on pyroelectric thin film arrays integrated on silicon.* in *Transducers—International Conf. on Solid- State Sensors and Actuators.* 1997. Chicago.

[77] B. Willing, M. Kohli, P. Muralt, N. Setter, and O. Oehler, Gas spectrometry based on pyroelectric thin film arrays integrated on silicon, *Sensors and Actuators A* 66, (1998)

[78] W. Lang, K. Kühl, and H. Sandmeier, Absorbing layers for thermal infrared detectors, *Sensors and Actuators A* 34, 243–248 (1992)

[79] D. Lienhard, F. Heepmann, and B. Ploss, Thin Nickel films as absorbers in pyroelectric sensorarrays, *Microelectronic Eng.* 29, 101–104 (1995)

[80] S. Bauer, S. Bauer-Gogenea, and B. Ploss, *Appl. Phys. B* 54, 544 (1992)

[81] K.C. Liddiard, Application of interferometric enhancement to self-absorbing thin film thermal IR detectors, *Infrared Phys.* 34, 379–387 (1993)

[82] C.M. Hanson, H.R. Beratan, and J.F. Belcher. *Uncooled infrared imaging using thin film ferroelectrics.* in *Infrared detectors and focal plane arrays VI.* 2000. San Jose: SPIE.

[83] R. Watton, P.A. Manning, M.J. Perkins, and J.P. Gillham. *Uncooled IR imaging: hybrid and integrated bolometer arrays*. in *Infrared technology and applications XXII*. 1996. Orlando: SPIE.

[84] R. Owen, J. Belcher, H. Beratan, and S. Frank. *Reproducibility advances in hybrid uncooled infrared devices II*. in *Infrared detectors and focal plane arrays IV*. 1996. Orlando, Florida: SPIE.

[85] H. Schopf, W. Ruppel, and P. Wuerfel, 16-element linear pyroelectric array with NaNO2 thin films, *Infrared Physics* 29, 101–106 (1989)

[86] M. Kohli, Y. Huang, T. Maeder, C. Wüthrich, A. Bell, P. Muralt, et al., Processing and properties of thin film pyroelectric devices, *Microelectronic Engineering* 29, 93–96 (1995)

[87] S. Fujii, T. Kamata, S. Hayashi, Y. Tomita, R. Takayama, T. Hirao, et al. *Pyroelectric linear infrared sensors made of La-modified PbTiO3 thin films and their applications*. in *SPIE*. 1995: SPIE.

[88] B. Willing, P. Muralt, T. Reimann, and O. Oehler. *High end pyroelectric thin film array detector system*. in *Transducers'99*. 1999. Sendai (Japan).

[89] B. Willing, M. Kohli, P. Muralt, and O. Oehler, Thin film pyroelectric array as a detector for and infrarted gas spectrometer, *Infrared Physics and Technology* 39, 443–449 (1998)

[90] R. Watton, F.W. Ainger, S. Porter, D. Pedder, and J. Gooding. in *Infrared technology X*. 1984: SPIE.

[91] C.M. Hanson, H.R. Beratan, J.F. Belcher, K.R. Udayakumar, and K.L. Soch. *Advances in monolithic ferroelectric uncooled IRFPA technology*. in *SPIE: Infrared detectors and focal plane arrays V*. 1998. Orlando, Florida: SPIE.

[92] R. Bruchhaus, D. Pitzer, R. Primig, M. Schreiter, W. Wersing, G. Hofmann, et al. *11 × 16 elementiges Pyrodetektorarray auf der Basis gesputterter selbstpolarisierter PZT-Dünnschichten*. in *5th Infrared Sensors and Systems*. 1997. Dresden: Dresden University Press.

[93] H.R. Beratan, C.M. Hanson, J.F. Belcher, and K.R. Udayakumar. *Thin film pyroelectric imaging array*. in *International Symposium on Integrated Ferroelectrics (ISIF)*. 1998. Monterey (CA).

[94] C.M. Hanson, H.R. Beratan, J.F. Belcher, K.R. Udayakumar, and K. Soch, *Next-generation ferroelectric uncooled IR detectors*. 1998, Rytheon Systems Company.

[95] J.A. Ruffner, P.G. Clem, B.A. Tuttle, C.J. Brinker, C.S. Sriram, and J.A. Bullington, Uncooled thin film infrared imaging device with aerogel thermal insulation: deposition and planarization techniques, *Thin Solid Films* 332, 356–361 (1998)

[96] P. Scheuerpflug, M. Hauck, and J. Fricke, Thermal properties of silica aerogels between 1.4 and 300 K, *J. Non-Cryst. Solids* 145, 196–201 (1992)

# Chapter 6

# RF Bulk Acoustic Wave Resonators and Filters

H.P. Loebl [a], C. Metzmacher [a], R.F. Milsom [b], P. Lok [c], A. Tuinhout [c]
[a] *Philips Research Laboratories, Weisshausstr.2, D-52066 Aachen, Germany*
[b] *Philips Research Laboratories, Cross Oak Lane; Redhill; Surrey, RH1 5HA, UK*
[c] *Philips Semiconductors, BL RF products, Gerstweg2, NL-6534 AE Nijmegen, The Netherlands*

Small sized, highly selective solidly mounted bulk acoustic wave (BAW) band pass filters are of great interest for mobile and wireless systems operating in the frequency range of 0.8 GHz up to more than 10 GHz. Applications for BAW filters include front-end filters for USPCS, UMTS, GSM, bluetooth and other standards. BAW filters can be fabricated on silicon or glass wafers using standard semiconductor production techniques. On-wafer packaging techniques allow to fabricate small, flip chip mountable BAW filter devices suited for System in Package (SiP) solutions.

The chapter explains the how BAW filters work, compares their performance with competing filter technologies (e.g. surface acoustic wave filters), describes the technique Philips uses to realize BAW filters and reviews briefly alternative technical approaches. BAW filters are based on electro-acoustic high Q resonators, which exploit the thickness extensional mode of a thin, highly oriented, piezoelectric AlN or ZnO film. One section of the chapter is devoted to the growth of piezoelectric AlN films, which have to be grown with the polar c-axis, oriented perpendicular to the substrate. It is shown that both the sputter deposition process and using a textured, well oriented electrode support excellent c-axis oriented growth of the AlN films. An example is the growth of AlN (0002) on Pt (111) oriented electrodes where we find an epitaxial relation. The growth mechanism of AlN films, which grow highly oriented as well on amorphous and smooth surfaces like SiO2 and the thickness dependence of their orientation is discussed. We show that AlN (0002) orientation is directly correlated with the electromechanical coupling coefficient $k_t$ of the films and therefore with the filter bandwidth.

The following section describes the modeling techniques we use to describe the filter and resonator response. A combination of a 1 dimensional analytical electro-acoustic model together with an electromagnetic model reveals optimal agreement between measurement and simulation. It is shown that the small details of the resonator and filter curves can be used to extract relevant material data. Embedding the electro-acoustic model into a commercial circuit simulator offers an excellent method for the designer to predict filter performance. The limits of this 1-D model are discussed briefly referring the reader to a more rigorous 2-D treatment of BAW resonators, which would allow to explain major acoustic loss mechanisms. Finally a few examples of filters operating in the range between 2 and 8 GHz are shown.

# 1. Introduction

Filtering of signals, which are closely separated in the frequency domain, is a key function in mobile radio front-ends (see Figure 1). The performance requirement means that this function must be implemented in a passive technology. 3G mobile standards like UMTS, where transmission and reception of signals is done simultaneously require narrow band filters (duplexers), which have flat pass- band with steep skirts and high out-of-band rejection. For that reason thin-film bulk acoustic wave (BAW) resonators, BAW ladder and lattice filters and acoustically coupled BAW filters have been studied intensively over the past 20 years for RF applications in the frequency range of 800MHz to 12 GHz [1–6, 14, 16, 20].

The use of BAW filters for radio front ends of mobile phones, in particular, has attracted a lot of interest, since surface acoustic wave (SAW) filters, which are well established at frequencies up to approx. 2GHz, have considerable drawbacks at higher frequencies: sub-micron lithography, dedicated small wafer-size expensive non silicon substrates such as $LiNbO_3$ or $LiTaO_3$, and poor power handling characteristics [27]. Ceramic resonator filters, which are also used in this frequency range, on the other hand, suffer from their relatively large size and low integration potential. BAW filters are remarkably small in size, can be processed on silicon or glass using standard silicon integration techniques. Furthermore, they can be designed with a low temperature coefficient of frequency (TCF), which is a very important requirement for accurate narrow band filters [24, 25]. This all makes BAW filters very attractive for today's semiconductor industry. The essential building blocks of these BAW filters are small sized BAW resonators (see Figure 2), which, in the technology of interest here, exploit the thickness extensional vibration mode of a thin piezoelectric film.

The resonance frequency $f_r$ is determined approximately by the thickness t of the piezoelectric film:

$$f_r = v_l/\lambda \approx v_l/2t \tag{1.1}$$

where $v_l$ is the longitudinal velocity of sound in the normal direction in the piezoelectric layer, t is the thickness of the piezoelectric film, and $\lambda$ is the acoustic wavelength. In

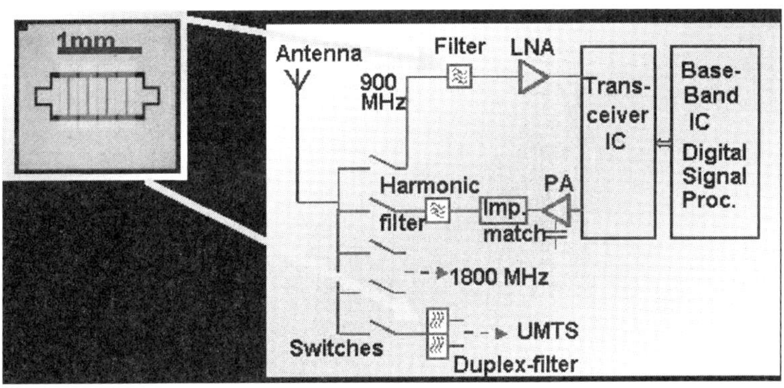


**Fig. 1.** Architecture of a mobile phone front end. Filters are required in e.g. GSM (900 MHz, 1800 MHz) and UMTS bands. Left: example of a die with a 4-section BAW ladder filter.

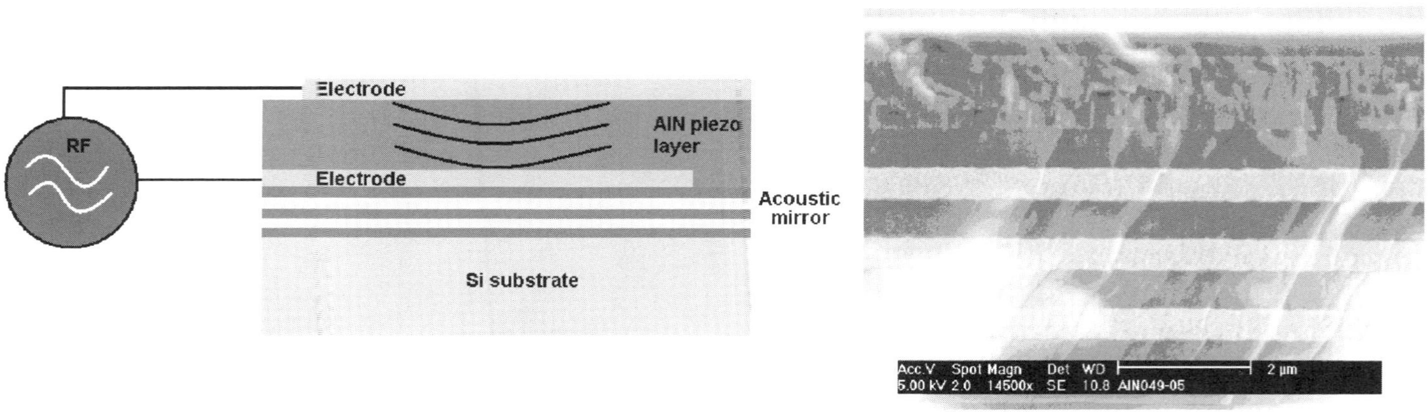


**Fig. 2.** Bulk acoustic wave resonator. Left: schematic showing acoustic Bragg reflector (acoustic mirror), electrodes, and piezoelectric layer; right: SEM cross-section of BAW resonator.

a real resonator device, of course, the frequency $f_r$ is different from eq. (1.1), since the acoustic properties of all other layers influence the resonator performance e.g. by the mass-loading effect of the resonator's electrodes. Despite the fact that equation (1.1) is only a crude approximation one can draw two important conclusions:

1. Since the velocity of sound is typically in the range between 3000–11000 m/s for most materials, we need a thickness of the piezo layer in the order of a micrometer.
2. The frequency of a resonator and thus the center frequency of a filter are mainly determined by the thickness t of the piezo layer. This poses also a strong challenge on the accuracy of thin film deposition. Despite the fact that the idea for high frequency BAW filters is already quite old, only the large improvements in the field of sputter deposition techniques over the past few years has made thin-film BAW resonators and filters technologically feasible.

For low loss BAW filters the acoustic energy has to be confined within the resonator. This is done by using an acoustic reflector, which reflects the energy back into the resonator. One implementation of such a reflector is realized using etching techniques adapted from micro electro mechanical systems (MEMS) technology. Here the thin film BAW resonator (TFBAR) is formed on a thin membrane of Si, $SiO_2$, or $Si_3N_4$. Since the acoustic impedance of air is negligible compared to the acoustic impedance of solids, almost all the energy is reflected back from the resonator/air interface. The TFBAR technology is applied industrially for PCS band duplex filters [15]. Another approach is the solidly mounted BAW resonator (SBAR). This reflector technique utilizes so-called Bragg mirrors consisting of a stack of thin layers of materials with alternate high and low acoustic impedance (acoustic impedance $= \rho \times v_l$, where $\rho$ is the density). Various material combinations exist for such acoustic interference coatings e.g. $SiO_2$/AlN [1, 2, 4, 25], $SiO_2$/W [20], $SiO_2$/metal-oxides [29]. Depending on the difference in acoustic impedance an effective reflector typically requires up to 5 pairs of layers. Figure 2 below shows a typical BAW resonator. The bottom electrode is made of a high conductivity material such as Al, Mo, W, or Pt. For the piezoelectric layer either AlN [1, 2, 4, 14, 17, 18], or ZnO [20, 22] is commonly used. Alternative materials including ferroelectrics are presently under discussion [7, 8, 9, 21]. Both AlN and ZnO have the hexagonal wurtzite type structure, i.e. in order to get a working device, the material has to be grown with the polar c-axis perpendicular to the substrate surface. Additionally the polarity may not change across the BAW resonator. The top electrode is again made of a high conductivity material.

In Figure 3 one can see the electrical impedance of a BAW resonator with its two characteristic frequencies. At the resonance frequency $f_r$ the electrical impedance $|Z|$ is very small (electrical field parallel to polarization, high current flowing) whereas at the anti-resonance frequency fa $|Z|$ is very large (electrical field anti-parallel to polarization, no current flowing). As mentioned above, filters are made by combining several resonators. Figure 3 shows a simple one-section ladder filter comprising a series and a parallel resonator (s,p). The parallel resonator p is shifted in frequency with respect to the series resonator s. When the resonance frequency $f_{r\_s}$ of the series resonator equals the anti-resonance frequency $f_{a\_p}$ of the parallel resonator, maximum signal is transmitted from input to output of the device. At the anti-resonance frequency $f_{a\_s}$ of the series resonator filter transmission is blocked, and at the resonance frequency

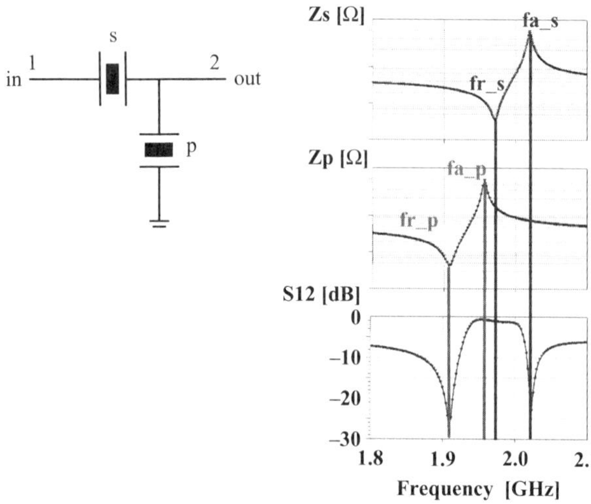


**Fig. 3.** Single section BAW filter consisting of one series and one parallel resonator. Right: Electrical impedance of series and parallel resonator. The bottom curve shows the transmitted signal $S_{21}$ revealing a band-pass filter characteristic.

of the parallel resonator $f_{r\_p}$ the filter input is connected to ground, so that the BAW filter also blocks signal transmission at this frequency. This results in the band-pass filter characteristic S21 shown in figure 3. The out-of-band rejection is typically 6 dB for a single filter section. The bandwidth of the filter is determined by the separation of $f_r$ and $f_a$ and is therefore proportional to the square of the electromechanical coupling coefficient $k_t^2$ (see eq. 1.2). Pulling the resonance frequency of the series resonator $f_{r\_s}$ and the anti-resonance frequency of the parallel resonator $f_{a\_p}$ apart, as shown in Figure 3, increases the bandwidth slightly. The dip in the center of the pass-band (see Figure 3 and Figure 12) can be recovered partially by adding small inductors at the filter ports.

$$k_t^2 = \frac{\pi^2}{4}\left(\frac{fa - fr}{fa}\right) \tag{1.2}$$

$k_t^2$ is approximately 6.25%–7% for AlN and 8.5% for ZnO. This appears to be sufficient for many mobile and wireless filter applications. Certainly ZnO would be preferable because it could offer a wider bandwidth. However, we have to consider the temperature dependence of the elastic, piezoelectric and dielectric constants and thermal expansion coefficient, which together result in a temperature coefficient of the resonance frequency TCF. For AlN and ZnO reported values are −25 ppm/K and −60 ppm/K respectively [23]. For the temperature dependence of the whole device the other layers (mass-loading layer, electrodes) also have to be considered. Reported values of TCF are −19 to −21 ppm/K for AlN SBAR filters, 0 ppm/K for compensated AlN TFBAR resonators [25], and −31 ppm/K for ZnO based SBAR filters [20]. The higher TCF for ZnO compared to AlN appears to be a drawback.

## 2. Experimental

In the following we describe typical experimental techniques for the fabrication of solidly mounted SBAR devices. Resonators and filters were fabricated on 6" silicon wafers and for comparison on glass wafers [7, 8, 9, 11]. We commonly used an acoustic reflector comprising $\lambda/4$ layers of high and low acoustic impedance materials. The thickness of the reflector layers was controlled optically. Electrode materials were Al, Mo and Pt. The piezoelectric layer was c-axis oriented AlN, which was deposited by pulsed reactive DC magnetron sputtering in a Unaxis CL200-cluster tool. The base pressure was in the low $10^{-8}$ mbar range. A pulse frequency of 20–100 kHz was used to avoid arcing. Chuck temperature was varied between 300 °C and 500 °C, and a RF substrate bias was applied to control stress and texture of the AlN films. The quality of the AlN films was routinely checked by X-ray diffraction $\theta/2\theta$ scans (Philips Fine Focus PW2213/20) and by measuring the full width at half maximum (FWHM) of the rocking curve of the AlN 0002 peak. The quality of the AlN films was excellent. A FWHM of 1.2–1.5° was achieved on Si substrates; approximately 1.5° was achieved on Pt electrodes. The AlN properties varied only slightly with the deposition conditions [9, 11]. Resonators and filters were measured electrically (one and two port measurements) on wafer level using wafer microprobes. The measurements were done with an R&S Vector Network-Analyzer in a frequency range between 10 MHz and 20 GHz with increased resolution in the vicinity of the resonance.

## 3. Microstructure of AlN

AlN and also ZnO have a hexagonal (6mm) wurtzite type structure (see Figure 4). When an electric field is applied along the polar c-axis, atoms and consequently the center of charge are displaced and the unit cell is strained. This effect represents the so-called inverse piezoelectric effect and is exploited in BAW resonators and filters. By applying an alternating RF field along the c-axis the BAW resonator is excited in its thickness extensional mode. Figure 4 reveals that the AlN film has to be grown c-axis

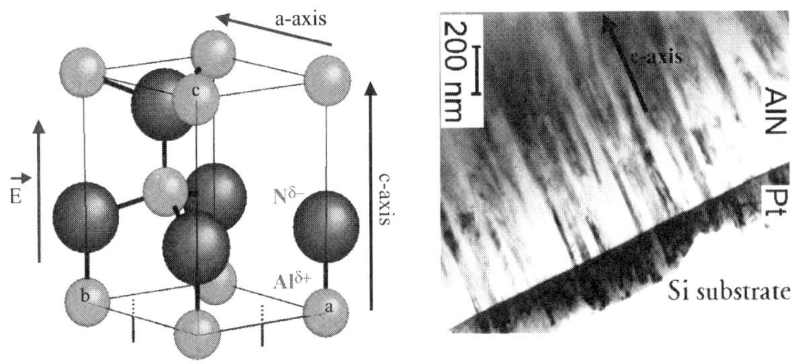


**Fig. 4.** Left: Schematic view of an AlN unit cell showing polar character. Right: TEM cross-section of a c-axis oriented polycrystalline AlN film on a Pt (111) electrode.

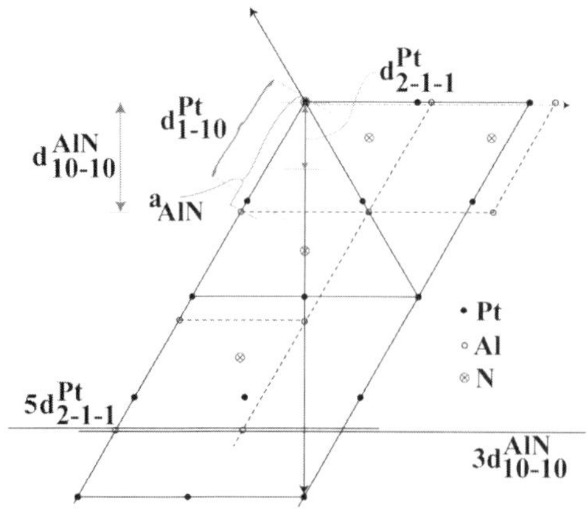


**Fig. 5.** Epitaxial relationship of AlN(0002) on Pt(111) in plan view, i.e. along AlN[0001] and Pt[111] direction. Dashed lines: unit cell of AlN, solid lines: 'hexagonal cell' of Pt.

oriented and that the a-axis is hereby randomly oriented in the substrate plane. The AlN film is a columnar grown and textured poly-crystal with a grain size (column width) of typically 35 nm.

C-axis oriented growth of AlN appears to be the naturally supported orientation, because it comes along with a with the hexagonal-closed packed (hcp) arrangement of N and Al atoms in the sequence of a double layer stacking according to ... AA'BB'AA'.... As long the growth of the AlN film is not disturbed, c-axis orientation will be the preferred orientation. Furthermore, the use of polycrystalline, textured electrodes may support c-axis oriented growth [8, 19]. In Figure 4 a TEM bright field cross-section of a Pt electrode with (111) orientation is shown. The TEM image indicates that the AlN growth occurs quasi-epitaxially and the Pt grain size and AlN grain-size match quite distinctly. Quasi-epitaxy can be deduced from the degree of texture of both AlN and Pt electrode and has been proved by selected area electron diffraction and micro diffraction in TEM in addition to macroscopic XRD. Indeed, an epitaxial orientational relationship exists (see Figure 5) which describes the growth of (0002) oriented AlN on Pt (111) geometrically according to the matching condition (3.1):

$$5 \cdot d_{2\bar{1}\bar{1}}{}^{Pt} = 3 \cdot d_{10\bar{1}0}{}^{AlN} \qquad (3.1)$$

Another important aspect of the growth of AlN is the thickness dependence of the AlN texture. As shown in Figure 6, AlN was grown in subsequent runs, each run depositing a AlN film thickness of approximately 1 μm. After each deposition an XRD rocking curve of the AlN0002 peak was measured. It can be seen that the rocking curve FWHM width of the (0002) peak gets smaller from run to run and that the net intensity of the 0002 peak becomes larger. The AlN c-axis orientation improves with thickness, if the growth is not disturbed. The TEM image reveals, that the orientation of the AlN crystallites in the beginning of the growth is misaligned with respect to perfect (0002) orientation. Since the (0001) oriented grains grow faster at the expense of the misaligned grains, we find eventually after a few 100 nm almost perfect c-axis

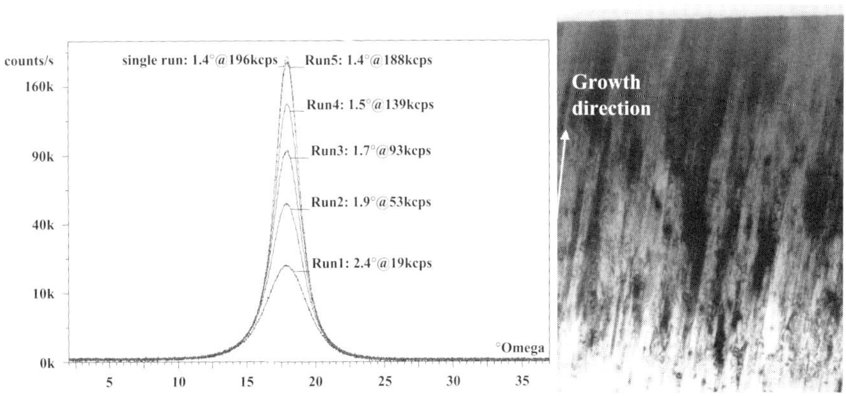


**Fig. 6.** Top left: XRD rocking curves of the AlN0002 peak of AlN films with different thickness. C-axis orientation improves from run to run. Right: TEM cross-section of thick AlN layer showing improvement of c-axis texture with thickness micro-structurally.

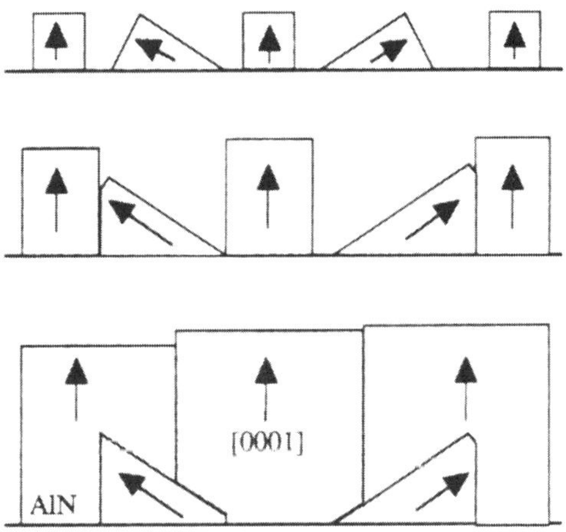


**Fig. 7.** Schematic depiction nucleation and growth mechanism of AlN according to ref. [26].

orientation of the AlN grains. The AlN growth mechanism is schematically depicted in Figure 7.

The dependence of the experimentally determined coupling coefficient $k_t$ on the rocking curve FWHM width of AlN0002 peak is shown in Figure 8 [11]. Films with poor c-axis orientation have significant lower values of the coupling coefficient $k_t$ with consequently reduced filter bandwidth. As can be seen in Figure 9, the sputter deposition process results in good oriented films both on $SiO_2$ and on Pt(111) substrate surfaces. The dependence of the AlN orientation on the sputter conditions is only small. Generally higher substrate temperature and moderate RF bias applied to the substrate support c-axis orientation most probably due the increased mobility of the sputtered N and Al ad-atoms at the substrate surface. In ref. [17,18,19] it was shown, that the

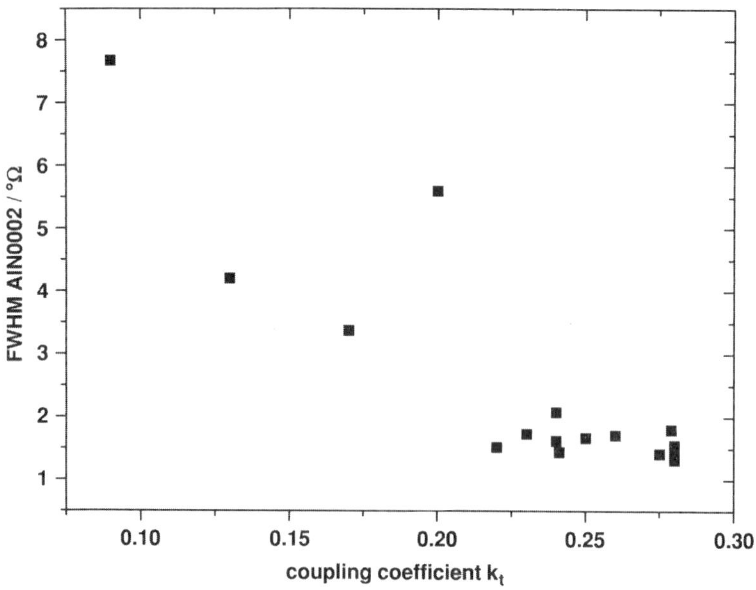


**Fig. 8.** Dependence of coupling coefficient $k_t$ on the rocking curve FWHM of the AlN0002 peak. Poorly oriented films show low coefficient $k_t$.

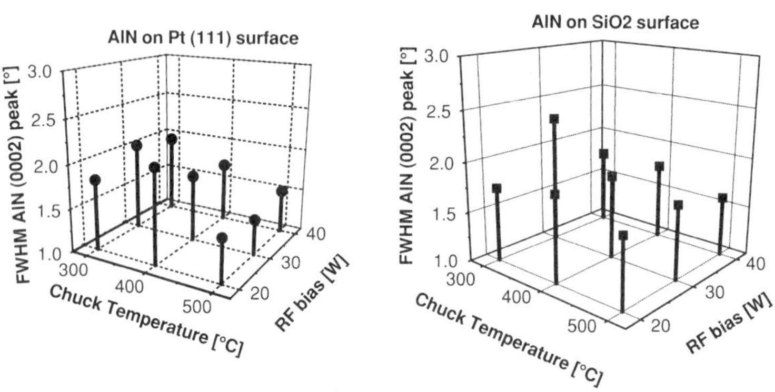


**Fig. 9.** Dependence of AlN c-axis orientation on process conditions. FWHM width of AlN(0002) peak depends only weakly from deposition conditions. Left: AlN on Pt(111) surface, right: AlN on thermal $SiO_2$ surface.

piezoelectric constant $d_{33}$ and thus $k_{33}$ or $k_t$ is influenced by the bias applied to the substrate.

## 4. Simulation

In order to meet the demanding specifications typical in BAW filter applications it is important to optimize all aspects of the resonator and filter performance, and therefore to consider both acoustic and electromagnetic (EM) contributions to the

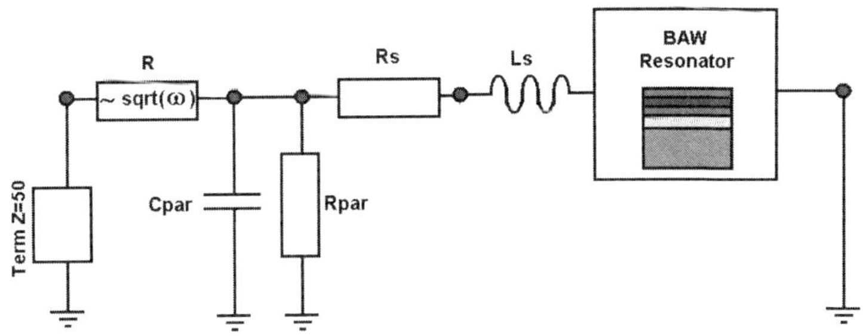


**Fig. 10.** Simulation of a BAW resonator: The electro acoustic 1-D resonator model described by the symbol right is imported in a commercial circuit simulator. The EM contributions are simulated as lumped elements.

filter behaviour. In particular, many mechanisms contribute to pass-band insertion loss. These include conductor loss, dielectric loss in the piezoelectric layer, substrate loss, electrical mismatch loss, acoustic loss in the materials and several additional acoustic loss mechanisms. On the other hand, rejection in the stop-band is typically limited by electrical (electromagnetic) effects. To help address related design issues, a method has been developed for simulating the different physical effects in a unified model. Computational efficiency is important in the design cycle, therefore a 1D electro-acoustic solution broadly equivalent to the model of Nowotny and Benes [10] is used to model the BAW resonators, while lumped-element equivalent-circuit models are used for modeling the EM part of the problem. In the first method, appropriate for modeling parasitics due to filter layout, lumped component values are extracted automatically from 2.5D EM simulation using the method described in [13]. The second method, appropriate for modeling single resonators, uses the simpler procedure: lumped elements describing parasitics are added either in series or parallel with the electro-acoustic model (see Figure 10).

The acoustic material loss of the resonator is described by the material quality factor $Q_m$ of each layer, which we introduce into the model through an imaginary component of the longitudinal sound velocity. Parasitic components due to interconnects are introduced into the model to match the measurements as closely as possible. Series resistance R and inductance L of the electrodes are added to the resonator model as lumped elements in the circuit simulator. Additionally a term proportional to $\sqrt{\omega}$ (square root of angular frequency) is included in the resistance. This approximately accounts for the RF-dependent redistribution of current in the conductors. A parasitic capacitor and resistor to ground model the stray electric field and associated loss in the Si substrate. Dielectric loss is characterized by introducing by an imaginary part into the dielectric constant of AlN.

We introduce additional loss (to account for otherwise unexplained measured loss) through an imaginary part of $k_t$. This is seen to improve the agreement with measurement. Alternatively it has been proposed to introduce an additional resistor in series with the static capacitance of the resonator [12]. Neither way of describing this additional loss mechanism seems to be justified on physical grounds. We suspect that the predominant additional loss mechanism may actually be acoustic radiation away from the resonator associated with imperfect energy trapping. A more rigorous treatment of this effect would require 2-D or even full 3-D modeling of the BAW resonators [28, 29].

## 5. Resonator and Filter Performance

By means of an acoustic Bragg reflector high Q factor resonators can be made. Figure 11 shows the electrical impedance Z and admittance Y of a series resonator (two identical resonators connected in series via the bottom electrode) for 1.95 GHz. The AlN thickness in this resonator was approx. 1900 nm; the resonator area was $268 \times 268\ \mu m^2$. For the bottom electrode we used (111) oriented Platinum, for the top electrode Aluminium was used. The acoustic reflector consisted of alternating $\lambda/4$ layers of $SiO_2$ and a metal oxide with low and high acoustic impedance respectively. The quality factor $Q_z$ at the anti-resonance was approximately 750, the quality factor at the resonance $Q_y$ was approx. 1014. The simulation was done using the combined 1-D electro-acoustic model (Nowotny & Benes theory) and the simpler of the two electromagnetic models described above. The acoustic material loss of the resonator is described by the material quality factor $Q_m$ of each layer, which we introduced into the model through an imaginary component of the longitudinal velocity of sound $v_l$ (see above). Parasitic components due to interconnections were introduced into the model to match the measurements accurately. A more fundamental treatment of parasitics based on an electromagnetic simulation of the filter layout is described in [13]. We found in our simulation $Q_m$ better than several 1000 for all layers (for this simulation we used a $Q_m$ of 6000 for all layers except for the Si substrate where we used $Q_m = 1000$ to account for acoustic scattering loss by the rough rear surface of the Si wafer). Velocity of sound and density of the layers were taken mainly from the literature. Dielectric loss was negligible for this resonator ($\tan\delta = 2\ 10^{-3}$ measured at 1 kHz). Series resistance $R_s$ and inductance $L_s$ of the electrodes were added to the resonator model as lumped elements ($R_s = 1.0\ \Omega$,

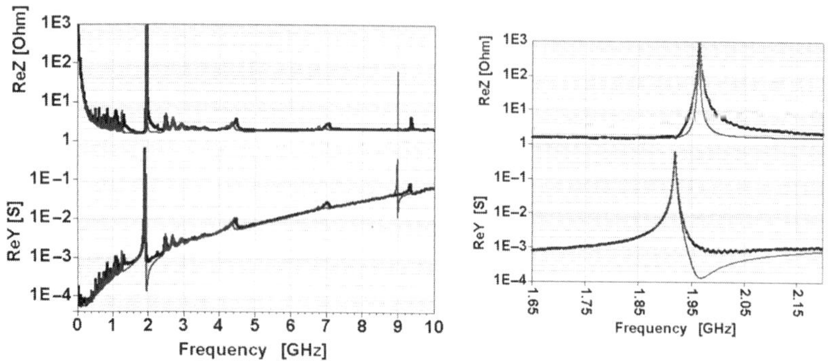


**Fig. 11.** Resonator at 1.95 GHz: Real part of impedance Z and admittance Y respectively. Measurement black curve, simulation blue curve. Right: anti-resonance and resonance on an enlarged scale.

$L_s = 0.061$ nH) in the circuit simulator. Additionally a resistance term proportional to $\sqrt{f}$ ($R = 1e^{-5}\sqrt{f}$) was included in the resistance. As explained above, this approximately accounts for the RF-dependent redistribution of current in the conductors, which flows closer to the edges at high frequencies. A parasitic capacitor $C_{par} = 0.19$ pF and resistor $R_{par} = 20$ kΩ to ground model the stray electric field and associated loss in the silicon substrate. Figure 11 shows that the agreement between measurement and simulation is excellent. Even the fine details of the resonator's response can be simulated and give valuable information, for example, concerning the centring of the Bragg reflector reflection band with respect to the resonator's centre frequency. The peaks in real part of the impedance (ReZ) below and above the fundamental anti-resonance peak correspond to transmission maxima of the acoustic reflector and can be used to determine the longitudinal velocity of sound in the reflector layers very accurately [13]. The small peak in the shoulder of the fundamental mode at 2.13 GHz, which is not predicted by the model, is identified as the fundamental TE mode of a parasitic resonator, which is formed between the reflector and the top electrode. Since this parasitic resonator has no bottom electrode, its resonance is at a higher frequency than that of the wanted resonator. The stray electric field excites this parasitic resonance. It can also be seen that the Bragg reflector, if well designed, suppresses higher harmonics and spurious modes very effectively. This results in a very clean resonator response and, as we will see below, in a clean filter stop-band.

The value of $k_t$ determined from our simulation is 0.26 ($k^2_{eff} = 6.2\%$), which is remarkably high and corresponds to the excellent orientation of the piezoelectric thin AlN film (see also Figure 7). In Figure 11 we see also that the quality factor in resonance $Q_r$ is simulated accurately ($Q_r$ is measured to be 1014) whereas the quality factor in anti-resonance $Q_a$ is predicted much higher than measured ($Q_a$ is measured to be 750). Generally we find the quality factor in resonance $Q_r$ a bit higher than the quality factor in anti-resonance $Q_a$. This effect is discussed in more detail in ref. [28, 29] and can be attributed to mode conversion into modes guided by the layered structure of the SBAR. Mode conversion occurs mainly at the electrode edges of the resonator. The guided modes traveling away from the resonator can be reflected by e.g. electrode edges and give rise to the ripple seen in Figure 11. The quality factor in anti-resonance $Q_a$ is therefore limited by mode-conversion, the quality factor in resonance $Q_r$ by the sheet resistance of the electrodes. For an adequate description of the effects of mode conversion which give rise to additional acoustic loss and ripple a 2D or even 3D simulation is needed [28, 29].

In the following Figure 12 we show a few examples of ladder-type filters, which have been realized using high Q BAW resonators with Bragg reflector. All these filters were made on silicon substrates. As discussed above, all filters reveal a clean stop band, which is due to the additional selectivity provided by the acoustic Bragg reflector (limited bandwidth of the reflector). Higher harmonics and spurious modes are therefore very effectively suppressed. The dips in the pass-band of the filters were made intentionally by pulling the resonance frequency of the non-mass-loaded resonators and the anti-resonance frequency of the mass-loaded resonators apart. Including small series inductors at the filter ports flattens this dip. Figure 12 shows a 4-section ladder filter for 1.95 GHz, a 2-section ladder filter for 2.85 GHz, and a 3-section ladder filter for

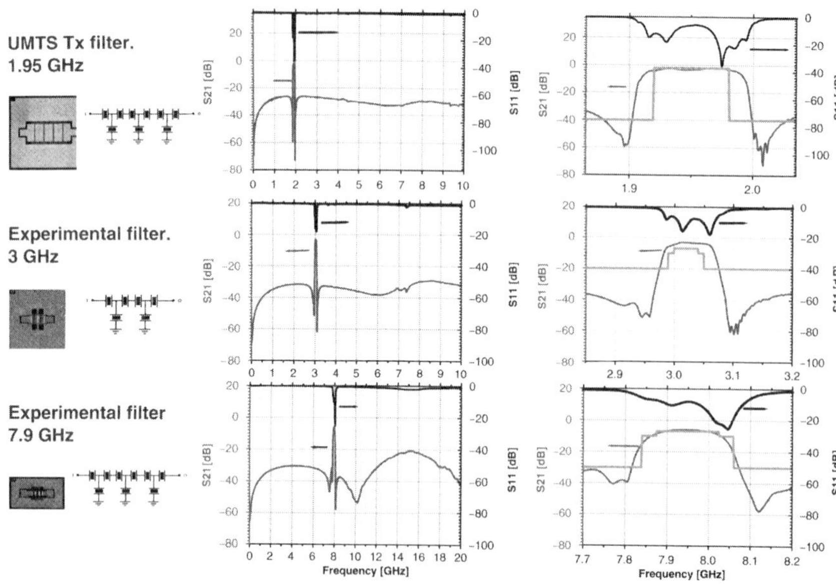


**Fig. 12.** 4-section, 2-section and 3-section BAW ladder-type band-pass filters for 1.95, 3, and 7.9 GHz. Left: typical layout; right: Filter curves depicting transmission S21 and reflection S11 in dB. Insertion loss is 1.9, 2.9 and 5.9 dB respectively. Bandwidth is 60 MHz, 79 MHz, and 105 MHz respectively.

8 GHz. The broad dip in the stop-band at 4.5, 6.5 and 12.5 GHz respectively is due to an LC resonance that depends on the details of the filter and package design.

In the last section we want to discuss briefly packaging of BAW filters. For BAW filters based on AlN films integration into standard CMOS processes is certainly an option, since most processing steps are compatible with standard semiconductor production. From yield and performance considerations, however, stand-alone flip chip mountable BAW filters are also attractive. As with SAW filters, there is clearly a trend to a wafer scale packages, which allows cost effective fabrication of BAW devices packaged on wafer level, which are small in area and height (Figure 13). In a System in Package (SiP) concept, which makes use of the optimal technology for each RF-function, these tiny BAW filter devices can be flip chip soldered on any carrier (e.g. on a laminate) and eventually molded into a RF front-end module [30].

## 6. Conclusion

Accurate BAW resonators and band-pass filters are made today in both SBAR and TFBAR technology for frequencies between 900 MHz and 12 GHz using either AlN or ZnO. In this paper we described bulk acoustic wave resonators and filters for frequencies between 2–8 GHZ on silicon wafers using c-axis oriented AlN and a Bragg acoustic interference reflector. The c-axis orientation of AlN improves with substrate temperature and with film thickness and can be influenced by the orientation

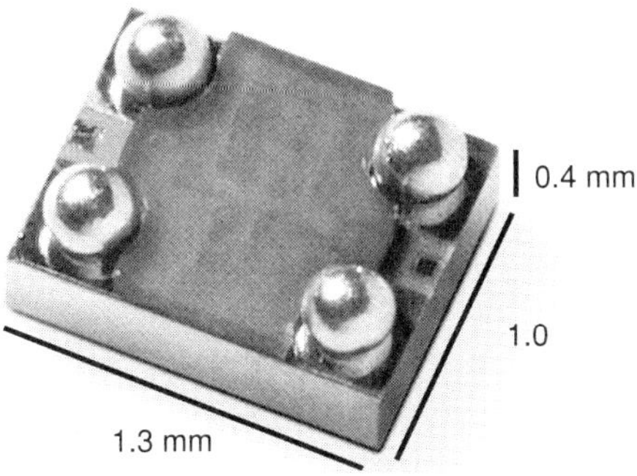


**Fig. 13.** Recent example of a packaged BAW filter for GSM 1900 [30].

of the underlying electrode. For Pt (111) electrodes a quasi-epitaxial relationship for the oriented growth of AlN can be found. These BAW filter devices show a remarkably high coupling coefficient $k_t$ and a clean stop-band. The responses of the electro-acoustic resonators and filters can be described by a combination of 1-D electro-acoustic plus electromagnetic modeling. To describe acoustic losses correctly a 2-D model is required.

## 7. Acknowledgement

We would like to thank our colleagues R.Kiewitt, W.Brand, R.Mauczok, H.D.Bausen, H.Nulens, G.Much, D.Wiechert, at the Philips Research Laboratories in Aachen and Eindhoven. This work was partially supported by the European Commission within the Project 'Microwave Electro-Acoustic Devices for Mobile and Land based Applications (MEDCOM, IST-1999-11411)'.

## References

[1] K.M. Lakin, G.R. Kline, K.T. McCarron, High Q microwave acoustic resonators and filters. IEEE Transactions on Microwave Theory a. Techniques, 41, 2139 (1993).

[2] K.M. Lakin, J. Belsick, J.F. McDonald, and K.T. McCarron, High Performance Stacked Crystal Filters for GPS and Wide Bandwidth Applications, 2001 IEEE Ultrasonics Symposium Atlanta, p. 833, 2001.

[3] R.C. Ruby, P. Bradley (SM), Y. Oshmyansky, A. Chien, Thin Film Bulk Wave Acoustic Resonators (FBAR) for Wireless Applications, 2001 IEEE Ultrasonics Symposium Atlanta, p. 833, 2001.

[4] K.M. Lakin, G.R. Kline, K.T. McCarron, Development of miniature filters for wireless applications, IEEE Transactions on Microwave Theory a. Techniques, 43, 2933 (1995).

[5] J.D. Larson III, R. Ruby, P. Bradley, Y. Oshmyansky, A BAW antenna duplexer for the 1900MHz PCS band. 1999 IEEE Ultrasonics Symposium Lake Tahoe, p. 887, 1999.

[6] K.M. Lakin, K.T. McCarron, J. Belsick, R. Rose, Filter Banks Implemented with Integrated Thin Film Resonators, 2000 IEEE Ultrasonics Symposium, Puerto Rico, p. 851, 2000.

[7] H.P. Löbl, M. Klee, R. Milsom, R. Dekker, C. Metzmacher, W. Brand, P. Lok, Materials for bulk acoustic wave (BAW) resonators and filters, Conference on Microwave Materials and their Applications MMA2000, Bled, Slovenia, September 2000, Journal of the European Ceramic Society, 21, 2633, (2001).

[8] H.P. Löbl, M. Klee, O. Wunnicke, R. Kiewitt, R. Dekker, E.v. Pelt, Piezoelectric AlN and PZT films for micro-electronic applications. 1999 IEEE Ultrasonics Symposium Lake Tahoe, p. 1031, 1999.

[9] Piezoelectric Materials for BAW Resonators and Filters, H.P.Lobl, M.Klee, C. Metzmacher, W.Brand, R. Milsom, P.Lok, F.van Straten 2001 IEEE Ultrasonics Symposium Atlanta, p. 807, 2001.

[10] H. Nowotny, E. Benes, General one-dimensional treatment of the layered piezoelectric resonator with two electrodes, J. Acoust. Soc. Am., 82, 513 (1987).

[11] H.P. Löbl, M. Klee, C. Metzmacher, W. Brand, R. Milsom, P. Lok, Piezoelectric Thin ALN Films for Bulk Acoustic Wave (BAW) Resonators, in Proceedings (SSMM) of the Asia Pacific Microwave Conference APMC 2001, Taipei, p. 37–42, 2001 and in Materials Science a. Engineering 9532, 1 (2002).

[12] J.D. Larson, P. Bradley, S. Wartenberg, R.C. Ruby, Modified Butterworth-Van Dyke Circuit for FBAR Resonators and Automated Measurement System, 2000 IEEE Ultrasonics Symposium, Puerto Rico, p. 865, 2000.

[13] R.F. Milsom, H.P. Löbl, D.N. Peligrad, J.W. Lobeek, A. Tuinhout, H.J. ten Dolle, Combined Acoustic Electromagnetic Simulation of Thin Film Bulk Acoustic Wave Filters, 2002 IEEE Ultrasonics Symposium Munich, proceedings, p. 963, 2002.

[14] H.P. Loebl, C. Metzmacher, D.N. Peligrad, R. Mauczok, M. Klee, W. Brand, R.F. Milsom, P.Lok, F.van Straten, A. Tuinhout, J.W Lobeek, Solidly Mounted Bulk Acoustic Wave Filters for the GHz Frequency Range, 2002 IEEE Ultrasonics Symposium Munich, proceedings, p. 897, 2002.

[15] P.D. Bradley, SM, R. Ruby, A. Barfknecht, F. Geefay, C. Han, G. Gan, Y. Oshmyansky, A 5 mm × 5 mm × 1.37 mm Hermetic FBAR Duplexer for PCS Handsets with Wafer-Scale Packaging, 2002 IEEE Ultrasonics Symposium Munich, proceedings, p. 907, 2002.

[16] K.M. Lakin, Coupled Resonator Filters, 2002 IEEE Ultrasonics Symposium Munich, proceedings, p. 879, 2002.

[17] M.A. Dubois, P. Muralt, Stress and Piezoelectric properties of AlN thin films deposited onto metal electrodes by pulsed direct current reactive sputtering, J. Appl. Phys. 89, 6389 (2001).

[18] M.A. Dubois, P. Muralt, Properties of AlN thin films for piezoelectric transducers and microwave filter applications, Appl. Phys. Lett. 74, 3032 (1999).

[19] G.F. Iriarte, J. Bjurström, J. Westlinder, F. Engelmark, I.V. Katardjiev, Synthesis of c-Axis Oriented AlN Thin Films on Metal Layers: Al, Mo, Ti, TiN and Ni, 2002 IEEE Ultrasonics Symposium Munich, proceedings, p. 300, 2002.

[20] J. Kaitila, M. Yililammi, J. Molaris, J. Ellä, T. Makkonen, ZnO base thin film bulk acoustic wave filters for EGSM band, 2001 IEEE Utrasonics Symposium Atlanta, p. 803, 2001.

[21] Q.X. Su, P. Kirby, E. Komoro, M. Imura, Q. Zang, R.W. Whatmore, Thin film bulk acoustic resonators using ZnO and lead zirconium titanate thin films, IEEE Trans. MMT-49, 749 (2001).

[22] S.L. Pinkett, W.D. Hunt, B.P. Barber, P.L. Gammel, Temperature characteristics of ZnO based thin film bulk acoustic wave resonators, 2001 IEEE Utrasonics Symposium Atlanta, p. 823, 2001.

[23] K.M. Lakin, J. Belsick, J.F. McDonald, K.T: Mc Carron, Improved bulk wave resonator coupling coefficient for wide band width filters, 2001 IEEE Utrasonics Symposium Atlanta, p. 827, 2001.

[24] K.M. Lakin, J.F. McDonald, K.T: Mc Carron, Temperature compensated bulk acoustic thin film resonators, 2000 IEEE Utrasonics Symposium Puerto Rico, p. 855, 2000.

[25] M.A. Dubois, P. Muralt, V.Plesky, BAW resonators based on aluminium nitride thin films, 1999 IEEE Ultrasonics Symposium, p. 907, 1999.

[26] A. Rodriguez-Navarro, W. Otano-Rivera, J.M. Garcia-Ruiz, R. Messier & L.J. Pilione, Development of preferred orientation of polycrystalline AlN thin films deposited by rf sputtering at low temperature, J. Mater. Res., 12, 1850–1855 (1997).

[27] J.D. Larson III, R.C. Ruby, P.D. Bradley, J. Wen, S. Kok, A. Chien, Power handling and temperature coefficient studies in FBAR duplexers for the 1900 MHz PCS band, 2000 IEEE Ultrasonics Symposium Puerto Rico, proceedings, p. 869, 2000.

[28] R.F. Milsom, H.P. Loebl, C. Metzmacher, P. Lok, A. Tuinhout, F.v. Straten, 2D Model of Solidly-Mounted and Membrane BAW Devices, 2003 IEEE IEEE Utrasonics Symposium Honolulu, Hawaii, p. 1802, 2003.

[29] H.P. Loebl, R.F. Milsom, C. Metzmacher, A. Tuinhout, P. Lok, F.v. Straten, Low Level Effects in SBARS and their Application to Device Optimization, 2003 IEEE IEEE Utrasonics Symposium Honolulu, Hawaii, p. 182, 2003.

[30] H.K.J. ten Dolle, J.W. Lobeek, A. Tuinhout, J. Foekema, Balanced Lattice-Ladder Bandpass Filter in Bulk Acoustic Wave Technology, International Microwave Symposium IMS, Fort Worth, TX, paper TU5D-6, 2004.

# Chapter 7

## High Frequency Tuneable Devices Based on Thin Ferroelectric Films

S. Gevorgian

*Department of Microtechnology and Nanoscience, Chalmers University of Technology, 412 96 Gothenburg, Sweden*

*Ericsson AB, Microwave and High Speed Electronics Research Centre, 431 84 Moelndal, Sweden*

Frequency and phase agile microwave technologies based on ferroelectrics are moving from laboratory demonstrators towards industrial applications. It is important, in this phase, to consider the problems which have not been addressed adequately while dealing with the materials aspects of ferroelectrics and developing laboratory demonstrators. Some of these problems are discussed in this chapter. The focus is concentrated on the ferroelectric varactors as the main component in tuneable microwave devices and systems. The I-V, C-V, and microwave performances of the ferroelectric varactors are considered as circuit components. It is shown that thin film ferroelectric varactors have performances comparable or better then competing commercial technologies (i.e. semiconductor), especially in the frequency range above 10–20 GHz, where there is a lack of high performance semiconductor varactors. The comparison of the basic tuneable technologies shows that thin film ferroelectric varactors are fast, low loss, low power consumption, and small size components useful for applications in power hungry and portable (including onboard and space) adaptable/reconfigurable microwave systems. Typical applications of ferroelectric varactors in devices (tuneable phase shifters, delay lines, and filters), and system demonstrators such as phased array antennas are discussed. Although some problems still need to be solved (hysteresis, industrial-scale reproducibility etc.), the performances of the ferroelectric varactors in subsystem and system level demonstrators indicate that the tuneable microwave technology based on ferroelectric films is mature enough to be considered for large scale industrial/commercial applications.

# 1. Introduction

Tuneable filters, phase shifters, and delay lines are widely used in modern microwave technology. Even more applications are envisaged in future phase/frequency agile, reconfigurable and adaptable microwave systems. Different physical phenomena in dielectrics (ferroelectrics [1], liquid crystals [2]), ferromagnetic materials, ferrites, and semiconductors [3] are used to utilise tuneable microwave components like filters, delay lines, phase shifters. Electric and/or magnetic fields (voltage, current), optical interactions or mechanical manipulations are used to achieve tuneability in components based on these materials. The ranges and the speed of tuneability, control power consumption, losses of microwave signal, power handling capability, potentials of integration, cost, and other parameters of the devices depend both on the materials used, the controlling mechanisms (magnetic, optical, electrical, mechanical), and the design. Selection of the type of a tuneable technology strongly depends on the specifics of the application.

In this Chapter a brief comparative analysis of tuneable microwave technologies is given to show the place of ferroelectrics in tuneable microwave devices. It is followed by a section reviewing ferroelectric varactors from the circuit applications point of view. The main features and implementation examples of analogue tuneable filters, delay lines and phase shifters based on ferroelectrics are summarised in Section IV, while Section V presents two typical phased array architectures using ferroelectric shifters and delay lines.

# 2. Tuneable Microwave Technologies

## 2.1. Mechanical

The mechanical tuning is the first tuning technique used in microwave technology. Early mechanically tuneable devices make use of coaxial lines or hollow metal waveguides and trimming screws/motors/stepper motors. These devices are simple to implement, offer low loss solutions; however, they are bulky, slow, not cost effective, sensitive to vibrations, require high control powers (with motors/stepper motors), or laborious (automated and manual) trimming. Quite recently bulk or thin film piezotransducers are used as actuators to develop relatively small size and relatively fast tuneable devices [4]. Bulk piezotransducer based devices are also simple to implement and offer low loss, low control power solutions. Nevertheless, bulk piezotransducers are still rather slow, sensitive to vibrations, and not suitable for cost effective integration. In contrast, the thin film bulk acoustic piezotransducers offer better performances in terms of speed and cost effective integration possibilities. These types of transducers are used as actuators in Microelectromechnical (MEM) devices.

Recently MEMs switches and varactors attracted considerable attention [5]. These devices offer low loss and low control power solutions, better integration possibility, and higher speeds than bulk mechanical devices. Although the idea behind MEMs devices is very simple, the large scale commercial applications of MEMs based tuneable (reconfigurable) devices turned to be not that simple. They require vacuum packaging;

still suffer from sticking/reliability problems. For analogue tuning (MEM varactor) the design and fabrication processes are rather complex. MEMs varactors have rather complex design in comparison with MEM switches, small Q-factor, and small range of tuneability (also in comparison with the semiconductor and ferroelectric) varactors.

Micromechanical/nanomechnical devices are gaining ground [6], [7], and most probably in the future they may become a major technology offering new tuneable devices with high integration possibilities and operation speeds comparable with semiconductor devices. No tuneable nanomechnical devices are reported so far. However, it seems that acoustic oscillations of micro/nano sized bars (cantilever, fixed ends) activated by electrostatic forces offer a possibility to develop tuneable resonators with extremely high Q-factors.

## 2.2. Magnetic

Tuneable devices using magnetic properties of materials have a rather long history of applications in microwave technology [3]. Magnetic properties of ferrites, magnetostaic (spin) waves, and magnetrostriction [8] has been considered for tuneable microwave applications. Devices based on ferromagnetic resonance make use of the external magnetic field induced (magnetic field dependent) resonances with rather high Q-factor (narrow bandwidth). In general, tuneable filters based on ferromagnetic resonance offer high selectivity and larger tuning ranges compared with other technologies. However the losses at resonant frequency are very high, limiting applications of these devices. Magnetically tuned devices require coils/currents and are rather slow. In spite of these disadvantages it seems that the tuneable devices making use of magnetic properties of the materials are given the second chance—Spintronics is a new and rapidly developing field opening up new possibilities in terms of new functionalities, integration possibilities, drastic size, control power and cost reduction [9].

## 2.3. Optical

Microwave photonics [10] makes use of optical fibres and optically controlled semiconductors for transmission, reception and processing/control of microwave signals Optically controlled devices have the advantage of remote control, inherent decoupling of control and microwave circuits. No decoupling networks for DC bias are requires, and they are potentially wide- and ultra wide band. The main disadvantage of the optically controlled/tuned devices is that they consume relatively large powers for tuning.

## 2.4. Liquid crystal

Liquid crystals also have been considered for microwave applications in the past. The tuneability of permittivity is associated with the anisotropy of the molecules, which was not large enough to get substantial tunings in microwave devices [11]. However, a recent report [2] shows that microwave devices with quite acceptable performances may be developed using specially synthesised liquid crystals with very large anisotropy. Besides integration and cost issues the main disadvantage of the liquid crystals is the low speed of tuning. Although the delay lines based on liquid crystals offer rather high figure of merit [2], the applications may be limited due to the issues like cost, speed, and temperature range.

### 2.5. Semiconductor

Semiconductor varactors (junction, heterojanction, Schottky, MOS etc.) and transistors are widely used in tuneable microwave devices [3]. They allow high density integration. Today the tuneable/switchable microwave devices are by far the most cost effective technology used in both commercial and defence systems. While the transistors based on Si, SiGe, and GaAs meet most of the strict requirements of modern microwave systems, there is a lack of adequate high Q-factor and high speed varactors, especially for frequencies above 10–20 GHz. Additionally, for large arrays, such as phased arrays with up to 10 000 and more radiating elements, the power consumption and heat sink are the main problems hindering applications of semiconductor devices. The desirable features of varactors for these systems include low power consumption, high tuning speeds and high tuneability at millimetre- submillimetre wave frequencies.

### 2.6. Ferroelectric

Thin film (epitaxial, ceramic/textured), thick (HTCC, LTCC) film, and bulk (single crystal, ceramics) ferroelectrics are considered for tuneable microwave applications. Ferroelectric varactors have substantially higher Q-factor at microwave and millimetre wave frequencies, higher tuning speed and lower power consumption. Ferroelectric varactors and tuneable devices based on them have potential (better, in comparison with other technologies listed above) of easy integration with standard Si and GaAs processes. Additionally, due to the high dielectric permittivity, the sizes of tuneable components based on ferroelectrics may be smaller in comparison with other technologies. Extra flexibility in terms of tuning, enhanced functionalities and performances of tuneable devices may be achieved by combination of ferroelectrics with ferrites [12] [13], ferroelectrics with semiconductors [14] or ferrites with semiconductors [15]. Devices based on such multifunctional materials offer dual (electric, magnetic) tuning possibility and extra flexibility in designing and shaping the device performances. For example, it is possible to tune the centre frequency of a filter and shape its pass/rejection independently. In the case of delay lines it is possible to tune the delay time while maintaining the input/output impedances on a desired level. Application of HTS electrodes, wherever it is possible, helps to reduce the losses, and, in some cases, to improve the overall performance of the devices due to the cooling. The main features of the analogue tuneable ferroelectric filters, delay lines and phase shifters are summarised in the sections below.

## 3. Ferroelectric Varactors

### 3.1. Basic designs

The varactors are the main components used in tuneable ferroelectric devices. Varactors used in microwave devices have two basic designs: parallel-plate and

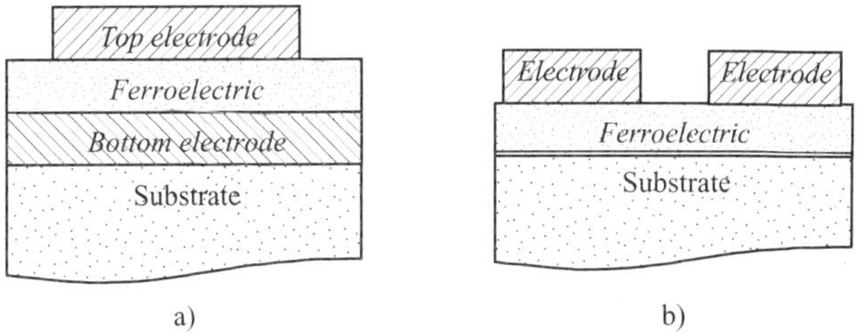


**Fig. 1.** Parallel-plate (a) and coplanar-plate (b) ferroelectric varactors on silicon substrate.

coplanar-plate, as shown in Fig. 1. In both parallel-plate and coplanar-plate designs the tuning (change in capacitance) is achieved by applying DC voltage to the plates, which causes a reduction of the permittivity of the ferroelectric film and hence the capacitance, provided that it has single crystal or polycrystalline structure. Amorphous ferroelectric films are not tuneable and they are used in high density capacitors. The main problem, in fabrication of both types of varactors, is in growing epitaxial (quasi-epitaxial) films with low microwave losses and high tuneability. The tuneability of the varactor is defined as relative change in the capacitance under applied DC field $E$:

$$T_c(E) = [C(0) - C(E)]/C(0) \tag{3.1}$$

where $C(0)$ and $C(E)$ are the capacitances under zero and E bias fields. The tuneability depends on the design of the capacitor, on the applied field, and the quality of the ferroelectric material used.

In comparison with the coplanar plate design the parallel-plate varactors have small sizes, require smaller tuning voltages, and have higher tuneability. The tuneability of the capacitance is practically the same as the tuneability of the permittivity, $T_\varepsilon(E) = [\varepsilon(0)\text{-}\varepsilon(E)]/\varepsilon(0)$. The coplanar-plate design, Fig. 1b, is simple to fabricate and integrate. However, due to the shunt connected partial capacitances of the substrate ($C_{Sub}$), air ($C_{air}$), the tuneability of the capacitance is smaller than that of the permittivity $T_\varepsilon(E)$ :

$$T_c(E) = [C_f(0) - C_f(E)]/[C_f(0) + C_{sub} + C_{air}] \tag{3.2}$$

The advantage of this design is that the tuneability may be traded against losses. For the given loss tangent of the ferroelectric film the total losses of the varactor may be reduced at the cost of reduced tuneability (increased tuning voltage) by increasing the gap width between the electrodes, Fig. 1b. The coplanar plate varactor may be designed to support higher voltages and higher microwave powers, while for parallel-plate varactor these parameters are (fixed) limited by the film thickness. When semiconductors are used as substrates (i.e. silicon [14]), the coplanar-plate design offers enhanced tuneability at low frequencies due to the action of MOS structure. Additionally, if the ferroelectric film is in polar phase its two polarisation states my be used to induce either accumulation (inversion) or depletion layers in the substrate at the ferroelectric/substrate interfaces. This effect may be used in non-volatile microwave switches.

In microwave devices the ferroelectric varactors are used as lumped element components where the sizes of the varactors are much smaller than the wavelength of the microwave signal in the ferroelectric [$\lambda_g = \lambda_o/\sqrt{(\varepsilon)}$, where $\lambda_o$ is the wavelength in

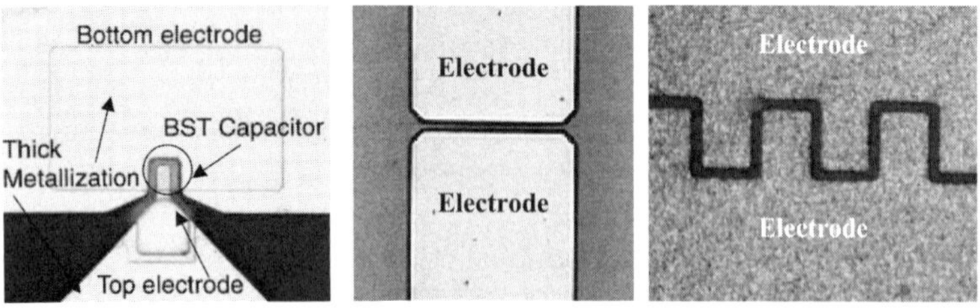


a) b) c)

**Fig. 2.** Photos of parallel-plate [16] (a) and coplanar-plate ferroelectric varactors with straight (a) and interdigital (c) electrodes.

the free space], and as distributed varactor structures. In the latter case the sizes of the ferroelectric components are comparable (as in resonators) or much larger (as in delay lines) then the wavelength $\lambda_g$. The coplanar plate design is more suitable in these applications since the impedance of the lines sections with ferroelectric films may be tailored by changing the gapwidth between the electrodes. Note that in a parallel-plate capacitor the eclectic field is perpendicular to the ferroelectric film, while in coplanar plate varactors it is predominantly in the plane of the film. Depending on the microstructure (columnar, granular etc.) of the film this relative orientation may play a certain role in device performance (losses, permittivity, tuneability), since the field may be normal or parallel to the grain boundaries.

### 3.2. Ferroelectric Films

From large number ferroelectric materials known today, only a limited number of them is considered for microwave applications. Both polar (ferroelectric) and paraelectric phases may be useful in tuneable microwave devices, provided they have low losses and reasonable tuneability [17]. However, often the paraelectric phase is preferred since in this phase there is no hysteresis associated with the domain walls. In this respect the quantum paraelectrics ($SrTiO_3$, $KTaO_3$, $CaTiO_3$, [18], [19]) have been given priority for some time, although they need to be cooled down to cryogenic temperatures in order to get reasonable tuneabilities at relatively low DC fields. On the other hand, for commercial applications room temperature operation is still preferable, and the solid solutions of $Ba_xSr_{1-x}TiO_3$ seem to be the material of choice for different reasons. It is the most studied at microwave frequencies composition, which allows microwave engineers to use the available data without getting too much involved in materials study. In addition, it allows controlling the Curie temperature by simple change of the *Ba* content. At room temperature paraelectric phases of $Ba_xSr_{1-x}TiO_3$ have low microwave losses ($tan\delta$) and substantial tuneability $T_\varepsilon(E) = [\varepsilon(0) - \varepsilon(E)]/\varepsilon(0)$ at relatively week DC fields.

Ferroelectrics have rather high permittivity (for thin films >100), allowing a substantial reduction in the sizes of microwave components. (In some cases high permittivity may be a negative feature, since it may cause matching problems in microwave circuits). The main advantages of ferroelectric films for microwave applications include frequency independent tuneability (up to 50% or more), high tuning speed (<1.0 ns), extremely small leakage currents and DC control power, high breakdown field, and radiation hardness. A useful figure of merit to characterise ferroelectric material was introduced in [20], which takes into account the DC field induced changes both in permittivity and loss tangent, and may be represented as:

$$K(E) = \frac{\varepsilon(0)}{\varepsilon(E)} \frac{T_{\varepsilon}^{2}(E)}{\tan\delta(0)\tan\delta(E)} \tag{3.3}$$

Where $T_{\varepsilon}(E)$ is the tuneability of the permittivity as defined above, $\tan\delta(0)$ and $\tan\delta(E)$ are the loss tangents of the ferroelectric with and without DC bias.

## 3.3. Substrates

The substrate materials considered for tuneable ferroelectric devices include metal (e.g. *Pt*), glass [21], ceramics (e.g. alumina) [22], semiconductors (*Si*, *GaAs*) [23], [24], crystalline dielectrics $MgO$ [25], [26], $Al_2O_3$ (sapphire) [21], [27], $LaAlO_3$ [25], [28] etc. Crystalline substrates ($MgO$, $LaAlO_3$, Sapphire) have been extensively considered for fabrication of tuneable ferroelectric devices, since they allow (hetero)epitaxial growth of ferroelectric films such as $SrTiO_3$ and $Ba_xSr_{1-x}TiO_3$ . In some cases even alumina [22], [29] is considered, which allows growth of polycrystalline ferroelectric films. Advantages of tuneable ferroelectric devices based on crystalline substrates (e.g. $MgO$, $LaAlO_3$, $Al_2O_3$) include good crystalline quality of ferroelectric films, low losses ($tan\delta < 0.01$), and high tuneability (>20%). However, they are not cost effective to use, partly due to the small sizes and cost of the available crystalline wafers. On the other hand, the large-scale applications of ferroelectric components and devices depend on the integration possibilities of microwave ferroelectric components with standard semiconductor processes. In this respect the ferroelectrics are not new in silicon technology. In recent years considerable efforts have been concentrated on the development *CMOS* based memory cells for Dynamic Random (*FERAM*) and nonvolatile memory applications, where ferroelectric capacitors are integrated with *MOS* transistors. Most of the material problems associated with the integration ferroelectrics with standard silicon processes is solved. In addition, there is a considerable progress in developing *CMOS* based *MMIC*s. Thus, there seems to be no major problems in the development of *SiMMIC*s integrating ferroelectric films, where some useful features of ferroelectrics are utilized.

## 3.4. Performance

The performance of a ferroelectric varactor depends not only on the composition of the film, but also on the strain, defects, electrode/ferroelectric interface chemistry, fabrication method, design etc. A simple scanning of the recent publications shows that depending on these factors varactors based on the same ferroelectric composition (i.e. $Ba_xSr_{1-x}TiO_3$) have performance ranging from rather good to totally unacceptable. Recently a substantial progress has been achieved with parallel-plate varactors, and Agile Materials [27] markets a number of tuneable microwave components.

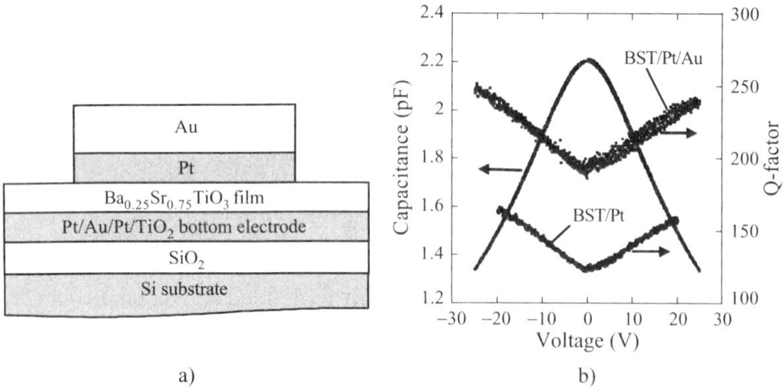


**Fig. 3.** Cross section (a) and C-V dependence (b) of a parallel-plate varactor with diameter of the top plate 30 μm.

### *3.4.1. C-V performance*

Figure 3 shows the typical C-V and Q-factor of a 30 μm BST/Pt/Au varactor at 1.0 MHz [30]. The varactor is fabricated on an n-type silicon [(100)Si, $\rho = 5\ \text{k}\Omega{\cdot}\text{cm}$] substrate with adhesive $TiO_2$ (15 nm) and Pt (100 nm) layers. BST films (300 nm thick) are grown by laser ablation of $Ba_{0.25}Sr_{0.75}TiO_3$ target using KrF excimer laser operating at 10 Hz and 1.5 $\text{J}{\cdot}\text{cm}^{-2}$ The bottom electrode and top electrodes are made of a 0.5 μm thick Au film and 50 nm Pt layer. For comparison, Pt(200 nm)/$TiO_2$/$SiO_2$/Si structures are used as substrate to form BST/Pt varactors. Use of thicker Au bottom electrode substantially increases the Q-factor of the varactor in comparison with a varactor with only Pt bottom electrode, provided that the fabrication processes optimised to grow high quality ferroelectric films, Fig. 3b. Both C-V and Q-V curves in Fig. 3b are given for one cycle of full reversal of the DC bias field. As it is seen no hysteresis is observed, which is quite important for device applications. On the other hand, if the fabrication process is not optimised and the film is in polar, ferroelectric phase, the C-V have two peaks ("butterfly" shape [24]), and the capacitance at given DC bias depends on the prehistory of biasing, which is not desirable in many device applications. The polar phase can be easily detected by measuring the P-V performance. In some cases the C-V dependence may have the form shown in Fig. 4a. In this case the capacitance does not return to its initial value after a full cycle bias reversal. There is no polarisation related hysteresis in P-V performance and the P-V is essentially linear, Fig. 4b. This performance is associated with charged defects in the film and not desirable for device applications.

### *3.4.2. I-V performance*

The I-V performance of ferroelectric varactors depends on the potential barriers at the electrode/ferroelectric interfaces and the conduction mechanism in the bulk of the ferroelectric film, Fig. 5. The I-V performance is closely correlated to defects in the film and at the interfaces and microwave losses [31] and in this respect is a powerful

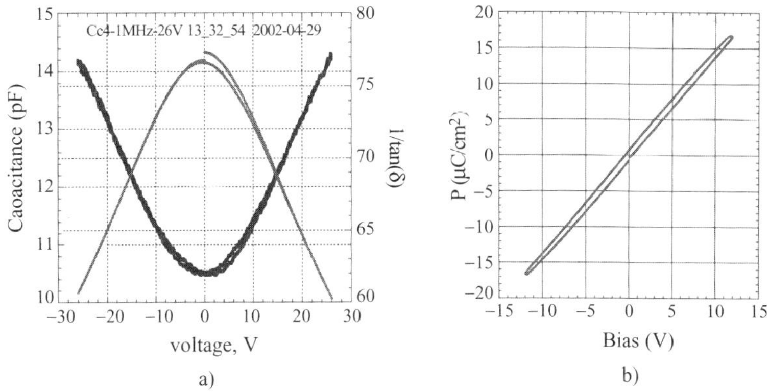


**Fig. 4.** The effect of charged defects on C-V performance of a varactor (a), and its P-V performance (b).

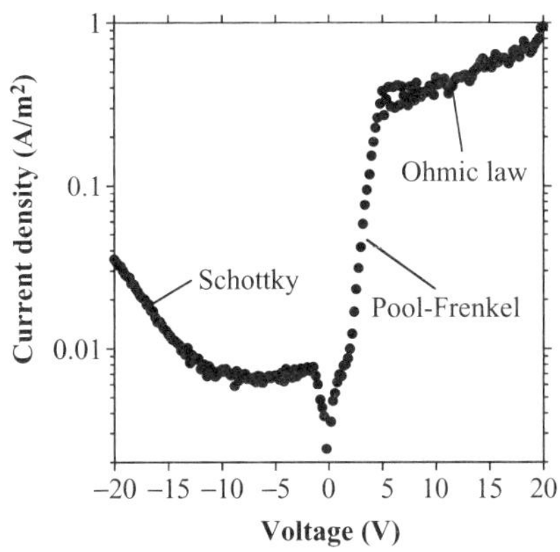


**Fig. 5.** Typical I-V performance of parallel-plate varactors with Pt/BST interfaces [31].

spectroscopic tool, along with the microstructure analysis, helping to interpret the performance of the varactor and use the information in optimisation of the fabrication technology and the design of the varactor. These mechanisms will not be considered here. Instead the main effects of the DC currents on the device performance will be discussed briefly.

Typically the (leakage) currents in ferroelectric varactors are much smaller in comparison with the competing semiconductor and magnetic technologies, and this is one of the heaviest advantages of tuneable ferroelectric technology. In terms of device performance the small leakage currents mean small control power consumption-an important issue for application in portable and space based microwave systems, and in systems where large arrays of the devices are densely packed and heating and heat sink is a serious issue. For example, current through our 10 μm BST/Pt/Au varactor at 20 V is about $5 \cdot 10^{-11}$. A which is 3 orders of magnitude less than reverse current through dual Schottky diode at 5V.

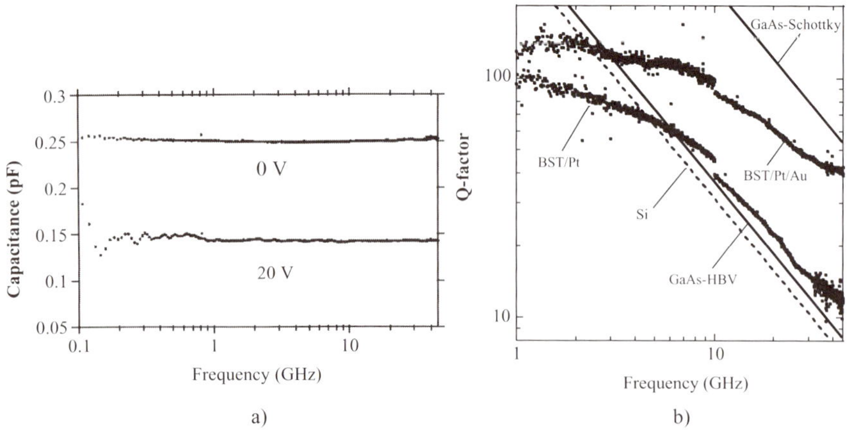


**Fig. 6.** Frequency dependencies of the capacitance (a) and Q-factor (b) of parallel-plate varactors.

No publications dealing with the noise performance of ferroelectric varactors are available. However, it follows from general considerations one may expect that at least 1/f noise, directly related to the leakage currents, is much smaller than that in semiconductor analogues.

### *3.4.3. Microwave performance*

Typical frequency dependencies of the capacitance and Q-factor of parallel-plate ferroelectric varactors are shown in Fig. 6. The design parameters of the varactors are given in Section III.4a, and the diameter of the top plate is 10 μm. As it can be seen the capacitance and the tuneability are fairly frequency independent. The tuneability of BST varactors at 25 V, $T_c(25)$, is more than 40% in the frequency range 1 MHz-45 GHz. In the frequency range 5–40 GHz the Q-factor of BST varactors fabricated on thick bottom Au/Pt electrode is more then 50 which is comparable or larger then Q-factors of the best semiconductor analogues, Fig 6.b [30]. With these parameters the BST varactors may outcompete semiconductor varactors in tuneable microwave devices above 10–20 GHz. Additionally, ferroelectric varactors have symmetric C-V characteristic, as Heterojunction Barrier Varactors (HBV [32]), and extremely small leakage current. Finally, the analysis of experimental results show that the permittivity of paraelectric $B_{0.25}Sr_{0.75}TiO_3$ films is about 200, and the loss tangent is an order of magnitude larger than in single crystals (e.g. $SrTiO_3$), indicating that there is a room for further improvement of the films/varactors.

## 3.5. Measurement and modelling of the varactors

Accurate measurement of the ferroelectric varactors, especially above 10–20 GHz is a challenging problem. On wafer microbe measurement is the main method used today for characterisation of varactors in a wide frequency band. Usually the test structures

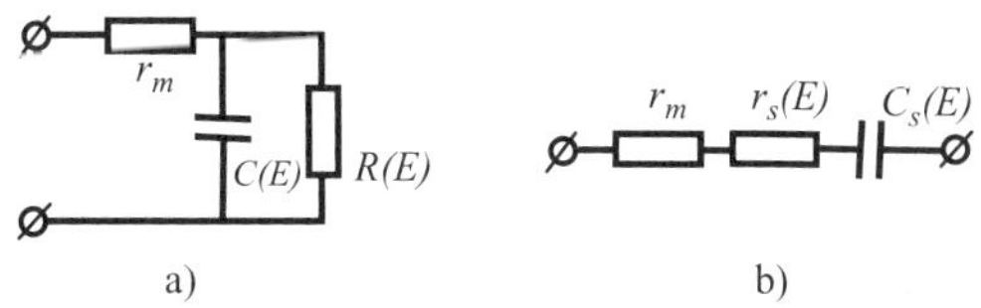


**Fig. 7.** Equivalent circuit representations of a ferroelectric varactor.

(varactors) similar the ones shown in Fig. 2 are used for microwave characterisation of the varactors and extraction of the dielectric properties of the ferroelectric films.

One of the main tasks of the measurements of the varactors as circuit components is to develop scalable circuit models for them. Low frequency (<100 MHz) C-V measurement give some information about usefulness of the varactors, but usually they are not adequate for development of circuit models useful for microwave frequencies. In a simplified equivalent circuit (Fig. 7) of a ferroelectric varactor $r_m$ is the resistance of the metal plates, $C(E)$ is the capacitance and, and shunt resistor $R(E)$ represents losses in the ferroelectric film. Both $C(E)$ and $R(E)$ are field dependent, while $r_m$ is bias independent. All these parameters are frequency dependent.

The parallel branch in Fig. 7a represents the ferroelectric film with permittivity $\varepsilon(E)$ and loss tangent $tan\delta$. For a simple parallel-plate varactor $C(E) = \varepsilon(E)\varepsilon_o A/t$, $R(E) = t/[2\pi f\varepsilon(E)\varepsilon_o tan\delta(E)\ A]$, where $A$ is the area of the smallest (top) plate, $t$ is the thickness of the ferroelectric film. The losses in the ferroelectric film are characterised by Q-factor or loss tangent: $Q_p(E) = 1/tan\delta(E) = 2\pi fR(E)C(E)$. Sometime a series equivalent circuit, Fig. 7b, is used, and the relationship between these two equivalent circuit elements is given by:

$$r_s(E) = \frac{R(E)}{1 + [2\pi f C(E) R(E)]^2} \tag{3.4}$$

$$C_s(E) = \frac{1 + [2\pi f C(E) R(E)]^2}{C(E)\,[2\pi f\, R(E)]^2} \tag{3.5}$$

The Q-factor of the varactor, $Q_s(E) = 1/\{2\pi f[r_s + r_m(E)]C_s(E)\}$:

$$Q_s(E) = \frac{Q_p(E)R(E)}{(1 + Q_p^2(E))r_m + R(E)} \approx \frac{Q_p(E)R(E)}{Q_p^2(E)r_m + R(E)} \tag{3.6}$$

In an extreme case, where the contact (electrode) resistance $r_m = 0$, one has $Q_s(E) = Q_p(E) = 1/tan\delta(E)$, and the frequency dependence of the varactor Q-factor nearly repeats the frequency dependence of the loss tangent, as it is seen in Fig. 6b for a varactor with thick *Au/Pt* electrodes. In another extreme case, where the losses in the electrodes dominate, $Q_p^2(E)r_m \gg R(E), Q_s(E) = R(E)/(Q_p(E)r_m) = 1/(2\pi f r_m C(E))$. In this case usually the thickness of the electrodes is much smaller than the skin depth, and the resistance $r_m$ is frequency independent. Since the permittivity, and hence capacitance (Fig. 6a) are frequency independent, the Q- factor of the varactor is characterised by inverse frequency dependence, $Q_s(E) = 1/(2\pi f r_m C(E))$, as it is seen for a varactor with thin *Pt* electrodes. This is the case also for semiconductor varactors (Fig. 6b), where the losses (resistance $r_m$) are given mainly by heavily doped semiconductor region(s) constituting the plate(s) of the varactor.

In actual device applications a ferroelectric varactor is subject to both high frequency and DC voltages and the circuit models should reflect the dependence of the varactor parameters on size, frequency, and field (DC and microwave) dependence. Due to the inherent and fast nonlinearity of the ferroelectric film, even a microwave signal causes changes in permittivity. It is supposed that in small-signal regime these changes are smaller than a limit tolerated by specifics of varactor applications. When the amplitude of microwave signal is large and/or the thickness (gapwidth in a coplanar-plate design) is small these changes cannot be ignored. Typically, the microwave signal reduces the permittivity the same way as the DC bias, however, the loss tangent increases with increasing amplitude of the microwave signal. This increase is associated with the higher order harmonic generation [33]. Hence, in measurements, where the extraction of complex permittivity of the ferroelectric film is concerned, care should be taken that the measurements are carried out at small signal regime. On the other hand, where the development of circuit models of the varactor are concerned, the measurements should be carried out at different levels of microwave power (large signal regime), and taken into account in circuit models.

The development of the design of the varactors, and microwave circuits based on them may be carried out using commercially available software (Momentum, HFSS, Sonnet etc.). However, these tools are rather time consuming, especially where complex circuits with large aspect ratio are considered. In some cases, especially for simple components like varactors, transmission line with ferroelectric films, the use of the conformal mapping [34], [35] technique may be more efficient in terms of time and cost. Additionally, this technique allows easy extraction of dielectric properties from impedance measurements.

## 3.6. Problems

There are several problems to be addressed where the ferroelectric varactor based tuneable microwave devices are considered for commercial and military applications. Some of them will be discussed in this section.

### *3.6.1 Temperature dependence*

The inherent temperature dependence of permittivity of ferroelectrics is a major concern for device applications. The temperature dependence may be compensated electronically at the expense of reduced tuneability. Nevertheless, there are some simple ways to address this issue. Figure 8 demonstrates one of the possible ways [36]. Two ferroelectrics with permittivity temperature dependences shown in Fig .7a arranged in a capacitor in a parallel form so that in the desired temperature range the resultant effective permittivity (capacitance) has no temperature dependence. At the same time in this temperature range the Q-factor and the tuneability are high. In an experiment with coplanar-plate electrodes, Fig .8b, the gapwidth is mach larger than the thickness of the ferroelectric layers. Due to this fact, and high permittivity of the films, the electric field

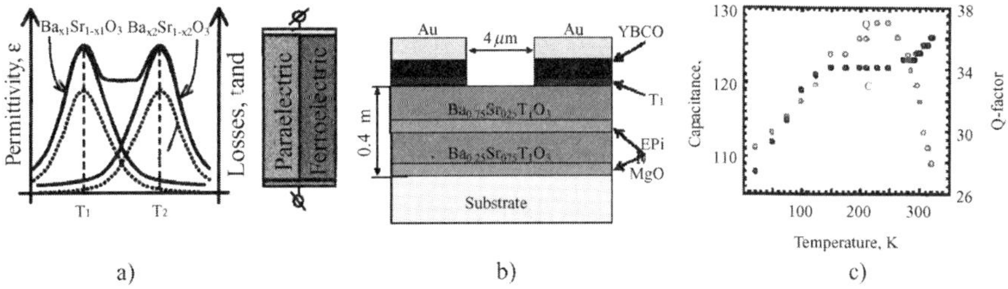


**Fig. 8.** Temperature compensated coplanar-plate varactor.

between the plates is practically in the plane of the films, as it is suggested in Fig. 7a. The measured performance clearly demonstrates the expected result, Fig. 8c.

### *3.6.2. Hysteresis*

Domain related and trapped charge related hysteresis loops are observed in tuneable ferroelectric devices, where the varactor is not optimised. Fig. 8a shows C-V performance of a coplanar-plate varactor based on $Na_{0.5}K_{0.5}NbO_3$ (NKN) films prepared by pulsed laser ablation [24]. The substrate used in this experiment is a 0.35-mm thick n-type silicon with resistivity $\rho g \approx 7.7$ kΩ cm and with a thin native $SiO_2$ layer. The average thickness of the films is about 0.5 μm. Au (≈0.5 μm)/Ti (10 nm) films are vacuum evaporated on top of $Na_{0.5}K_{0.5}NbO_3$ films. In this case the NKN film is in polar phase. Figure 9b shows the P-V performance of the varactor where polarisation hysteresis is observed with a coercive voltage about 1.5 V. The C-V performance, as it is expected, has peaks at the same bias voltages. However, the standard "butterfly" shape of the curve is distorted. The capacitance, after a full cycle of voltage reversal, is lower then the original value, which is a clear indication of presence of deep traps in ferroelectric film. Note that due to the specifics of microwave measurements it takes about 30 minutiae to produce a dependence shown in Fig. 8a, which indicates that the time constant of the traps is very large. Most probably the oxygen vacancies

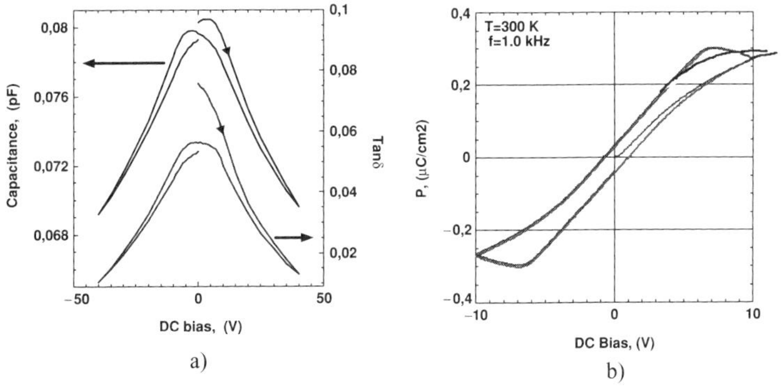


**Fig. 9.** DC dependence of he capacitance measured at 40 GHz (a) and P-V dependence (b) of a coplanar-plate varactor based on NKN films.

in the film and/or film/electrode interface act as traps. Usually a proper annealing in oxygen atmosphere eliminates or substantially reduces the charge/trap related effects. On the other hand, even in paraelectrics one may have strain induced polar phase [37], and associated hysteresis. The strain and the oxygen vacancies usually are in dynamic equilibrium [97]. Most probably, the oxygen vacancies may be completely eliminated in strain released films by proper selection of the buffer layers and the substrates.

In some cases, where the electrode/ferroelectric interfaces are not symmetric, the peak of the C-V curve shifts from $V = 0$ position, which is similar to the imprint in memory cells. In general, the higher tuneability and low losses [higher $K(E)$,(3.3)] are the main parameters one needs in tuneable device applications, and, ideally, a completely symmetric C-V with no hysteresis and/or charging effects is preferable. Any asymmetricity, charging processes, and temperature dependence cause problems for control electronics.

### 3.6.3. Power handling capability and nonlinearity

The experimental work carried out by [30] shows that the ferroelectric varactors have substantially higher power handling capability than semiconductor analogues, and even MEMs varactors/switches. Additionally, there are different ways to increase the power handling capability of ferroelectric varactors and devices based on them. The simplest way is to increase the thickness of the ferroelectric film in a parallel-plate varactor, or the gapwidth in a coplanar plate varactor, so that the changes in the permittivity induced by microwave field are much smaller (the limits are specified by specifics of the applications) then the changes induced by applied DC field. One way to increase the power handling capability [39] of a coplanar plate varactor is to use extra high resistivity DC biasing strips in the gap between the main electrodes of the capacitor, Fig. 10. In the proposed design the gap between the main (microwave) electrodes is selected based on power handling requirements and can be rather large, while the gap between the high resistivity (DC) electrodes is rather small, allowing to substantially reducing the required DC control voltage. These extra films are transparent at microwave frequencies and do not contribute in the losses of the varactor.

In alternative applications the microwave field dependent dielectric properties of ferroelectrics may be used in harmonic generators [40], mixers and pulse compressors

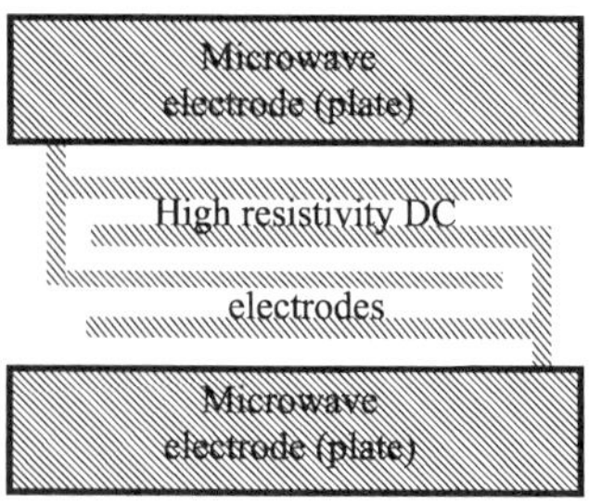


**Fig. 10.** Extra high resistivity DC bias electrodes in the gap of coplanar-plate varactor [39].

[41]. In this case the varactors need to be designed so that the microwave field induced changes in capacitance are as high as possible.

## 4. Tuneable Ferroelectric Devices

### 4.1. Tuneable filters and matching networks

Tuneable microwave filters used today in commercial and defence microwave systems are mechanical, or based on magnetic (ferrite, YIG) and semiconductor varactors. Mechanically tuneable filters are useful for simple trimming where no fast tuning is required or the adjustment (trimming) of the centre frequency of the filter is done rarely. Although MEM switches have relatively high Q-factor, the analogue tuned MEM varactors and filters based on them still have high losses, they are rather slow, have small tuning range and more suitable for frequencies below 10 GHz [5].

Commercially, semiconductor varactor based filters are available below several GHz [42]. They have rather high losses and the selectivity (number of poles, steepness of the skirts) is not high due to the low Q-factor of the semiconductor varactors. The selectivity (Q-factor) of YIG filters is rather high, but unfortunately the losses are also high. In some applications YIG filters with High Temperature Superconductors (HTS) [43] have substantially smaller losses; however, the filters need to be cooled to below 90 K.

In terms of power handling capability mechanically tuned (bulk) filters have no competitors. YIG filters are next to handle rather high power levels. Both these technologies are quite complex and not cost effective. Ferroelectrics offer relatively simple solutions.

Thin film ferroelectric tuneable resonators, matching networks, and filters are one of the most discussed, but perhaps relatively less successful components. The reason is that for most of applications the loss requirements are quite strict and practically no tuneable components meet these requirements. In most of the system applications filters should have very low losses in the passband and high selectivity. i.e. steep skirts, which is not easy to fulfil, especially for narrow band (bandwidth less than 5%) filters.

Resonators are the main building blocks of filters. In general, for a bandpass filter the fractional bandwidth $BW$, the insertion losses $IL$, and the unloaded $Q$-factor of the resonators used are related as [Kohn]:

$$IL_{dB} = 4.34 \frac{\sum g_i}{BW \cdot Q} \tag{4.1}$$

where $g_i$ represents capacitances (scaled) and inductance in the low frequency prototype of the filter. For a given Q-factor of the resonators (and hence varactor included in the resonator) this formula sets the limits of achievable losses and filter bandwidth.

Both lumped element (i.e. capacitor and inductor coils with sizes much smaller than the wavelength of microwave signal) and waveguide resonators incorporating ferroelectric varactors have been considered in tuneable filters. In lumped element resonators [27], [45], [46], the ferroelectric varactor is the main part of the resonator capacitor. The $Q$-factor of the resonator may be given as $1/Q = 1/Q_L + 1/Q_v$, where

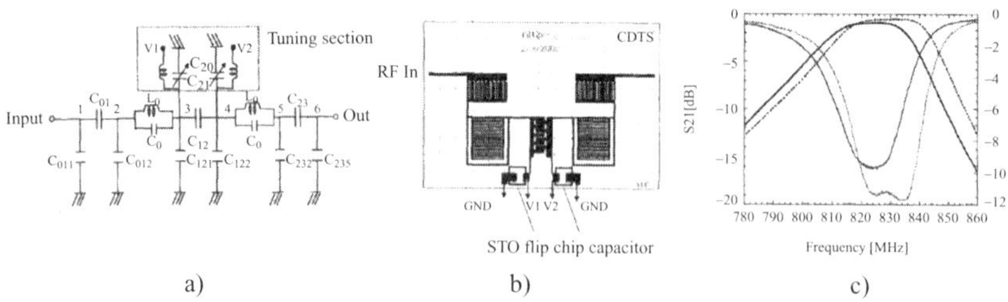

**Fig. 11.** Equivalent circuit (a), layout (b) and performance (c) of a superconductor filter with interdigital (coplanar-plate) ferroelectric (STO) varactors at $T = 65\ K$.

$Q_L$ and $Q_v$ are the $Q$-factors of the inductor coil and ferroelectric varactor respectively. Below 10.0 GHz the Q factor of the varactors is usually higher (> 100) than the Q-factor of thin film lumped inductor coils ($Q_L < 100$), and the performance of the resonators, matching networks, filters etc. based on lumped element ferroelectric LC resonators is limited by the losses in metals strips of the coils. A possible way to reduce the losses is to use superconductor strips instead of normal metal. Figure 11 shows equivalent circuit, layout ant the performance of a superconducting filter with STO varactors [46]. The flip-chip varactors are made of 2.0 mm thick STO on $LaAlO_3$ substrate. Both STO and superconducting ($YBa_2Cu_3O_7$) films are prepared by reactive co-evaporation technique. The bandwidth is about 3.5%. The tuning, about 1.0% at E < 2.0 V/$_m$m field, may be increased by increasing the bias DC field.

A third order lumped element tuneable ferroelectric filter using parallel-plate BST varactors (*Pt* bottom electrodes 0.2 µm) operating at room temperature has been demonstrated in [45]. The centre frequency of the filter tuned from 176 to 276 GHz (57% 9 under DC bias 6.0 V. The insertion loss of the filter is 3.0 dB, of which only 1.5 dB is attributed to the STO varactors.

The Q-factor of waveguide (Microstrip, CPW, hollow) resonators incorporating ferroelectric varactors $1/Q = 1/Q_r + [1/(kQ_v)]$, depends on the varactor inclusion factor $k < 1.0$. The unloaded Q-factor of waveguide resonators $Q_r$ is usually higher than the Q-factor of the varactors. Typically, for Microstrip/coplanar type resonators it is larger than several hundred, while for hollow waveguide resonators it is even higher and, in contrast to lumped element resonators, in this case the Q-factor of tuneable resonator (i.e. resonator with a varactor) is limited by the Q-factor of varactors. In waveguide based tuneable filters the varactor may be only a part of the resonators. For applications in low loss, narrow band filters with high selectivity (i.e. multipole filters) the Q-factor of the tuneable resonator may be kept rather high by keeping the inclusion factor very small, $k \ll 1$. In this case the varactor acts as a perturbation and the changes in the resonant frequency (tuning) are small, although the tuneability of the varactor may be high. In other words one can trade the high Q-factor for reduced tuneability. Room temperature hollow waveguide tuneable filters with commercially acceptable performances have been demonstrated recently. Tuneable filters with centre frequencies 14 GHz and 32 GHz, bandwidths about 0.9%, and tunings about 1.8% had insertion losses 3dB and 4.5 dB respectively [47]. The varactors used in these filters have coplanar-plate design and require rather high tuning voltages (about 300 V) due the large gap between the electrodes. Tuneable filters with ferroelectric films parley filling the gap between the coupled microstrip lines have been demonstrated in [48]. In most of the reported cases the applied DC field changes not only the centre frequency of the filter, but also the bandwidth, which is a reflection of the fact that the applied field changes the impedance of the resonators and the coupling between them. An original method to design tuneable ferroelectric filters with constant fractional bandwidth is proposed in [49].

In system applications the size/cost of the filters is a critical issue. While selecting the filter design this issue has to be addressed properly. For example, for communications applications tuneable filters based on lumped elements are preferable since they offer integration and size/cost reduction possibilities, especially at relatively low microwave frequencies (typically below 10 GHz). In this frequency range the filter performance is limited by the low Q-factor of the varactor and it is not beneficial to use bulky waveguides (hollow or integrated-planar). It seems more reasonable to use lumped element topology where the filter performance may be limited by Q-factor of lumped inductor coils. For frequencies above 10 GHz the waveguide designs offer low loss and highly selectivity at the cost of limited tuneability, although these types of filters usually are not cost effective.

### 4.2. Tuneable delay lines

A few publications report analogue tuneable delay lines, however a large number of them deal with delay line type phase shifters (true-time delay lines). Delay lines have considerable applications in modern microwave and digital technologies. Delay Locked Loops (DLL) [50], and Voltage Controlled Oscillators (VCO), [51], feed-forward amplifiers [52], are typical examples, not to mansion phased array antennas and the radar, and measurement [52] systems. Delay lines are also used in digital circuits such as time to digital converters [53], and computer printed circuit boards to minimise the skew of the clock [54].

Substantial delay times (up to 10 ns at 2.0 GHz) have been achieved in delay lines based on 3dB hybrids loaded by series resonant circuits with ferroelectric varactors. These types of delay lines have relatively small size; however, due to the used resonator circuits they have narrow bandwidth (about 5%). Small size ultra-wideband delay lines are easily to implement using coplanar and coplanar strip line sections of length $l$ incorporating ferroelectric films (distributed varactors), Fig. 12a. In the equivalent circuit of the device (section dx), Fig. 11b, $L$ and $C$ are the per-unit line inductance and capacitance of the line without ferroelectric film, and $C_c$ is the per-unit length capacitance

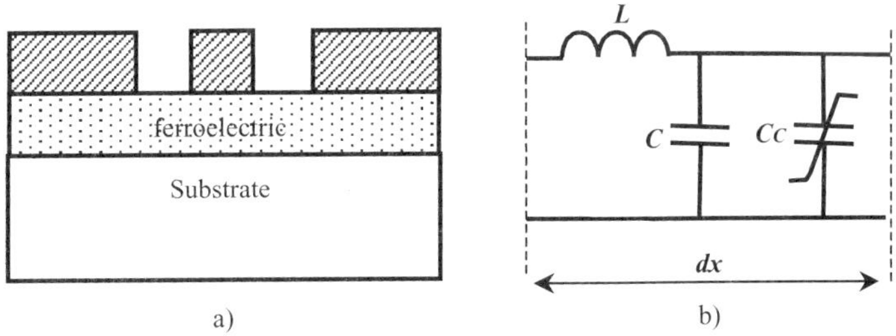


**Fig. 12.** Coplanar waveguide ferroelectric delay line (a) and its equivalent circuit (b).

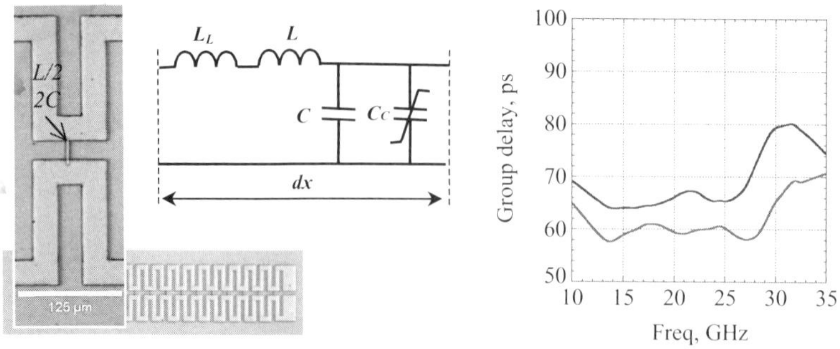


a) b)

**Fig. 13.** Synthetic coplanar strip waveguide delay line. Zoomed unit cell and its equivalent circuit (a), and wideband performance at 0 and 30 V DC bias. The total length o the device is 2.0 mm [56].

associated with the ferroelectric film (distributed varactor). Typically $C$ represents the partial capacitances due to the substrate and air, and is much smaller than $C_c$. As a first approximation the delay time is given as:

$$\tau(E) = \frac{l}{c_o}\sqrt{\varepsilon_{eff}(E)} = \sqrt{LC_c(E)} \tag{4.2}$$

where $\varepsilon_{eff}(E)$ is the DC field dependent effective permittivity, and $l$ is the length of the line section with the ferroelectric film. $C_c$ and the effective dielectric permittivity may be easily calculated using conformal mapping technique [34]. The tuneability is given by (3.2), the capacitances are assumed to be given for per unit length. The disadvantage of this type of delay line is that to reduce the tuning voltages, increase the delay time and its tuneability the width of the slot between the ground plane and signal line need to be as small as possible. The reduction of the slot width inevitably leads to increased losses both in metal strips and ferroelectric film. To reduce the metal losses superconductive electrodes [25], [55] may be used if the cryocooling is acceptable. Obviously, the losses with normal metal electrodes are even higher, making this design less attractive. For room temperature applications with normal metal (Au) electrodes the delay line biased on a synthetic coplanar strip waveguide with periodically loaded ferroelectric varactors offers better performance in terms of losses (dB) per unit delay time (ns). Photographs of such a delay line and its zoomed unit cell (including its equivalent circuit) are shown in Fig. 13a. Typically the per unit length inductance $L$ and capacitance $C$ are much smaller than the lumped inductance $L_L$ and varactor capacitance Cc, and the delay time and impedance of the synthetic line are give by $\tau \approx \sqrt{(L_L C_c)}$, $Z \approx \sqrt{(L_L/C_c)}$. About 10% frequency independent tuneability is achieved in the frequency band 45 MHz-25 GHz with a figure of merit 45dB/ns [56].

### 4.3. Tuneable phase shifters

The main phase shifter topologies and technologies are summarised in detail in [3]. Ferroelectric varactors are essentially analogue components and most of the analogue phase shifter topologies considered in [3] may be realised on ferroelectric varactors. Nevertheless, an attempt has been made to develop digital phase shifters based on ferroelectric varactors [57], [47]. Rat–rate [58], coupled microstrip [28], and Lange coupler [59] phase shifters are based on sections of transmission lines and have sizes not acceptable in some microwave systems.

Periodically loaded line type phase shifters are essentially delay line type phase shifters. In this case a standard transmission line (i.e. coplanar, microstrip) is periodically loaded by ferroelectric varactors. The unit cell equivalent circuit of the devices is similar to the one shown in Fig. 13a, where the length $dx$ should be replaced by $l_{u.c.}$, i.e. length of the unit cell, $L_L = 0$, $L$ and $C$ are the per unit length inductance and capacitance. Below Bragg frequency, $f_B = 1/[\pi\sqrt{L_L(C + C_c})]$, the differential phase shift may be approximated as:

$$\Delta\phi(E) = 2\pi fn\sqrt{L\,[C + C_c(E)]} \tag{4.3}$$

where $n$ is the number of the cells. The impedance $Z \approx \sqrt{[L/(C + C_c)]}$. The periodically loaded line type phase shifters have linear phase frequency dependence with substantial change in the differential phase shift in a bandwidth of interest. On the other hand for many applications, especially in phase shifter based wide band phased array antennas, the differential phase shift within operation frequency bandwidth should be constant. Careful selection of the topology and special design efforts are require for achieving such performance [47], [60].

From integration/size perspective phase shifters based on lumped LC components [59], [61] are more desirable. Lumped element inductors and ferroelectric varactors allow substantial size reduction of phase shifters based on periodically loaded synthetic lines [59], [27], especially for low frequency applications. Figure 14a shows a prototype synthetic line ferroelectric phase shifter designed for operation at about 5.0 GHz. The overall length of the device is 2.4 mm, and the figure of merit at 4.7 GHz is about 60°/dB [59]. Currently Agile Materials & Technologies [27] market phase shifters of this type for frequencies 1.8–2.2 GHz, with overall sizes 1.0 × 6.0 mm$^2$ and figure of merit 90°/dB. The all-pass network based lumped element phase shifters proposed recently, Fig. 14b, are small size and potentially wide band.

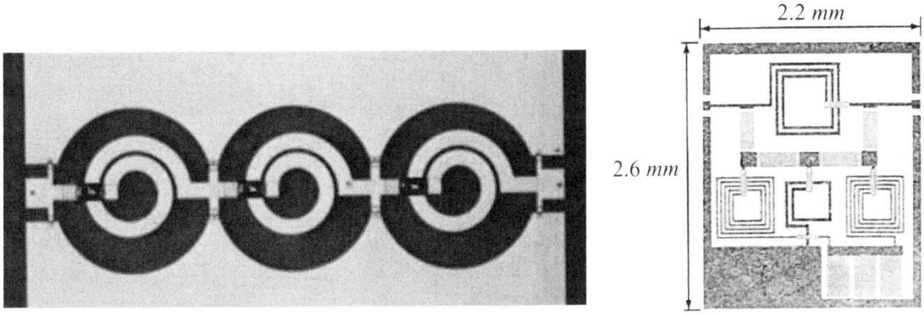


**Fig. 14.** Lumped element synthetic transmission line [59] (a) and all-pass network [61] (b) phase shifters.

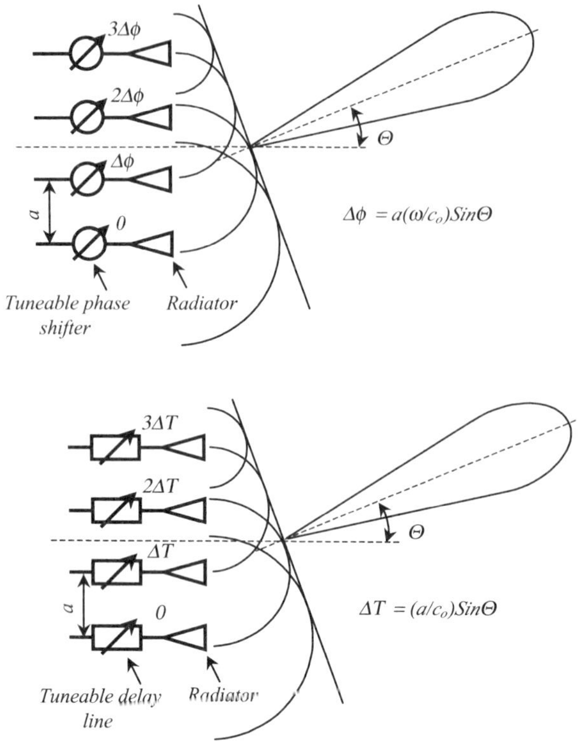


**Fig. 15.** 4 element 1D phased arrays based on phase shifters (a) and delay lines (b).

## 5. Subsystem and System Applications of Tuneable Ferroelectric Devices

Phase shifters and delay lines have a wide range of applications in microwave systems, including oscillators, amplifiers, etc. Voltage Controlled Oscillators (VCO) base on thin film ferroelectrics have been first considered in [63]. Phased array antennas are more typical applications, where these components are used in very large numbers [3]. Simplified (4 element 1D array) schematic diagrams of phased arrays based on phase shifters and delay lines are shown in Fig. 15. For military and space applications 2D phased array may consist of up to several thousand phased shifters or delay lines. For such arrays the speed, power consumption and sizes of the phase shifters and delay lines are critical issues. In this respect ferroelectric phase shifters and delay lines have no competitors. It seems that the future commercial microwave communications systems, including cellular (mobile) systems, tend to be reconfigurable and adaptable. In these systems antennas should be with tuneable beamshapes and scan angles. At present there are reports on phased array antennas for space [64] and commercial [65], [47] applications based on ferroelectric technology.

In phase shifter based beam scanner, Fig. 15a, the scanning of the beam is achieved by changing the phase shift. The maximum required phase shift for full $\Theta = 90^\circ$

scan angle is $2\pi$, which is easily achievable using ferroelectric technology. The delay line type phase shifters with linear phase frequency response may be used in phased arrays operating in narrow frequency band. For wide band (narrow pulse) phased array antennas the phase shift needs to be frequency independent in the required frequency band of operation.

In time delay scanning systems, Fig. 15b, the delay lines should provide non-dispersive (i.e. frequency independent) delay time in a wide frequency range. For example, for a 1D array consisting of 8 elements with the inter element spacing $a = \lambda_o/2$, $\lambda_o$ is the free space wavelength the total length of the array is about 4 $\lambda_o$. At $f = 23$ GHz and scan angle $\Theta = 90°$, the required incremental delay time is $\Delta T = a/c_o = 1/(2f) =$ 0.02 ns. Hence the maximum delay time to be provided by a delay line should be about ng $\Delta T = 8{\cdot}0.02 = 0.16$ ns, which is easy to achieve in the delay lines discussed above. However, the required tuning range for full 90° scanning is almost 100%, which is vary hard to achieve by using ferroelectric varactors with reasonably low loss. For small scan angles, requiring less than 75% tuning, the ferroelectric varactor based delay lines may have quite acceptable performances.

## 6. Conclusions

Thin film ferroelectric varactors have performances comparable or better then competing commercial technologies (e.g. semiconductor, MEM). The comparison of the basic tuneable technologies shows that thin film ferroelectric varactors are fast, low loss, low power consumption and small size components useful for applications in power hungry and portable (including onboard and space)adaptable/reconfigurable applications. Although some problems still need to be solved (hysteresis, industrial-scale reproducibility etc.), the performances of the ferroelectric varactors in subsystem and system level demonstrators indicate that the tuneable microwave technology based on ferroelectric films is already mature enough to be considered for large scale industrial/commercial applications.

## References

[1] O. G. Vendik, "Microwave Tuneable Components and Subsystems Based on Ferroelectrics : Physics and Principles of Design", Integrated Ferroelectrics, vol. 49, pp. 181–190, 2002

[2] S. Muller, C. Weil, P. Scheele, Y. Kryvoshapka, and R. Jakoby, "Novel Liquid Crystals for Tuneable Microwave Components", "Tuneable Ferroelectric Microwave Devices", IMS'2004 Workshop on New Technologies for Frequency-or Phase-Agile Microwave Circuits and Systems

[3] Koul S. K., Bhat B., Microwave and Millimeter Wave Phase Shifters, Artech House, 1991

[4] Gevorgian S., "Tuneable ferroelectric/piezoelectric devices", Chapter in book Piezoelectric Materials and Devices (N. Setter Ed.), Lausanne, Switzerland, 2002

[5] G. Rebeiz, RF MEMS Theory, Design, and Technology, Wiley, 2003

[6] Nanoelectronics and Nanoscale Processing. Special Issue, Proc. IEEE, vol. 91, No 11, 2003

[7] Nanotubes hold promise for RF, Wireless Europe, p. 13, October 2003

[8] Hristoforou E., Hauser H, and Niarcos D, Magnetostrictive delay line characterisation", Journal of Magnetism and Magnetic Materials, 242–245, 2002

[9] Spintronics Technology. Special Issue, Proc. IEEE, vol. 91, No 5, 2003

[10] A. Viulcot, B. Cabon, and J. Chazelas (Editors), Microwave Photonics, Kluwer Publ., 2003

[11] Kuki T., Fujikake H., and Nomoto T., "Microwave Variable Delay Line Using Dual-Frequency Switching mode Liquid Crystals", IEEE Trans. Microwave Theory Techn., Vol. 50, pp. 2604–2608

[12] Demidov V. E., Kalinikos B. A, Karmanenko S. F., Semenov A. A., and Edenhofer P., "Electrical tuning of disperssion characteristics of surface electromagnetic-spin waves propagating in ferrite-ferroelectric layersed structure", IEEE Trans., Microwave Theory Techn., vol. 51, pp. 2090–2096, 2003

[13] Jia Q. X., et al. "Integration of nonlinear dielectric barium strontium titanate with polycrystalline yttrium iron garnet", Appl. Phys. Lett., vol. 74, pp. 1564–1566

[14] Gevorgian S., Abadei S., Berg H., and Jacobsson H., "MOS Varactors with Ferroelectric Films", IEEE MTT-S2001 Dig., vol. 2, pp. 1195–1197, 2001

[15] Fiallo H. H., et al. "Transmission properties of metal-semiconductor-relaxor microstrip lines", IEEE Trans. Microwave Theory Techn., vol. 42, pp.1176–1182

[16] Seriacco J., Acikel B., Hansen p. , Taylor T., Xu H., Speck J. S., and R. York, "Tuneable passive integrated circuits using BST thin films", Integrated Ferroelectrics, vol. 49, pp. 161–170, 2002

[17] S. Gevorgian, and E. Kollberg, "Do we really need ferroelectrics in paraelectric state only in electrically controlled microwave devices"? IEEE Trans. Micr. Theory Techn., Vol. 49, pp., 2117–2124, 2001

[18] S. Gevorgian, "Quantum Paraelectrics in Tuneable Microwave Devices: Promises, Current Status and Problems", IEEE Int. Microwave Symposium Workshop on Ferroeletric Materials and Devices for Microwave Applications, Philadelphia, June 2003

[19] Lemanov etc., Solid State Communic. 110, 611–614 (1999)

[20] O. G. Vendik, I. B: Vendik, and E. Kollberg, "Commutation Quality Factor of Two State Switchable Devices" IEEE Trans. Microwave Theory Techn., vol. 48, pp. 802–808, 2000

[21] Acki B. Liu Y., Naga A. S., Taylor T.R., Hansen P. J., Speck J. S., and R. York, "Phase shifters using (BaSr)$TiO_3$ thin films on sapphire and glass substrates", IEEE MTT-S'2001, pp.

[22] Kozyrev A. B., Gaidukov M. M., Gagarin A. G., Tumarkin A. V., and Razumov S. V., "A finline 60- GHz phase shifter based on a (Ba,Sr)TiO3 ferroelectric thin film", Technical Physics Letters, vol. 38, pp. 239–241, 2002

[23] Erker E. G., Nagra A. S., Liu Y., Periaswamy P., Taylor T. R., Speck J., and York R. A., "Monolithic Ka-band phase shifter using voltage tuneable BaSrTiO3 parallel-plate capacitors", IEEE Microwave and guided Weave Letters, vol. 10, pp. 1012, 2000

[24] S. Abadei, S. Gevorgian, C.-R. Cho, and A. Grishin, "Low frequency and Microwave Performances of Laser Ablated Epitaxial $Na_{0.5}K_{0.5}NbO_3$ Films on High Resistivity $SiO_2$/Si Substrates", J. Appl. Phys, Vol. 91, pp. 2267–2276, 2002

[25] Carlsson E. F., Petrov P. K., Chakalov R. A., Larsson P., Ivanov Z., and Gevorgian S., "Experimental study of thin film HTS/ferroelectric CPW phase shifters for microwave applications", Inst. Phys. Conf. Ser. No158 (UK), pp. 339–342, 1997

[26] S. W. Kirchoefer, J. M. Pond, A.C. Carter, W. Chang, K. K. Agarwal, J. S. Horwitz, and D. B. Chrisley, Microwave and Optical Technology Letters, **18**, 168 (1998)

[27] http://agilematerials.com

[28] Van Keuls F. W., Romanofsky R. R., Boham D. Y., Winters M. D., Miranda F. A., Mueller C. H., Treece R. E., Rivkin T. V., and Galt D., "YBCO,Au)/STO/LAO thin film conductor/ferroelectriccoupled microstrip phase shifters for phased array applications", Appl. Phys. Lett., vol. 71, pp. 3075–3077, 1997

[29] S. Delpat, M. Ouaddari, F. Vidal, M. Chaker, and K. Wu, "Voltage and Frequency Dependnet Dielectric Properties of BST-0.5 Thin films on Alumina Substrate", IEEE Microwave and Wireless Components Letters, vol. 13, pp. 211213, 2003
[30] A. Vorobiev, P. Rundqvist, K. Khamchane, S. Gevorgian, "Silicon substrate integrated high Q-factor parallel-plate ferroelectric varactors for microwave/millimeterwave applications", Appl. Phys. Letters, Vol.83, p. 3144, 2003
[31] A. Vorobiev, P. Rundqvist, K. Khamchane, and S. Gevorgian, "Microwave Loss Mechanisms in $Ba_{0.25}Sr_{0.75}TiO_3$ Thin Film Varactors", Accepted for publication, Appl. Phys. Letter, 2004
[32] E. Kollberg, J. Stake, and L. Dillner, "Heterostructure barrier varactors at submillimetre waves", Phil. Trans. R. Soc. Lond., A, vol. 354, pp. 2383–2398, 1996
[33] P. Rundqvist, A. Vorobiev, S. Gevorgian, "Large Signal Circuit Model of Parallel-Plate Ferroelectric Varactors", Proc. EuMC'2004
[34] S. Gevorgian, H. Berg, H. Jacobsson, and T. Lewin, "Basic Parameters of Coplanar-Strip Waveguides on Multilayer Dielectric/Semiconductor Substrates. Part 1: High Permittivity Superstrates", IEEE Microwave Magazine, pp. 60–70, June 2003
[35] S. S. Gevorgian, T. Martinsson, P. Linnér, E. Kollberg, "CAD Models for Multilayered Substrate Interdigital Capacitors", IEEE Trans. Microwave Theory Techn., vol. 44, No. 6, pp. 896–904, 1996
[36] S. Gevorgian, P. K. Petrov, Z. Ivanov, and E. Wikborg, "Tailoring the temperature coefficient of capacitance in ferroelectric varactors", Appl., Phys. Lett, Vol. 79, pp. 1861–1863, 2001
[37] N. A. Pertsev, Tagantsev, and N. Setter, "Phase transitions and strain induced ferroelectricity in $SrTiO_3$ epitaxial films", Phys. Rev. B, vol. 61, pp. R825–R829, 2000-II
[38] A. Kozyrev, A. Ivanov, T. Samiolova, O. Soldatenkov, and A. Astafiev, "Nonlinear Response and Power Handling Capability of Ferroelectric $Ba_xSr_{1-x}TiO_3$ Film Capacitors and Tuneable Microwave Devices", J. Appl. Phys., vol. 88, pp. 5334-53-42, 2000
[39] Y.-K. Yoon, D. Kim, M. G. Allen, J. S. Kenney, and A. T. Hunt, "A reduced Intermodulation Distortion Tuneable Ferroelectric Capacitor-Architechture and Demonstration", IEEE Trans. Microwave Theory Techn., vol. 51, pp. 2568–2576, 2003
[40] T. B. Samoilova, K. F. Astafev, T. Rivkin, and D. S. Ginley, "Frequency conversion in coplanar waveguide based on $Ba_xSr_{1-x}TiO_3$ film", J. Appl. Phys., vol. 90, pp. 5703–5707, 2001
[41] A. T. Findikoglu, Q. X. Jia, D. W. Roger, and X. D. Wu, Electron. Lett., vol. 31, pp. 1814–1816, 1995
[42] Tunable Filters, Pole/Zero Corp
[43] Oats D. E., and Dionne J. F., "Magnetically tuneable resonators and filters", IEEE trans. Appl. Supercond., vol. 9, pp. 4170–4175, 1999
[44] G. L. Matthaei, L. Young, and E. M. T. Jones, Microwave Filters, Impedance Matching Networks, and Coupling Structures, McGraw-Hill, N. Y., 1964
[45] A. Tombrak, J.-Paul Maria, T. Ayguavives, Z. Jin, G. T. Stauf, A. I. Kingon, and A. Mortazawi, "Voltage-Controlled RF Filters Employing Thin-Film Barium-Strontium-Titanate Tuneable Capacitors", IEEE Trans. Microwave Theory and Techn., vol. 51, pp. 462–576, 2003
[46] B. H. Moeckly, and Y. Zhang, "Strontium Titanate Thin Films for Tuneable $YBa_2Cu_3O_7$ Falters", IEEE trans. Appl. Supercond., vol. 11, pp. 450–453, 2001
[47] A. Kozyrev, O. Basov, A. Gagarin, A. Ivanov, D. Kosmin, S. Krasilnikov, A. Pavlov, and A. Zemtsov, "Tuneable ferrelectric Microwave Devices", IMS'2004 Workshop on New Technologies for Frequency- or Phase-Agile Microwave Circuits and Systems
[48] F. Miranda, C. H. Mueller, F. W. Van Keuls, G. Subramanyam, and S. Vignesparamoorthy, "Performance enhancement of Tuneable Bandpass Filters using Ferroelectric Thin Films", Integrated ferroelectrics, vol. 50, pp. 121–131, 2002

[49] I. Vendik, O. Vendik, V. Pleskachev, a. Svischev, and R. Wördenweber, "Design of Tuneable Ferroelectric Filters with a Constant Fractional Bandwidth", IEEE Int. Microwave Symposium, pp. 1461–1464, 2001

[50] Jung Y. J., Lee S. W. Shim D., Kim W., Kim C., and Cho S. I., "A dual loop delay locked loop using multiple voltage controlled delay lines", IEEE J. Solid State Circuits, vol. 35, pp. 784–791, 2001

[51] Koo R. C., and Long J. R., "An inductively tuned quadrature oscillator with extended frequency control range", Proc. 2003 IEEE Int. Symposium on Circuits and Systems, vol.1, pp. 709–712

[52] Seo C., "Novel phase shift line in feedforward circuit using photonic bandgap", Microwave and Optical technology lett., vol. 38, pp. 357–359, 2003

[53] Rubin B. J., and Singh B., "Study of meander line delay in circuit boards", IEEE Trans. Microwave Theory. Techn., vol. 48, pp. 1452–1460, 2000

[54] Dudek P., Szczepanski S., and Hatfield J. V., "A high resolution CMOS time-to-digital converter utilizing a vernier delay line", IEEE J. Solid Stae Circ., vol. 35 (2), pp. 240–247, 2000

[55] DeGroot D. C., Beall A. J., Marks R. B., and Rudman D. A., "Tuneable Microwave properties of YBCO/STO thin film transmission lines", IEEE Appl. Supercond., Vol. 5, pp. 2272–2275, 1995

[56] D. Kuylenstierna, A. Vorobiev, P. Linnér, and S. Gevorgian, "Ferroelectrically Tuneable Delay-lines", Proc. EuMC 2004

[57] Sherman V., Astafiev K., Setter N., Vendik O., Vendik I., Hoffman-Eifert S., Böttger U., and Waser R., "Digital reflection type phase shifter based on a ferroelectric planar capacitor", IEEE Micr. Wireless Comp. Letters, vol. 11, pp. 407–409, 2001

[58] Kozyrev A., Ivanov A., Keis V., Khazov M., Osadchy V., Samoilova T., Soldatenkov O., Pavlov A., Koepf G., Mueller C., Galt D., and Rivkin T., "Ferroelectric films: Nonlinear properties and applications in microwave devices", Dig. IEEE MTT-S'1998, pp. 985–988

[59] J. Serraiocco, B. Acikel, P. Hansen, T. Taylor, H. Xu, J. S. Speck, and R. A.York, "Tuneable Passive Integrated circuits Using BST Thin Films", Integrated Ferroelectrics, vol. 49, pp. 161–170, 2002

[60] A. Deleniv, S. Abadei, and S. Gevorgian, "Tuneable Ferroelectric Filter-Phase Shifter", Dig., IEEE Int. Microwave Symposium, Vol. 2, pp. 2003

[61] D. Kim, Y. Choi, M. Ahn, M. G. Allen, J. S. Kenney, and P. Marry, "2.4 GHz Continuously Variable Ferroelectric Phase Shifters Using All-Pass Networks", IEEE Microwave and Wireless Comp. Letters., vol. 13, pp. 434–436, 2003

[62] B. Romanofsky, "Broadband, Low-Loss K- and Ka Band Phase Sifters based on Thin Ferroelectric Films", IMS'2004 Workshop on New Technologies for Frequency- or Phase-Agile Microwave Circuits and Systems

[63] F. A. Miranda, G. Subramanyam, F. W. Van Keus, R. Romanofsky, J. D. Warner, and C. H. Mueller, "Design and Development of Ferroelectric Tuneable Microwave Components for Ku- and K-Band Satellite Communication Systems", IEEE Trans. Microwave Theory Techn., vol. 48, pp. 1181–1189, 2000

[64] R. Romanofsky, J. Bernhard, F. Van Keuls, F. Miranda, G. Washington, and C. Caney, "K-band Phased Array Antennas Based on Ba0.6Sr0.4TiO$_3$ Thin Film Phase Shifters", IEEE Trans. Microwave Theory Techn., vol. 48, pp. 2504–2510, 2000

[65] http://www.globalspec.com/ParatekMicrowave

# Chapter 8

# MEMS for Optical Functionality

S.-G. Kim,[1,3] G. Barbastathis[1,3] and H. L. Tuller[2,3]

[1]*Department of Mechanical Engineering*
[2]*Department of Materials Science and Engineering*
[3]*Microsystems Technology Laboratories*
*Massachusetts Institute of Technology, Cambridge, MA 02139, USA*

We discuss key features of MEMS technology enabling new functionality in microphotonic devices, including "arrayability", i.e. the ability to make massively parallel optical devices in a small form factor, "reconfigurability," the ability to change optical properties spatially and temporally, and "nano positioning," the ability to position micro-scale devices with nanometer accuracy. Our goal, in this chapter, is to examine a few select optical MEMS devices, emphasizing the emergence of these functionalities and their impact on optical product viability. In section 2, we begin with a historical introduction, tracing the origins of today's optical MEMS devices to solid state research performed in the 70's and 80's, followed by classification of MEMS-based photonic devices in section 3. In section 4, we describe in some detail select MEMS products, and argue why the three functional requirements of arrayability, reconfigurability and nano positioning result in unique system's properties. We conclude, in section 5, by discussing future promises and challenges facing MEMS-based optical devices.

# 1. Introduction

The field of optical micro-electro-mechanical systems (optical MEMS or MOEMS) has experienced rapid growth in the recent decade, especially in the late part of the 90's. The primary application domains, ordered by approximate precedence in embracing MEMS technology, have been: projection displays, components for optical fiber communications (sources, switches, cross-connects, routers, etc.) and optical sensing and imaging. The strong—albeit unsteady—market demand, coupled with challenging intellectual problems sparked extensive research and development activities in both academia and industry. In tune with these efforts were major funding initiatives, first by governments, and subsequently by venture capital and other private investors. Many novel designs, devices, and systems resulted, and some products are in wide commercial use today, especially in the projection display business.

One aspect of MEMS usage in optical systems is miniaturization. This is important in many cases from the point of view of footprint and materials costs. However, miniaturization alone could not have been sufficient to make optical MEMS technologies commercially successful in most cases. The key success factors of many optical MEMS products have been the new optical functionalities enabled by the use of MEMS. Among many benefits of MEMS technology that were reviewed as core success factors, the ones that stand out are:

(1) "*arrayability*," which is the term that we use for the ability to place tiny devices in large-scale arrays;
(2) "*reconfigurability*," *i.e.* the ability to reconfigure the optical properties, spatially and temporally, via localized micro-actuation and/ or deformation; and
(3) nanoscale precision control of position and alignment for microscale devices, which we will refer to as "*nano positioning*."

We present an overview of cases where a combination of these features has led to commercial successes by creating new optical functionalities, and discuss materials-related challenges and future trends for optical MEMS research and commercialization.

# 2. Historical Background

The first demonstration of a miniature (~2mm × 2mm) torsional mirror was by Kurt Petersen, while at IBM [1]. Quoting from this paper, Peterson predicts that "... silicon micromechanics may eventually find a practical implementation for displays (especially if silicon-driving circuitry can be integrated on the same chip, matrix addressing the two-dimensional array of mirrors—all electronically)." This statement proved to be truly prophetic, given the current success of MEMS display products such as Texas Instruments' Digital Micromirror Device (DMD) [2]. It can also be seen that micro-mirror array based displays are one of the oldest threads in optical MEMS technology [2–5]. The trend led to numerous other applications for micro-mirror arrays, as we will see later in this paper.

The key idea of a DMD based display is to provide an array of bistable micro-mirrors to intercept the light path so that one of the mirror positions directs the light onto a screen, whereas the other position directs the light onto a block where it is absorbed. Therefore, each mirror modulates a different pixel. By dithering the mirror, gray-scaling can be accomplished as well. Color displays can be obtained by standard techniques such as a color wheel (most compact and cheapest, and therefore most common) or three mirror arrays (for red-green-blue) combined with dichroic optics and projected onto the same screen. This technology is discussed in more detail in section 4.1.

An alternative MEMS display design was proposed in 1994 by a Stanford group, led by David M. Bloom [4]. The modulation concept, as opposed to micro-mirrors, was to modulate the light via diffraction from miniature gratings, one per pixel on the screen. This concept was termed the "grating light valve" (GLV) by its inventors. The structure and micro-mechanics of the GLV are discussed in section 2.2. In the most common GLV implementation, the light is sent through a *Schlieren* system which blocks the $0^{th}$ diffraction order and lets the $\pm 1^{st}$ diffraction orders through. So, when the grating is inactive there is no light transmitted to the screen, whereas upon actuation of the grating the corresponding pixel is illuminated by the interference pattern of the $+1^{st}$ and $-1^{st}$ orders. The key difference between the GLV and micro-mirror based displays is that the GLV actuators are long ($\sim$0.5mm) thin ($\sim$few μm) beams alternating between two vertical positions and, hence, can be actuated faster by orders of magnitude (10s of MHz) than the bulkier torsional micro-mirrors (typically kHz). Therefore, typically the GLV needs not be a full 2D pixel matrix, but is instead implemented as a single row, rapidly scanned in the vertical direction to produce the full display. For the same reason, the GLV row can have many more pixels than the linear dimension of a micro-mirror based display of the same generation. However, there is a caveat: since the modulation of light is based on diffraction, the efficiency is reduced, and high-power laser sources are required to obtain sufficient brightness (whereas micro-mirror based displays work with a simple light bulb.) Therefore, the GLV is naturally targeted towards high-end projection markets that can afford the price and space taken by the bulky laser sources in return for excellent resolution of thousands of pixels per dimension over a large projection area, e.g. digital cinema. On the other hand, micro-mirror displays have been successful at the lower end mass market of computer and video projectors, where XGA resolution suffices and price/footprint/weight are critical factors.

The comparison of DMDs and GLVs is interesting because it points out how the intricate relationship between optical principle (reflection *vs* diffraction) and moving parts (tilting rectangular plates *vs* thin beams moving vertically) can influence seemingly extraneous factors (light source selection, resolution, footprint, price) and radically change the outlook of an optical MEMS product. Interestingly, it is now public information that Texas Instruments considered alternative markets for its DMD product, such as laser printers, before discarding them in favor of the display market [5].

A natural extension of the micro-mirrors theme was what is now known as the "silicon micro-optical bench," where optical arrangements were miniaturized to the surface of a silicon chip, whereas with standard optics they would normally take several square feet on an optical table [6]. Several ingenious concepts were demonstrated for dealing with light at these small scales, including pop-up micro-mirrors, diffractive micro-lenses (Fresnel zone plates) and on-chip sources [7]. Such optical systems have great potential in a number of applications such as optical processing [8] and sensing [9]. However, the major application that emerged from these efforts was optical networking, because it was in that area that the need for miniaturization was the most pressing. Primarily, the technology driver was the desire to reduce the size of a router for a dense wavelength division multiplexing (DWDM) node from that of a large refrigerator to a briefcase, figuratively speaking. That need was important not only because such nodes were anticipated to be placed at central spots in metropolitan areas, where real estate is expensive, but also because the capacity demand was projected to increase at exponential rates.

## 3. Classification of Optical MEMS Devices

Walker and Nagel [10] studied and reviewed MEMS technologies for optical applications and classified them depending on the types of micro-optics used and optical functions performed, such as: sources, detectors, free space optics, wave guided optics, transmissive optics, reflective optics, diffractive optics and interferometric optics as shown in Table 1.

Micromirrors have been used to steer the optical beam path through the waveguide, in and out of optical fibers and to redirect light out of coaxial paths. Many of the reflective mode MEMS devices have been developed on bare silicon [11], metallic surfaces, and differential index multilayers. The VCSEL (vertical cavity surface emitting laser) is a typical MEMS device built on differential-index multilayers to achieve high reflectivity [12]. The mirror array has been the most successful MEMS application in optics and has been used for projection displays [2–5], free space optical switching [13] among many others. Diffractive optics has enabled planar microlenses, Fresnel zoneplate lenses and gratings. Efforts have been made to make them tunable in digital [14] or analog manners

**Table 1.** Key optical devices developed based on MEMS technology

| | Devices | Products |
|---|---|---|
| Reflective Mode Devices | Micro-mirror array for information displays, Phase correction piston arrays | Texas Instrument's DMD, Daewoo's TMA |
| Diffractive Mode Optics | Fresnel zone plate<br>Tunable Gratings | Silicon Light Machine's GLV |
| Free space Optics | 3-D optical switches<br>2-D optical switches | Optical Micro Machines, Lucent Lambda router, Xros cross-connect |
| Waveguide optics | Planar waveguide<br>Fiber inline alignment | |
| Interference Mode Devices | Tunable Fabry-Perot external cavity lasers | |
| Detectors | Bolometer<br>micro-spectrometer | Honeywell-bolometer array<br>Polychromator |
| Transmission Devices | Optical micro shutters and choppers | |
| Adaptive optics | Deformable mirrors for aberration correction | |
| Sources | VCSEL External Cavity Laser | |

[15]. In the aforementioned devices, light propagates in free space before reaching the actuator, which deflects or diffracts it. An alternative for handling light is to confine it in a waveguide. It then becomes impossible to interact with it directly; however, one can still modify light propagation by bringing a MEMS actuator close enough to the waveguide that light from the waveguide couples evanescently with the actuator [16, 17, 23]. Typically, evanescent coupling results in phase delay, which in turn can be used for interferometric switching, etc.

Optical MEMS research has demonstrated a large number of new devices and systems. Some of the optical MEMS devices have become great commercial successes, while many have not. Miniaturization alone is inadequate to create new markets and/or to become more competitive. The most successful optical MEMS devices commercially have been those, which combine at least two of the three functional requirements among "arrayability," "reconfigurability" and "nano positioning." Some of these successful optical MEMS cases are listed in Table 1.

## 4. MEMS Products: Case Studies

The study of optical MEMS technology reveals that the same basic optical MEMS technology has often been used for radically different applications, e.g. micro-mirrors for displays versus cross-connects in telecommunications systems. The reason this broad sharing has been possible is that the applications share at least some of the functional requirements mentioned earlier, namely arrayability, reconfigurability, and nano-positioning. In this section, we overview some optical MEMS products, which emphasize the significance of these attributes for commercial success.

### 4.1. Digital Micromirror Devices (DMD) for Digital Light Processing (DLP)

In the digital information era, more multimedia applications require brighter and larger-screen displays as well as compact and personalized displays. Conventional film projectors cannot display digital multimedia information. Many electronic projection display technologies have been developed since the first CRT projection display was demonstrated by RCA in the 1940's. At present, there are two major projection display technologies competing in the rapidly growing market: DLP (Digital Light Processing) and high-temperature Poly LCD projection. DLP is very strong in the expensive large venue projector market and low-cost ultra portable projector market, while LCD projection is strong in the middle arena.

The large venue projectors can project more than 5,000 ANSI lumens on to a screen and are very expensive, heavy and bulky, and need special care for installation and maintenance. 3-chip DLP technology is dominant in this category given its ability to deliver up to 10,000 lumens of light at much lower cost than existing CRT projectors or ILA (Illuminated Light Amplifier) projectors from Hughes-JVC. The heat generation by light absorption and photo-degradation still limits the use of high-power light sources for LCD. The biggest market segment is the so-called portable projector group and about 95% of projectors in the world market fall within this category. Currently, the mainstream projection technology for this category is the one-chip DLP. The one-chip

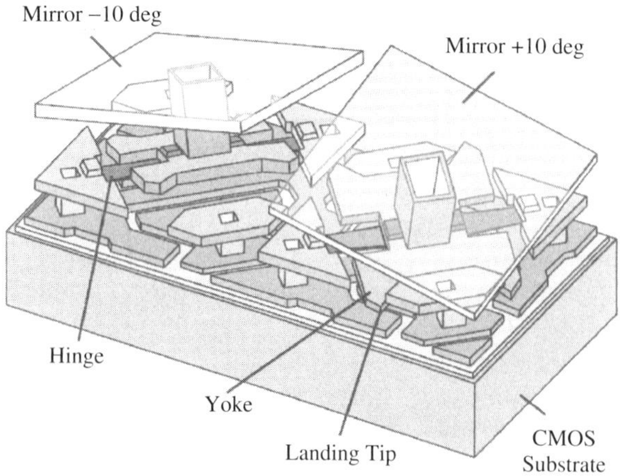


**Fig. 1.** Two DMD pixels (mirrors are shown as transparent.) in opposite tilting states [2].

DLP engines tend to be more compact than 3-panel LCD engines, and dominate the market with the light weight mobile projectors under 3-pounds.

At the core of a digital light-processing projector is Texas Instrument's patented Digital Micromirror device (DMD) [2], a MEMS chip with an array of micromirrors mounted on its SRAM surface. Therefore, *arrayability* is the key optical functionality offered by the DMD device. Each of these tiny mirrors tilts in response to electrostatic charges on the mirror's mounting substrate. Depending on the degree of tilt (+10°, 0°, −10°), individual mirrors reflect on/off control of light from the projection lamp. Controlling the duty ratio of on/off stages, 10 bit (1024) grayscales can be created digitally for each primary color. The color image is then accomplished with either rotating color wheels with one DMD chip or dichroic mirrors and combining prisms with three DMD chips for red, green and blue. Incoming video or graphic signals are turned into a digital code of binary data (0s and 1s) that tilts the DMD mirrors accordingly. By throwing the modulated light through the projection lens with precise digital signal processing, a large screen full color image is projected on to the screen.

The DMD chip has micromirrors on top of a completed CMOS memory circuit as shown in Fig. 1. The DMD chip is monolithically fabricated by CMOS compatible processes. Each micromirror switch has an aluminum mirror, sized 16 μm by 16 μm. The mirror is tilted along the hinge in diagonal direction, through electrostatic attraction produced at the air gap under the mirror and on top of the CMOS. The key technical concern on the DMD chip has been the issue of mechanical reliability. Due to the high speed PWM (pulse width modulation), the torsional hinge requires reliability of more than $10^{13}$ cycles through its operating life. Hinge failure and hinge memory problems have been solved by continuous design improvements over the last ten years.

### 4.2. The Thin film Micromirror Array (TMA)

The Thin-film Micromirror Array (TMA) is a reflective type spatial light modulator fabricated with MEMS technology [3]. Micromachined thin-film piezoelectric actuators

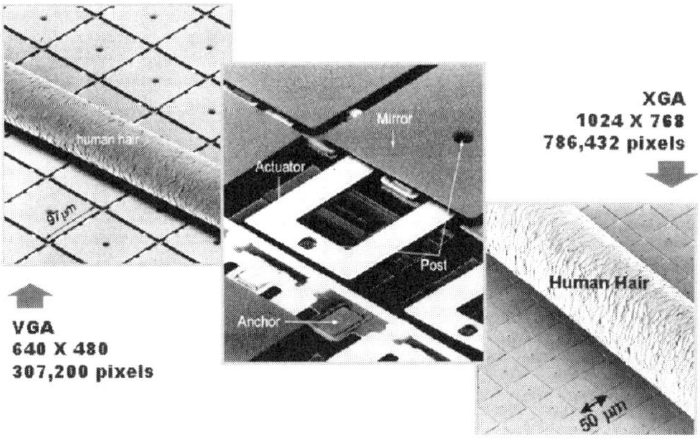


**Fig. 2.** Piezoelectric cantilever-actuated micromirrors for TMA display [3].

are used to control the tilt angle of each micromirror, which defines the gray scale of the matching screen pixel. The light reflected on a mirror surface forms a rectangular shaped image on the projection stop plane and this image moves along the horizontal axis as the tilt angle changes. When a mirror does not tilt, all of the light reflected by the mirror is blocked and absorbed by the projection stop and the pixel image on the screen is in the darkest state (black). When the mirror is fully tilted, all of the light reflected by the mirror goes out through the projection stop and the pixel image is in the brightest state (white). The amount of light that passes through the projection stop is linearly proportional to the tilt angle of each mirror. The precise control of the tilt angle can generate gray levels on the screen in between the brightest and the darkest state. Therefore, the TMA shares the arrayability functional requirement with the DMD device, with the additional functional requirement of nano positioning of the miniature mirrors to control grayscale. The application of piezoelectric actuation to array MEMS technology was the key factor for achieving these functional requirements.

Each pixel consists of a dual layered structure, a mirror layer and an actuator layer. Mirror-tilting is performed by thin-film piezoelectric actuators in the form of micro-cantilevers. As shown in Fig. 2, a mirror is connected to the underlying cantilevers through a support post. The cantilevers are anchored to the underlying PMOS substrate. A cantilever consists of the non-stoichiometric silicon nitride ($SiN_x$) supporting layer, Pt bottom electrode, PZT layer, and top Pt electrode. When an electric field is applied between the two electrodes, the piezoelectric layer shrinks in the horizontal direction and expands in the vertical direction. Since the neutral plane of the cantilever shifts toward the bottom electrode due to the thickness of the supporting layer, the mechanical contraction of the piezoelectric layer causes an upward vertical deflection of the cantilever and, consequently, the tilting of the mirror on top of it.

The TMA chip is monolithically fabricated over a PMOS active matrix by surface micromachining techniques. The active matrix is a transistor array that addresses the video signal to each pixel. The size of each mirror is 49 μm × 49 μm for XGA format TMA chips. The mirror layer is made of aluminum and is sputter-deposited on the second sacrificial layer. The aluminum layer is patterned to make mirror shapes by a dry etching process. The sacrificial layers are removed to form the air gaps. Since two different materials are used as sacrificial layers, the release process is performed in two steps. First, the second sacrificial layer is removed through the openings between mirrors by plasma etching. After completely removing the second sacrificial layer, the first sacrificial layer is exposed to air and removed by $XeF_2$ vapor etch process. Both release methods show fast lateral etching rates, leave no residues, do not etch or damage the other layers such as Al and PZT. Figure 2 presents the SEM photographs of the completed VGA and XGA format TMA chips after the release process. The 800,000 finished mirrors in an XGA format TMA chip have initial tilting positions of $0^\circ \pm 0.03^\circ$. A working projector prototype of 5,400 true ANSI lumen was presented at the Asian Display 1998 with three TMA chips and a 1 kW Xenon lamp. The prototype showed overall light transmitting efficiency of 22%.

TMA's PZT microcantilever actuators have been used not only for display, but also for fast-scanning AFM tip arrays [18] and strain-tuning of photonic devices [19].

### 4.3. The Grating Light Valve (GLV)

The GLV is a unique MEMS product that acts as a dynamic, tunable grating to precisely vary the amount of laser light that is diffracted or reflected [4, 14]. The Grating Light Valve is comprised of a series of ribbons on the surface of a silicon chip as shown in Fig. 3, which can be moved up or down over very small distances by controlling the magnitude of electrostatic forces between the ribbons and the substrate. The ribbons are arranged such that each element is capable of either reflecting or diffracting light. This allows an array of elements, when appropriately addressed by control signals, to vary the level of light reflected off the surface of the chip. This control of light can be analog (variable control of light level) or digital (switching light on or off). Because GLV devices utilize the principle of diffraction to switch, attenuate and modulate light, they are highly accurate, simple to manufacture, and have relatively high power handling capabilities. However, because the projected light is obtained from diffraction orders rather than direct reflection, some loss of incident power is inevitably incurred.

Each GLV element consists of six dual-supported parallel ribbons formed on silicon nitride and coated with a reflective top metal layer. The top is used also as an electrical conductor to create electrostatic attraction to the bottom common electrode plane, as

**Fig. 3.** Grating Light Valve ribbons to form a digitally tunable diffraction grating [4].

shown in the figure. The on/off switching speed of GLV is as fast as 20 nanoseconds, which is much faster than that of existing tilting mirrors. Current GLV devices are claimed to have a diffraction efficiency near the theoretical maximum of 81%, fill factor of 95% and top layer reflectivity of 91%, with an overall device efficiency of about 70%. This optical efficiency corresponds to an insertion loss of about 1.5 dB.

The large screen projection display has been the primary target application of GLV technology. It has been demonstrated that a front projection system for a 1,920 × 1,080 HDTV image can be made with the scanning of a linear GLV array of 1080 pixels. However, due to the keen competition between DLP and LCD and the consequent price war in the consumer market, the need of the additional cost of scanning mechanism and the speckled image resulting from the laser beam, the commercialization of GLV projection display has been significantly delayed at the present time. However, the digital tunability of GLV has successfully introduced many reconfigurable optical devices, such as reconfigurable blocking filters, dynamic gain equalizers and GLV printers.

### 4.4. The Polychromator

The Polychromator is an electronically programmable, dark-field correlation spectrometer system based on a programmable MEMS diffraction grating developed in a joint effort of Honeywell, Massachusetts Institute of Technology, and Sandia National Laboratory [20]. Rather than serving as means for projecting images, the Polychromator is a newly developed sensor capable of distinguishing between a number of different gaseous species remotely by utilizing a combination of optics and MEMS. This device provides sufficient sensitivity and selectivity to detect very small amounts of gaseous species for security and biological imaging applications.

The Polychromator chip does not need a reference cell, which many conventional gas analysis techniques require. Instead, the reference spectrum for each correlection measurement is formed by modulating the Polychromator grating. The latter consists of thousands of mechanical beam elements on a silicon wafer. The beams and actuators are manufactured with standard thin film process techniques. Each grating element is 10 microns wide and one centimeter long and is designed to move up and down as shown in Fig. 4. Light from the environment (e.g., a suspicious "gas cloud") is directed onto the chip after being collected by an optical relay, such as a telescope or binoculars. The grating is then programmed to "match" the incoming spectrum with a reference spectrum. One thereby determines whether the gas cloud contains the substance that the grating was programmed to match or not. Evidently, the key functional requirement in this device is reconfigurability. The Polychromator replaces a bank of gas tanks containing the reference chemicals, which would otherwise have to be used to implement correlation spectroscopy.

### 4.5. Uncooled IR Bolometers

Infrared sensors allow objects to be identified and imaged according to their body temperatures. The ability to image by temperature allows for remote thermometry and the ability to see through darkness, smoke and inclement weather. Until recently, high performance cameras have required expensive cooling to cryogenic temperatures

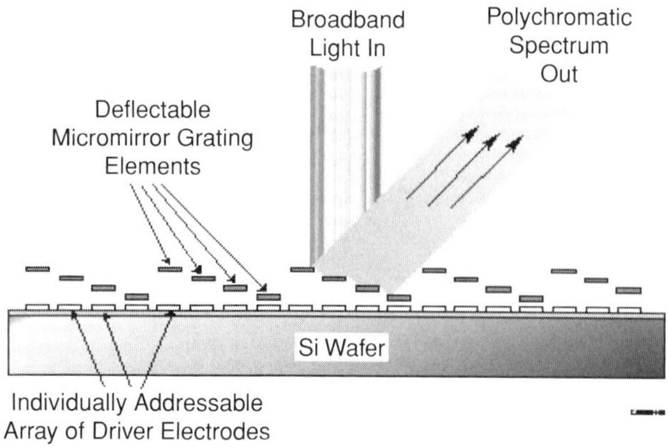


**Fig. 4.** Polychromator developed by MIT, Sandia and Honeywell [20].

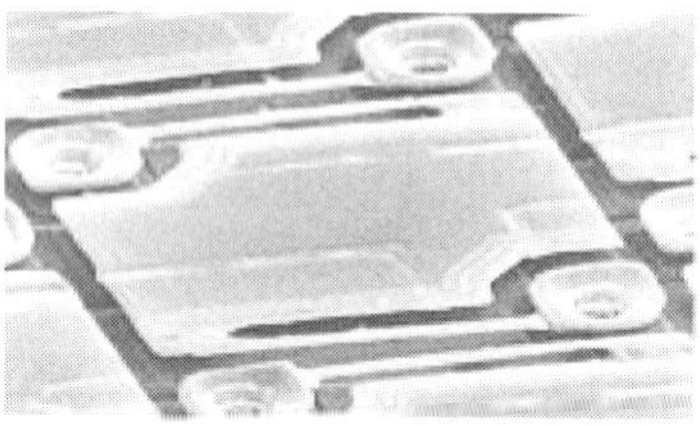

**Fig. 5.** Uncooled bolometer pixel (coplanar design by Honeywell) [21].

to detect photons and to suppress thermal noise. More recently, detectors operating at room temperature have been developed using micromachined bolometers and pyroelectric detectors. Uncooled infrared FPA (focal plane array) structures can image infrared scenes generated by humans and objects by transforming the induced temperature changes in the detectors into electrical signals. Uncooled bolometers utilize the temperature-induced changes in electrical resistance, polarization, and dielectric properties of the detector materials.

Detectors on uncooled FPAs need high thermal isolation from the substrate to achieve high sensitivity. MEMS technology plays a key role in enabling the fabrication of very low thermal mass detectors with exceptionally high thermal isolation. Wood and colleagues [21] fabricated a 2-dimensional array of resistive bolometers with the thermal isolation structure as shown in Fig. 5. The silicon nitride membrane, supported by two tethers, floats above the silicon substrate. The size of the tether was adjusted to provide a thermal conductance of $8 \times 10^{-8}$ W/°K. The thermal mass of the plate was $8 \times 10^{-10}$ J/°K and the thermal response time was 10 msec. Operating at 30 frames per second with f/1 optics, the measured NETD (noise equivalent temperature difference) was 0.04 °C with vanadium oxide as the resistive material. Honeywell has commercialized this design into products. Cole and colleagues developed a monolithic 2D array using the pyroelectric properties of lead titanate with a similar thermal isolation scheme as the one developed by Honeywell. The predicted NETD of this device was 0.01 °C and has been adopted for many military applications. A tri-level design was fabricated

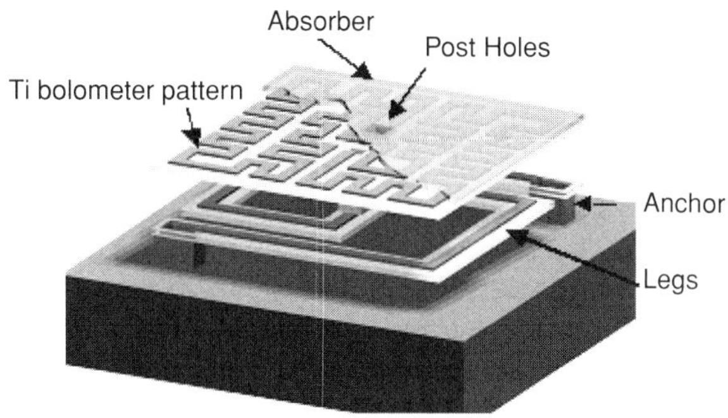


**Fig. 6.** Tri-level uncooled bolometer structure by Kim, et al. [22].

by Kim and colleagues [22] as shown in Fig. 6. The key objectives of this design were to increase the fill factor and to minimize the thermal conductance. Kim et al achieved a fill factor of over 92% compared to a fill factor of 62% for Honeywell's coplanar design for a similar thermal conductance.

## 4.6. MEMS for integrated optics

The optical MEMS examples given so far can be classified as "free space optics," since the moving structures (mirrors, grating beams, etc.) interact with light in its free space propagation (radiation) mode. By contrast, light in integrated optics is confined to optical waveguides laid out on top of a proper optical material, as shown in Fig. 7. In this case, it is clearly difficult to control light propagation by means of mirrors or diffraction gratings in direct contact with the light beam. However, an alternative means of controlling light via mechanical structures in integrated optics still exists, based on evanescent coupling between the guided light and dielectric structures such as bridges or cantilevers suspended over the waveguide [23]. These structures are basically phase modulators, and they work by altering the effective refractive index of the waveguide and, hence, the group velocity of light. The demonstration reported in reference [23]

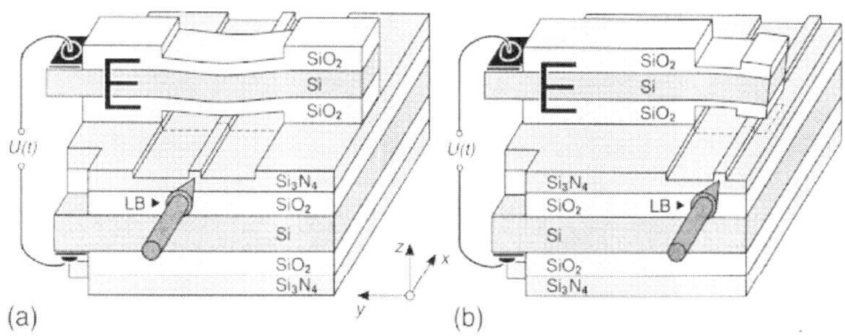


**Fig. 7.** MEMS for switching and tuning of integrated optics. (a) A bridge structure and (b) a cantilever suspended over an optical waveguide change the mode shape, resulting in phase shift of the guided light [23].

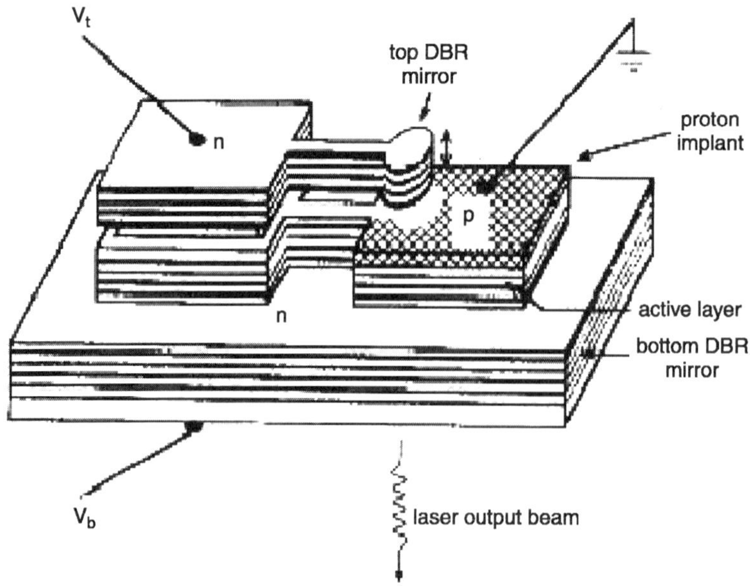


**Fig. 8.** Tunable VCSEL structure (after Ref. 24).

was a 2 × 2 optical switch of the Mach Zehnder interferometer type, where the phase delay was controlled by the MEMS device via the aforementioned phase modulation property, as opposed to the typical thermo-optic effect.

### 4.7. Tunable Fabry-Perot cavities

The "nano positioning" functionality is critical for applications such as tunable laser cavities, where the cavity length specifies the lasing wavelength. This requirement led to a successful marriage between the positioning capabilities of MEMS and the Vertical Cavity Surface Emitting Laser (VCSEL) technology [24]. VCSELs had been invented earlier [25–27] as a means to get solid state lasers to emit light in the vertical direction with respect to the substrate, which is preferable for numerous applications in illumination, optical interconnects, etc. In the original design, the top reflector of the VCSEL device was a distributed Bragg reflector (DBR) structure. In the MEMS implementation [24], the reflector was attached to a movable cantilever structure which was capable of tuning the resonance wavelength by as much as 15 nm with a low actuation voltage, approximately 5.7 V (Fig. 8). Since these first attempts, tunability has improved to cover the entire 1530–1620 nm range of dense wavelength division multiplexing (DWDM) communications [25].

### 4.8. Strain-Tunable Optical Devices

Tunable gratings such as the GLV and the Polychromator can be described as "digital," since actuation results in change of the grating period and modulation profile by

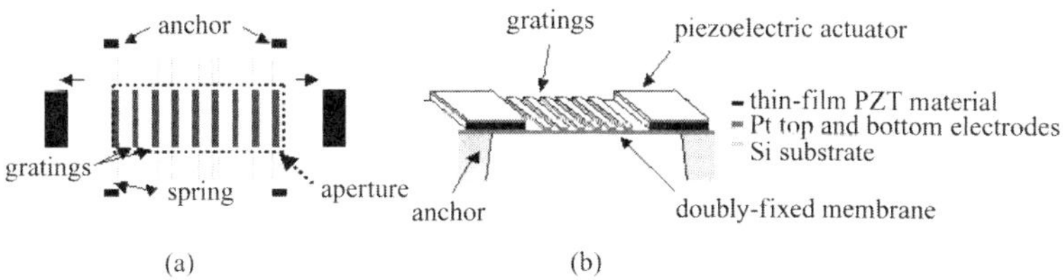


**Fig. 9.** Analog-tunable Gratings (a) Electrostatic comb drive actuated grating with flexured ends [15], (b) Piezoelectric thin film (PZT) actuated grating on a solid membrane [19].

discrete amounts, equal to the size of the micromachined ribbons. Therefore, reconfigurability is limited by the smallest feature size that the available lithography can deliver in a structurally sound configuration. To cover the range of reconfigurability at ultra small scales, a new class of "analog" tunable gratings has been developed [15, 19]. Nano-positioning is an additional functionality offered by these devices, necessary for reconfiguration to be as accurate as necessary for optical applications. However, another aspect of reconfigurability is sacrificed: e.g., these analog devices cannot perform the polychromator function in a correlation spectrometer. Analog tuning of a diffractive grating is achieved by transverse actuation of the grating structure using thin-film piezoelectric actuators [15, 19] or electrostatic comb drives with flexured gratings [15], as illustrated in Fig. 9.

To achieve transverse actuation, the grating grooves are defined on a floated deformable membrane. Then, the membrane is mechanically stretched by thin-film piezoelectric actuators. The piezoelectric actuators, a lead zirconate titanate (PZT) layer sandwiched between top and bottom electrodes located at both ends of the membrane, are capable of producing sufficient force to stretch the membrane up to 0.3% strain, or equivalently, 0.3% diffracted angle change. This is accomplished through the application of an electric field across the PZT film to result in its shrinkage, due to reorientation of crystallographic domains. This shrinkage on the PZT film correspondingly strains the membrane in the transverse direction. An alternative method of transverse tuning of the grating period involves using electrostatic comb-drives on a more compliant grating structure [15]. Based on the fine voltage control previously demonstrated on a thin-film piezoelectric material, the resolution of the grating period change can be better than 0.5 Å (corresponding to angular resolution better than 2.1 μradians), although the current metrology limits the observable resolution to the order of 1 nm. Out-of-plane membrane tilt and bending, asymmetrical rotation of the membrane during actuation, thermal disturbances, modal vibrations, and optical detection techniques may limit the lowest achievable resolution. The demonstration of the tunable diffractive grating shows tuning of the first diffracted order angle up to 486 μradians at 10 V, with minimum observable grating period displacements at approximately 0.6 nm at 1 V.

Reconfigurability combined with nano positioning leads to some interesting application capabilities for the tunable grating devices. For example, using ultra-precise tunable gratings one can compensate for thermal detuning in wavelength-sensitive optical networking elements such as wavelength multiplexers/demultiplexers and routers. Another intriguing possibility is the concept of "optical diversity," which we have developed to compensate for the aperture limitation in grating-based spectrometers. This is described in more detail in the next paragraph. It should also be noted that arrayability is possible for the grating devices, especially the piezoelectric version where the actuators occupy smaller real estate surrounding the active grating area, compared to the electrostatic version.

Optical diversity is motivated by a challenge that miniaturization often poses to optical devices: the limited optical aperture. Briefly, the problem is posed as follows: the spectral resolution in a grating spectrometer is determined by the number of periods $N$ that the grating is composed of. Since the available area in a miniature spectrometer is, by definition, small, the acquired spectra can get "blurred." In other words, the measured value of spectral density at a given wavelength returned by the spectrometer does not correspond to the true spectral density at that wavelength of the specimen that is being tested (e.g. a potentially dangerous chemical) but also contains contributions from other neighboring wavelengths. The mathematical operation describing the blurring of the measured spectrum is a *convolution* of the true spectrum with the impulse response of the grating. The impulse response turns out to depend on $N$, as well as the quality of the collector optics that relay light from the specimen to the grating. The effect of the convolution can be undone by the inverse operation, namely a *deconvolution*, which can be performed digitally by post-processing the measured spectra. However, the deconvolution process is highly sensitive to noise that is unavoidably present in any measurement. It has been discovered that, by tuning the grating and repeating the spectral measurement, we can recover the blurred spectra with much less sensitivity to noise and, consequently, much higher accuracy than a single blurred measurement would allow.

Optical diversity works as follows: the grating period is first set to an initial value and a blurred measurement is taken. Then the grating is actuated so that the period changes by a small amount, and a new measurement, also blurred, is taken. The process is repeated, depending on the amount of noise present in the measurements—the lower the signal-to-noise ratio, the larger the number of necessary measurements. The ensemble of blurred measurements thus collected is subsequently processed using a class of mathematical techniques called "regularized pseudo-inverse" [29]. After this step, the final de-blurred spectrum is acquired. The quality of the spectrum de-blurring process is determined by the noise present in the measurements and also by the precision and accuracy of grating actuation. Thus, optical diversity is a good example of how reconfigurability and nanopositioning help solve a significant optical problem stemming from miniaturization of an optical spectrometer system.

## 5. Materials Challenges in Optical MEMS Devices

Micromachining enables the fabrication of miniature movable structures ideally suited for applications in the optical domain. Small mechanical displacements, e.g. a quarter wavelength in an interferometer can produce on-off switching or create modulation effects stronger than in conventional electrooptic or free carrier activated devices. Likewise, arrays of micromirrors can serve as the basis of displays and of optical cross-connects as discussed above. At the same time, these miniature MOEMS structures must satisfy certain dimensional and structural constraints which may be less important for other MEMS applications. For example, surface roughness determines mirror reflectivity while stressed membranes distort reflected optical images. Waveguides integrated within MOEMS must exhibit reproducible indices of refraction and optical losses. Actuators, whether electrostatic, piezoelectric or thermal-mechanical, require low voltage operation, power dissipation and chip real estate, all the while operating reproducibly over many switching cycles. Other chapters in this book focus, in some detail, on MEMS processing issues. In the following, we mention aspects of particular importance for MOEMS devices.

A prime example of the special materials and fabrication requirements of MOEMS is the DMD projection display chip developed by Texas Instruments (TI). Instead of utilizing the conventional surface micromachining process utilizing phosphosilicate glass for the sacrificial layers and polysilicon for the micromechanical elements, TI developed a low temperature process utilizing conventional positive photoresist and sputter deposited aluminum alloys [30]. This enables fabrication of all the underlying electronic and micromechanical components beneath the micromirrors thereby imparting a high fill factor and small mirror gaps. To insure high reliability, special attention was directed towards minimizing stiction between contacting parts, hinge memory and sensitivity to particulates. For the former, a self-assembled monolayer lubricant is applied to lower adhesion forces. In addition, feedback electronics are utilized to monitor and control the movement of the mirrors to eliminate overshoot and oscillations to further protect against stiction [31]. A low creep Al alloy was developed to minimize shifts in micromirror angle with time. Packaging was assigned to class 10 rather than class 10,000 clean rooms to eliminate failures due to foreign particles, keeping in mind that these devices contain more than one million micromechanical elements and that the eye can readily detect even a few faulty pixels within the display.

Microcantilever beams, membranes and microbridges are often stressed, particularly when thin films are deposited onto these structures. The stresses may be of an intrinsic nature (e.g. lattice mismatch) or due to thermal-mechanical stresses resulting from differences in thermal expansion coefficients. This forces structures out of plane, often at unpredictable angles. Indeed, Chen et al. [32] proposed a $2 \times 2$ optical switch based on a polysilicon microcantilever beam with attached hinged micromirror which is normally out of plane due to a Cr-Au layer which places the polysilicon under compressive stress. The switch is actuated by applying a voltage between the cantilever and the substrate thereby bringing the mirror into the optical path length of the integrated optical fibers. The micromirrors used in telecommunication for signal routing are typically coated with gold for its excellent reflecting properties in the IR. However, since Au serves as a deep trap in silicon, great care must be taken to avoid poisoning the electronic devices integrated into the structure [33].

Silicon nitride is often used to fabricate optical waveguides by a variety of techniques including sputtering, CVD, PECVD and LPCVD. Both the general method used and the detailed processing conditions contribute to variations in index of refraction $n$, optical loss and the magnitude and nature of residual stresses. For example, utilizing PECVD, one finds $1.9 < n < 2.2$ and stresses range from tensile to compressive depending on the Si/N ratio and the deposition conditions [32]. Thus, both the choice of materials and the processing conditions must be carefully controlled to ensure reproducible and stable structures.

In addition to passive optical materials such as $Si_3N_4$, active materials such as electrooptic and magnetooptic materials are of interest for optical modulation, pyroelectric materials for IR sensing and fluorescent materials as light sources. Other active materials of interest include piezoelectric materials such as PZT. The majority of such materials are complex oxides, often with the perovskite or garnet structure and normally, in the bulk state, require very high temperatures to achieve desired phase purity and microstructure. To enable integration with silicon based microelectronic and/or MEMS chips, these materials are prepared as thin films either by vapor phase techniques (e.g. sputtering, PLD, MOCVD, MBE) or chemical routes (e.g. sol gel). In general, the degree of crystallinity and orientation, microstructure, stoichiometry and built in stress depend sensitively on the deposition conditions, type and temperature of substrate during deposition and subsequent annealing conditions. In general, the films show reduced activity compared to their bulk counterparts. For example, an electrooptic modulator prepared by deposition of $BaTiO_3$ films onto an MgO single crystal substrate exhibited an effective electrooptic coefficient of 50 pm/V [34], less than an order of magnitude lower than the maximum value measured for single crystals [35]. Similar effects are often noted for piezoelectrics with reduced effective piezoelectric coefficients. These are suspected to be related to constraints imposed by the substrate, and by defects and other interfaces. Buffer layers, between silicon and the active oxides are essential to prevent interdiffussion and oxidation and also serve as seed layers to aid in the nucleation of the appropriate phases and crystallographic orientations. An excellent review of many of these issues may be found in the article by Muralt [36].

## 6. Conclusions and Discussion

Optical functionality offered by MEMS (MOEMS) has expanded from light emission and detection to optical amplification, switching, spatial modulation and routing, as well as basic optical signal processing [37]. The performance of optical MEMS devices has been improved significantly together with the advancements in materials for optical MEMS devices as well as a systems approach to MEMS design. However, many of the optical MEMS devices have not become commercially successful, while a few have become great successes. Miniaturization alone is insufficient to create new markets and ensure commercially competitive. The collection of MEMS devices discussed in this article was not meant to be exhaustive but rather indicative of the possibilities that MEMS technology offers to optical systems via the novel functionalities of "arrayability," "reconfigurability," and "nano positioning" and that the most commercially successful optical MEMS devices have been those, which combine at least two of the three functionalities.

The future information technology requires faster and smarter communication networks, which may only be obtained through all optical networking with more sophisticated optical functionalities, such as intelligent optical signal processing and adaptive optics. Integration of optical devices with MEMS technology will accomplish this goal if they are designed with well-defined optical functionality and the proper use of materials and processes.

## References

[1] K. E. Petersen, "Silicon as a mechanical material," *Proc. IEEE* 70(5):420–457, 1982.

[2] P.F. Van Kessel, L. Hornbeck, R. E. Meier, and M. R. Douglass, "A MEMS-based Projection Display", *Proc. Of the IEEE*, Vol. 86, No. 8, 1998.

[3] S.G. Kim, K.H. Hwang, J. Hwang, M.K. Koo, and K.W, Lee, "Thin-film Micromirror Array (TMA)—A New Chip-based Display Device for the Large Screen Display", *Journal of the Society of Information Display*, Vo. 8, No. 2, 2000.

[4] R. B. Apte, F. S. A. Sandejas, W. C. Banyai, and D. M. Bloom, "Deformable grating light valves for high resolution displays," *Proc. Solid State Sensor and Actuator Workshop*, 1994.

[5] S. Horsley, "DMD: From R&D to a Profitable Business," oral presentation at the 1st International Symposium on Nanomanufacturing, Cambridge, MA, 2003; http://nanoman.mit.edu.

[6] M. C. Wu, L.-Y. Lin, S.-S. Lee, and K. S. J. Pister, "Micromachined free-space integrated micro-optics." *Sensors and Actuators A: Physical* 50(1–2):127–134, 1995.

[7] L. Y. Lin, S. S. Lee, K. S. J. Pister, and M. C. Wu, "Micro-machined three-dimensional micro-optics for integrated free-space optical system," *IEEE Photonics Tech. Lett.* 6(12):1445–1447, 1994.

[8] M. E. Motamedi, M. C. Wu, and K. S. J. Pister, "Micro-opto-electro-mechanical devices and on-chip optical processing," *Optical Eng.* 36(5):1282–1297, 1997.

[9] J. M. Kahn, R. H. Katz and K. S. J. Pister, "Mobile Networking for Smart Dust", ACM/IEEE Intl. Conf. on Mobile Computing and Networking, Seattle, WA, August 17–19, 1999.

[10] S. J. Walker and D. J. Nagel, Optics and MEMS, NRL report NRL/MR/6336-99-7975, May 15, 1999, http://code6330.nrl.navy.mil/6336/moems.htm.

[11] M. A. Chan, S. D. Collins and R. L. Smith, "A micromachined pressure sensor with fiber-optic interferometric readout," *Sensor and Actuators* A, vol, A43, pp. 196, 1994.

[12] E.C.Vail, G.S. Li, W. Yuen, et al, "High performance micromechanical tunable vertical cavity surface emitting lasers," *Electronic Letters*, vol. 32, pp. 1888, 1996.

[13] D.T. Neilson, et al., "Fully Provisional 112x112 Micro Mechanical Optical Crossconnect with 35.8 Tb/s Demonstarted capacity," Technical Digest of Optical Fiber Communication Conference 2000, PD-12, March 2000.

[14] R. Corrigan, R. Cook and O. Favotte, "Silicon Light Machines-Grating Light Valve Technology Brief", white paper, Silicon Light Machines company website, http://www.siliconlight.com/htmlpgs/glvtechframes/glvmainframeset.html.

[15] C. Wong, W. Shih, Y. Jeon, S. Desai, S. Kim and G. Barbastathis, "Analog tunable gratings with sub-nanometer resolution", *Proc. of Solid-State Sensor and Actuator Workshop*, Hilton Head, South Carolina, June, 2002.

[16] W. Lukosz, "Integrated optical nanomechanical devices as modulators, switches, and tunable frequency filters, and as acoustical sensors," presented at Integrated Optics and Microstructures, Boston, MA, 1993.

[17] W. Lukosz, "Integrated optical chemical and direct biochemical sensors", Sensors and Actuators B, 29, 1995, pp. 37–50.

[18] Y. K. Kim, J. M. Bae, S. Y. Son, J. H. Choi, and S. G. Kim, "High Speed Atomic Force Microscope Cantilevers with Built-in Piezoelectric Actuator", *Proc. of MOEMS '99*, Mainz, Germany, September 1999.

[19] Chee-wei Wong, Yongbae Jeon, G. Barbastathis and Sang-Gook Kim, "Analog tunable gratings driven by thin-film piezoelectric microelectro mechanical system actuators", *Applied Optics*, Vol. 42, No. 4, 2003.

[20] E.S. Hung and S. D. Senturia, "Extending Travel Range of Analog-tuned Electrostatic Actuators", *J. of Microelectromechanical Systems*, V. 8, N. 4, 1999.

[21] R.A. Wood, C.J. Han, and P.W. Kruse, "Integrated uncooled infrared detector imaging arrays," *Proc. IEEE Solid-State Sensor and Actuator Workshop*, 1992.

[22] H.K. Lee, J.B.Yoon, E. Yoon, S.B. Ju, Y.J. Yong, W. Lee and S.G. Kim, "A High Fill-Factor Infrared Bolometer using Micromachined Multilevel Electrothermal Structures", *IEEE Trans. on Electron Devices*, Vol. 46, No. 7, July 1999.

[23] R. Dangel and W. Lukosz, "Electro-nanomechanically actuated integrated-optical interferometric intensity modulators and 2 × 2 space switches," Opt. Commun. 156:63–76, 1998.

[24] M. S. Wu, E. C. Vail, G. S. Li, W. Yuen, and C. J. Chang-Hasnain, "Tunable micromachined vertical cavity surface emitting laser," *Electron. Lett.* 31(19):1671–1672, 1995.

[25] S. Kinoshita, K. Morito, F. Koyama, and K. Iga, "Reproducible fabrication of AlAs/GaAs circular buried heterostructure (CBH) surface-emitting lasers with low threshold," *Electron. Lett.* 24(11):699–700, 1988.

[26] J. L. Jewell, A. Scherer, S. L. McCall, Y.-H. Lee, S. Walker, J. P. Harbison, and L. T. Florez, "Low-threshold electrically pumped vertical-cavity surface-emitting microlasers," *Electron. Lett.* 25(17):1123–1124, 1989.

[27] C. J. Chang-Hasnain, "1.55- μm tunable VCSEL for metro-WDM applications," *Proc. SPIE* 4580: 40–45, 2001.

[28] W.-C. Shih, C. Hidrovo, S.-G. Kim, and G. Barbastathis, "Optical diversity by nanoscale actuation," IEEE Nanotechnology conference, San Francisco, CA, August 2003.

[29] M. Bertero and P. Boccacci, "Introduction to inverse problems in imaging," Institute of Physics Publishing, 1998.

[30] L. J. Hornbeck, "The DMD (TM) projection display chip: A MEMS-based technology," *MRS Bulletin*, vol. 26, p. 325, 2001.

[31] H. Hogan, "Microelectromechanical systems: A little shake, rattle and roll," *Photonics Spectra*, vol 36, no. 10, p. 68, 2002.

[32] R.T. Chen, H. Nguyen and M.C. Wu, "A high-speed low-voltage stress-induced micromachined 2 × 2 optical switch," *IEEE Photonics Tech. Lett.* Vol. 11, no. 11, p. 1396, 1999.

[33] M. Tabib-Azar, Integrated Optics, Microstructures and Sensors, Kluwer Academic Publ., Boston, MA, 1995, p. 171.

[34] B.W. Wessels, "Ferroelectric oxide epitaxial thin films: synthesis and non-linear optical properties," *J. Crystal Growth*, 195, 706 (1998).

[35] R.L. Holman, L.M. Althouse Johnson and D.P. Skinner, "The desireability of electrooptic ferroelectric materials for guided-wave optics," Proc. 6th IEEE Int. Symp Appl. Ferroelectrics, Lehigh Univ. Bethlehem, PA, 1986, p. 32.

[36] P. Muralt, "Ferroelectric thin films for micro-sensors and actuators: a review," J. Micromech. Microeng. 10, 136 (2000).

[37] G. Chik, "Optical Component Evolution/Revolution," International Workshop on Future Trends in Microelectronics: the Nano Millennium, Ile de Bendor, France (2001).

# Part B

## Materials, Fabrication-Technology, and Functionality

# Chapter 9

## Ceramic Thick Films for MEMS

R.A. Dorey and R.W. Whatmore
*Nanotechnology Group, School of Industrial and Manufacturing Science, Cranfield University, Cranfield, Bedfordshire, UK, MK43 OAL*

Many applications of ceramic thick films for MEMS have been highlighted in part A: applications and devices. While scaling of devices would appear to be a smooth transition, there are often fundamental changes in the nature of the fabrication processes used to realise devices with different thicknesses of ceramic film. Together with the chapters on thin films, this chapter aims to provide the reader with an understanding of some of the processes, and constraints, involved in processing of ceramic films for MEMS.

This chapter deals, specifically, with the fabrication of thick films where the film thickness is typically between 1 and 100 μm. The first part of this chapter provides an overview of the issues of interfacial reactions and constrained sintering which are typically associated with processing thick films. Interfacial reactions are often of concern due to the typical processing temperatures employed (600–1200 °C) and the associated high diffusion speeds of atomic species at these temperatures. The role of diffusion barriers in preventing interfacial reactions is discussed.

Following the overview of processing issues associated with thick films, a range of deposition techniques are reviewed. Screen printing, composite sol gel, electrophoretic deposition, thick film sol gel, direct writing, and ceramic paint are discussed.

The chapter concludes by offering a brief summary of typical thick films properties and thick film specific patterning techniques to compliment the chapters on material properties and processing.

# 1. Introduction

Micro-electromechanical systems (MEMS) containing active electroceramic components are of great importance in the fields of sensors, actuators and transducers [1, 2, 3]. There exist a number of active materials that exhibit useful piezoelectric responses. The most commonly used of these piezoelectric materials is lead zirconate titanate (PZT) [4] and many of the other systems also contain significant levels of lead. There are many issues that affect the processing of Pb rich materials such as the volatility and reactivity of Pb. It is the authors' intention to provide the reader with an overview of the different techniques for, and issues associated with, the production of functional oxide ceramic thick films/features. This article will also highlight some of the issues specifically associated with the production of Pb containing thick films due to the abundance of functional materials with high Pb contents.

Thick films are considered to be films ranging in thickness between 1 μm and 100 μm. They occupy a technological region between the processing capabilities of thin film deposition techniques and machining of bulk ceramics [1]. The ability to produce films in this range of thicknesses is of great importance due to the conflicting desires for greater miniaturisation, system integration and device power/sensitivity [5, 6].

Systems such as sensors, actuators and transducers require a certain thickness of active material in order to be sufficiently powerful/sensitive [7, 8] to be of commercial interest. Such thicknesses of material are not easy to fabricate using thin film production techniques such as sol-gel, sputtering, pulsed laser deposition (PLD), physical vapour deposition (PVD) or chemical vapour deposition (CVD) due to the slow deposition rates and high levels of stress generated during processing [9] which can lead to cracking of the film. Conversely the production, machining and bonding of bulk ceramics is not practical due to the expense, waste and difficulty in handling the ceramics [10]. Thick film fabrication routes allow the possibility of directly integrating the films onto the substrate so eliminating the difficulty of handling thin fragile ceramic components.

The two main issues associated with the processing of ceramic thick film materials are those of material compatibility (interfacial reactions) and constrained sintering. These are unique problems faced by thick film processing techniques due to the high processing temperatures and the presence of the substrate.

# 2. Common Issues with Thick Film Production

## 2.1. Interfacial reactions

### *2.1.1. Substrate materials*

Thick film fabrication, like bulk ceramic processing, is based on powder processing routes and as such requires high processing temperatures in order to be able to densify the films. One of the issues associated with this need for high processing temperatures is the selection of an appropriate substrate material to support the film and form the MEMS. The easiest solution to this problem is to use a thermally and chemically stable substrate such as polycrystalline alumina or sapphire. However, these materials are

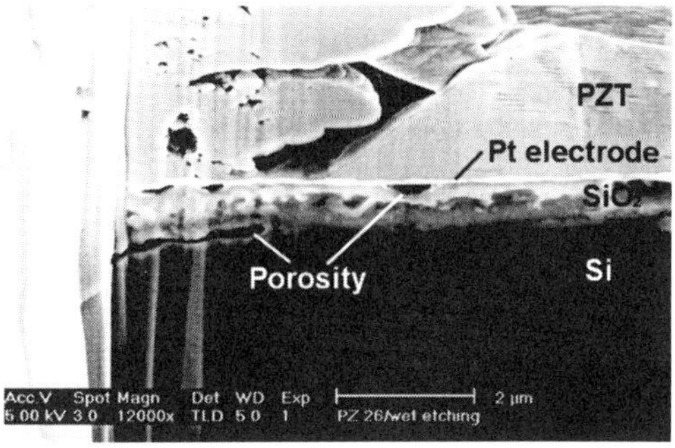


**Fig. 1.** Example of reaction between Si and Pb (within the $SiO_2$ layer) during the processing of a PZT thick film.

difficult to process (they are hard and chemically inert), can be expensive and there is little micromachining know-how relating to them. Instead there is a considerable drive to integrate the functional oxide ceramic thick films with silicon [11, 12], silica (glass) [13, 14] and metals [9, 15] such as steel, iron, chrome, nickel, copper and cobalt. These materials are of great interest as they are readily available, inexpensive, easy to machine and, in the case of metals, ductile [15]. The use of these materials in conjunction with thick film oxide ceramics is, however, limited by their upper processing temperature capabilities of approximately 900 °C [16, 17] and undesirable reactions between the substrate and film (i.e. lead and silicon) [18, 19].

### *2.1.2. Diffusion barriers*

Diffusion barriers are often deposited onto the surfaces of substrates to prevent interdiffusion of atomic species between the substrate and film. Although silicon can be processed at temperatures as high as 900 °C it has been shown that deleterious reactions can occur between silicon and lead [7, 20, 21] at much lower temperatures (figure 1). During the sintering of Pb containing films on silicon, Pb was found to diffuse readily through $SiO_2$ or $Si_3N_4$ layers [22] on the surface of silicon, leading to the formation of lead silicate compounds at the silicon / PZT interface, delamination, and increases in the conductivity of the $SiO_2$ layer [22, 23]. Such diffusion of the lead species is often enhanced by the presence of a liquid phase [17, 21].

The use of diffusion barriers has been employed to separate the silicon and lead containing materials and to prevent the migration of atomic species. The requirements of a diffusion barrier are that it should resist the passage of atoms though it, it should not be a weak link through poor adhesion to the substrate or film, and it should be as thin as possible so as not to affect the performance of the device as a whole. Examples of successfully implemented thin diffusion barriers on silicon include 70 nm yttria stabilized zirconia (YSZ)/30 nm $TiO_2$ [24], 120 & 150 nm $ZrO_2$ [25, 19], 350 nm $TiO_2$[26], 300 nm MgO [19], Rh/TaN [27]. Thick film diffusion barriers have also been employed to yield satisfactory results. i.e. 1 µm columnar YSZ [28], 1 µm $SiO_2$/50 nm $TiO_2$/500 nm Pt [7].

### *2.1.3. Electrode materials*

There are a limited number of suitable materials available for use as back electrodes. These electrodes are deposited onto the substrate prior to the deposition of the active electroceramic component and must therefore be able to withstand the processing temperatures used during the processing of the electroceramic and remain electrically conducting. Pt, Au, and Au/Pt/Pd [15], Ag/Pd [29] alloys have been shown to be effective back electrodes as they do not oxidise at the high processing temperatures. However, reactions between Si and Au, with the formation of a Si-Au eutectic [23], have been reported. This effectively limits the use of Au in conjunction with silicon substrates.

The alloy systems tend to be applied using a screen printing process and are themselves relatively thick (1–2 μm) which may limit their use in certain applications. The thickness of such an electrode does in itself act as an effective diffusion barrier. The use of Pt electrodes is limited to below 850 °C as the electrodes tend to degrade due to the large thermal expansion mismatch between silicon and Pt and the enhanced creep characteristics of Pt at elevated temperatures which result in the formation of islands and/or nodules [15, 30]. To enhance the adhesion between the substrate and the Pt electrode a thin (5–20 nm) Ti or $TiO_2$ adhesion layer is often used.

A reduction in processing temperature to below 800 °C could increase the range of materials that would be suitable for use as back electrodes. Possible alternative metals could include less noble metals such as Cr, Al, Cu [15]. An alternative approach to using noble metals as electrodes is to utilise a conducting ceramic such as $RuO_2$. [31].

## 2.2. Constrained sintering

### *2.2.1. Shrinkage*

One of the key issues associated with the production of any coating on a rigid substrate is that of shrinkage. Constrained shrinkage occurs when the coating undergoes a reduction in volume while the dimensions of the substrate remain unchanged. This occurs initially during the drying of the film when fluid between the ceramic particles is removed through evaporation. A further volume reduction occurs during the sintering of the ceramic particles when pores are eliminated. In an unconstrained body the volume reduction is accomplished via isotropic shrinkage. However, when the film is supported on a rigid substrate the shrinkage in the plane of the film ($x + y$) is prevented [32]. This leads to the generation of in-plane tensile stresses within the film. The principles dictating constrained shrinkage during drying and sintering are comparable. The primary difference being that during the early stage of drying there is sufficient fluid present to allow a degree of particle reorientation to relieve stress. Once a critical volume of fluid has been removed no further particle reorientation is possible and the situation can be likened to that of a film being sintered at very low temperature.

The general sintering behaviour of a 'green' film supported on a rigid substrate and subjected to a constant heating rate has been modelled by Zhao and Dharani [33] using a viscoelastic finite element simulation. Material properties, Young's modulus, and Poisson's ratio were assumed to be functions of density while the viscosity was taken to be a function of grain size, density and temperature. Density and grain size data

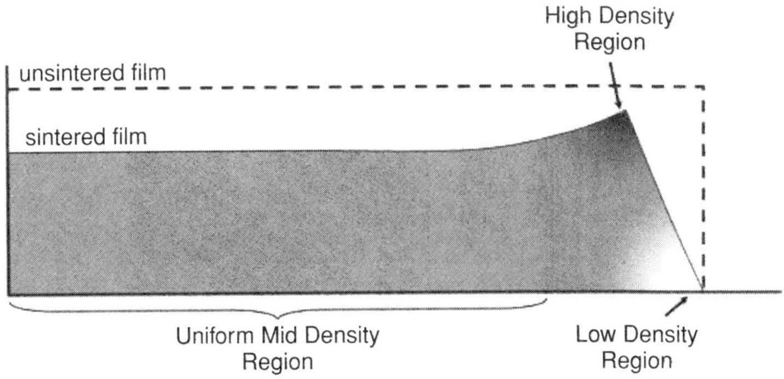


**Fig. 2.** Schematic of the general shrinkage behaviour of a constrained film during sintering.

were obtained from observations of unconstrained sintered samples. Zhao and Dharani modelled the effect of a range of films of different thicknesses fully constrained on a rigid substrate and found that the general sintered shape and density variations were consistent. A generalised schematic of these results is shown in figure 2.

The main body of a film was found to have a uniform density approximately midway between that of the unconstrained edge (highest density) and constrained edge (lowest density). For a 100 μm thick film the mean relative density of the constrained films was found to reach a maximum of 0.72 while that of unconstrained films was approximately 0.98. Although the form of the density variation was found to be comparable across the thickness ranges modelled, it was found that the maximum density attainable was lower for thinner films. The edge effect density variations predicted may prove to be an issue for MEMS processing where small features are required. This study was conducted assuming a constant heating rate and as such did not account for any effects of sintering time on the densification behaviour. Stech *et al.* [34] showed that for a $TiO_2$ film an appreciable degree of densification occurred during the first 150 minutes with higher densities attained at higher sintering temperatures. This observation indicates that relative densities greater that 0.72 could be attained if longer sintering times were employed. Although promising, care should be taken when applying these findings to the sintering of PZT thick films, as the use of longer sintering times and higher temperatures could lead to greater lead loss and a reduction in functional properties of the film. This is of particular concern as the surface area to volume ratio of films is higher that that of bulk ceramics and as such small levels of lead loss could have significant effects on the resultant functional properties. Kosec *et al.* [35] have demonstrated that the use of a Pb rich sintering atmosphere can be used to counter the evaporation of Pb during processing.

A finite element simulation of the effect of sintering time has shown that enhanced densification can be achieved if some form of stress relaxation can occur [36]. In a constrained film the generation of tensile in-plane stresses also give rise to shear stresses. Flow of material under the action of these shear stresses is able to bring about a reduction in the level of tensile stress (analogous to creep stress relaxation) as shown in figure 3. A non-dimensional relaxation parameter (ß—given by the ratio of the rate

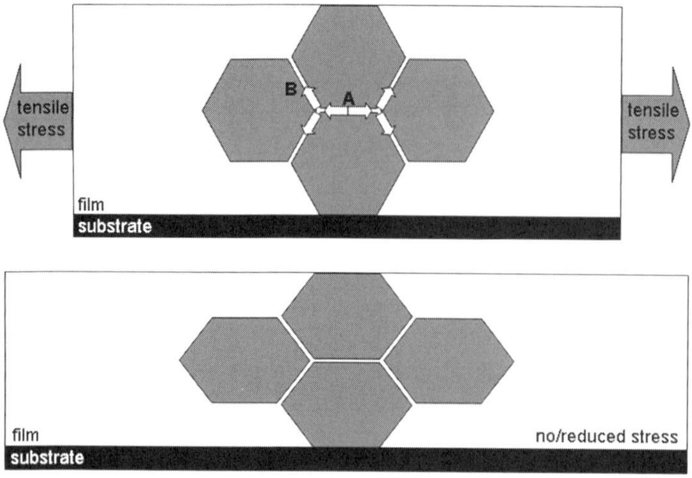


**Fig. 3.** Schematic of effect of a liquid phase sintering aid during sintering of thick films.

constant for shear diffusion and the rate constant for densification) was proposed to account for this stress relaxation.

The fundamental driving force for densification is given by the sintering pressure p:

$$p = -\frac{(\sigma_x + \sigma_y + \sigma_z)}{3} \tag{2.1}$$

In an unconstrained film $\sigma_x = \sigma_y = \sigma_z = -p_o$. When the film is constrained, $\sigma_x$ and $\sigma_y$ are acted upon by the opposing constraint stress $\sigma$ such that

$$\sigma_x - \sigma = \sigma_y - \sigma = -p_o \tag{2.2}$$

And equation 2.1 becomes:

$$p = p_o - 2/3\sigma \tag{2.3}$$

As the non-dimensional relaxation parameter, ß, → 0 (i.e. no shear relaxation), $\sigma \to 3/2$ $p_o$ and $p \to 0$ (i.e. no sintering). Conversely, if ß $\gg 1$ then $p \to p_o$ and the substrate has no effect on sintering. In reality ß is likely to be between the two extremes.

This analysis does provide a good indication of how the sintering of constrained films can be enhanced by encouraging the shear stress relaxation. If the diffusion controlled shear deformation can be enhanced then ß will increase and the degree of sintering will be greater. If densification and stress relaxation are controlled by grain boundary and lattice diffusion respectively then increases in the grain size and sintering temperature can lead to an increase in ß. However, care should again be taken if applying this to the PZT as discussed earlier. The analysis does provide an indication as to one possible way in which the application of sintering aids is beneficial to the sintering thick films. The sintering aid acts as a fast diffusion path for atomic species and also encourages the dissolution of material from areas of low stress (A) and the subsequent deposition of material at areas of high stress (B) (figure 3). A further route for stress relaxation is that of particle slippage whereby individual powder particles are able to move relative to one another. This action would be enhanced by the presence of a liquid phase sintering aid [35] which acts as a lubricant allowing particles to slide over each other more easily. In a real system, both techniques are likely to be active. Particle slippage is likely to dominate during the early stages of sintering while the particles are still free to move. During the latter stages of sintering (and in the absence of sufficient liquid phase), when particles are less free to move, stress relaxation can be accomplished through diffusion controlled shear relaxation.

Reduced sintering rates are not the only effect that can occur due to constrained shrinkage. If the tensile stress generated during sintering (or drying) exceeds the fracture strength of the material then the film will crack. This is of particular importance if the density of the film is low (i.e. during drying, early stages of sintering and at constrained free edges) as the strength of the film will be severely reduced [32, 37]. Studies of the liquid phase sintering behaviour of $ZrO_2$ (with $Bi_2O_3$ sintering aid) and the solid state sintering of alumina have shown that the $ZrO_2$ which was able to undergo a degree of stress relaxation due to the liquid phase sintering aid did not crack during sintering [38].

Closely related to the effect of constrained shrinkage is that of thermal expansion mismatch which can also cause the film to crack. Large differences in the thermal expansion coefficients of the substrate and film will lead to the generation of tensile stresses in the film. These stresses will either arise during heating prior to sintering ($\alpha_{substrate} > \alpha_{film}$) or during cooling after sintering ($\alpha_{substrate} < \alpha_{film}$) [33]. Tensile stresses within the unsintered film are of greater concern due to the lower strength of the film. Hence care must be taken when using substrates with high thermal expansion coefficients (i.e. metals). Silicon and PZT, near the morphotropic phase boundary, both have thermal expansion coefficients in the region of $2 \times 10^{-6}$ K$^{-1}$ at temperatures below 500 °C [39, 40, 41] which helps to minimise the thermal expansion mismatch. At temperatures above 500 °C, and with non MPB composition PZT, significant thermal strains may arise.

### 2.2.2. *Sintering aids*

In solid state sintering of ceramic particles (in the absence of sintering aid) diffusion is slow even at high temperatures. The processing temperatures required to sinter a ceramic powder can be dramatically reduced through the addition of a liquid phase sintering aid. The liquid phase sintering aid initially facilitates the reorientation of the ceramic particles (acting as a lubricant) to allow a better degree of packing and increasing the density of the film. The liquid phase sintering aid also acts as a fast diffusion path for atomic species and encourages dissolution/reprecipitation of the ceramic so encouraging the densification process to occur.

The sintering temperature of bulk PZT is between 1100–1300 °C. Due to the high surface area to volume ratio of films it is very important to control the loss of volatile compounds. For this reason excess lead oxide and lead rich atmospheres are often employed when processing lead containing films. PbO has a melting point of approximately 850 °C and as such has been used as a sintering aid [27, 29, 42, 43]. Other examples of sintering aids used to reduce the processing temperatures to between 700 and 850 °C include PbO-$Cu_2O$ [21, 44, 45], $Pb_5Ge_3O_{11}$ [46, 47, 48], $LiBiO_2$-CuO [49],

$PbO\text{-}PbF_2$ [10, 50, 51], $Bi_2O_3\text{-}B_2O_3\text{-}CdO$ [10, 11, 52], Borosilicate Glass [11, 20, 22, 43], Li/PbO [53] and $PbO/TiO_2$ [54, 55, 56].

Although sintering aids significantly reduce the sintering temperature and enhance densification the presence of significant levels of non/low-function material within the thick film can reduce the resultant functional properties. Furthermore, as sintering aids enhance the degree of solution-reprecipitation of the major phase (i.e. PZT) there is a large degree of atomic mixing. This may affect the electromechanical properties of the ceramic. i.e. if a soft doped PZT is the major phase, but the sintering aid contains a significant proportion of hard dopants then the sintered PZT material will behave as a hard doped material. Care should therefore be taken when selecting the appropriate sintering aid.

# 3. Thick Film Deposition Techniques

## 3.1. Screen printing

Screen printing is the most widely used thick film deposition technique. During the printing process an ink is forced though a fine mesh to deposit it onto the desired substrate. The film is then dried and sintered at elevated temperatures to yield a dense thick film. One of the advantages associated with this technique is the ability to directly pattern the film by selectively masking certain areas of the mesh. This direct patterning eliminates the need to pattern (typically by etching) the film at a later stage. Limitations in the maximum resolution are imposed by the size of the mesh used; hence this technique may not be suitable for producing very fine features.

The screen printing ink is composed of 4 types of constituent which are thoroughly mixed to yield a homogenous product [54]. The 4 constituents are:

*Functional Phase*—this is composed of the material to be deposited (i.e. PZT).

*Binder phase*—this is typically a liquid phase sintering aid and is added in relatively small amounts (5–15 vol% relative to the functional phase). In addition to acting as a sintering aid it also helps to bind the functional phase together and to the substrate.

*Carrier phase*—this acts as support for the solid powders during the printing process. It typically represents between 20 and 40 vol% of the ink and is composed of volatile components (solvents) and non-volatile components (polymers). The solvents are removed during the initial drying stage to leave the polymers to temporarily bind the film together. During the burn out stage the polymers are removed to leave only the functional and binder phases in place.

*Modifiers*—these are added in small quantities to modify the rheological properties of the ink. These additions are again removed during the drying and burnout stages.

The mesh is constructed of stainless steel wire, or nylon, and is stretched across a frame. The pattern is produced by coating the mesh with a photosensitive emulsion and selectively exposing it to UV light. The unwanted material can then be removed to yield the patterned mesh. The mesh is positioned above (but not in contact with) the substrate and the ink placed on the mesh in front of a squeegee. The rheological properties of the ink are such that it does not pass though the mesh at this stage. The squeegee is then drawn across the mesh forcing it into contact with the substrate. The action of shear stresses on the ink reduce its viscosity allowing it to flow though the open sections of the mesh. Once the squeegee has passed the mesh springs back from the surface of the substrate leaving the ink in place. The ultimate resolution of the screen printing process is limited by the resolution of the mesh and the flow of the ink once printed.

Once the ink has been printed it is dried to remove the solvents. At this stage further layers can be deposited to increase the film thickness if required. Once the film has been dried the organic components (i.e. polymers & modifiers) can be removed by a thermal treatment between 350 and 600 °C. A final sintering treatment of between 850 and 950 °C is then performed to densify the film. The burnout and sintering treatments are usually performed at the same time.

### 3.2. Composite film technology

Although thick films can be deposited by screen printing the technique is still based on the sintering of micron size ceramic particles which limits the minimum temperature that can be used to achieve dense products. To overcome this limitation, ceramic powder processing can be combined with low temperature sol-gel processing to yield a hybrid technique capable to producing thick films at low temperatures [44, 57, 58, 59, 60]. The process entails producing a composite slurry consisting of a ceramic powder and an oxide ceramic producing sol. The composite slurry is then deposited using a spin coating process whereby the electroded substrate wafer is coated with the slurry and spun at high speed to yield a thick film. The film is pyrolysed to convert the sol to an oxide ceramic and remove the organic components of the slurry. Successive layers can be deposited to increase the film thickness. A final sintering treatment above 600 °C is used to sinter the film and develop the functionality (i.e. perovskite phase in PZT). The resultant properties of the film are typically lower than those of the corresponding bulk ceramic (see table 1) due to the presence of significant levels of porosity. The incorporation of low melting point sintering aids [21, 44, 47] and repeated sol infiltration and pyrolysis [21] have led to significant improvements in the functional properties of these films to the point where the relative permittivity of films sintered at 710 °C are comparable to that of the bulk ceramic. The use of repeated sol infiltrations and low melting point sintering aids have allowed high density films to be produced at low temperatures [21] as shown in figure 4.

This technique has also been applied to a number of other deposition techniques including screen printing [25], dip coating [61], spray coating [62, 63], and interfacial polymerisation [64]. A variation of the composite sol-gel route is to use conventional thick film processing followed by infiltration of the oxide ceramic producing sol to increase the green density of the film [65]. Subsequent sintering is then enhanced due to the presence of the highly sinterable sol-gel derived nanophase.

### 3.3. Electrophoretic deposition

Electrophoretic deposition (EPD) is a simple, fast and inexpensive deposition technique for obtaining thick ceramic films. One of the advantages of EPD over other

**Table 1.** Comparison of values of relative permittivity for thick films and bulk ceramics (valued for tan δ have also been included where reported in the literature)

| Thick film processing route | Processing temperature | ε (& tan δ) @ 1kHz Film | Bulk | Ref |
|---|---|---|---|---|
| Screen print | 1000 °C | 204 (0.019) | 560 (0.085) | 50 |
| Screen print | 1050 °C | 840 (0.0089) | 2010–2050 (0.004) | 10 |
| Screen print | 950–1030 °C | 600–750 | 2100 | 6 |
| Screen print | 750–950 °C | 400–750 (0.05) | *1100** | 65 |
| Screen print | 750–950 °C | 620–1024 (0.05) | *1100** | 65 |
| Screen print | 900 °C | 1500 (0.06) | 2650 (0.04) | 17 |
| Composite film (screen) | 710 °C | 600 | 1200 | 25 |
| Composite film (spin) | 710 °C | 1100 (0.02) | 1200 (0.003) | 21 |
| Composite film (dip) | 630 °C | 1300 (0.2) | 3240 (0.021) | 61 |
| Composite film (dip) | 550 °C | 2500 | 3240 | 100 |
| Composite film (interface) | 950 °C | 771 (0.052) | 1200 | 64 |
| Microstereo lithography | 700–850 °C | 200 (0.01) | | 85 |

**-estimated bulk values*

**Fig. 4.** Scanning electron microscope micrograph of a fracture cross-section of a high density composite film produced at 710 °C with sintering aid and repeated sol infiltration.

thick film deposition techniques is the ability to coat complex geometries [66]. The three stages to the process are the formation of a charged suspension, the deposition of charged particles onto an electrode under the action of a DC voltage and the final sintering [51, 53, 67].

The charged suspension can either be formed using electrosteric (dispersants) or electrostatic (pH control) stabilisation. In both cases a low concentration suspension is formed with a typical solids loading below 1 vol% [52]. The substrate is then set as a voltage relative to the reference electrode and the charged particles migrate under the action of the DC field. The field is either maintained at a constant value or adjusted so as to keep the current density constant [66]. Although conditions of constant current density are preferred, as this has the effect of maintaining a constant field in the suspension (and hence deposition rate), care must be taken not raise the voltage too high so as to avoid the evolution of gasses and/or heat at the electrodes which can lead to defects in the films [68].

Electrostatic suspension of PZT particles can readily be accomplished though the use of either acetic acid [53] or $HNO_3/NH_3$ [66]. Maintaining the pH of the suspension between 3 and 6 prevents the agglomeration of the PZT particles and associated reduction in deposition rate and production of poor quality films [66].

In order to deposit thick films using EPD it is necessary for the substrate to be conducting or coated with an electrode. This may act as a limitation for the use of this technique in the production of MEMS where often the ceramic film must also cover non-electroded (non-conducting) areas.

A further limitation is the high density of Pb containing compounds as it is very difficult to maintain large (i.e. heavy) particles in suspension. Tassle and Randall [53] found that PZT particles greater than 0.4 μm settled out of the suspension with those particles greater than 2 μm is size settling very rapidly. A consequence of this settling is the practice of operating EPD in a sideways or upwards deposition so that particles settling out of suspension do not contaminate the substrate.

A thorough review of the electrophoretic deposition technique, as applied to many ceramic oxide films, is given by Sarkar and Nicholson [69].

### 3.4. Thick film sol gel

While processing of oxide ceramic particle films does represent the main approach to realising thick films, conventional sol gel thin film chemistries have been modified to allow thick sol gel films to be fabricated. Thick sol gel processing makes use of alternative solvents such as diols [70, 71] which result in less rigid structures that are able to undergo stress relaxation with greater ease. Alternatively, additives which act both as strengtheners and plasticizers may be added to the system. Examples of such additives include polyvinyl pyrrolidine (PVP) [3, 72] and polyvinyl butral (PVB) [73]. These additives increase the strength by acting as 'long range' links holding the structure together during drying, and promote stress release by breaking up the polymeric gel network making the structure more flexible and able to deform plastically. With increased levels of additive the achievable thickness increases. However, increasing the levels of these additives also significantly increases the amount of organic material present. Hence, the structure may initially be capable of withstanding the shrinkage but it may fail when further shrinkage occurs when the additives are removed. These two conflicting effects on increasing the amount of plasticizer mean that there is an optimum level of addition where the greatest crack resistance is achieved. This optimum level of additions has been shown to be between 1 and 1.5 mol% [72] which yields a single layer thickness of between 700 and 1500nm. This is much greater than that achievable using conventional sol gel techniques.

The ability to deposit such thick films brings with it its own issues. When the individual layers are this thick it is difficult to effectively remove the vapour/gases released as the film dries and the organic phase breaks down during pyrolysis. With standard thin film processing, the diffusion distances for gas to escape are relatively small. However, as the thickness increases so do the diffusion distances. This means that processing has to be conducted at a slower rate to allow time for the gas/vapour to escape. If the gas is unable to escape then pores will be formed which will decrease the properties of the film [74].

In thick sol gel films pores are generally formed at the bottom of the individual layers due to the greater diffusion distances required to effectively remove the gas and the constraining effects of the substrate (or layer below). When multiple layering is employed the resultant structure becomes one of alternating high and low density zones [75] with the pores trapped at the interface between individual layers. Despite the formation of these pores there is still considerable interest in developing sol gel materials that can be used to deposit films suitable for piezoelectric devices using a minimum number of layers.

A further complication associated with the processing of thicker films is that of compositional inhomogeneity arising from crystal growth. Ti rich PZT tends to nucleate in preference to Zr rich PZT [76, 77]. Hence, as the PZT crystal grows Zr is expelled from the perovskite lattice leading to an increase in concentration of Zr ahead of the crystal growth front. This Zr is gradually incorporated into the crystal lattice as the grain grows, leading to a variation in the ratio of Ti and Zr. Furthermore, local levels of lead may be reduced ahead of the growth front resulting in the formation of pyrochlore in place of perovskite PZT. This latter effect can be reduced by incorporating a slight (<10 at%) lead excess in the CSD film. The type of compositional variation differs depending on the manner in which the PZT nucleates and grows. If the PZT nucleates on the back electrode the resultant structure will be characterised by a rising Zr content as the distance from the back electrode increases. If intermediate crystallisation is carried out a 'saw toothed' distribution of Zr/Ti can be observed where a Ti rich material is observed at the base of each layer annealed [8, 76]. Depending on the level of Pb excess, an increase in Pb concentration may also be observed at the surface of the layers. This is caused by the expulsion of excess lead ahead of the crystal growth front [74, 76]. When the PZT nucleates at the back electrode and at the surface of the film, the middle of the layer is found to be Zr and Pb rich [74]. This is a result of the two growth fronts meeting in the centre of the film.

## 3.5. Direct writing

Recent developments in the field of direct writing have enabled a range of microscale devices to be created. Application of such technology has the benefit of eliminating the need to pattern the devices after deposition. Furthermore, the fact that the patterns are printed directly under computer control eliminates the need to make expensive photolithographic masks and screens, as well as increasing the flexibility of the system [78, 79]. Many of the direct writing technologies have been developed through the continued refinement of existing macroscopic solid freeform fabrication techniques.

A number of micro-scale features have been produced using the ink jet printing technique. Ink jet printing is divided into two categories—continuous flow and drop on demand [80]. In the continuous flow process a steady stream of ink flows though a fine aperture. Rayleigh instability (assisted by the superimposition of a mechanical oscillation) causes the stream of ink to break up into single drops which can be deflected by a pair of charged plates. This technique has a faster deposition rate, but the need to deflect the ink drops elecrostatically requires that the ink be conductive [80, 81]. In drop on demand printing the ink drops are ejected from the print head when required (acoustic, mechanical or thermal ejection) and are positioned on the substrate by the mechanical movement of the print head relative to the substrate [80]. Both approaches require careful control of the ink to ensure good dispersion with no agglomerates or debris. Commercially available ink jet printers have a resolution of 1200 dpi ($\sim$20 μm).

The inks used for ink jet printing typically have a low solids loading ($<$5 vol%). The drying behaviour of the ink must be carefully controlled in order to maintain a high printing resolution. If the solvent evaporation is too slow the ink will spread on impact with the substrate leading to a loss of resolution. However if the ink dries too quickly then the printing nozzle may clog [82]. Typically a conventional ceramic ink drop with a diameter of 60 μm will spread to give a feature 600 μm wide and 1 μm high. Greater control of the feature size can be accomplished by reducing the volume of the ink drop [81] or by using an ink with a higher solids loading. Seerden *et al.* [80] have demonstrated that a ceramic ink with up to 40 vol% solids can be successfully printed using a drop on demand print head. By mixing the ceramic powder with a thermoplastic wax it was possible achieve alumina features with walls thinner than 100 μm. Other examples of features produced using drop on demand printing include 3 mm high $ZrO_2$ walls with widths of 340 μm [83] and $ZrO_2$ pillars 100 μm in diameter and 2 mm high [84].

Direct writing of patterns on substrates has also been accomplished using a micropen tool [78, 79] whereby a ceramic ink with 25–35 vol% solids loading is extruded through a pen. Related to the micropen is the technique of robocasting which again used a directable extrusion system to directly write a pattern on the substrate. The main difference between the systems is that robocasting using a more highly loaded ink (55–60 vol%) to produce lines approximately 575 μm wide. Both production techniques require a high temperature burn-out stage and a sintering stage to produce finished the product. Both robocasting and micropen rely on mechanically moving the delivery system or substrate and will therefore suffer the same mechanical control limitations imposed by the drop on demand ink jet printing technique.

Jiang et al. [85] have successfully produced PZT microfeatures on silicon substrates using a microstereolithographic process. Features between 10 and 150 μm thick and with line widths of between 20 and 100 μm were produced. This technique has also been used to fabricate small alumina gears 20 μm thick with diameters of approximately 400 μm and 1 mm [86]. To produce the PZT structures a PZT powder mixed with a UV curable monomer (1,6-hexabediol diacrylate (HDDA)), a photoinitiator (benzoin ethyle ether), solvent and dispersant to form a UV sensitive suspension with approximately 33 vol% solids loading. This suspension is then drawn across the substrate to leave a uniform layer which is then selectively polymerised using a x-y scanning focused UV laser beam (Ar laser) with spot size of 1–2 μm. Control of the z position of the substrate allows further layers of suspension to be deposited and patterned. Once the required thickness has been obtained the unexposed areas were removed by washing with water and IPA. The PZT microfeatures were then subjected to a heat treatment designed to burn out the organic polymer (650 °C/4 hours) and sinter the PZT (700–850 °C/ 1–2 hours). The final electrical properties of the features were comparable to those of screen printed thick films. Structures with comparable feature sizes were produced using an alternative microstereolithographic process whereby a liquid crystal dynamic lithographic mask was placed in the path of the a columnated UV beam [87]. The resultant image is then reduced in size, using a lens, and focused onto the surface of the UV curable suspension. The thickness of the components is increased, as before, through successive deposition of further layers. The smallest feature size attainable was 50 μm with a resolution of approximately 5 μm.

A close relation of microstereolithography is gelcasting whereby a mould is first built up by selectively exposing a polymeric monomer to UV light as with microstereolithography. Once the mould has been produced a suspension of ceramic particles in a polymer are cast into the mould and allowed to cure. The ceramic moulding is then heated to remove the organic components and to cause the ceramic to sinter. PZT microcomponents (with dimension ~1cm in size) have been fabricated using this route [88]. The applicability of this technique for producing MEMS type components is questionable due to the difficulty in ensuring thorough filling of the micro-scale moulds.

To date the use of direct write fabrication of MEMS systems has not become established due to the difficulties in operating at such small scales. Direct write fabrication techniques are readily capable of producing structures with feature sizes in the order of 100's of μm. For the direct fabrication techniques to be viable alternatives to the more established techniques it will be necessary to demonstrate the capability of producing feature sizes in the order of 10's of μm.

### 3.6. Paint

Related to the production of true ceramic thick films is the work being conducted on PZT thick film paints [89] which require no sintering process. Instead PZT is held in suspension in a liquid polymer which sets without the need for heat. The resultant properties of the thick films are significantly lower that those of conventional thick film materials, but there may be applications where piezoelectric thick films are required to be deposited at low temperatures.

## 4. Electrical Properties of PZT Thick Films

Due to the low processing temperatures and addition of significant levels of sintering aid the dielectric properties of PZT thick films are often significantly below those of the corresponding bulk ceramics. Table 1 gives details of the values of relative permittivity of thick films relative to those of the bulk ceramic (values of tan δ are also included where presented in the literature). Due to the large dependence of relative permittivity on composition—in particular dopant additions—it is not practical to draw comparisons between the properties of different materials. Instead only those papers that reported the thick film relative permittivity along with those of the bulk ceramic are reported. Values of the $d_{33,f}$ piezoelectric coefficient which were reported were between 1/4 and 1/3 those of the bulk ceramic [6, 10, 65, 85]. It should be noted that the value of $d_{33,f}$ is not only affected by the processing conditions, level of porosity and presence of sintering aid but also by the rigid substrate. The effect of the substrate clamping is to reduce the piezoelectric coefficient [90, 91] and to reduce the degree to which the film can be poled [91]. It is not the author's intention to go into greater detail. Instead the readers are directed to the appropriate references.

Although easy to measure and useful for determining the piezoelectric properties of the thick films, the $d_{33}$ coefficient can give a misrepresentative idea of the piezoelectric capabilities of the thick film. Often thick films are used in bending mode devices where the $d_{31}$ piezoelectric coefficient dictates the effectiveness of the material. Due to the small scale of the thick films it is not possible to measure the $d_{31}$ piezoelectric coefficient directly using a standard Berlincourt type meter. A number of systems have been developed to measure the $d_{31}$ piezoelectric coefficient by inducing a bending motion in a whole wafer [92]. The disadvantage of these systems is that the wafer must be fully coated and that only an average value of the piezoelectric coefficient can be obtained. An alternative approach [93] uses a modified Berlincourt meter, requires much smaller samples and can obtain localised values of $d_{31}$. Values of $e_{31,f}$ (strain equivalent of $d_{31,f}$) reported for a high density composite sol-gel film were between $-4.5$ and $-8$ C/m$^2$ [21] where calculated value of $e_{31,f}$ for the bulk ceramic was approximately $-16$ C/m$^2$. These values compare with those from a screen printed material which were typically less than $-1$ C/m$^2$ [25]. The enhanced properties were attributed to the high density and low level of non-functional material in the composite sol-gel films.

## 5. Patterning Thick Films

### 5.1. Material removal

The production of a uniform thick film, although of central importance for the manufacture of MEMS, does not represent the complete processing issues. As with the production of MEMS devices from thin film materials, the use of thick film materials also requires the shaping (patterning) of films. The primary differences between the patterning of thin and thick films arise due to two factors. The first, most obvious, is the increased volume of material that must be removed during patterning. The second difference is due to the different microstructures usually associated with the thick films.

Thin films are primarily etched using a form of reactive ion etching [94] allowing high side wall angles to be obtained due to the anisotropic nature of the etching. The disadvantage of these reactive ion etching techniques is the slow speed and need for dedicated machines when etching PZT (to prevent cross-contamination) [94]. This slow etch rate effectively prohibits the use of many of these thin film etching techniques for etching thick films on the grounds of the extremely long processing times required and the associated expense [94].

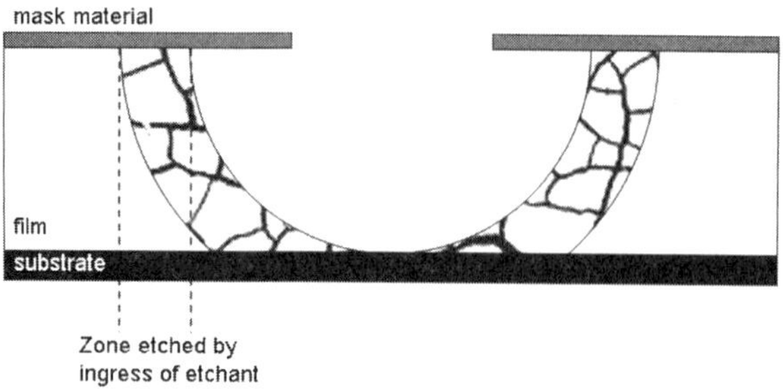


**Fig. 5.** Schematic representation of how porosity can lead to accelerated undercutting during wet etching.

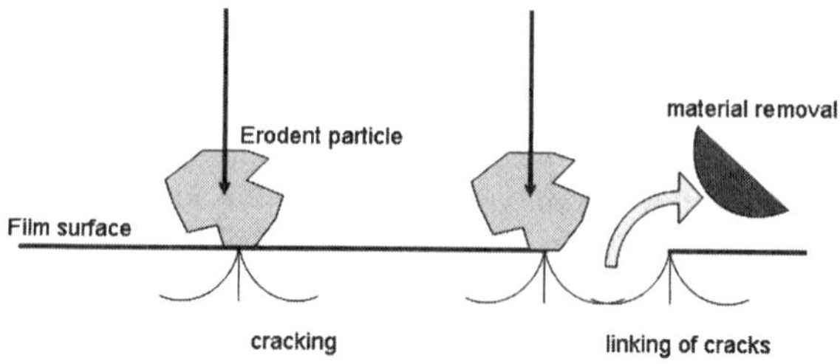


**Fig. 6.** Action of high speed particle on brittle substrate.

Thin films can also be etched using a wet chemical process. There are a number of etch recipes for etching PZT [94]. They are typically based around a HCl/HF mix where the HCl attacks the Ti and Pb components while the HF attacks the Ti and Zr components of the PZT. A number of further additions can also be added to the etchant to remove the insoluble Pb reaction products.

The high density of thin films, when made by sol-gel, PLD, sputtering, etc, lends itself to good edge control during wet etching. Due to the isotropic nature of wet etching it is only possible to obtain a 1:1 ratio between undercut and thickness [94]. The degree of undercutting for wet etching of thick films is generally greater due to the different microstructures of the thick films. These films, made using a route involving the consolidation of ceramic particles through the action of heat, usually have a higher degree of residual porosity. During etching, etchant can enter this porosity and accelerate the rate of undercutting (figure 5).

An alternative patterning technique that has successfully been employed for the production of thick film MEMS devices is that of powder blasting [28, 95]. Fine ceramic particles (3–30 μm) are accelerated under the action of high pressure gas and directed at the surface of the thick films. The particles impact the surface of the films at speeds of up to 290 m/s. When the high speed ceramic particles impact the surface of films the area directly under the particle is cracked (figure 6).

Repeated impacts lead to a network of cracks which link together allowing material removal to occur. The process is fast with relatively inexpensive equipment costs. Detailed patterns with features as small as 30 μm can be produced using this technique

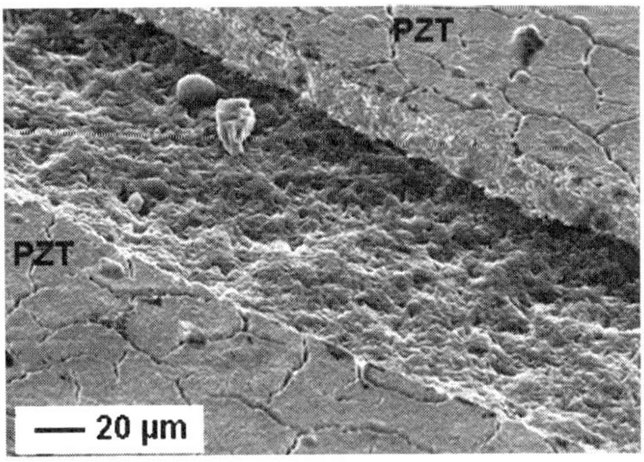


**Fig. 7.** SEM micrograph of an 80 µm wide trench powder blasted through a PZT thick film into a Si substrate [97].

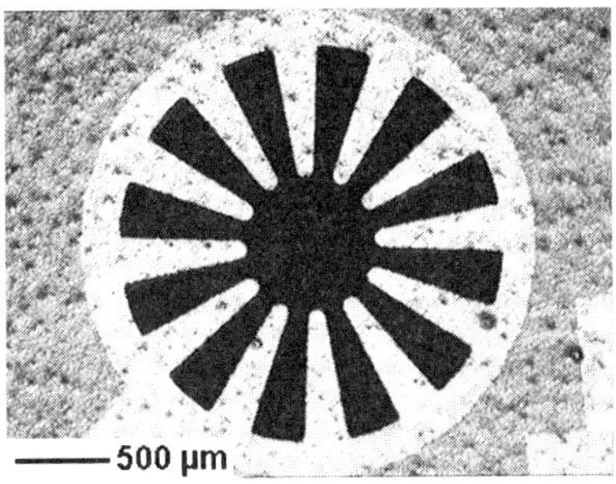


**Fig. 8.** A 1mm spoke test actuator formed by powder blasting into a PZT-on-silicon composite layer [28].

by selectively masking the surface of the films with a metal or elastomer mask [28, 96]. Powder blasting is particularly effective at eroding hard brittle materials and is well suited for patterning ceramics. However, there is limited selectivity between such hard materials. Due to the anisotropic nature of the process high aspect ratio holes can be produced. It is not possible to get perfectly straight side walls due to the interaction of the erosion particles with the edges of the mask. Figures 7 and 8 show features in a 10 µm thick composite sol-gel film obtained through powder blasting.

### 5.2. Direct patterning

The approaches adopted to yield a sol gel film that can be patterned are akin to those used with photoresist polymers. Sol gel chemistries have been modified so that the solubility of the system decreases [98, 99] when exposed to UV light. A decrease in solubility can be achieved through the use of beta-diketones with UV sensitive chelating rings which break on exposure to UV light [98] or through the use of photosensitisers or photoinitiators which cause enhanced crosslinking of the gel phase to occur on exposure to UV light [3, 99]. Such techniques have been used to produce patterned features with thicknesses in excess of 1 µm using a single deposition stage.

## 6. Conclusions

A number of established and emerging thick film deposition techniques have been presented. Although providing many ways to deposit thick films onto substrates the processes are all limited by the common reliance on oxide ceramic processing technology. The very nature of oxide ceramic processing necessitates the use of high processing temperatures to achieve effective densification. Very high processing temperatures can be avoided through the use of liquid phase sintering aid, small ceramic particles and repeated sol infiltration with the consequence that thick film processing can routinely be conducted in the range of 600–900 °C. This reduction in processing temperature is likely to increase the range of materials that can be successfully integrated into MEMS. It has already been demonstrated that the low temperature processing routes can be successfully used to deposit oxide ceramics onto low temperature substrates such as silicon and steel.

There exist two effective ways to pattern the thick film materials—wet etching and powder blasting. However, to obtain very high aspect ratio features for MEMS structures it will be necessary to further develop the direct patterning techniques for the deposition of these thick films. The current direct patterning technologies such as screen printing, stereolithography, ink jet printing are capable of producing features with dimensions in the order of 100 μm. This size range is at the upper border of MEMS processing and future developments within this field will increase the usefulness of these techniques within the filed of MEMS. The ability to realise feature sizes in the order of 10's of microns with comparable resolution will be the goal of future developments of direct write fabrication technologies.

There are still a number of challenges to be faced in the producing of thick film structures, but the continued refinement of current techniques, adaptation of unrelated technologies and development of wholly new approaches will be able to deliver the requirements needed for commercial thick film MEMS production.

## 7. Acknowledgements

R. Dorey is a Royal Academy of Engineering/EPSRC postdoctoral research fellow. The authors would like to acknowledge the contribution to this article of work from the EC framework V 'Parmenide' project and EPSRC grant GR/N05970/01

## References

[1] R.W. Whatmore, *Ferroelectrics*, **225**, 179, (1998)

[2] N. Setter, *J. Euro. Ceram. Soc.*, **21**, 1279, (2001)

[3] R.W. Whatmore, Q. Zhang, Z. Huang, and R.A. Dorey, *Materials Science in Semiconductor Processing*, **5**, 65, (2003)

[4] M. Koch, A. Evans, and A. Bunnschweiler, Microfluidic, *Technology and Applications* (Research Studies Press Ltd., Baldock, Hertfordshire, England), p137

[5] Y. Jeon, Y.G. Seo, S.-J. Kim, and K. No, *Integrated Ferroelectrics*, **30**, 91, (2000)

[6] L. Simon, S. Le Dren, and P. Gonard, *J. Euro. Ceram. Soc.*, **21**, 1441, (2001)
[7] P. Glynne-Jones, S.P. Beeby, P. Dargie, T. Papakostas, and N.M. White, *Meas. Sci. Technol.*, **11**, 526, (2000)
[8] N. Ledermann, P. Muralt, J. Baborowski, S. Gentil, K. Mukati, M. Cantoni, A. Seifert, and N. Setter, *Sensors and Actuators A*, **105**, 162, (2003)
[9] Q.F. Zhou, H.L.W. Chen, and C.L. Choy, *Thin Solid Films*, **375**, 95, (2000)
[10] S. Le Dren, L. Simon, P. Gonnard, M. Troccaz, and A. Nicolas, *Mat. Res. Bull.*, **35**, 2037, (2000)
[11] E.S. Thiele, and N. Setter, *J. Am. Ceram. Soc.*, **83**, 1407, (2000)
[12] N. Setter, and R. Waser, *Acta mater.*, **48**, 151, (2000)
[13] E. Belloy, S. Thurre, E. Walckiers, A. Sayah, and M.A.M. Gijs, *Sensors and Actuators*, **84**, 330, (2000)
[14] D. Mülsenberg, *Microelectronics Journal*, **28**, 419, (1997)
[15] S.M. Spearing, *Acta mater.*, **48**, 179, (2000)
[16] E.S. Thiele, D. Damjanovic, and N. Setter, *J. Am. Ceram. Soc.*, **84**, 2863, (2001)
[17] M. Kosec, D. Murko, J. Holc, B. Malic, M. Ceh, T. Haike, and H. Beige, Z. Metallkd., **92**, 97, (2001)
[18] O.M. Kanunnikova, F.Z. Gilmutdinov, and A.A. Shankov, *International Journal of Hydrogen Storage*, **27**, 783, (2002)
[19] Y. Jeon, J. Chung, and K. No, *J. Electroceram.*, **4**, 195, (2000)
[20] S.P. Beeby, A. Blackburn, and N.M. White, *J. Micromech. Microeng.*, 9, pp. 218–229, (1999)
[21] R.A. Dorey, S.B. Stringfellow, and R.W. Whatmore, J. Euro. Ceram. Soc., **22**, 2921, (2002)
[22] R. Maas, M. Koch, N.R. Harris, N.M. White, and A.G.R. Evans, *Materials Letters*, **31**, 109, (1997)
[23] M. Koch, N. Harris, R. Maas, A.G.R. Evans, N.M. White, and A. Brunnschweiler, *Meas. Sci. Technol*, **8**, 49, (1997)
[24] Y.B. Kim, T.S. Kim, K.S. Choi, and D.J. Choi, *Integrated Ferroelectrics*, **35**, 199, (2001)
[25] R.A.Dorey, R.W. Whatmore, S.P. Beeby, R.N. Torah, and N.M. White, Integrated Ferroelectrics, Integrated Ferroelectrics, **54**, 651, (2003).
[26] F.F.C. Duval, R.A. Dorey, and R.W. Whatmore, *Thin Solid Films*, in press
[27] T. Futakuchi, K. Nakano, and M. Adachi, *Jpn. J. Appl. Phys.*, **39**, 5548, (2000)
[28] S. A. Wilson, R. D. Haigh, J. E. A. Southin, R. A. Dorey, and R. W. Whatmore, *Proceedings of Euspen 03*, Aachen, Germany, 19–20 May, (2003)
[29] V. Farrari, D. Marioli, A. Taroni, and E. Ranucci, *Sensors and Actuators B*, **68**, 81, (2000)
[30] P. D. Hren, S.H. Rou, H.N. Al-Shareef, M.S. Ameen, O. Auciello, and A.I. Kingon, *Integrated Ferroelectrics*, **2**, 311, (1992)
[31] B. Wang, K.W. Kwok, HJ.L.W. Chan, C.L. Choy, K.Y. Tong, E.Z. Luo, J.B. Xu, and I.H. Wilson, *Materials Characterisation*, in press
[32] S.-Y. Tzeng, and J.-H. Jean, *J. Am. Ceram. Soc.*, **85**, 335, (2002)
[33] Y. Zhau, and L.R. Dharani, *Thin Solid Films*, **245**, 109, (1994)
[34] M. Stech, and P. Reynders, and J. Rodel, *J. Am. Ceram. Soc.*, **83**, 1889, (2000)
[35] M. Kosec, J. Holc, B. Malic, and V. Bobnar, *J. Euro. Ceram. Soc.*, **19**, 949, (1999)
[36] R.K. Bordia, and R. Raj, *J. Am. Ceram. Soc.*, **68**, 287, (1985)
[37] R.K. Bordia, and A. Jagota, *J. Am. Ceram. Soc.*, **76**, 2475, (1993)
[38] T. J. Garino, and H.K. Bowen, *J. Am. Ceram. Soc.*, **70**, C315, (1987)
[39] T. Mitsui, M. Marutake, E. Sawaguch, *in Landolt-Börnstein Numerical data and functional relationships in science and technology, Volume 9, Suppliment and Extansion to volume 3, Ferro- and antiferroelectric substances*, ed. K.-H. Hellwege and A.M. Hellwege (Springer-Verlag, Berlin, 1975) p. 309

[40] D. Berlingcourt, H.H.A. Krueger, and C. Near, *Technical publication TP-226, Properties of Piezoelectricity Ceramics* (Morgan Electro Ceramics, www.morgan-electroceramics.com)

[41] Y. Okada, and Y. Tokumara, *J. App. Phys.*, **56**, 314, (1984)

[42] G. De Cicco, B. Morton, D. Dalmonego, and M. Prudenziati, *Sensors and Actuators*, **76**, 409, (1999)

[43] M. Prudenziati, B. Morten, and De Cicco, *Microelectronics international*, **38**, 5, (1995)

[44] D.L. Corker, Q. Zhang, R.W. Whatmore, and C. Perrin, *J. Euro. Ceram. Soc.*, **22**, 383, (2002)

[45] D.L. Corker, R.W. Whatmore, E. Ringgaard, and W.W. Wolney, *J. Euro. Ceram. Soc.*, **20**, 2039, (2000)

[46] P. Tran-Huu-Hue, F. Levassort, F.V. Meulen, J. Holc, M. Kosec, and M. Lethiecq, *J. Euro. Ceram. Soc.*, **21**, 1445, (2001)

[47] T. Hayashi, T. Inoue, and Y. Akiyama, *J. Euro. Ceram. Soc.*, **19**, 999, (1999)

[48] F.F.C. Duval, R.A. Dorey, Q. Zhang, and R.W. Whatmore, *J. Euro. Ceram. Soc.*, J. Euro. Ceram. Soc., 23, 1935, (2003).

[49] X.X. Wang, K. Murakami, O. Sugiyama, and S. Kaneko, *J. Euro. Ceram. Soc.*, **21**, 1367, (2001)

[50] C. Lucat, F. Menil, and R. Von Der Muhll, *Meas. Sci. Technol.*, **8**, 38, (1997)

[51] T. Sweeney, PhD Thesis, Cranfield University, UK, (1998)

[52] M.-C. Wang, M.-S. Huang, and N.-C. Wu, *J. Euro. Ceram. Soc.*, **21**, 695, (2001)

[53] J. Van Tassel, and C.A. Randall, *J. Euro. Ceram. Soc.*, **19**, 955, (1999)

[54] M. Prudenziati, *Thick Film Sensors* (Elsevier, NL, 1994, ISBN 0444897232), p113

[55] Y. Jeon, Y.G. Seo, S.-J. Kim, and K. No, *Integrated Ferroelectrics*, **30**, 91, (2000)

[56] K. Kakegawa, T. Kato & Y. Sasaki, *J. Euro. Ceram. Soc.*, **20**, 1599, (2000)

[57] B. Kindl, D.J. Carlsson, Y. Deslandes, and J.M.A. Hoddenbagh, Ceramics International, **17**, 347, (1991).

[58] D.A. Barrow, T.E. Petroff, R.P. Tandon, and M. Sayer, *J. Appl. Phys.*, **81**, 876, (1997)

[59] D.A. Barrow, T.E. Petroff, and M. Sayer, Surface and Coatings Technology, **76–77**, 113, (1995)

[60] M. Lukacs, M. Sayer, and S. Foster, *Integrated Ferroelectrics*, **24**, 95, (1999)

[61] A.L. Kholkin, V.K. Yarmarkin, A. Wu, P.M. Vilarinho, and J.L. Baptista, *Integrated Ferroelectrics*, **30**, 245, (2000)

[62] T. Olding, M. Sayer, and D. Barrow, *Thin Solid Films*, **398–399**, 581, (2001)

[63] M. Kobayashi, T.R. Golding, M. Sayer, and C.-K. Jen, *Ultrasonics*, **39**, 675, (2002)

[64] T. Tsurumi, S. Ozawa, G. Abe, N. Ohashi, S. Wada, and M. Yamane, *Jpn. J. Appl. Phys.*, **39**, 5604, (2000)

[65] H.J. Kim, Y.-B. Kim, J.-Y. Kang, and T.S. Kim, *Integrated Ferroelectrics*, **50**, 11, (2002)

[66] J. Ma, and W. Cheng, *J. Am. Ceram. Soc.*, **85**, 1735, (2002)

[67] T.G. Sweeney, and R.W. Whatmore, *Ferroelectrics*, **187**, 57, (1996)

[68] B. Su, C.B. Ponton, and T.W. Button, *J. Euro. Ceram. Soc.*, **21**, 1539, (2001)

[69] P. Sarkar, and P.S. Nicholson, *J. Am. Ceram. Soc.*, **79**, 1987, (1996)

[70] N.J. Philips, S.J. Milne, *J. Mater. Chem.*, **1**, 893, (1991).

[71] R.W. Schwartz, T.L. reighert, P.G. Clem, D. Dimos, D. Liu, *Integrated Ferroelectrics*, **18**, 275, (1997).

[72] H. Kozuka, S. Takenaka, *J. Am. Ceram. Soc.*, **85**, 2696, (2002).

[73] Y. Takahashi, *Thin Solid Films*, **370**, 5, (2000).

[74] S. Sporn, S. Merklein, W. Grond, S. Seifert, S. Whal, A. Berger, *Microelectronic Engineering*, **29**, 161, (1995).

[75] K. Miyazawa, K. Ito, J. Mayer, M. Ruhle, *J. Am. Ceram. Soc.*, **81**, 2333, (1998).

[76] S.A. Impey, Z. Huang, A. Patel, R. Beanland, N.M. Shorrocks, R. Watton, R.W. Whatmore, *Journal of Applied Physics*, **83**, 2202, (1998).

[77] H.J. Willems, D.J. Wouters & H.E. Maes, *Microelectronic Engineering*, **29**, 217, (1995).

[78] B.H. King, D. Dimos, P. Yang, S.L. Morissette, *J.Electroceramics*, **3**, 173, (1999)

[79] S.L. Morissette, J.A. Lewis, P.G. Clem, J. Cesarano III, and D.B. Dimos, *J. Am. Ceram. Soc.*, **84**, 2462, (2001)

[80] K.A.M. Seerden, N. Reis, J.R.G. Evans, P.S. Grant, J.W. Halloran, and B. Derby, *J. Am. Ceram. Soc.*, **84**, 2514, (2001)

[81] P.F. Blazdell, and J.R.G. Evans, *J. Mater. Proc. Tech.*, **99**, 94, (2000)

[82] J.A. Lewis, *Current Opinion in Solid State and Materials Science*, **6**, 245, (2002)

[83] X. Zhao, J.R.G. Evans, and M.J. Edirisinghe, *J. Am. Ceram. Soc.*, **85**, 2113, (2002)

[84] X. Zhao, J.R.G. Evans, M.J. Edirisinghe, and J.H. Song, *J. Mater. Sci.*, **37**, 1987, (2002)

[85] X.N. Jiang, C. Sun, Z. Zhang, B. Xu, and Y.H. Ye, Sensors and Actuators, **87**, 72, (2000)

[86] X. Zhang, X.N. Jiang, and C. Sun, *Sensors and Actuators*, **77**, 149, (1999)

[87] M. Farsari, F. Claret-Tournier, S. Huang, C.R. Chatwin, D.M. Budgett, P.M. Birch, R.C.D. Young, and J.D. Richardson, *J. Mater. Proc. Tech.*, **107**, 167, (2000)

[88] D. Guo, K. Cai, C. Nan, L. Li, and Z. Gui, *Scripta mater.*, **47**, 383, (2002)

[89] J.M. Hale, and B. de Poumeyrol, *Proc. Ferroelectrics UK 2000*, 205, (2000), ISBN 1-86125-135-1

[90] M.-A. Dubois, and P. Muralt, *Sensors and Actuators*, **77**, 106, (1999)

[91] R.A. Dorey, and R.W. Whatmore, *Integrated Ferroelectrics*, **50**, 111, (2002)

[92] J.F. Shepard Jr, P.J. Moses, S. Trolier-McKinstry, *Sensors and Actuators A*, **71**, 133, (1998)

[93] J.E.A. Southin, S.A. Wilson, S. Schmitt, and R.W. Whatmore, *J. Phys. D:Appl. Phys.*, **34**, 1446, (2001)

[94] R.A. Miller, and J.J. Bernstein, *Integrated Ferroelectrics*, **29**, 225, (2000)

[95] H. Wensink, and M.C. Elenspoek, *Sensors and Actuators A*, **102**, 157, (2002)

[96] H. Wensink, H.V. Jansen, J.W. Berenschot, and M.C. Elwenspoek, **10**, 175, (2000)

[97] R.D. Haigh (2003), Private Communication

[98] N. Tohge and Y. Takama, *J. Mater. Sci.*, **10**, 273, (1999).

[99] S. Marson, R.A. Dorey, Q. Zhang, and R.W. Whatmore, *J. Euro. Ceram. Soc.*, **24**, 1925, (2004).

[100] A.L. Kholkin, V.K. Yarmarkin, A. Wu, M. Avdeev, P.M. Vilarinho, and J.L. Baptista, *J. Euro. Ceram. Soc.*, **21**, 1535, (2001)

# Chapter 10

## Thin Film Piezoelectrics for MEMS

S. Trolier-McKinstry
*Materials Research Institute and Materials Science and Engineering Department, Penn State University*

P. Muralt
*Ceramics Laboratory, Swiss Federal Institute of Technology EPFL, Lausanne, Switzerland*

Thin film piezoelectric materials offer a number of advantages in microelectromechanical systems (MEMS), due to the large motions that can be generated, often with low hysteresis, the high available energy densities, as well as high sensitivity sensors with wide dynamic ranges, and low power requirements. This chapter reviews the literature in this field, with an emphasis on the factors that impact the magnitude of the available piezoelectric response. For non-ferroelectric piezoelectrics such as ZnO and AlN, the importance of film orientation is discussed. The high available electrical resistivity in AlN, its compatibility with CMOS processing, and its high frequency constant make it especially attractive in resonator applications. The higher piezoelectric response available in ferroelectric films enables lower voltage operation of actuators, as well as high sensitivity sensors. Among ferroelectric films, the majority of the MEMS sensors and actuators developed have utilized lead zirconate titanate (PZT) films as the transducer. Randomly oriented PZT 52/48 films show piezoelectric $e_{31,f}$ coefficients of about $-7$ C/m$^2$ at the morphotropic phase boundary. In ferroelectric films, orientation, composition, grain size, defect chemistry, and mechanical boundary conditions all impact the observed piezoelectric coefficients. The highest achievable piezoelectric responses can be observed in $\{001\}$ oriented rhombohedrally-distorted perovskites. For a variety of such films, $e_{31,f}$ coefficients of $-12$ to $-27$ C/m$^2$ have been reported.

## 1. Introduction

The field of MEMS is a large and growing one, with numerous means reported for both sensing and actuation on-chip. Given the plethora of mechanisms by which the environment can be detected and/or useful responses made, it is worth considering the impetus for integrating piezoelectric thin films into MEMS devices (i.e. what advantages offset the need to introduce new materials into the cleanroom environment?). As usual, the answer to such a question depends significantly on the device or function in question. However, a couple of attributes come to the fore in promoting the use of piezoelectric devices in MEMS applications. These include:

1) The relatively straightforward manner in which high frequency resonant structures, with good temperature stability, can be implemented. At present there is a substantial research effort in developing MEMS devices with high electrical quality factors for rf circuits. Low frequencies are straightforward to obtain via MEMS techniques [1, 2]. While considerable progress has been made recently in the development of electrostatically actuated devices with high resonant frequencies [3, 4], these devices are inherently rather small, require sophisticated patterning techniques, and are susceptible to mass-loading changes on environmental exposure. In contrast, piezoelectric resonators with resonant frequencies in the high MHz—GHz range are widely used in scanning acoustic microscopes [5], and have seen considerable development in thin film bulk acoustic resonators (FBAR) [6, 7].
2) Piezoelectric sensors do not require power themselves (although of course any associated electronics such as charge or voltage amplifiers, etc, will need to be powered). As a result, piezoelectric MEMS are interesting for low power requirement sensors. Indeed, in situations where the sensor is operated only on an intermittent basis, it is also possible to visualize using any such sensor in a mechanically noisy environment as a power source the remainder of the time. Such energy harvesting schemes have been implemented in bulk and thick film piezoelectrics [8, 9, 10, 11]. Moreover, as sensors, it is possible to design piezoelectric devices with broad dynamic range and low noise floors.
3) The ability to perform large amplitude actuation with lower drive voltages and low hysteresis. The preponderance of MEMS literature utilizes electrostatic actuation of flexural structures. Electrostatics is relatively easy to implement, and offers the possibility of large amplitude actuation, though typically at the cost of large driving voltages and substantial hysteresis. Current-based actuation approaches, such as those utilized in many thermal and magnetically driven devices, typically require high power to operate, and in some cases are inherently slow (e.g. due to thermal time constants). In contrast, the piezoelectric effect can be utilized to drive large displacements in MEMS structures at modest voltages, low powers, and with low hysteresis.
4) The fact that piezoelectricity shows good scaling with size. That is, the energy density available for actuation remains high, even as device sizes drop. Poor scaling is, of course, one of the principal reasons that electromagnetic motors are not attractive at MEMS size scales.

5) The straightforward ability to provide electrical signals to drive or sense the device. Much like the case of electrostatics, piezoelectrics need only electrical contact for sensing or actuation. From the point of view of electronic circuit design, piezoelectric elements are CMOS compatible. The resulting signals are easily processed on chip.

With these advantages, there are some application areas where integration of piezoelectric films into the device becomes practical. Considerable progress has been made in this area over the last 20 years, with devices such as filters, micromotors, micropumps, micro-sonar arrays, scanning force microscopy tips, accelerometers, etc. [12–17] having been demonstrated.

## 2. Important Piezoelectric Coefficients for MEMS

Parameters needed for the description of the piezoelectric effect in a body with homogeneous properties on the scale of its dimensions are the strain tensor $x_i$ (where $i = 1$ to 6 in reduced notation [18]), the stress tensor $\sigma_i$ ($i = 1$ to 6), as well as the electric field $E_i$ and the electric displacement field $D_i$ ($i = 1$ to 3) vectors. Piezoelectricity means that there is a linear relation between the D-field and strain/stress:

$$D_i = \sum_k d_{ik}\sigma_k \quad or \quad D_i = \sum_k e_{ik}x_k \tag{2.1}$$

These equations describe the so-called direct effect. The piezoelectric coefficients $d_{ik}$ and $e_{ik}$ are the components of a 3$^{rd}$ rank tensor, which in reduced tensor notation corresponds to a $3 \times 6$ matrix. The $d$ and $e$ coefficients are related to each other through the stiffness tensor $c_{ij}^E$: $d_{ik} = e_{ip}c_{pk}^E$. The converse effect is described by the same set of piezoelectric coefficients:

$$x_i = \sum_k d_{ki}E_k \quad or \quad \sigma_i = -\sum_k e_{ki}E_k \tag{2.2}$$

Piezoelectricity (see [19]) occurs in crystal classes containing no center of inversion (it is also forbidden in crystal class 432). In some piezoelectric crystal classes there exists a polar axis. All the piezoelectrics used in thin film form belong to this material group. (Quartz, which has no polar axis, is only a useful piezoelectric in single crystalline form). It is conventional to assign the coordinate index 3 to the polar axis direction, and to define $d_{33}$ as a positive quantity.

In MEMS technology, most of the piezoelectric thin films are and will be polycrystalline materials. The piezoelectric effect is averaged over all the grains. At this point one has to distinguish between ferroelectric and non-ferroelectric polar materials. The latter do not allow reorientation of the polar axis. In this case, the material growth process has to provide for a textured structure that includes the alignment of the polar directions. In ferroelectric materials, the polar axis can be reoriented by an electric field. In these materials, an internal electrical polarization, i.e. the spontaneous polarization, is observed. By means of poling, i.e. the application of a large electric field able to switch the polarization, the polarization component parallel to the poling field is aligned. As a result, there is a net piezoelectric effect. In ceramics, it is conventional to assign the

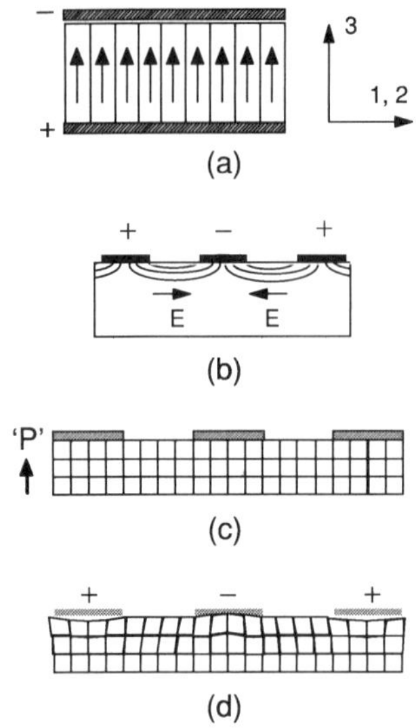


**Fig. 1.** Electrode systems for driving piezoelectric films: (a) Planar capacitor structure with top and bottom electrodes. (b) Electric field with interdigital electrodes. (c) Case of polar film with uniform perpendicular polarization. (d) Schematic drawing of strains induced by $d_{15}$ between the electrodes, and $d_{33}$, $d_{31}$ below the electrodes.

index 3 to this poling direction. For the case of vertically stacked capacitor structures (see Fig. 1a), direction 3 is perpendicular to the film plane, using the same convention as for polycrystalline ferroelectrics. The piezoelectric coefficient $d_{33}$ thus describes the piezoelectric effect in the out-of-plane direction, as e.g., the thickness change of the free film. The directions 1 and 2 are therefore in the plane of the film. In a polycrystalline film, these directions are equivalent. This means that the in-plane strains due to $E_3$ are isotropic and $d_{31}$ and $d_{32}$ are the same. The shear strain coefficients $d_{15}$ and $d_{14}$ are equal as well: $x_5 = d_{15}E_1$, $x_4 = d_{14}E_2$. The polycrystalline thin film geometry thus recovers the cylindrical symmetry of poled ceramics, yielding the same three independent elements of the piezoelectric tensor as for tetragonal symmetry ([18]), i.e. $d_{33}$, $d_{31}$, $d_{15}$. In the case of non-ferroelectric materials, the polar crystal axis is by convention the "3" axis as well. In order to profit from a maximal longitudinal and transverse response ($d_{33}$ and $d_{31}$), the film orientation must be (001) or (00$\overline{1}$). The two possibilities have opposite piezoelectric responses! Uniform texture of either of the two possibilities has thus to be achieved. In figure 1(a), the arrows symbolize the direction of the polar axis in a non-ferroelectric system such as AlN and ZnO, or the average polarization per grain in a poled ferroelectric material. The in-plane directions are again equivalent if the film is polycrystalline, or if the polar axis is a threefold or higher rotation axis. In both cases, the resulting symmetry is thus again the same as for poled ceramics. Polar films also offer the possibility to work with shear strains (if the film is ferroelectric it needs to be poled; using the same electrodes for poling and operation, the field is always parallel to the polarization). In between interdigital electrodes [20] the field lies in the film plane (electric field in 1 and 2 directions: $E_1$, $E_2$), thus perpendicular to the polar axis. The resulting shear strains add to the strains of $d_{33}$ and $d_{31}$ below the electrodes, as shown in figure 1. The result is an alternating compression and dilatation of the surface layer. At a given frequency, surface acoustic waves are excited whose wavelength is defined by the period of the interdigital electrode. The nature of the wave depends on a large number of parameters of the piezoelectric film, other involved layers and the substrate. The main interest in the use of piezoelectric films is to excite SAW waves in non-piezoelectric materials with high SAW velocities such as sapphire [21] and diamond [22, 23] in order to realize high frequency RF filters. To some extent it is also possible to sputter polar films with an oblique incidence of the atom flux onto the substrate. The polar axes consequently grow inclined to the substrate normal [24] and shear strains are excited even in the planar capacitor geometry. Shear coefficients are more difficult to access in ferroelectric films, as the polarization direction follows always the poling field, which has the same geometry as the operating field when the corresponding electrodes are the same.

The main difference between thin film and bulk materials lies in the fact that thin films are used in a composite structure, where the total elastic properties are often dominated by the other part of the structure. This other part may be a silicon cantilever, a silicon oxide or a nitride membrane, for instance. The interaction with the substrate is very anisotropic. At the interface along the in-plane directions (indices 1,2), the piezoelectric thin film and the substrate have identical strains. Perpendicular to the film plane, the thin film is free to move, i.e., $\sigma_3 = 0$. As a consequence, there is no deformation mode in which only one piezoelectric coefficient is involved. The complete equation of state needs to be analyzed. For illustration, a thin film clamped to a much thicker substrate is considered (i.e. a piezoelectric laminated plate). When a field is applied, the strains in the plane stay at zero ($x_1 = x_2 = 0$); in-plane stresses and an out-of-plane strain are developed. The equation of state using the compliance tensor and $d$ coefficients reads now as (for the average polar direction parallel to E-field: no shear components):

$$\begin{aligned} x_1 &= \left(s_{11}^E + s_{12}^E\right)\sigma_1 + d_{31}E_3 = 0 \\ x_3 &= 2s_{13}^E\sigma_1 + d_{33}E_3 \end{aligned} \tag{2.3}$$

The first line of equation 3 allows derivation of the in-plane stresses, that is an effective $e_{31,f} = -\sigma_1/E_3$. The second line leads to the definition of an effective $d_{33,f} = x_3/E_3$:

$$\begin{aligned} e_{31,f} &= \frac{d_{31}}{s_{11}^E + s_{12}^E} \\ d_{33,f} &= d_{33} - \frac{2s_{13}^E}{s_{11}^E + s_{12}^E}d_{31} \end{aligned} \tag{2.4}$$

Identical expressions can be derived using the $e$-coefficients. Both effective coefficients can be measured directly. It turns out that $d_{33,f}$ is always smaller than $d_{33}$ and that the absolute value of $e_{31,f}$ is always larger than that of $e_{31}$.

The input and output parameters for actuator and sensor applications of piezoelectric laminated plates is schematically described in Fig. 2. In an actuator, application of a voltage leads to a piezoelectric in-plane stress causing a deflection of the structure, whereas the piezoelectric thin film is strained in the out-of-plane direction. In the sensor

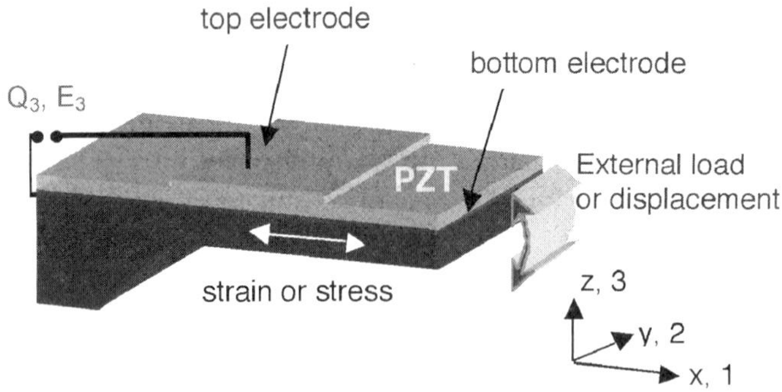


**Fig. 2.** Input ($E_3$, load) and output parameters (displacement, $Q_3$) in actuator and sensor applications for piezoelectric laminated plates.

mode, in-plane strains create the piezoelectric charges that record the deformation of the flexible structure. In addition, the film is sensitive to out-of-plane stress (which also causes in-plane strain!).

A cantilever structure can be modelled by analytical methods [25, 26].) In the actuator mode, a constant curvature is established at a given voltage. The deflection at the end of the beam is proportional to the square of the beam length. Excursions of 10 to 20 μm with 500 μm long beams have been obtained [27]. More complicated structures must be analyzed by means of finite element calculations [28].

The piezoelectric coefficients are not the only material parameters of interest. In addition to force, deflection and piezoelectric charge (current), other performance issues are important. In resonating structures the coupling coefficient ($k^2$) is essential. It describes the effectiveness of energy transformation from electrical to mechanical energy. Part of the electrical energy is dissipated and transformed to heat. The dielectric loss angle (tanδ) is involved in this phenomenon. The dielectric loss also generates a noise current or voltage, which imposes limits to the resolution, or more precisely, to the signal-to-noise-ratio of sensors. Whether the intrinsic noise due to tanδ limits the performance or not depends on the application, electronics, and external noise levels. Table 1 summarizes the figures of merit for the various criteria. Note that the figure of merit for intrinsic signal-to-noise ratio does not depend on whether voltage or current is detected.

**Table 1.** Figures of merit for piezoelectric sensors (limited by signal to noise ratio)

| Physical parameter | Figure of merit |
|---|---|
| deflection force, piezoelectric charge of deflected piezoelectric laminated structure | $e_{31,f}$ |
| piezoelectric voltage in deflected piezoelectric laminated structure | $e_{31,f}/\varepsilon_0\varepsilon_{33}$ |
| coupling coefficient for plate wave | $e_{31,f}^2/\varepsilon_0\varepsilon_{33}$ |
| coupling coefficient for thickness wave | $e_{33}^2/(\varepsilon_0\varepsilon_{33}c_{33}^D) \approx d_{33,f}^2 \cdot c_{33}^E/\varepsilon_0\varepsilon_{33}$ |
| Signal-to-noise ratio | $e_{31,f}(\varepsilon_0\varepsilon_{33}\tan\delta)^{-1/2}$ |

## 3. AlN and ZnO

Both AlN and ZnO are wurtzite structured materials which show a piezoelectric response along [0001]. For MEMS applications, ZnO and AlN are commonly sputter deposited ([29, 30]). ZnO is preferentially deposited at room temperature in order to obtain a high resistivity. AlN does not exhibit conductivity problems and good films are reported to grow between 100 °C and 900 °C. A convenient temperature for sputter tools as well as for applications is the range of 200 to 400 °C [31]. Essential for achieving good piezoelectric coefficients is the homogeneous nucleation of grains with (0001) texture of the same polar direction. It appears that a more perfect {0001} texture, as determined by X-ray diffraction (no peaks of other orientations, small rocking curve width) includes as well a better selectivity between the (0001) and $(000\bar{1})$ versions. In other words, the same growth mechanisms leading to a more perfect (0001) texture and crystalline quality lead also to a preferential growth of one of the polarizations. A first indication of such phenomena was found by Hickernell and coworkers [32] when they observed that the SAW transducer loss was smaller when the X-ray peak width was smaller. More recently, a clear correlation was established between the coupling coefficient of a BAW resonator and the rocking curve width of the AlN film [33]. Evaluating $d_{33,f}$ of AlN as a function of various deposition conditions, it was observed that the most important parameter influencing the $d_{33,f}$ was the ion self bias of the substrate holder, meaning the ion energy and flux of the bombarding species ($Ar^+$, $N_2^+$) [34]. Besides process influence, a distinct influence of the substrate was observed as well. The crystalline order of the latter may influence AlN nucleation in favor of or against (0001) orientation. According to Table 2, AlN and ZnO thin films exhibit quite similar piezoelectric properties. The transverse coefficient is almost equal, while the longitudinal effect is somewhat larger in ZnO. The ZnO thin film properties given in Table 2 are close to single crystal values. In the case of AlN there are no single crystal reference data available.

AlN has two major advantages over ZnO. First, it is perfectly compatible with silicon semiconductor technology, whereas Zn, as a fast diffusing ion, is problematic. Secondly, AlN is a large band gap (6 eV) material with a large resistivity, whereas ZnO is really a semiconductor (3.0 eV) with the inherent risks that off-stoichiometry might lead to doping (as e.g. Zn interstitials [35]) and thus to an increased conductivity. It is,

**Table 2.** Thin film piezoelectric and dielectric properties

| Coefficients / figures of merit | ZnO [40, 41] | AlN [31, 42] | PZT (1–3 μm) [43] |
|---|---|---|---|
| $e_{31,f}$ ($Cm^{-2}$) | −1.0 | −1.05 | − 8 .. −12 |
| $d_{33,f}$ (pm/V) | 5.9 | 3.9 | 60 .. ..130 |
| $\varepsilon_{33}$ | 10.9 | 10.5 | 300 .. 1300 |
| $e_{31,f}/\varepsilon_0\varepsilon_{33}$ (GV/m) | −10.3 | −11.3 | −0.7 ..−1.8 |
| $e^2_{31,f}/\varepsilon_0\varepsilon_{33}$ (GPa) | 10.3 | 11.9 | 6 .. 18 |
| $\tan\delta$ (@1 to 10 kHz, $10^5$ V/m) | 0.01 .. 0.1 | 0.003 | 0.01 .. 0.03 |
| $e^2_{31,f}/sqrt(\varepsilon_0\varepsilon_{33}\tan\delta)$ ($10^5$ $Pa^{1/2}$) | 3 .. 10 | 20 | 4 .. 8 |
| $c^E_{33}$ (GPa) (PZT52/48 ceramic) | 208 | 395 | 98 |
| $d^2_{33,f}\cdot c^E_{33}/\varepsilon_0\varepsilon_{33}$ | 7.4% | 6.5% | 7 % .. 15 % |

in fact, difficult to obtain ZnO with a high resistivity. The dc conductivity translates into a high dielectric loss at low frequencies, which is especially harmful for sensors and actuators working at frequencies below about 10 kHz. The fact that more MEMS studies have been performed with ZnO than with AlN is explained by the better availability of ZnO films, the less demanding vacuum conditions for ZnO, and some early negative experiences with mechanical stresses in AlN thin films. This is, however, not a problem when using modern magnetron sources and careful tuning of ion bombardment impact. The two materials are compared in Table 2 on the basis of measured thin film properties. An accurate method to determine $d_{33,f}$ consists of utilizing a double side laser interferometer, which measures the thickness change unambiguously and avoids errors due to sample deflections [36]. The $e_{31,f}$ coefficient is determined by means of deflecting or deflected structures. The various methods are listed in [37] and [38]. The data on ZnO are based on SAW experiments. MEMS type deflective devices where the ZnO is grown on metal layers typically show smaller $e_{31,f}$ values of typically $-0.6$ C/m$^2$ [39]. Table 2 shows that AlN is better suited for deflection devices (especially as sensors), while ZnO should yield larger coupling coefficients for longitudinal bulk acoustic wave generation.

## 4. Ferroelectric Thin Films

In other devices, maximizing the piezoelectric coefficient is of considerable importance in reducing the drive voltage or increasing the speed or sensitivity of a MEMS device. For example, in a piezoelectrically-actuated switch for rf signals, the overall speed of the device is limited by the device size, with smaller dimensions yielding higher resonance frequencies. However, since it is imperative that a certain standoff be maintained, in order to maintain a good "on–off" ratio at high frequencies, while the device must be laterally small, it should still be possible for a vertical gap (ideally of several microns) to be maintained [44]. This can be achieved at modest driving voltages if the actuator piezoelectric coefficient is large.

For applications where large thin film piezoelectric coefficients are required, ferroelectric compositions are extremely attractive. However, this comes at some cost in terms of complexity of the actuator composition (e.g. the necessity for 2–5 cations), need to develop film deposition routes that maintain stoichiometry, phase, and in some cases, orientation, and challenges associated with bringing new materials sets into cleanrooms. Thankfully, the piezoelectric MEMS community can benefit from the tremendous infrastructure development resulting from ferroelectric nonvolatile access memories, which should speed implementation of ferroelectrics-based MEMS devices. This section will detail the available literature on the piezoelectric properties of ferroelectric thin films, with emphasis on the importance of morphotropic phase boundaries and high transition temperatures for MEMS applications.

In ferroelectric materials, there are several contributions to the available piezoelectric response. The intrinsic piezoelectric response is the response that would be made due to application of an electric field to an appropriately averaged ensemble of single crystals. The extrinsic response is typically associated with motion of non-180° domain walls or of phase boundaries. In thin films, the extrinsic contribution to the piezoelectric coefficient is considerably reduced relative to bulk materials [45, 52, 46]. This limits the magnitude of the piezoelectric coefficient that is available in randomly oriented films. However, it has been demonstrated that the intrinsic piezoelectric coefficients are dependent on crystallographic orientation [47]. In particular, many rhombohedrally-distorted perovskites, when poled and driven along the crystallographic [001], have extremely high piezoelectric coefficients [48]. In contrast, in some of the tetragonal perovskites, the piezoelectric response is maximized when measured along the polar axis. As a result, it is extremely interesting to compare the properties of randomly oriented to textured films to see if the same enhancement of the piezoelectric coefficient transfers to thin film samples. Orientation can be controlled by deposition conditions, heating rate, and substrate choice, among other factors.

### 4.1. Morphotropic phase boundaries

The most widely utilized ferroelectric thin films for piezoelectric applications are based on lead zirconate titanate (PZT). This system illustrates many of the important features of the other ferroelectric compositions, so emphasis will be placed on this family. The lead zirconate titanate phase diagram is dominated by a rhombohdrally distorted ferroelectric region at low Zr contents, a tetragonal region at high Ti concentrations, separated by a morphotropic phase boundary (MPB), with a sliver of monoclinically distorted perovskite at lower temperatures near the MPB [49]. In bulk ceramics, maxima in the piezoelectric d and e coefficients are generally observed at the MPB. The same behavior is often [50–54], but not universally [55, 56] reported in thin films (See Fig. 3). At present there does not appear to be a strong correlation between factors like deposition method (e.g. not all films deposited by chemical solution deposition show comparable composition dependence), film orientation, or film thickness with whether or not strong enhancement of properties is observed at the MPB. Perhaps some of these differences can be ascribed to the effect of grain size, or the relative poling efficiency in films of different compositions.

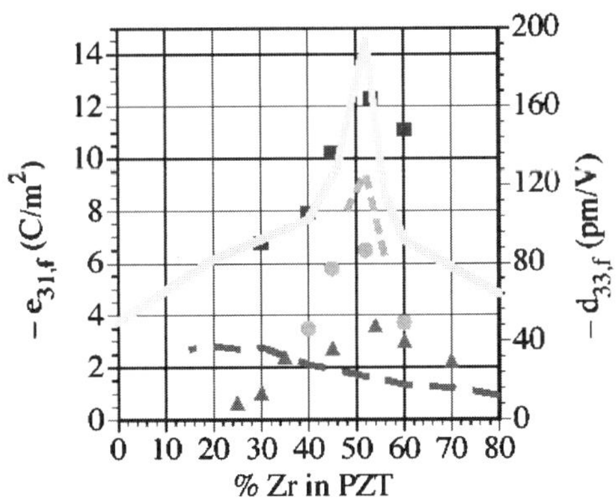


**Fig. 3.** Composion dependence of the relative piezoelectric response in PZT thin films. Data from [50, 51, 52, 54, 56]. Symbols are for $e_{31,f}$ data, lines are for $d_{33,f}$ data.

For good piezoelectric response, it is important not only that a system display a morphotropic phase boundary, but what the transition temperature at the morphotropic phase boundary is. Of course, the magnitude of the piezoelectric coefficient can be enhanced if a system with a low transition temperature is utilized. However, this comes at the expense of heightened temperature dependence in the piezoelectric response, increased tendency towards depoling, and often limitations on the available drive levels due to low coercive fields [57]. Consequently, it is interesting to consider systems that show morphotropic phase boundaries with high transition temperatures. In the lead-based perovskites, the highest transitions temperatures at known morphotropic phase boundaries occur in the PZT and $PbYb_{1/2}Nb_{1/2}O_3$-$PbTiO_3$ systems. In both cases, the transition temperatures are near 360 °C at the MPB. Even higher transition temperature MPB compositions have been reported recently in $BiMeO_3$-$PbTiO_3$ solid solutions. High piezoelectric coefficients ($d_{33} = 465$ pC/N) have been measured in bulk ceramic $BiScO_3$-$PbTiO_3$ 36/64 with transition temperatures of 450 °C [57, 58]. While the transition temperatures of thin films can differ from those of bulk ceramic materials, as a rule of thumb, for the large film thicknesses required in many MEMS devices, the transition temperatures are often reasonably close (often within ~50 °C) to those observed in bulk ceramics and single crystals. Consequently, the high $T_c$ MPB ceramics offer a good guide as to the materials that will be interesting for high $T_c$, high piezoelectric activity thin films.

### 4.2. Effect of film orientation

There are discrepancies in the literature concerning whether or not there is a strong orientation dependence of the piezoelectric properties in ferroelectric thin films. For example, various authors have reported maxima for {111}, {100}, or randomly oriented PZT films near the morphotropic phase boundary [51, 52, 59]. It is possible that the differences are related to changes in the film domain state, perhaps due to residual stresses. Certainly, the piezoelectric response does appear to correlate with the magnitude of the remanent polarization; films with low remanent polarizations have low piezoelectric coefficients. This may also indicate the importance of grain size in ferroelectric thin films, especially for films under tensile stress. As has been shown by Tuttle and co-workers [60], in many thin films, the domain state is governed substantially by the stress experienced by the film during cooling through the phase transformation temperature. Consequently, perovskite structured thin films, when exposed to tensile stresses during the cooling process, tend to have their polar vectors in orientations approximately parallel to the substrate (of course constrained by the crystallography of the system); films under compressive stresses tend to have the polarization oriented more nearly perpendicular to the substrate. Because the mobility of non-180° domain walls in ferroelectric thin films tends to be limited compared to bulk materials, once the ferroelastic domain orientation is established, it is difficult to change at room temperature [45, 61]. Larger grained films tend to show more complicated domain structures within a single grain [60], and so may be less susceptible to reduction in the remanent polarization and piezoelectric response.

PZT film processing offers the possibility to achieve textured films. Growth of PZT is strongly nucleation controlled, meaning that the nucleation can be manipulated to

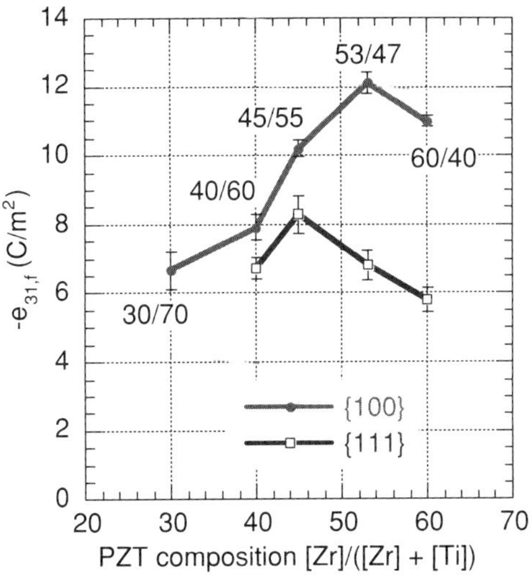


**Fig. 4.** Transverse piezoelectric response in {111} and {100} textured PZT films as a function of composition. All of the films were deposited on Pt-coated Si substrates. Data from [43].

achieve a desired orientation. On a perovskite single crystal substrate or electrode, PZT grows almost inevitably in the same orientation as the template. On platinum this is less evident. Platinum films are most often deposited with a (111) texture. As the lattice constant of Pt is rather close to that of PZT (2–3 % mismatch), one could expect that PZT tends to grow with a (111) orientation on Pt. This is, however, not necessarily so. A mixture of all major orientations is obtained in the general case. Control of the orientation by using thin buffer layers has been reported to be successful in sputtered films on Pt-coated substrates. For example, $PbTiO_3$ will often adopt a (100) orientation on Pt, and serve as a template for well-oriented PZT films, while a thin $TiO_2$ buffer layer can yield (111) oriented perovskites [62]. During sol-gel film processing, the concentration of excess lead, the pyrolysis temperature, as well as the ramp speed used in the rapid thermal annealing all play a substantial role for orientation control [63]. The temporary formation of a $Pt_3Pb$ interface layer has been identified as a mechanism promoting (111) orientation [64, 65]. Texture control is important for obtaining optimal properties. PZT {100} films show substantially larger piezoelectric responses than those of {111} films, as shown in Fig. 4.

Similar orientation—induced enhancements in the piezoelectric response have been reported in a number of perovskite solid solutions. In particular, high piezoelectric coefficients have been reported in {001} films (on the rhombohedral side of the morphotropic phase boundary) in $PbMg_{1/3}Nb_{2/3}O_3$-$PbTiO_3$ (PMN-PT) [66, 67] $PbYb_{1/2}Nb_{1/2}O_3$-$PbTiO_3$ [78] and $BiScO_3$-$PbTiO_3$ [78, 79]. For example, in PMN-PT 70/30 films on Si, the ratio of the piezoelectric response for {100} oriented films, relative to {111} ranged from 1.6–1.9 [66], with a maximum $e_{31,f}$ coefficient of $-5.5C/m^2$. In both PYbN-PT 50/50 and 60/40 epitaxial films on $SrRuO_3/SrTiO_3$, $e_{31,f}\{100\}/e_{31,f}\{111\}$ was ~3.6 (See Fig. 5). For the {100} films, $e_{31,f}$ values of $-14C/m^2$ to $-19C/m^2$ were obtained in films of 1 and 3 microns thickness,

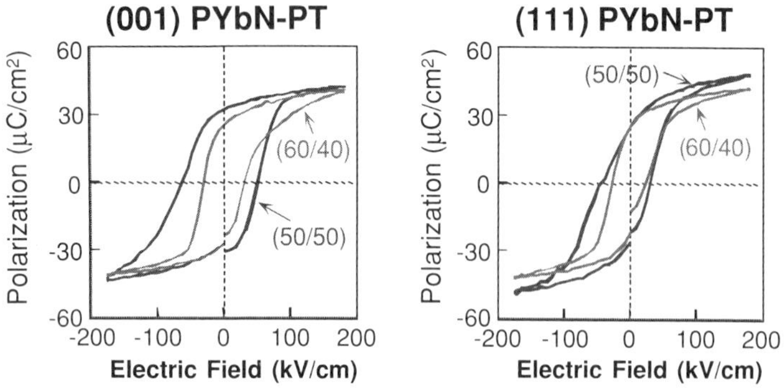


| (001) PYbN-PT | (50/50) | (60/40) |
|---|---|---|
| $P_r$ ($\mu C/cm^2$) | 30 | 26 |
| $\varepsilon_r$ | 1000 | 1000 |
| **$e_{31,f}$ ($C/m^2$)** | **-12** | **-9** |

| (111) PYbN-PT | (50/50) | (60/40) |
|---|---|---|
| $P_r$ ($\mu C/cm^2$) | 25 | 24 |
| $\varepsilon_r$ | 1300 | 1300 |
| **$e_{31,f}$ ($C/m^2$)** | **-3.3** | **-2.5** |

**Fig. 5.** Piezoelectric anisotrophy in PYbN-PT epitaxial films. Data from [78].

respectively [78, 68]. For the $BiScO_3$-$PbTiO_3$ films, $e_{31,f}$ of $-9$ to $-12$ C/m$^2$ have been achieved in {100} epitaxial films [78, 79]. This is especially encouraging, because the high piezoelectric response is coupled with a high transition temperature of ~460 °C (~100 °C higher than at the PZT morphotropic phase boundary). In general, for films that show enhanced piezoelectric response with orientation (of the type predicted based on phenomenology [47, 69]), the achieved anisotropy in directly related to the quality of the orientation. Thus, many of the epitaxial films show stronger anisotropy than fiber textured films (where the percentage of oriented material may be substantially reduced—see for comparison [78, 70, 71]. This type of orientation-enhanced piezoelectric response is especially useful in thin films, as ferroelastic domain wall motion is so heavily clamped.

It is critical to note, however, that orientation itself, does not guarantee large piezoelectric response. This is clear in a comparison of lead magnesium niobate—lead titanate films grown on Si with those grown on most single crystal oxide substrates. Films grown on Si generally have low remanent polarization values (~10 μC/cm$^2$), and consequently, low $e_{31,f}$ values when poled through the film thickness. This is largely because the tensile stress in the films resulting from the mismatch in thermal expansion coefficients rotates the hysteresis loop clockwise [72]. The result is low remanent piezoelectric response. In contrast, films grown on single crystal oxide substrates such as $SrTiO_3$ or $LaAlO_3$, show much squarer hysteresis loops, and larger piezoelectric coefficients [73, 67]. It is important to note that this is not just a consequence of crystal quality or epitaxy. Thus, when the elastic constraints imposed by the substrate are removed, the remanent polarization and $e_{31,f}$ coefficients of epitaxial PMN-PT films on Si increase substantially [67]. While some of the relaxor-based materials such as PMN-PT appear to be especially sensitive to imposed stresses, similar stress-induced rotations of the hysteresis loops have been observed in PZT [74]. Thus, it is likely that the behavior is rather general.

In addition to orientation, the piezoelectric properties of ferroelectric materials including PZT are strongly influenced by composition, grain size, defect chemistry, and mechanical boundary conditions [75, 76, 77]. This is one area where film deposition conditions and substrate choice can strongly impact the quality of the resulting materials. For example, it has been suggested that small grain size can lead to grains where a limited number of domain variants are observed [60]. The result for films on Si can be lower remanent polarizations, and suppressed extrinsic contributions to the available piezoelectric response [60, 45]. In many polycrystalline films, the lateral grain size is controlled by the grain size of the underlying Pt electrode. Since the latter is often quite fine, it is not uncommon for film grain sizes to be in the range of 50–200 nm. If, however, the deposition conditions have been optimized so that nucleation is not dominated by the electrode crystallites, then larger grain sizes can result.

A second important factor is substrate-dominated changes in the observed piezoelectric response. As explained in the section on piezoelectric coefficients, thin films are mechanically clamped by a massive substrate. The composition of the substrate would be expected to be important both due to differences in the thermal expansion coefficient (and hence the resulting film stress) and due to the importance of the substrate elastic moduli in controlling the mechanical boundary conditions of the film. There has been relatively little literature on the importance of the substrate in controlling the magnitude of the piezoelectric response. This may be in part because it is so difficult to control the microstructure of the ferroelectric film independent of the substrate. Work by Dubois suggests that for a given film microstructure and poling state, the magnitude of the $e_{31,f}$ coefficient increases with applied static tensile strains, and decreases for applied compressive strains [38]. Figure 6 shows a comparison of the piezoelectric responses

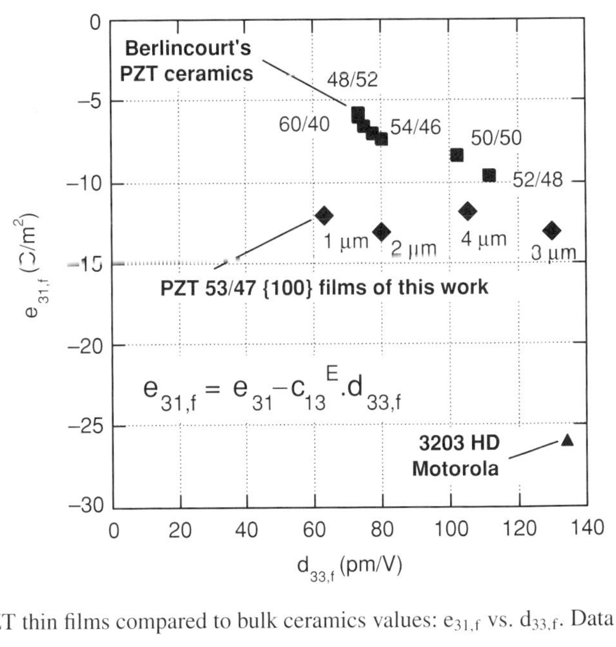


**Fig. 6.** PZT thin films compared to bulk ceramics values: $e_{31,f}$ vs. $d_{33,f}$. Data from [43].

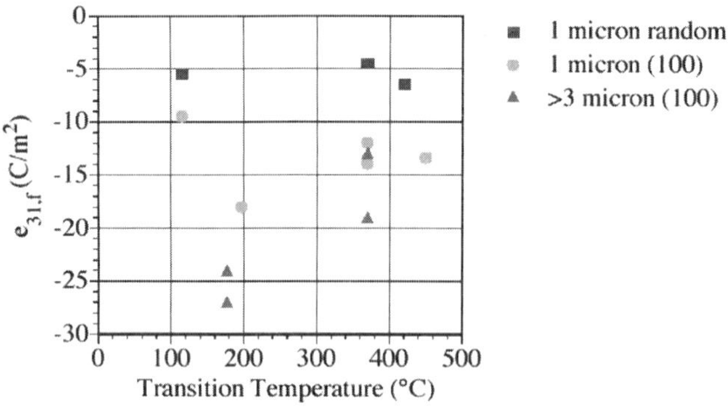


**Fig. 7.** Piezoelectric coefficients of a number of morphotropic phase boundary ferroelectric thin films plotted as a function of the measured transition temperature. Data from [15, 52, 67, 73, 78, 79].

of thin film and bulk PZT ceramics, highlighting the importance of elastic boundary conditions. 3 to 4 μm thick PZT films are close to the performance calculated for a standard hard ceramics, the PZT 4D. However, their response falls below the $e_{31,f}$ of the optimized, rather soft ceramics (Motorola 3203D) by a factor of two. It is noteworthy that an increase in film thickness increases $d_{33,f}$ significantly, but does not improve $e_{31,f}$. Such behavior could be explained if porosity increases with thickness: substrate effects are reduced—increasing $d_{33,f}$—and in-plane stiffness is reduced as well—decreasing $e_{31,f}$. However, TEM and SEM inspections do not support the idea of increased porosity. This effect is thus very likely related to a mechanism of elastic domains that can move more easily in thicker films, and that give rise to out-of-plane strain, but not to in-plane strain (rhombohedral symmetry).

Figure 7 shows a comparison of the available data in the literature on morphotropic phase boundary ferroelectrics. The low transition temperature compounds (<200 °C) are PMN-PT materials, the films near 350 °C–420 °C are either PZT or PYbN-PT, while the highest transition temperature data are for $BiScO_3$-$PbTiO_3$ films. The substrates range from single crystal oxides (e.g. $LaAlO_3$ or $SrTiO_3$) to Si. In all cases where data were reported, the measured transition temperatures (rather than the bulk numbers) were used in the plot. This was done since the transition temperatures in films depend on the levels of stress the film is under and the composition (including A:B site ratio). It is clear that the highest piezoelectric coefficients are shown in oriented thin films on the rhombohedral side of the morphotropic phase boundary, suggesting that the domain rotation argument proposed for bulk single crystals applies to thin films as well. In addition, it is clear that the magnitude of the piezoelectric coefficient depends somewhat on film thickness, with the highest piezoelectric constants being reported for thicker films.

## 5. Conclusions

This paper reviews the current state of piezoelectric thin films for microelectromechanical systems. Over the last 15 years, considerable progress has been made in optimizing the deposition conditions for thin films to improve the available piezoelectric activity. Control of the growth process is essential in the wurtzite-structured materials, since the polarization direction cannot be switched following deposition. The resulting materials show good temperature stability in the response, coupled with smaller piezoelectric coefficients. Much larger $e_{31,f}$ values can be achieved in many of the ferroelectric compositions. The resulting properties are dependent on film grain size, thickness, and orientation, with the best responses reported to date in $\{001\}$ oriented films near the morphotropic phase boundary.

## References


[1] F. D. Bannon, J. R. Clark, C. T. C. Nguyen, *IEEE J. Sol. State Circuits* **35** (2.4), 512 (2000).
[2] L. W. Lin, R. T. Howe, A. P. Pisano, *J. MEMS* **7** (2.3), 286 (1998).
[3] L. Pescini, H. Lorenz, R. H. Blick, *Appl. Phys. Lett* **82** (2.3), 352 (2003).
[4] M. A. Abdelmoneum, M. U. Demirci, C.T.-C. Nguyen, The Sixteenth Annual International Conference on Micro Electro Mechanical Systems, 2003. MEMS-03 Kyoto. IEEE, 698 (2003).
[5] Y. Ito, K. Kushida, K. Sugawara, H. Takeuchi, *IEEE TUFFC* **42** 316 (1995).
[6] M. A. Dubois and P. Muralt, *Appl. Phys. Lett.* **74**, 3032 (1999).
[7] S. H. Kim, J.H. Kim, H. D. Park, and G. W. Yoon, *J. Vac. Sci. Tech. B* **19**, 1164 (2001).
[8] M. Umeda, K. Nakamura, and S. Ueha, *Jpn. J. Appl. Phys. Part 1* **36** (5B): 3146 (1997).
[9] G. W. Taylor, J. R. Burns, S. M. Kammann, W. B. Powers, and T. R. Welsh, *IEEE J. Ocean. Eng.* **26** (2.4), 539 (2001).
[10] G. K. Ottman, H. F. Hofmann, G. A. Lesieutre, *IEEE Trans. Power Electron.* **18** (2.2), 696 (2003).
[11] P. Glynne-Jones, S. P. Beeby, N.M. White, *IEE Proc. Sci. Meas. Tech.*, **148** (2.2) 68 (2001).
[12] J. J. Bernstein, S. L. Finberg, K. Houston, L. C. Niles, H. D. Chen, L. E. Cross, K. K. Li, and K. Udayakumar, *IEEE Trans. UFFC*,**44**, 960 (1997).
[13] Y. Nemirovsky, A. Nemirovsky, P. Muralt, and N. Setter, *Sen. and Act., A* **56**, 239 (1996).
[14] P. Muralt, M. Kohli, T. Maeder, A. Kolkin, K. Brooks, N. Setter, and R. Luthier, *Sen. and Act., A* **48**, 157 (1995).
[15] P. Muralt, *IEEE Trans. UFFC* **47** (2.4) 903 (2000).
[16] D. L. Polla and L. F. Francis, *MRS Bulletin*, **21** (7) 59 (1996).
[17] L.-P. Wang, K. Deng, L. Zou, R. Wolf, R. J. Davis, and S. Trolier-McKinstry, *IEEE Electron Device Lett.* 23, 182 (2002).
[18] J. F. Nye, *Physical Properties of Crystals: Their Representation by Tensors and Matrices*, (Clarendon Press, Oxford 1979).
[19] T. Ikeda, *Fundamentals of Piezoelectricity*, (New York : Oxford University Press, 1996).
[20] R. M. White and V.W. Voltmer, *Appl. Phys. Lett.* **17**, 314 (1965).
[21] K. Tsubouchi and N. Mikoshiba, *IEEE Trans. Son. Ultrasonics* **32**, 634 (1985).
[22] H. Nakahata, H. Kitabayashi, T. Uemura, A. Hachigo, K. Higaki, S. Fujii, Y. Seki, K. Yoshida, and S. Shikata, *Jpn. J. Appl. Phys. Pt 1* **37** 2918 Sp. Iss. SI (1998).
[23] G. F. Iriarte, *J. Appl. Phys.* **93**, 9604 (2003).
[24] I. Cerven, T. Lacko, I. Novotny, V. Tvarozek, M. Harvanka, *J. Cryst. Growth* **131** (3–4): 546 (1993).
[25] J. G. Smits and W. S. Choi, *IEEE TUFFC* **38** (2.3), 256 (1991).
[26] Q. Meng, M. Mehregany, K. Deng, *J. Micromech. Microeng.*, **3**, 18 (1993).
[27] Ph. Luginbuhl, G.-A. Racine, Ph. Lerch, B. Romanowicz, K. G. Brooks, N.F. de Rooij, Ph. Renaud, and N. Setter, *Int. Conf. Solid-State Sens. Act., Proc.*, **1**, 413 (1995).

[28] T. Fabula, H. Wagner, B. Schmidt, S. Buttgenbach, *Sens. Act. A* **42** (1–3), 375 (1994).
[29] F. S. Hickernell, *Proc. IEEE* **64** 631 (1976).
[30] A. Rodriguez-Navarro, W. Otano-Rivera, J. M. Garcia—Ruiz, and R. Messier, *J. Mater. Res.* **12** 1850 (1997).
[31] M. A. Dubois and P. Muralt, *Appl. Phys. Lett.* **74** (20), 3032 (1999).
[32] F. J. Hickernell, R.X. Yue, F.S. Hickernell, *IEEE Trans. UFFC* **44**, 615 (1997).
[33] H. P. Loebl, C. Metzmacher, R.F. Milsom, P. Lok, F. van Straten, A. Tuinhout, *J. Electroceramics*, forthcoming issue.
[34] M.-A. Dubois, P. Muralt, *J. Appl. Phys.* **89**, 6389 (2001).
[35] A. Barker, S. Crowther, D. Rees, *Sensors & Actuators* **58**, 229 (1997).
[36] Kholkin AL, Wutchrich C, Taylor DV, Setter N, *Rev. Sci. Instrum.* **67** (5), 1935 (1996).
[37] J. F. Shepard, Jr., P. J. Moses, and S. Trolier-McKinstry, *Sens. Actuators A* **71** 133 (1998).
[38] M. A. Dubois and P. Muralt, *Sens. Act. A* **77** (2.2), 106 (1999).
[39] F. J. von Preissig, H. Zeng, E. S. Kim, *Smart Mater. Struct.* **7**, 396 (1998).
[40] J. G. Gualtieri, J.A. Kosinski, A. Ballato, *Trans. UFFC* **41**, 53 (1994).
[41] G. Carlotti, G. Socino, A. Petri, and E. Verona, in *Proc. 1987 IEEE Ultrason. Symp.* Oct. 1987, p. 295.
[42] K. Tsubouchi, K. Sugai, and N. Mikoshiba, *IEEE Ultrason. Symp.* Oct. 1981, pp. 375.
[43] N. Ledermann, P. Muralt, J. Baborowski, S. Gentil, K. Mukati, M. Cantoni, A. Seifert and N. Setter, *Sens. Act. A* **105**, 162 (2003).
[44] D. Peroulis, S. P. Pacheco, K. Sarabandi, and L. P. B. Katehi, *IEEE Trans. Microwav. Theory Tech.* **51** (10) 259 (2003).
[45] F. Xu, S. Trolier-McKinstry, W. Ren, and B. Xu, *J. Appl. Phys.* **89** (2.2), 1336 (2001).
[46] K. Saito, T. Kurosawa, T. Akai, T. Oikawa, and H. Funakubo, *J. Appl. Phys.* **93** (2.1) 545 (2003).
[47] X. H. Du, J. H. Zheng, U. Belegundu, and K. Uchino, *Appl. Phys. Lett.* **72** (19), 2421 (1998).
[48] S.-E. Park, and T.R. Shrout, *IEEE Trans. UFFC* **44** 1140 (1997).
[49] B. Noheda, D. E. Cox, G. Shirane, R. Guo, B. Jones, L. E. Cross, *Phys. Rev. B* **63** (2.1), 014103 (2001).
[50] H. D. Chen, K. R. Udayakumar, C. J. Gaskey, and L. E. Cross, *Appl. Phys. Lett.* **67** (23) 3411 (1995).
[51] A. Seifert, N. Ledermann, S. Hiboux, J. Baborowski, P. Muralt, and N. Setter, *Integr. Ferro.* **35** (1–4), 1889 (2001).
[52] F. Xu, R. A. Wolf, T. Yoshimura, and S. Trolier-McKinstry, *Proc. 11th Int. Symp. Electrets*, 386 (2002).
[53] R. Wolf and S. Trolier-McKinstry, *J. Appl. Phys.***95** (2.3), 1397 (2004).
[54] Haccart T, Soyer C, Cattan E, Remiens D, *Ferroelectrics* **254** (1–4), 185 (2001).
[55] I. Kanno, H. Kotera, K. Wasa, T. Matsunaga, T, Kamada, and R. Takayama, *J. Appl. Phys.* **93** (7), 4091 (2003).
[56] D.-J. Kim, J.-P. Maria, A. I. Kingon, and S. K. Streiffer, *J. Appl. Phys.* **93**, 5568 (2003).
[57] R. E. Eitel, C. A. Randall, T. R. Shrout, P. w. Rehrig, W. Hackenberger, and S. E. Park, *Jpn. J. Appl. Phys. Part 1* **40** (10) 5999 (2001).
[58] T. R. Shrout, private communication (2002).
[59] K. Kakimoto, H. Kakemoto, S. Fujita, and Y. Masuda, *J. Am. Ceram. Soc.* **85** (2.4), 1019 (2002).
[60] B. A. Tuttle, T. J. Garino, J. A. Voight, T. J. Headley, D. Dimos, M. O. Eatough, *Science and Technology of Electroceramic Thin Films*, ed. O Auciello, R Waser 117, Kluwer Academic Publishers, The Netherlands (1995).
[61] A. L. Kholkin, M. L. Calzada, P. Ramos, J. Mendiola, and N. Setter, *Appl. Phys. Lett.* **69** 3602 (1996).

[62] P. Muralt, *J. Micromech. Microeng.* **10**, 136 (2000).
[63] K.G. Brooks, I.A. Reaney, R. Klissurska, Y. Huang, L. Bursill, N. Setter, *J. Mater. Res.* **9** 2540 (1994).
[64] S.-Y. Chen and I.-W. Chen, *J. Amer. Cer. Soc.* **77** 2337 (1994).
[65] R.W. Whatmore, Q. Zhang, Z. Huang, and R.A. Dorey, *Materials Science in Semiconductor Processing* **5** 65 (2003).
[66] J. H. Park, F. Xu, and S. Trolier-McKinstry, *J. Appl. Phys.* **89**(2.1) 568 (2001).
[67] T. Yoshimura and S. Trolier-McKinstry, unpublished.
[68] T. Yoshimura and S. Trolier-McKinstry, *Integr. Ferroelectr.* **50**, 33 (2002).
[69] A. J. Bell, *J. Appl. Phys.* **89**(7), 3907 (2001).
[70] Q.Q. Zhang, Q. F. Zhou, and S. Trolier-McKinstry, *Appl. Phys. Lett.* **80**(18), 3370 (2002).
[71] Q. F. Zhou, Q. Q. Zhang, and S. Trolier-McKinstry, *J. Appl. Phys.* **94** (5) 3397 (2003).
[72] Z. Zhang, J-H. Park and S. Trolier-McKinstry, *MRS. Proc. 596 Ferroelectric Thin Films VIII*, editors: R. W. Schwartz, P. C. McIntyre, Y. Miyasaka, S. R. Summerfelt, and D. Wouters, 73–77 Materials Research Society, Warrendale, PA (2000).
[73] J.-P. Maria, J. F. Shepard, Jr., S. Trolier-McKinstry, T. R. Watkins, and A. E. Payzant, accepted, Int. J. Appl. Ceram. Tech. (2004).
[74] J. F. Shepard Jr., S. Trolier-McKinstry, M. Hendrickson, and R. Zeto, *MRS Proc. 459: Materials for Smart Systems II* 47–52 (1997).
[75] N. Kim, *Ph. D. Thesis*, The Pennsylvania State University (1994).
[76] F. Jona, and G. Shirane, *Ferroelectric Crystals*, Pergamon Press, New York (1962).
[77] T. M. Shaw, S. Trolier-McKinstry, and P.C. McIntyre, *Ann. Rev. Mater. Sci.* **30** 263 (2000).
[78] T. Yoshimura and S. Trolier-McKinstry, *J Appl. Phys.* **92**(7), 3979 (2002).
[79] J. Nino, T. Yoshimura and S. Trolier-McKinstry, *J. Mat. Res.* **19**(2.2), 568 (2004).

# Chapter 11

# Science and Technology of High Dielectric Constant Thin Films and Materials Integration for Application to High Frequency Devices

O. Auciello, S. Saha, W. Fan, B. Kabius, S. K. Streiffer, D. Y. Kaufman, J. Im[1] and P. Baumann[2]

*Argonne National Laboratory, Materials Science Division*
*9700 South Cass Ave, Argonne, IL-60439, USA*
[1] *present address: Agere Systems, Wafer Fab R&D, Alhambra, CA*
[2] *present address: Aixtron, Aachen, Germany*

The science and technology of $Ba_{1-x}Sr_xTi_{1+y}O_{3+z}$ (BST) thin films are of great interest due to potential applications in a broad range of high frequency and high dielectric constant thin film based devices. In this chapter, we discuss the synthesis and characterization of BST thin films and the development of material integrations strategies suitable for fabrication of BST-based capacitors for integration in devices. In section 2, we discuss the synthesis of BST films using RF-magnetron sputtering and characterization of composition and microstructure and how they affect the physical and electrical properties of the films. In section 3, we discuss the synthesis of BST films via MOCVD and the resulting microstructure and electrical properties. Section 4 is dedicated to a discussion of the effects of BST/electrodes interface contamination and structure on the electrical properties of BST-based thin film capacitors. Section 5 is dedicated to a discussion of the application of sputter-deposited BST thin films to the fabrication of capacitors and their integration into phase shifters developed by York's group at the University of California-Santa Barbara. In section 6, we discuss the first integration of BST thin films with copper-based bottom electrodes using a new diffusion barrier layer to inhibit the oxidation of copper layers when growing oxide BST films at high temperature in an oxygen environment. The chapter concludes with a discussion of the view to future work on the science and technology of BST thin films and integration into high-frequency devices.

---

This work was supported by the US Department of Energy, BES-Materials Sciences, under Contract W-13-109-ENG-38.

# 1. Introduction

It is well known that nonlinear dielectrics such as $Ba_{1-x}Sr_xTi_{1+y}O_{3+z}$ (BST), exhibit large variation in permittivity, $\varepsilon$, as a function of the electric field applied to the material. Therefore, BST is a suitable candidate for high-frequency voltage-tunable microwave devices such as resonators, filters, and phase-shifters [1–6]. For these applications it is desirable to produce BST thin films, which have maximum tunability with a minimum loss factor. A theoretical parameter, K (figure of merit) can be defined to reflect the quality of the film in terms of tunability and loss as:

$$K = \frac{\left[\%tunability \equiv \dfrac{\varepsilon_{max} - \varepsilon_{min}}{\varepsilon_{max}}\right]}{\tan\delta(\%)}$$

where $\varepsilon_{max}$, $\varepsilon_{min}$ are the maximum and minimum dielectric permittivity, respectively, over the investigated voltage range, and the dissipation factor ($\tan\delta$) given as a percentage. The relationship above shows how the loss factor along with the tunability contributes to the device performance. In order to obtain high K values it becomes necessary to understand the composition-microstructure-electrical property relationships of polycrystalline BST thin films. We have performed extensive studies of issues such as the effect of composition, thin film deposition parameters, electrode influence on the electrical properties of BST thin films in order to understand composition-microstructure-property relationships. Different levels of optimization are highlighted in each section that show how carefully chosen parameters result in film property improvement in each stage.

As part of the process optimization for integration of BST films into high frequency devices, we studied properties of BST solid solution films deposited by rf-magnetron sputtering, MOCVD, and the fabrication of BST capacitors based on Cu-electrodes. RF-magnetron sputtering provided easy control of stoichiometry while MOCVD BST films provided superior composition control and conformal coverage of device structures. Along with growth issues, we have addressed the effects of surface contamination on the BST electrical properties. In this review we report results from our extensive research program focused on integrating thin film BST into current microwave technology. We also highlight results from our collaborative work with York et al. of UCSB, whose group used our BST films to fabricate phase-shifters based on their circuit design in order to demonstrate the viability of BST thin film technology for use in microwave circuits.

# 2. Synthesis and Characterization of $Ba_{1-x}Sr_xTi_{1+y}O_{3+z}$ Thin Films by RF-Magnetron Sputtering

## 2.1. A. Synthesis of BST solid-solution films

$Ba_{1-x}Sr_xTi_{1+y}O_{3+z}$ thin films were synthesized using RF-magnetron sputtering in Ar-$O_2$ gas mixtures [7]. The substrate and target (a sintered stoichiometric $Ba_{0.5}Sr_{0.5}TiO_3$ disc) were positioned parallel to each other in an on-axis configuration with 10 cm separation. BST films were deposited at 650 °C on Pt(120nm) / $TiO_2$(40nm)/ $SiO_2$(300nm)/Si or Pt(100nm)/$SiO_2$(300nm)/Si substrates. High resistivity silicon (>10 kΩ-cm) substrates were also used in anticipation of microwave circuit applications.

Patterned Pt top electrodes were deposited through a shadow mask onto the BST films at 350 °C, using electron beam evaporation. Subsequent to top electrode deposition, samples were annealed in air at 550 °C for 30 min. This was followed by a 700 °C anneal for 60 min, to improve the structure and chemistry of both the bulk BST film and the top electrode/BST interface, with the aim of reducing bulk and interfacial contributions to the dielectric loss tangent. RBS, XRD, field emission SEM (FESEM), and atomic force microscopy AFM were used to investigate the thickness, composition, crystallographic orientation, microstructure, and roughness of the BST films, respectively. Dielectric properties (relative permittivity, ε, and dielectric loss, tan δ) were measured as a function of applied electric field at 10 kHz and 0.1 $V_{rms}$. oscillation level, using a HP 4192A impedance analyzer. Capacitance and loss were also characterized at 1MHz using a Keithley C-V meter. Leakage current was measured using a Keithley SMU 237 measurement unit.

### 2.1. B. synthesis of BST multilayered structure

BST multilayer structures were fabricated using RF- magnetron sputtering in Ar-$O_2$ gas mixtures with the same geometry as mentioned above. First, a thin BST layer (~5nm) with a (Ba+Sr)/Ti ratio of 0.73 was deposited at room temperature on top of the Pt bottom electrode. A low temperature (room temperature) BST film deposition process for the initial layer improved the device yields of BST capacitors, presumably by reducing roughening and hillocking of the Pt [6]. Subsequently, a 60 nm main BST layer with a (Ba+Sr)/Ti ratio of 0.9 was deposited at 650 °C onto the initial layer. Finally, another Ti-rich [~5nm, (Ba+Sr)/Ti = 0.73] BST interfacial layer was deposited onto the heterostructure at 650 °C. All three BST layers were deposited without breaking vacuum, using a single ceramic stoichiometric $Ba_{0.5}Sr_{0.5}TiO_3$ target. Different compositions and microstructures for the layers were achieved by changing the total process gas ($Ar+O_2$) pressure, which affects the relative incident fluxes of the Ba, Sr, and Ti species via gas-phase collisions [7].

### 2.2. Composition control and physical properties

Figure 1 (a) shows the variation of the (Ba+Sr)/Ti and Ba/Sr ratios in BST films as a function of the total ($Ar+O_2$) deposition pressure with the $O_2$/Ar ratio fixed at 1/5. The ratio of (Ba+Sr)/Ti changed from 0.73 to 0.98 for films grown with total gas pressure from 22 to 58 mTorr. The growth rate of the BST films depends on the total process pressure (Figure 1 (b)). Variation of the $O_2$/Ar ratio in the 1:1–1:5 range did not affect the metallic composition of the BST films.

BST films with (Ba+Sr)/Ti ratios ≥ 0.9 exhibited a polycrystalline structure characterized by (100), (110), and (111) peaks in the XRD spectra [7], which disappeared for (Ba+Sr)/Ti ratios below 0.85 [7]. This, coupled with the relatively high permittivities still found for these samples, suggests that high-Ti BST films are nanocrystalline,

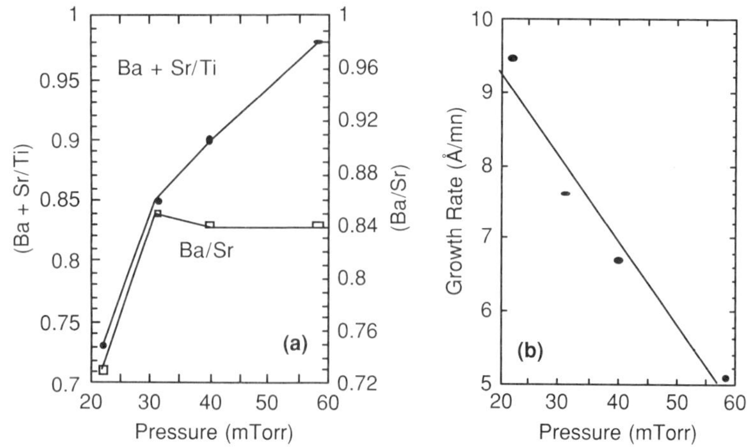


**Fig. 1.** (a) Variation of (Ba+Sr)/Ti and Ba/Sr ratios for BST films deposited at 650 °C from a stoichiometric target with a Ba/Sr ratio of 50/50, and (b) growth rate as a function of the total pressure (Ar+$O_2$) [7].

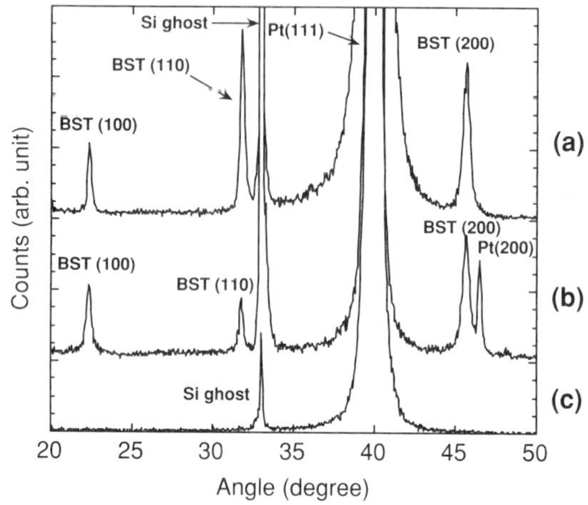


**Fig. 2.** XRD spectra of BST films without (a) and with (b) (Ba+Sr)/Ti = 0.73 interfacial layers. (c) XRD spectrum of a BST film with a (Ba+Sr)/Ti = 0.73 composition.

BST films with (Ba+Sr)/Ti ratios > 0.9 exhibited a dense and granular microstructure, while BST films with (Ba+Sr)/Ti ratios < 0.85 exhibited a featureless microstructure. The root-mean-square (rms) roughness measured by APM was approximately 3 nm for BST films with (Ba+Sr)/Ti $\geq$ 0.9 and 2 nm for BST films with (Ba+Sr)/Ti $\leq$ 0.85, correlating with the structural changes indicated by XRD spectra.

For multilayered BST films, the X-ray diffraction spectra with interfacial layers are shown in Fig. 2. The XRD spectrum for the layered BST film including the interfacial layers shows a general decrease in the intensity of the BST peaks and a change in the relative intensities of the (100) and (110) peaks (Fig. 2 (b)). An XRD spectrum of a

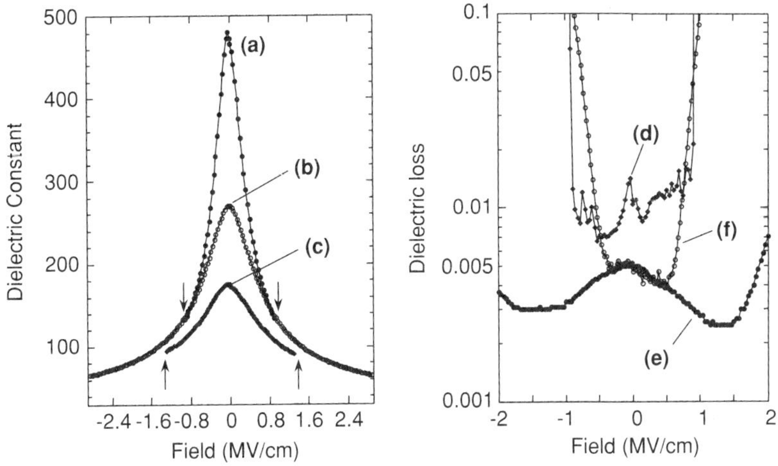


**Fig. 3.** Permittivity of BST films without (a) and with (b) (Ba+Sr)/Ti = 0.73 interfacial layers. (c) Permittivity of a BST film with (Ba+Sr)/Ti = 0.73. Also, dielectric loss of BST films without (d) and with (e) (Ba+Sr)/Ti = 0.73 interfacial layers. (f) dielectric loss of a BST film with (Ba+Sr)/Ti = 0.73. The arrows in Fig. 3 (a) and (c) indicate the breakdown field of the BST films.

BST film with the same composition as the Ti-rich interfacial layers is shown in Fig. 2 (c), as a reference. Although transmission electron microscopy would be necessary to confirm this hypothesis, the absence of diffraction peaks from the XRD spectrum of the Ti-rich layer is indicative that the layer is likely nanocrystalline, as an amorphous film would show negligible tunability, which is not the case as described latter on this paper. The reduction in the BST peaks intensity of the layered BST film can be interpreted as an indication that the Ti-rich interfacial layer reduces the overall crystallinity of the layered BST film.

### 2.3. Electrical properties

In Fig. 3 curves (a) and (d) shows the field-dependent dielectric properties of a single 70 nm BST layer ((Ba+Sr)/Ti = 0.90). This capacitor exhibited zero-bias permittivity of 480, 72% tunability at breakdown, and a dielectric loss $\tan\delta = 0.013$ at zero bias and room temperature. The K value was calculated as 55. In comparison, a BST capacitors with a Ti-rich composition [(Ba+Sr)/Ti ratio of 0.73] showed zero-bias permittivity of 170, 47% tunability and a dielectric loss $\tan\delta = 0.0047$ (curves (c) and (f) in Fig. 3) at zero bias, which yielded a K value of 100. Improvement in the multilayer structure is shown in Fig. 3, curves (b) and (e), the field-dependent dielectric properties of a BST capacitor with a (Ba+Sr)/Ti = 0.90 BST layer sandwiched between two (Ba+Sr)/Ti = 0.73 BST interfacial layers. This heterostructure exhibited low loss, almost identical to that of the interfacial BST layer itself (~0.005) at zero bias and room temperature, yet maintained an intermediate zero-bias permittivity of 270, and high tunability (76%) of the primary BST layer, corresponding to a K value of 150.

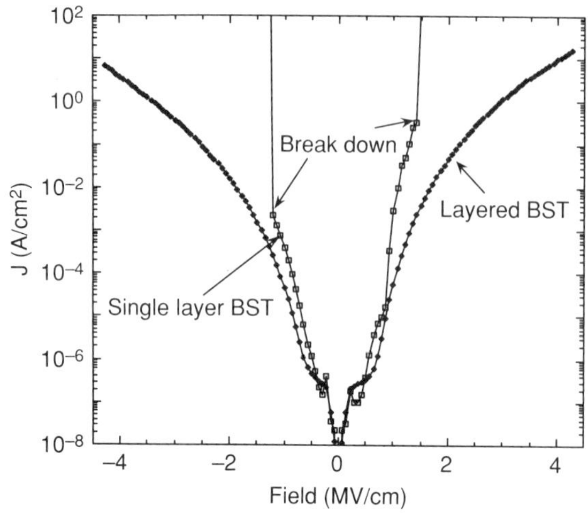


**Fig. 4.** Leakage current density vs. electric field curves for a capacitor with Pt/(Ba+Sr)/Ti = 0.73/(Ba+Sr)/Ti = 0.9/(Ba+Sr)/Ti = 0.73/Pt layered structure and a capacitor with a single layer BST ((Ba+Sr)/Ti = 0.9).

The data presented in Fig. 4 indicate that a single layer near-stoichiometric BST film exhibited a breakdown field of ~1 MV/cm [see arrows in Fig. 3 curves (a) and (c)]. On the other hand, BST capacitors with bottom and top (Ba+Sr)/Ti = 0.73 interfacial layers exhibited substantially improved breakdown fields up to ~4 MV/cm, as shown in dependence of leakage current on applied field, Fig. 4. The mean breakdown field is 4.2 MV/cm with a standard deviation of 0.2, resulting from measurements on 10 layered BST capacitors. This increase in breakdown field results in a tunability comparable to that of the near-stoichiometric single layer film, despite some reduction in zero-bias permittivity.

## 3. Synthesis and Characterization of MOCVD $Ba_{1-x}Sr_xTi_{1+y}O_{3+z}$ Thin Films

### 3.1. Synthesis

In the second approach $Ba_{1-x}Sr_xTi_{1+y}O_{3+z}$ thin films were fabricated using a large area vertical MOCVD system. Metalorganic precursors of $Ba(thd)_2$, $Sr(thd)_2$, and Ti $(O\text{-}iPr)_2\text{-}(thd)_2$ with polyamine adducts were introduced using high purity nitrogen as a carrier gas into the MOCVD reactor, via a temperature-controlled flash-vaporizer and a computer-controlled liquid delivery system (ATMI LDS-300B) that provides good composition control and reproducibility of the delivered precursor mixture. The temperatures of the delivery lines were controlled to avoid condensation or premature reaction of the precursors prior to introduction into the MOCVD reactor. The precursors were thoroughly mixed with high purity reactive gases ($O_2$ and $N_2$) in a showerhead designed to provide deposition of BST films with uniform composition and thickness over

**Table 1.** Deposition and processing parameters for MOCVD synthesis of BST thin films.

| | |
|---|---|
| Substrates: | Pt(1000Å)/$SiO_2$(1000Å)/Si |
| Substrate heater temperature: | 650 °C |
| Reactive gases: | $O_2$ and $N_2O$ |
| Reactive gas flow rate: | 250–1000 SCCM |
| Reactor pressure: | 1.5–2.7 Torr |
| Top electrodes: | e-beam evaporated Pt (1000Å) |
| Post electrode anneal: | 550 °C for 0.5 hrs |
| Electrical characterization: | HP4192A at 1MHz and 0.1 V rms |

large area substrates. The film deposition and processing conditions are summarized in Table 1. The (Ba+Sr)/Ti ratios were varied between 0.96 and 1.05 while the Ba/Sr ratio of the BST thin films was kept at 70/30. Further details of the deposition system have been reported previously [8]. We report the properties of BST films deposited in a thickness range of 90 to 480 nm.

### 3.2. Microstructure and crystallinity of BST films

$Ba_{1-x}Sr_xTi_{1+y}O_{3+z}$ thin films deposited by MOCVD are polycrystalline and show a highly (100) oriented growth. Figure 5 (a) shows the x-ray diffraction pattern of a BST film grown at 675 °C with a (Ba+Sr)/Ti ratio of 0.93, while Fig. 5 (b) shows a SEM picture revealing the grain structure of the BST film. Films grown with a total reactive gas flow rate of 750 sm exhibited strong (100) fiber texture. Excellent film uniformity was obtained over a 4-inch wafer. Initial growth rates were ≈12 Å/min, but higher rates of ≈50Å/min were achieved after optimization of the reactor showerhead configuration and vaporizer temperature uniformity (± 5 °C). It is well-established that the electrical properties of BST film-based capacitors depend on the (Ba+Sr)/Ti

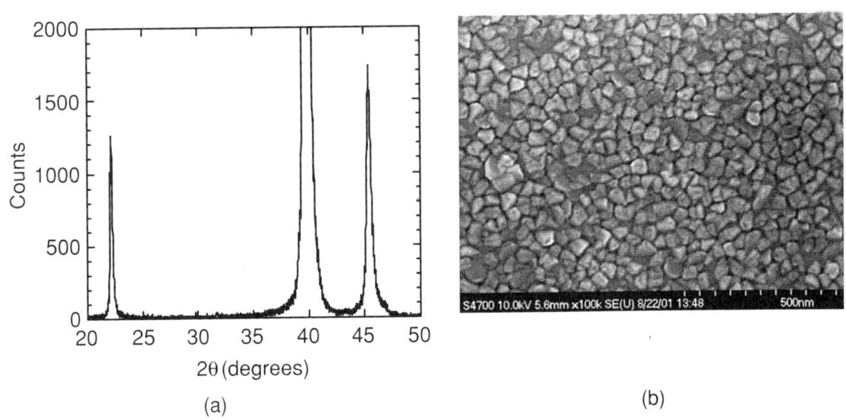


**Fig. 5.** (a) XRD spectrum of a BST film grown by MOCVD at 675 °C with a (Ba+Sr)/Ti = 0.93. (b) SEM picture showing the microstructure of the BST film which XRD is shown in (a).

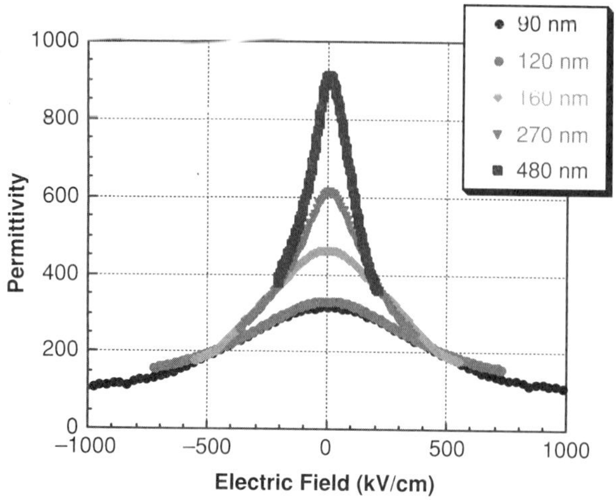


**Fig. 6.** Relative permittivity vs. electric field characteristics as a function of BST film thickness.

ratio [7–10]. Slightly Ti rich samples with (Ba+Sr)/Ti about 0.96 were studied here. A broader range of (Ba+Sr)/Ti ratios is still under investigation.

### 3.3. Electrical properties

Figure 6 shows the room temperature relative permittivity vs. electric field characteristics as a function of BST film thickness. The zero-field dielectric constant increases systematically with increasing film thickness, while at higher fields, the permittivities become thickness independent, as consistent with previous observations [1, 11]. Losses for all samples were less than 0.5% at room temperature and zero bias.

Figures 7 show the leakage current characteristics of a 90 nm BST film as a function of temperature and applied electric field. The leakage current behavior can be divided into low-field and high-field regimes. The low-field regime is characterized by weak field dependence and a monotonic increase in leakage with increasing temperature. The high-field regime, in contrast, shows a strong dependence on applied electric field, and a distinct crossover point beyond which the leakage current is found to decrease with increasing temperature. A detailed analysis of the leakage characteristics can be found elsewhere [12].

## 4. $(Ba_xSr_{1-x})Ti_{1+y}O_{3+z}$ Interface Contamination and Its Effect on Electrical Properties

In order to fabricate high quality BST/electrode heterostructures for such devices, it is important to understand the effect of contamination at the film-electrode interfaces. For example, carbon- and hydrogen-containing contaminants at interfaces are generally known to affect the Schottky barrier height, which may be one of contributing factors to leakage current of BST capacitors. Hydrocarbon contaminants are generally adsorbed

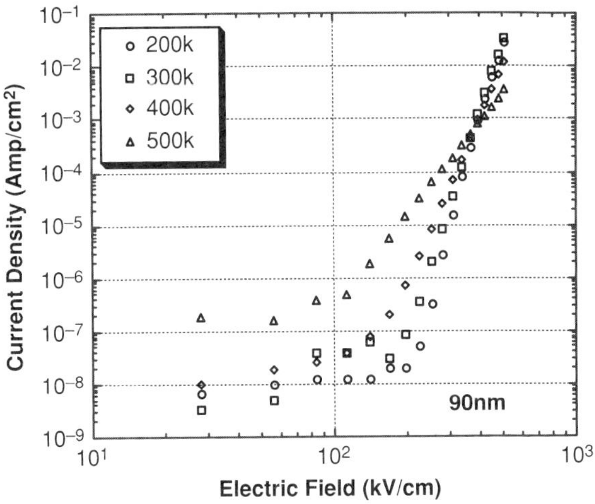


**Fig. 7.** Leakage current characteristics of a 90 nm BST film as a function of temperature and applied electric field.

on the surface of BST layers or metal electrodes when substrates are transported through air from one growth chamber to another in a research environment. This problem should be overcome in an industrial environment where it is expected that the fabrication of BST capacitors will be done in integrated cycles without exposing the capacitors layers to air. In any case, we performed systematic studies on the extent of contamination of the surfaces of BST and Pt thin films using *in-situ*, real time mass spectroscopy of recoiled ions (MSRI), coupled with thermal decomposition/desorption in an oxygen ambient for the cleaning of the surfaces. Surface compositional analysis was performed using the MSRI technique, which permits compositional analysis of the top surface layer (<3 monolayers) of thin films in backfill sputtering and reactive gas environments.

Figure 8 (a) shows the result from *in situ*, real-time MSRI analysis performed during heating of BST films in vacuum or oxygen atmospheres up to 700 °C, in order to determine the conditions that result in the elimination of C and H species from the surface of the BST film. Complete removal of C and H species from the surface of BST films was accomplished only when heating the films at ≥550 °C in ≥1 mTorr of oxygen 5 min [13, 14], which indicates that the oxygen atmosphere is particularly critical for achieving a clean BST film surface prior to top electrode deposition. The elimination of C and H species from the BST surface is due to chemical reactions between oxygen and C and H species that result in the formation of volatile CO and/or $CO_2$ and $H_2O$ molecules activated by the high substrate temperature. A gradual decrease in surface contamination is observed at low $O_2$ pressure (0.5 mTorr and vacuum) over the entire temperature range studied. In contrast, upon heating in 1 mTorr of oxygen, the magnitude of the carbon peak starts to decrease above about 400 °C (Fig. 8 (a)), while the magnitude of the hydrogen peak starts to decrease above 300 °C (Fig. 8 (b)); both fall below the limits of detectability between 500 and 550 °C.

The effect of C and H contamination on the electrical properties can be observed in the leakage properties shown in Fig. 9 (a) and (b). Before the standard post-capacitor

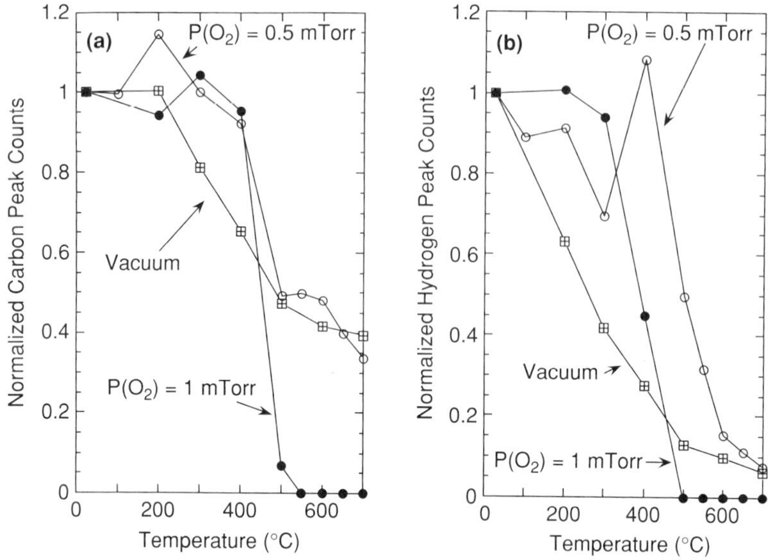


**Fig. 8.** MSRI peak count evolution for (a) carbon and (b) hydrogen during heating of BST films in vacuum and in oxygen ambient of $5 \times 10^{-4}$ Torr and $1 \times 10^{-3}$ Torr, respectively.

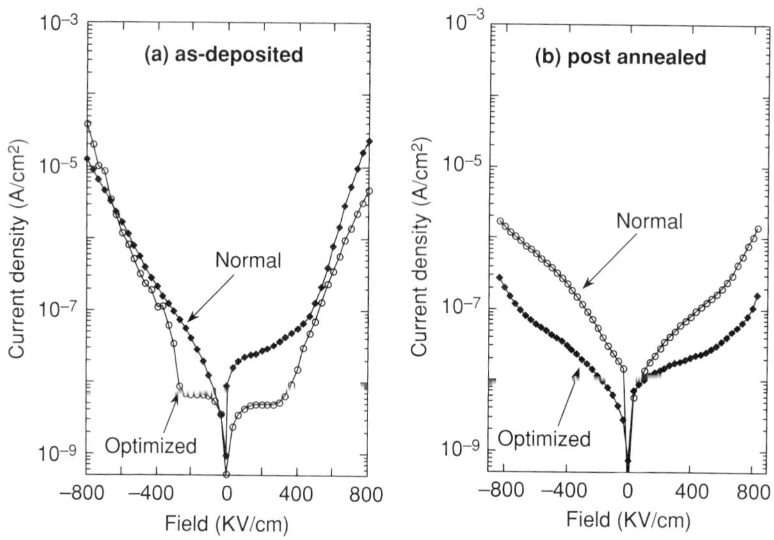


**Fig. 9.** Current density vs. electric field for Pt/60nm BST/Pt heterostructures with Pt top electrodes deposited either at 350 °C in vacuum (marked as "normal") or deposited under conditions yielding a clean top interface (marked as "optimized"), for (a) as-deposited electrodes and (b) after annealing in air for 30 minutes at 550 °C.

fabrication anneal, a strong polarity dependence is observed in the leakage of the capacitor with top electrodes deposited at 350 °C in vacuum and therefore with top interface contamination, compared to the sample with Pt electrodes deposited under optimized conditions yielding a clean top interface (Fig. 9 (a)). It is important to notice that the vacuum achieved in non-ultrahigh vacuum deposition systems, including ours, used to produce electrode layers, is in the range of $10^{-6}$ – $10^{-8}$ Torr. Under these vacuum conditions, C and H species from the background gas re-adsorb on the film surface quickly if the latter is cooled down below the 500 °C range. This can be observed if the system has a suitable surface analytical technique such as our MSRI method capable of monolayer analysis resolution. Therefore, in order to produce a clean Pt/BST interface, we annealed the BST film in 1 mTorr at 550 °C, and then deposited the first 50Å of Pt under these same condition. This was followed by deposition of the remaining top Pt electrode (950Å) at 350 °C in vacuum. In view of the BST processing described above, the asymmetric leakage, observed also in BST capacitors with the top Pt electrode deposited at 350 °C in vacuum, can be correlated with differences in the behavior of the top and bottom electrode interfaces, and therefore to some degree to the presence of C and H contaminants at the top Pt/BST interface. In addition, BST capacitors with top Pt electrodes deposited under the optimized conditions shown in Fig. 8, and therefore with clean top and bottom interfaces, exhibited almost no polarity dependence (Fig. 9 (a)) and reduced dielectric loss (0.012) but similar capacitance, compared to BST capacitors with Pt top electrode deposited at 350 °C (dielectric loss = ~0.02).

The conclusions presented above are reinforced by the behavior of BST capacitors after annealing of the whole Pt/BST/Pt capacitors at 550 °C in air, following top Pt electrode deposition at 350 °C in vacuum. Fig. 9 (b) shows the leakage current densities after annealing of the as-fabricated capacitors, at 550 °C for 30 minutes in air. Substantial reduction and a relatively symmetric behavior of the leakage are found for BST capacitor fabricated both with the optimized top Pt electrode process and the standard top Pt process [13, 14]. After the post-capacitor fabrication anneal, both BST capacitors show similar values of the dielectric loss (~0.01).

## 5. Phase Shifters Based on Optimized Magnetron Sputter-Deposited BST Films

Optimized BST thin films described above have been incorporated into monolithic microwave integrated capacitors in thin-film phase-shifters circuits in work performed by York et al. based on their unique devise design [15, 16]. The approach taken by York's group work was to fabricate monolithic parallel-plate capacitors using standard IC processes, and integrate these with microwave coplanar-waveguide transmission-line structures. Figure 10 illustrates a simple parallel-plate capacitor structure integrated in shunt along a coplanar transmission-line (CPW) (a), an array of capacitors loading a transmission line (b), and a typical capacitance vs. voltage curve (c) for the device fabricated at UCSB using BST films produced at Argonne. For lowest possible loss, the silicon substrates can be micromachined in the CPW gap region as shown in

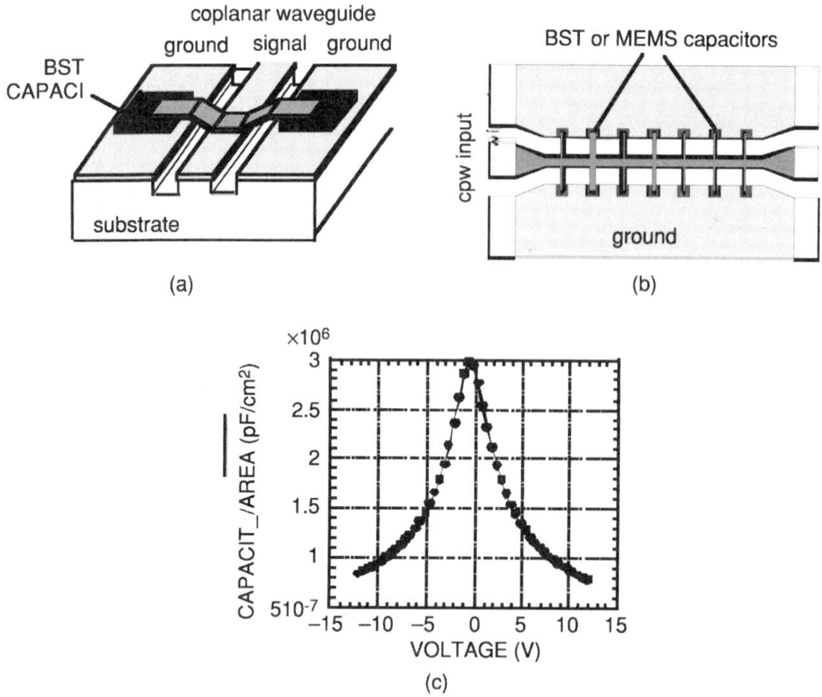


**Fig. 10.** a) Integrated thin—film capacitor structure on a coplanar waveguide, b) distributed circuit phase shifter using a periodically-loaded coplanar waveguide and c). Low-frequency (1 MHz) C-V curve of a typical UCSB/ANL device showing 4:1 capacitance (work done by York's group at UCSB).

Fig. 10 (b). This also has the advantage of reducing the capacitance per unit length on the unloaded-line, allowing for larger capacitive loading in a circuit.

The monolithic delay lines shown in Fig. 10 (b) are distributed circuits using a coplanar waveguide periodically loaded with thin-film BST varactors. When designed correctly, this structure is a synthetic transmission line with a phase velocity that can be controlled by changing the value of the external loading capacitors. The parallel plate capacitor topology utilizes the tunability of the BST film effectively and requires lower control voltages than interdigitated electrode designs. Conductor losses are low in this topology.

York et al. have recently used this topology to demonstrate analog phase shifter circuits using monolithic GaAs varactors loading a coplanar waveguide (CPW) transmission line in K-band, resulting in $<4$ dB insertion loss at 20 GHz and a continuously programmable delay of 0–360 degrees at this frequency [15, 16]. Excellent performance of the phase shifters was the result of careful modeling of the circuit, especially with respect to RF losses, which enabled the optimization of the structure for low insertion loss and high performance. An especially important parameter is the loading factor 'x', which is the ratio of the variable capacitor per unit length to the unloaded line capacitance per unit length. For a given variable capacitor technology and substrate

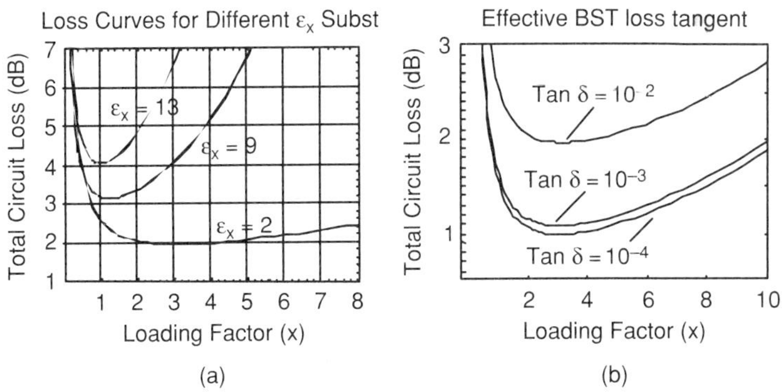


**Fig. 11.** Predicted losses for BST-based circuits like that of Fig.11. (a) Influence of substrate dielectric constant (tan $\delta = 0.02$, or Q = 50); (b) Influence of loss tangent ($\varepsilon_r = 2$) (work done by York's group at UCSB).

dielectric constant there exists an optimum-loading factor, which results in a circuit with the lowest possible insertion loss. This has been experimentally verified by measuring the losses on GaAs varactor loaded transmission lines, and it was found that losses for the optimally loaded transmission line are dominated by the losses in the variable capacitor. Thus, further efforts to reduce the losses of the circuit must concentrate on reducing losses in the variable capacitor, which favors the use of thin-film BST technology.

To demonstrate the potential of BST thin film varactors for distributed phase-shifters, a loss analysis was carried out by York's group for different substrate dielectric constants and for different values of loss tangent of the BST film. The results are displayed in Figure 11 (a), which shows that there is a certain optimum loading factor for minimization of the total circuit loss. The total circuit loss decreases with a decrease in the substrate dielectric constant, due to a subtle interplay between conductor loss and characteristic impedance. Since the BST technology enables the use of different substrates, it should be possible to achieve substantially lower loss than with GaAs-based circuits. Another set of calculations performed by York's group was to determine the effect of the BST thin film loss tangents on the total phase shifter losses as shown in Figure 11 (b). At loss tangents of $10^{-3}$ or lower (specified at 10 GHz) the performance is limited by conductor loss only, indicating that significant improvements over diode-based circuits can be obtained with material improvements, particularly on electrode materials once the BST material has been fully optimized.

York et al. have recently demonstrated a BST-based periodically loaded line phase shifter [17] as shown in Figure 10 (b). The phase shifter was capable of producing up to 160 $^\circ$ of phase shift at 30 GHz. The phase shift increases linearly with frequency as expected for a variable time delay element. The preliminary results of this phase shifter are shown in Figure 12. Although this phase shifter has not yet been fully optimized, the results confirm the potential of thin film BST technology for application to high frequency devices. Advances in materials and materials integration strategies are required to further improve on these results, particularly on electrode materials, which is the topic of the next section.

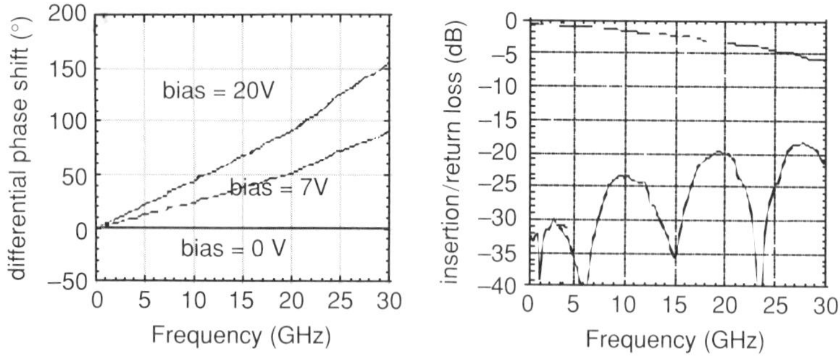


**Fig. 12.** Measured performance of a periodically loaded line phase shifter fabricated at UCSB (work done by York's group at UCSB).

## 6. Synthesis and Properties of $(Ba_xSr_{1-x})TiO_3$ Thin Film Capacitors with Layered Cu-Based Electrode

Copper has drawn much attention [18, 19] as an interconnect material for integrated sub-micron circuit technology due to its low resistivity and high electro- and stress-migration resistance, which are superior to that of Al and Al-alloy-based IC interconnection technologies, and this is the reason why Cu has now been introduced into IC fabrication. In addition, compared to Pt (widely used as the electrode material in ferroelectric and complex oxide thin film devices), copper's low cost, high conductivity, and relatively easy reactive ion etching properties, makes it an appealing candidate to replace Pt for non-volatile ferroelectric memories and other ferroelectric and complex oxide thin film-based devices. However, there are several issues to be solved before Cu can be successfully integrated as an electrode for oxide thin films. We have developed layered structures of TiAl/Cu/Ta to solve the Cu diffusion, adhesion, and oxidation problem encountered in the fabrication of high-k thin film-based capacitors for application to high frequency devices [20]. Tantalum (Ta) has been extensively investigated and demonstrated to be an excellent diffusion barrier for Cu integration with $SiO_2$/Si substrates [21, 22], due to its high melting temperature and thermodynamic stability with respect to Cu and Si. The TiAl layer, working as oxygen diffusion barrier, exhibits very good barrier properties against oxygen penetration at elevated temperature [23, 24]. Polycrystalline $(Ba_xSr_{1-x})TiO_3$ (BST) thin films, which is a high-dielectric oxide material, were subsequently grown on the Cu-based electrode as discussed in our prior publication [20].

### 6.1. Synthesis of Cu-based electrodes

Stacked layers of TiAl(20–50 nm)/Cu(100–200 nm)/Ta(20 nm) were produced on thermally oxidized Si(100) substrate by using ion-beam sputter deposition. The as-deposited Cu-based electrode exhibited much low resistivity (2.7 μΩ·cm) than the bulk resistivity of platinum (10.7 μΩ·cm), which may greatly reduce the RC constant and boost the switching speed of the BST thin film devices for high-frequency application. To understand Cu diffusion and oxidation mechanism and to develop materials integration strategies for producing a reliable Cu-based electrode technology for high-k capacitors, TiAl/Cu/Ta structures were heated up to 600 °C in conventional furnace filled with constant flow of 99.9% oxygen. Electrical measurement, XPS depth profile, RBS, and XRD analyses revealed that the layered Cu-based electrode was able to maintain its low electrical resistivity (2.7 μΩ·cm) and structure integrity in oxygen up to 600 °C. BST thin films were deposited on the Cu-based electrodes substrates by radio-frequency magnetron sputter deposition at temperature from 400 to 600 °C in an mixture of $O_2$ and Ar (1:5). For the first time, polycrystalline BST thin films with perovskite structure were obtained on Cu electrode, using TiAl and Ta as barriers. The layered BST/TiAl/Cu/Ta/$SiO_2$/Si structure with a very smooth BST surface was studied using cross-sectional SEM and TEM. No interdiffusion was detected in the sample grown at $\leq 500$ °C.

### 6.2. Composition and microstructural characterization of layered Cu-based electrodes

Figure 13 (a) shows a STEM bright-field image of a cross-section BST/Cu-based electrode sample, while Fig. 13 (c) shows line-scans of the elements Al, Ti, O and Cu measured across all the layers. These line-scans were derived from elemental maps calculated from energy-filtered TEM images using the three-window-method [25]. The imbedded image (Fig. 13 (b)) gives an example of O-K map. The pixel-size of these images is 1.2 nm and the CCD-camera was operated with 2× binning. The intensity was integrated for a line-scan width of 10 pixels to improve the signal-to-noise ratio. The width of the energy selection aperture used for recording the individual energy filtered images was 10 eV for the Ti-$L_{23}$ edge, O-K edge, and Cu-$L_{23}$ edge and 30 eV for the Al-K edge. These line-scans across all layers show steep concentration gradients at the interfaces with a width of only a few nanometers. Very low oxygen level has been detected within the Cu layer and practically no Cu atom diffused into Si substrate, which indicate the excellent barrier properties of TiAl and Ta.

The optimized process for fabrication of BST thin film capacitors with Cu-based electrode involves sputter deposition of BST thin films at 400 to 450 °C followed by 2-min rapid thermal annealing (RTA) at 700 °C in oxygen atmosphere [26]. With this method, the low-k interfacial layer formed between BST and TiAl barrier and interface roughening caused by Cu-recrystallization at high temperature can be greatly suppressed. As the result, large dielectric constant, $\varepsilon_r = 280$, and low dielectric loss, $\tan\delta = 0.007$, were achieved for BST capacitors with bottom Cu-based electrodes. The dielectric constant of the 160-nm BST layer as a function of applied electric field is plotted in Fig. 14 (a). The positive and negative signs of the field were decided by the voltage applied on the Pt (top-electrode) with respect to the TiAl/Cu/Ta (bottom-electrode). The field-dependent permittivity shown in Fig. 14 (a) indicates a tunability of 53%. Fig 14 (b) shows the leakage current density as a function of the applied electric field. The asymmetric shape of the I-V curve is primarily due to the asymmetric electrode structure using Pt at top and TiAl at bottom. Leakage below $2 \times 10^{-8}$ A/cm$^2$ at 100 kV/cm was achieved with a breakdown field of 450 kV/cm.

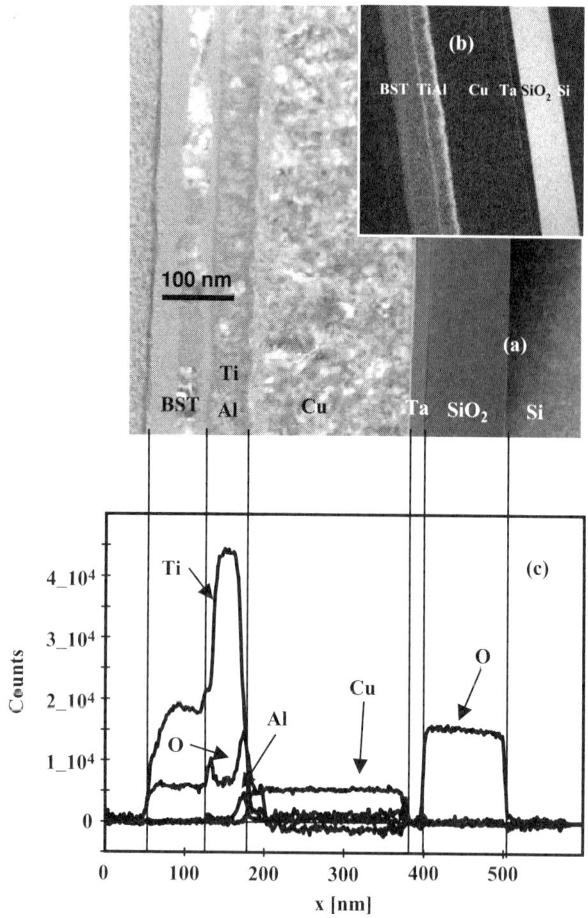


**Fig. 13.** (a) Cross-sectional STEM bright field image, (b) O-K map of BST/TiAl/Cu/Ta sample, and (c) line-scan of Al, Ti, Cu, and O element distribution across the layer, showing that Cu layer is free of oxygen and is well confined within the TiAl and Ta barriers.

## 7. Conclusions

In conclusion, we have shown that solid solution and multi-layered BST films with high tunabilities, low losses, and high dielectric breakdown fields can be synthesized using RF-magnetron sputter deposition with judiciously chosen process parameters. Specifically, control is demonstrated of the composition (i.e., the (Ba+Sr)/Ti ratio) of BST films grown from a single stoichiometric target by use of tailored target-substrate geometry and deposition pressure. MOCVD BST films were also grown to demonstrate the capability of producing very good quality large area films which show excellent electrical properties. The issues related to interface contamination have been touched upon, where we demonstrated that post top electrode deposition annealing treatments can lead to large improvement in electrical properties of the BST films, specifically

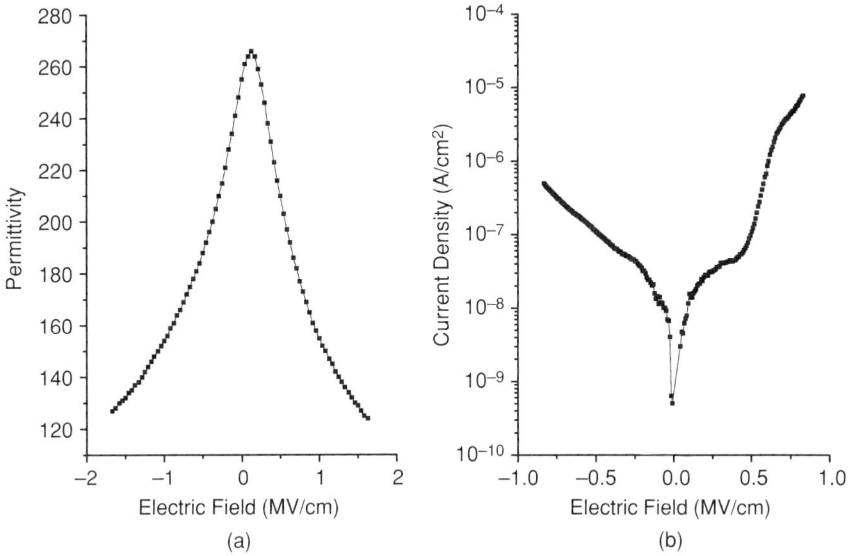


**Fig. 14.** (a) C-V and (b) I-V curves of BST thin film deposited at 450 °C followed by 2-min RTA.

reducing loss and leakage current. The improvements in the electrical properties have been attributed to the elimination of C, H impurities from electrode/BST interface through a thermal annealing process.

Periodically loaded line phase shifters with up to 160 ° of phase shift at 30 GHz were fabricated by York *et al.* using BST films produced in our program. The phase shift increased linearly with frequency as expected for a variable time delay element. The phase shifters produced by using our BST films are not fully optimized yet. However, the result presented in this paper indicate the potential of thin film BST technology for high frequency devices.

A successful attempt was made to replace Pt with Cu as the bottom electrode for BST films. To overcome the problems of Cu integration with high-k oxides, we have investigated the oxidation and diffusion resistance of TiAl/Cu/Ta multilayered structure, which showed high electrical conductance and excellent thermal stability in oxygen environment up to 600 °C. The electrical properties of these heterostructured films show good electrical properties, while further work is being pursued towards improvements of the film properties and successful integration into microwave and other potential devices.

## 8. Acknowledgments

This work was supported by the U. S. Department of Energy, FreedomCAR and Vehicle Technologies Program and Basic Energy Sciences—Materials Sciences under contract W-31-109-ENG-38, and by the DARPA-FAME program. The authors would like to acknowledge the contributions of Prof. R. York's group within the

DARPA-FAME program, particularly on the integration of ANL thin films into high quality devices designed and fabricated at UCSB and measurements of the electrical properties shown in Fig. 10, 11, and 12, which were done at UCSB.

## References

[1] D. Flaviis, N.G. Alexopolous, and M. Staffsudd, IEEE Trans. Microwave. Tech. **45**, 963 (1997).

[2] V.K. Varadan, D.K. Ghodgaonker, V.V. Varadan, J.F. Kelly and

[3] P. Glikerdas, Microwave Journal, **35**, 116 (1992).

[4] J.M. Ponds, S.W. Kirchoefer, W. Chang, J.S. Horwitz, and D.B. Chrisey, Integr. Ferroelectr., **22**, 317 (1998).

[5] F.A. Miranda, F.W. Van Keuls, R.R. Romanofsky, and G. Subramanyam, Integr. Ferroelectr. **22**, 269 (1998).

[6] A.I. Kingon, S.K. Streiffer, C. Basceri, and S.R. Summerfelt, MRS bull. **21**, 18 (1995).

[7] J. Im, O. Auciello, P.K. Baumann, S.K. Streiffer, D.K. Kaufman, and A.R. Krauss, Appl. Phys. Lett. **76**, 625 (2000).

[8] P.K. Baumann, D.Y. Kaufman, J. Im, O. Auciello, S.K. Streiffer, R.A. Erck, and J. Giumarra, Integrated Ferroelectrics **34,** 255 (2001).

[9] G.W. Dietz, M. Schumacher, R. Waser, S.K. Streiffer, C. Basceri, and A.I. Kingon, J. Appl. Phys. **82**, 2359 (1997).

[10] C. Basceri, S.K. Streiffer, A.I. Kingon, R. Waser, J. Appl. Phys. **82**, 2497 (1997).

[11] S.K. Streiffer. C.Basceri, C.B. Parker, S.E. Las, and A.I. Kingon, J. Appl. Phys. **86**, 4565 (1999).

[12] S. Saha, D.Y. Kaufman, S.K. Streiffer and O. Auciello, Appl. Phys. Lett. **83**, 1414 (2003).

[13] O. Auciello, A.R. Krauss, J. Im and J.A. Schultz, in Annu. Rev. Mater. Sci., edited by O. Auciello and R. Ramesh, **28**, 375 (1998).

[14] J. Im, A.R. Krauss, Y. Lin, J.A. Schultz, O. Auciello, D.M. Gruen, and R.P.H. Chang, Nucl. Instrum. Meth., **B 118**, 772 (1996).

[15] P. Padmini, T.R. Taylor, M.J. Lefevre, A.S. Nagra, J.S. Speck and R.A. York, Appl. Phys. Lett. **75**, 3186 (1999).

[16] A. Nagra, T.R. Taylor, P. Periaswamy, J.S. Speck, and R.A. York, MRS Symp. Proc.: "Materials Issues for Tunable RF and Microwave Devices", **vol. 603**, 37 (2000).

[17] R. York, A. Nagra, P. Periaswamy, O. Auciello, S.K. Streiffer, and J. Im, Integrated Ferroelectrics **34,** 177 (2001).

[18] H. Takasago, K. Adachi, M. Takada, J. Electron Mater. **12**, 319 (1989)

[19] J. Li, G. Vizekelethy, P. Revesz, J.W. Mayer, K.N. Tu, J. Appl. Phys. **69**, 1020 (1991).

[20] W. Fan, S. Saha, J.A. Carlisle, O. Auciello, R.P.H. Chang, and R. Ramesh, Appl. Phys. Lett. **82**, 1452 (2003).

[21] K. Holloway, P.M. Fryer, C. Cabral, Jr., J.M.E. Harper, and P.J. Bailey, J. Appl. Phys. **71,** 5433 (1992).

[22] P. Catania, J.P. Doyle, and J.J. Cuomo, J. Vac. Sci. Technol. A**10**, 3318 (1992).

[23] S. Aggarwal, A.M. Dhote, H. Li, S. Ankem, and R. Ramesh, Appl. Phys. Lett. **74,** 230 (1999).

[24] A.M. Dhote, O. Auciello, D.M. Gruen, and R. Ramesh, Appl. Phys. Lett. **79**, 800 (2001).

[25] *Energy-Filtering Transmission Electron Microscopy*, edited by L. Reimer (Springer Berlin Heidelberg , 382 (1995).

[26] W. Fan, B. Kabius, J.M. Hiller, S. Saha, J.A. Carlisle, O. Auciello, R.P.H. Chang, and R. Ramesh, J. Appl. Lett. **94**, (Nov. 15, 2003).

# Chapter 12

## Permittivity, Tunability and Loss in Ferroelectrics for Reconfigurable High Frequency Electronics

A.K. Tagantsev, V.O. Sherman, K.F. Astafiev, J. Venkatesh and N. Setter

*Ceramics Laboratory, Swiss Federal Institute of Technology EPFL, 1015 Lausanne, Switzerland*

Miniaturized tunable capacitors open interesting perspectives for new *active* rf components: Miniaturized tunable filters, resonators, phase-shifters, antennas, etc. can potentially be manufactured with the desirable characteristics of very high Q (quality factor), low power consumption, low insertion loss, high isolation, and high speed replacing bulky, multi-component systems. The field of rf-MEMS (we refer to the wide definition of MEMS as microsystems not necessarily including moving parts) is being intensively developed for this end.

The permittivity of ferroelectrics, especially in temperatures not too far from the ferroelectric phase transition is strongly electric field dependent. Therefore a ferroelectric capacitor can in principle serve as a varactor, voltage dependent capacitor, fulfilling the same function as rf-MEMS switch or a semiconductor varactor. In comparison to rf-MEMS capacitor switches, tunable ferroelectric capacitors do not have mechanical parts and their response is faster. While the materials are not the standard IC materials and require development, the technology is rather simple albeit not yet worked in details. Tunable ferroelectrics are attractive above $\approx$10 GHz where the relatively high microwave loss of pin diodes makes this solution less attractive. MEMS switches which are also being developed at present possess a relatively slow communication speed. For this reason ferroelectric tunable devices are being developed for various applications as discussed earlier (chapter 6).

Indeed in last decade intensive development efforts have been made in this direction. At present several small companies manufacture thin film based tunable devices, and R&D is on going in a number of large electronic companies and in academic laboratories in Europe, America and Asia. The work on miniaturized systems shows a number of differences between properties of thin film and those of corresponding bulk materials (crystals and ceramics), some of the differences are understood by now and some are still under investigation. In addition, both experimental results and models show that multi-phase materials (composites) can result in improvement or deterioration of performance, depending on the microstructural details.

In this chapter a review of the properties of ferroelectric materials that are relevant to microwave tunable devices is presented: we discuss the theory of dielectric response of tunable bulk materials and thin films; the experimental results from the literature and from own work are reviewed; the correspondence between the theoretical results and the measured properties of tunable materials is critically analyzed; nominally pure, real (defected), and composite bulk materials and thin films are addressed. In addition, techniques for characterization of tunable ferroelectrics and applications of these materials are briefly presented.

---

Remark: This chapter is adapted from J. Electroceramics, vol. 11, pp. 5–66, 2003.

## 1. Introduction

The recognition of the potential usefulness of ferroelectric materials in tunable high-frequency devices dates back over 40 years [1]. However, due to various reasons related to both device electronics and materials technology, it is only in the past decade that intensive development efforts are being made in this direction. The main driving force for this resurgent interest is the potential for substantial miniaturization of microwave (MW) components and systems (accompanied also by a large cost reduction) and the potential for integration with microelectronic circuits due to the development of thin and thick film ferroelectric technology. A number of review papers [2–5] have appeared recently, covering different aspects of tunable ferroelectric materials and their applications. However, the recent theoretical, experimental and technical developments (e.g. investigations of tunable composites) necessitate revisiting the subject. In the present work, we present a comprehensive updated discussion of the main issues of tunable ferroelectric materials.

The desired material properties are quite well known: a high tunability and low dielectric losses are required. The underlying physics has been already developed (see e.g. [2, 6]), but its applicability to present day problems is not straight forward; While materials are being developed, questions arise such as: Is it preferable to use the tunable material in its paraelectric phase, in its ferroelectric phase or at about the phase transition temperature? What is the interplay between dielectric constant and tunability? Is it possible to have a material with an intermediate permittivity yet a high tunability? Can a material be of a high permittivity yet of low losses? Are high tunability and low loss contradictory? What is the low-loss limit one can expect and how will it vary with frequency and temperature? Can mixing of a low loss dielectric and a ferroelectric result in an overall low loss without deteriorating strongly the tunability of the composite? Why are thin films often more lossy than bulk materials of similar compositions? Are loss mechanisms similar in thin films and in bulk single crystals? Why do films often exhibit lower permittivity and tunability than bulk materials of a similar composition and is this phenomenon general and inevitable? Other questions are related to the significance of the measurement results: Is the measurement of losses at kHz frequencies useful for the evaluation of the performance of the material at GHz frequencies? Is the measured tunability at kHz frequencies indicative of the tunability at MW frequencies? Are losses measured at zero bias field indicative of losses under field?

In the following review some considerations and answers to the above questions are presented: First, we summarize the principal applications of tunable ferroelectrics in order to estimate the needed performance and through it the details of the needed properties. Next, the theory of dielectric response in tunable ferroelectrics (permittivity, tunability and dielectric losses) without and under dc field is presented and discussed, first for ideal ferroelectrics, then for real materials (including defects) and for multi-phase materials (composites). The evaluation methods of the relevant properties are described and the significance of the obtained results is assessed. Experimental work on properties of modern tunable ferroelectrics is summarized, based on results reported in the literature and on results at the authors' laboratory. These results are discussed in the context of the theory.

## 2. Applications

The applications of tunable ferroelectrics in microwave devices are summarized below in order to estimate the performance that is required from the materials. A detailed discussion of designs and applications of ferroelectric based microwave components can be found in Refs. [2, 3, 5, 7, 8].

For a microwave engineer the main attraction of ferroelectric materials is the strong dependence of their dielectric permittivity $\varepsilon$ on the applied bias electric field $E_0$. This characteristic is commonly described by the *tunability* $n$ defined as the ratio of the dielectric permittivity of the material at zero electric field to its permittivity at some non-zero electric field, as expressed by Eq. (2.1). The *relative tunability* ($n_r$) is defined by Eq.(2.2).

$$n = \frac{\varepsilon(0)}{\varepsilon(E_0)} \tag{2.1}$$

$$n_r = \frac{\varepsilon(0) - \varepsilon(E_0)}{\varepsilon(0)} = 1 - \frac{1}{n} \tag{2.2}$$

An example of dielectric constant—electric field dependence for a bulk ferroelectric material ((Ba, Sr)$TiO_3$ [9]) is presented in Fig. 2.1.

The dielectric loss in ferroelectrics is not as small as that of common microwave dielectric materials and the loss tangent ($\tan\delta$) is an important characteristic of the material, which should be taken into account in the device design. The temperature dependence of the dielectric permittivity over the operating temperature interval is another important issue: this is of particular concern in the vicinity of the ferroelectric transition temperature, in which the material exhibits a high but strongly temperature dependent tunability. The trend "the higher the dielectric constant, the higher the

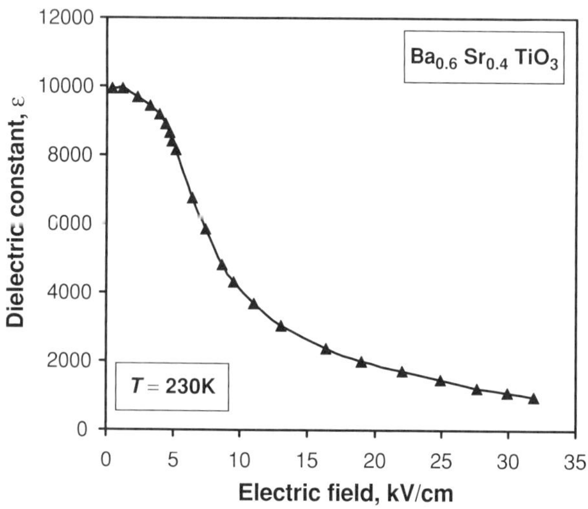


**Fig. 2.1.** Typical field dependence of the dielectric permittivity of a tunable ferroelectric material $Ba_{0.6}Sr_{0.4}TiO_3$ [9].

**Table 2.1.** Permittivity, temperature coefficient of permittivity and quality factor of some linear dielectrics and ferroelectrics used as the microwave resonators. (1) Values taken from the datasheet of Morgan Electro Ceramics company. (2) Values calculated based on the data reported in Refs. [12] [13] (3) Values calculated based on the data reported in Ref. [14]. (4) Values are taken from Ref. [10]. (5) Values taken from Ref. [2]. The temperature dependence of the dielectric permittivity is given in term of $TCE = \varepsilon^{-1}\partial\varepsilon/\partial T \times 10^6$

| Material | $\varepsilon_r$ @ 25 °C | TCE (ppm/K) @ 25 °C | $Q = 1/\tan\delta$ @ 25 °C | Loss measured @ $f$(GHz) |
|---|---|---|---|---|
| Ba-Zn-Ta-O (1) | 30 | −18...−26 | 12600 | 10 |
| Zr-Sn-Ti-O (1) | 37 | −7...−45 | 5800 | 9 |
| Ba-Nd-Ti-O (1) | 88 | −8...−46 | 1100 | 5 |
| $TiO_2$ (4) | 100 | −900 | 14500 | 3 |
| $SrTiO_3$ | 300 (2) | −2000 (2) | 1000 (5) | 10 (5) |
| $Ba_{0.6}Sr_{0.4}TiO_3$ | 4000 (3) | −40000 (3) | 50 (5) | 10 (5) |

tunability, loss, and temperature dependence of the dielectric permittivity" is observed for many dielectrics [10, 11]. This trend can be traced across different materials, it is illustrated in Table 1

The correlation between the tunability and the loss tangent forces designers to choose the material with the optimal trade-off between these two parameters for best device performance. This optimal trade-off is found by an integral parameter called the Commutation Quality Factor ($K$). The commutation quality factor was originally introduced to characterize the switching properties of semiconductor two-state microwave switches (p-i-n diodes [15] and field effect transistors). It was shown that this parameter does not depend on the characteristics of the passive network that is used together with the switchable element in the microwave circuit designs. In other words, the factor $K$ can be considered as an invariant that characterizes the tunable performance of the material. Later this theory was adapted to other switchable and tunable components including ferroelectrics [16]. The Commutation Quality Factor for ferroelectrics is given by:

$$K = \frac{(n-1)^2}{n \cdot \tan\delta\,(U_{\min}) \cdot \tan\delta\,(U_{\max})}, \tag{2.3}$$

where $U_{\max}$ and $U_{\min}$ are the voltages applied in the two states of the ferroelectric capacitor and $n$ is the tunability of the capacitor determined as the ratio $\varepsilon\,(U_{\min})\,/\,\varepsilon\,(U_{\max})$.

In theory, all forms of ferroelectric materials, bulk single crystals, bulk ceramics, thin films, and thick films can be used in RF and microwave tunable elements. Each of these forms has its advantages and shortcomings.

## 2.1. Bulk ferroelectrics

Cylinders or cubes of ferroelectric single crystals and ceramics can be used in *tunable dielectric resonators* and *tunable filters* [17]. The linear dimensions of these filters and correspondingly their mass are proportional to the length of the electromagnetic wave propagating in the ferroelectric medium which in turn is related to the dielectric

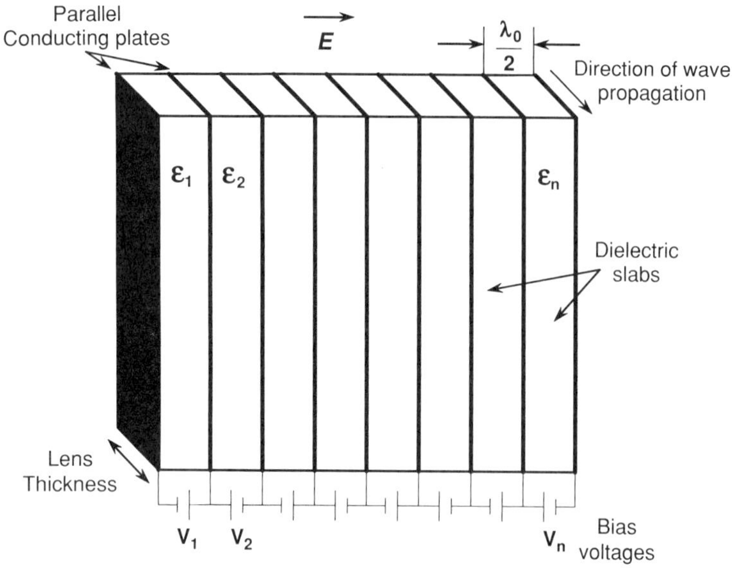


**Fig. 2.2.** Sketch of ferroelectric lens antenna, [18].

permittivity of the material according to the relation.

$$l = \frac{c}{f_0 \cdot \sqrt{\varepsilon}} \tag{2.4}$$

where $c = 3 \times 10^8$ m/s is the speed of light in vacuum, and $f_0$ and $l$ are the frequency and length of the electromagnetic wave, respectively. The potential reduction in dimensions and mass from using a ferroelectric material is proportional to $\sqrt{\varepsilon}$.

These elements can be used at radio frequencies (RF) as parallel plate *tunable capacitors* or *varactors*. Presently, the high loss of ferroelectrics (compared to that of semiconductor varctors at these frequencies) and mainly the high voltage needed for tuning limit the use of this kind of bulk varactors in comparison to the semiconductor ones.

Bulk ferroelectrics can also be used in *lens antennas* [8, 18]. Shown in Fig. 2.2, a lens antenna is a stack of electroded ferroelectric slabs [18]. The dielectric constant of each slab is controlled by varying the dc electric field applied. As a result an electromagnetic beam can be steered by passing through the antenna.

The use of ferroelectric single crystals or bulk ceramics as dielectric substrates for tunable *microstrip or strip line structures* is limited because their high and field dependent permittivity requires an adaptive impedance matching to 50 Ω in addition to the high operation voltage. Some of these problems can be minimized by an appropriate design as shown by Flaviis *et al.* [19].

At microwave frequencies, the loss in ferroelectrics can be significantly lower than that in semiconductor varactors. However, the use of relatively thick bulk ferroelectric materials still requires very high voltages for the tuning, of the order of hundreds of volts to tens of kilovolts. This limits the applicability of bulk ceramics and single crystals in tunable devices.

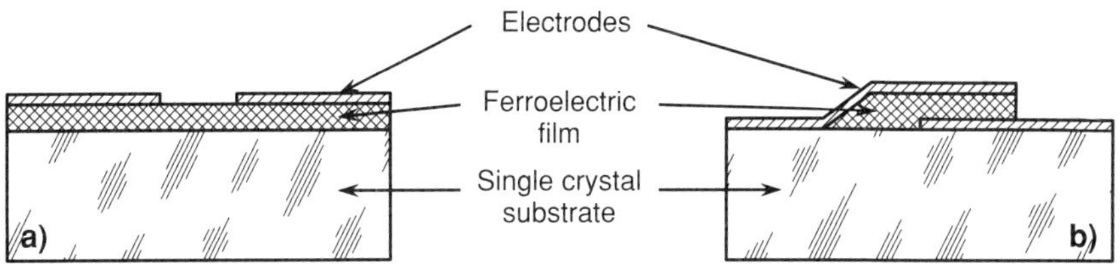


**Fig. 2.3.** Planar (a) and trilayer (b) planar capacitors based on ferroelectric film [2, 20].

## 2.2. Ferroelectric thin films

Thin films are very attractive for microwave tunable applications due to the low tuning voltages and the relatively low production cost. Sapphire, MgO, and $LaAlO_3$ substrates which are frequently used in microwave circuits are also used as substrates in ferroelectric film based components. These substrates provide good lattice matching to the perovskite ferroelectrics whereas the standard Si wafers, which are attractive substrates when the requirement of 100% compatibility with IC technology is imposed, do not provide lattice matching, resulting in inferior properties of the deposited ferroelectric films. Another problem with the use of Si for substrates for tunable devices is too high conductivity of poly-Si.

The simplest tunable passive component based on thin ferroelectric films is a *varactor* which can be made as a planar structure, the co–called "planar capacitor" (Fig. 2.3a), or as a tri-layer "sandwich" structure (Fig. 2.3b) [2, 20].

For the tri-layer capacitor the dc voltage is applied across the film thickness and as a result relatively small dc voltages (1–20 V) are sufficient for effective tuning. However, because of the high dielectric permittivity and the small thickness, the varactors have large capacitance which limits their high frequency applications. Moreover, in the parallel plate configuration, the electrodes contribute to the loss tangent of the device much more than in the case of the planar structure.

For the planar structures, the electric field is applied along the film across the gap, which is typically more than an order of magnitude larger than the film thickness. This makes the capacitance smaller, however, the voltage needed for tuning increases.

*Microwave ferroelectric phase shifters* are the most widely studied tunable ferroelectric components. Their importance stems from the role they could play in phased array antennas. A phased array antenna consists of thousands of radiating elements which should be served by thousands of phase shifters. The phase shifters are used to modify and control the width and angle of the steered radar beam. At present each phase shifter is a housed microwave semiconductor module. The use of ferroelectric films enables the integration of the phase shifters with the microwave circuits on one substrate thus substantially reducing the size, mass, and cost of the antennas.

A simple coplanar line structure patterned on a ferroelectric thin film coated substrate makes a phase shifter (Fig. 2.4a) in which the phase velocity of the electromagnetic wave that passes through the line is controlled by the applied dc electric field, $U_{dc}$ (via changing the permittivity of the film, which in turn controls the wave velocity) [21]. It is also possible to tune the phase velocity of the electromagnetic wave passed through the coplanar line by loading this line with tunable ferroelectric capacitors (in contrast to

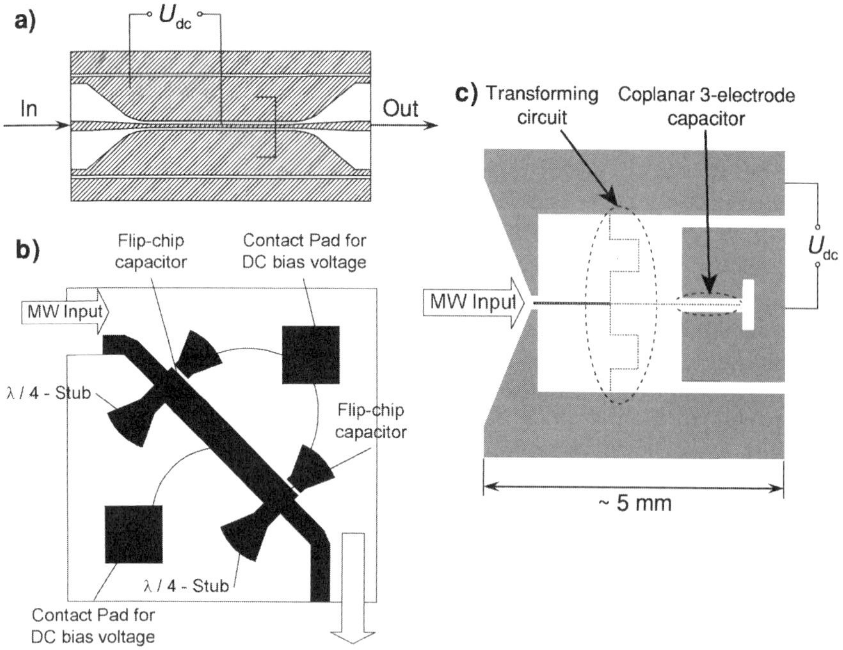


**Fig. 2.4.** Ferroelectric microwave phase shifters: (a) coplanar line analog phase shifter [21], (b) periodically loaded line phase shifter [23], (c) reflection type digital phase shifter [24].

the previous case where the coplanar line is homogeneously loaded with a ferroelectric film) [22].

Another possibility is the so-called digital phase-shifter which is used only at two values of the tuning voltage providing a fixed phase shift (e.g. 45° or 90°). Examples of this kind of phase shifters are presented in Fig. 2.4b and c. Figure 2.4b shows a layout of a periodically loaded line digital phase shifter with two planar capacitors connected in parallel to the line [23]. Another solution is presented in Fig. 2.4c [24] showing a phase shifter that consists of a transforming circuit that is terminated by a coplanar 3-electrode capacitor. The advantage of the digital phase shifters compared to those based on the simple coplanar line is that the former can be better optimized. Namely, better impedance matching with the 50 Ω external wave impedance and lower amplitude modulation distortions at switching can be reached in the digital phase shifters.

The Figure of Merit ($F_{\rm ph}$) of microwave phase shifters is the ratio of the phase shift produced by the phase shifter ($\Delta\phi_{\rm ph}$) to its insertion loss ($L_{\rm dB}$). The upper bound of the Figure of Merit $F_{\rm ph}$ for an optimally designed one-bit digital phase shifter (not taking into account conductor and interconnections losses) can be determined as [25, 26]:

$$F_{PH}^{\max} = 6.6\,\frac{\Delta\varphi/2}{\sin\left(\Delta\varphi/2\right)}\cdot\sqrt{K}, \tag{2.5}$$

where $K$ is the commutation quality factor of the ferroelectric capacitor used in the circuit and $\Delta\varphi$ is the desired phase shift.

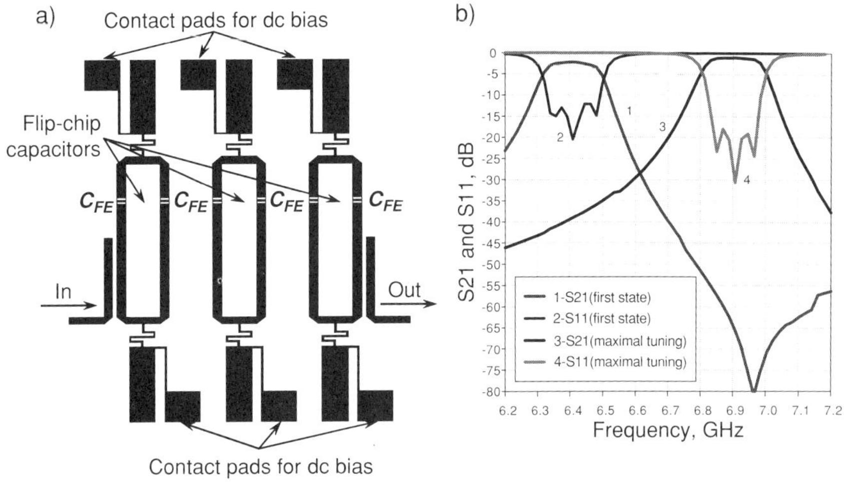


**Fig. 2.5.** Layout of a tunable 3-pole filter based on ferroelectric planar capacitors (a) and its simulated characteristics at different values of dc bias voltage (b) where S21 and S11 are the transmission and reflection coefficients, respectively [31].

Values of $F_{ph}$ in the range of 30–100 (for frequencies of 1–30 GHz) were reported [19, 21–24, 27, 28] for phase shifters based on ferroelectric films with tunability 1.5–2, loss tangent 0.02–0.05, and with copper or gold as the conducting layers.

*Microwave tunable filters* based on ferroelectric thin films have also been recently demonstrated. [29, 30] The Figure of Merit for the tunable band pass filter ($F_{BPF}$) is characterized by the ratio between the shift of the central frequency due to the tuning and the average bandwidth of the filter. The Figure of Merit of the tunable $N$-pole filter as a function of the $K$-factor can be found in the following form [31]:

$$F_{BPF} = \frac{\omega_{02} - \omega_{01}}{\sqrt{\Delta\omega_1 \Delta\omega_2}} = \frac{1}{2N^2} \cdot \sqrt{K}, \tag{2.6}$$

where $\omega_{01}$, $\omega_{02}$, $\Delta\omega_1$, and $\Delta\omega_2$ are the central frequencies and the band widths of the filter in two inter-switchable states. The layout of a 3-pole filter and its simulated frequency characteristics are shown in Fig. 2.5 [31].

The designs of other microwave active and passive ferroelectric thin film based components such as microwave *voltage controlled oscillator* [27], *frequency modulators* [32], *parametric amplifiers* [9] and *tunable power divider* [33] were also reported recently.

## 2.3. Ferroelectric thick films

Screen printed or tape cast thick films are regarded as promising ferroelectric elements for hybrid RF/microwave circuits because of their low production costs. The principles of design and operation of ferroelectric thin film based microwave components can be applied to the thick film devices with only minor modifications [34]. However, to date, high loss and technological limitations on the linear dimensions

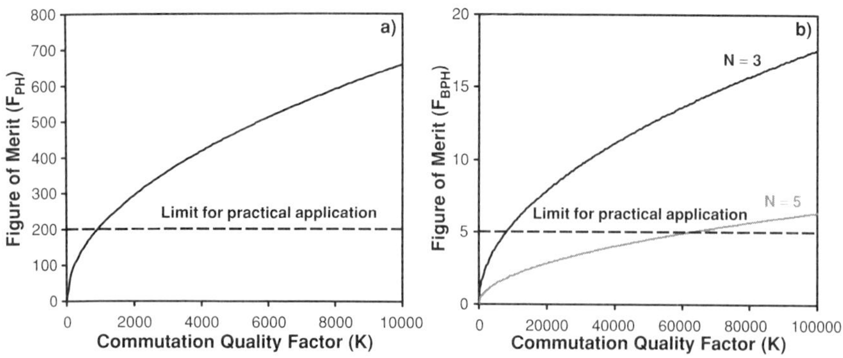


**Fig. 2.6.** Calculated dependences of the figure of merit of (a) phase shifters and (b) band-pass filters on the commutation quality factor of switchable component. ($N$ in figure (b) corresponds to the order of filter).

associated with the thick film technology still impede the use of ferroelectric thick films in tunable microwave devices. In addition, the planar microwave devices made of thick ferroelectric films are expected to have worse tunability compared to thin film devices. The reason is that only a thin upper layer of the thick ferroelectric film contributes to the effective tuning whereas the contribution of the rest of the film can be regarded as that of a non-tunable capacitance connected in parallel with the upper tunable part of the film.

## 2.4. Requirements imposed on the material by the device performance needs

Generalizing the above discussion, it is of interest to estimate the needed properties of the ferroelectric varactors for various potential applications.

Figure 2.6a shows the calculated dependence of $F_{\text{ph}}$, the figure of merit for a phase shifter, on the commutation quality factor of the material. For practical applications, in order to be competitive with semiconductor devices, $F_{\text{ph}} \geq 200$ is required, which implies $K > 900$. Of course this requirement depends also on the frequency used and on the specifics of the application of the device.

Figure 2.6b shows the dependence of $F_{\text{BPF}}$ (figure of merit for band-pass filter) on the commutation quality factor. One can see that, for example, to have a filter which can be tuned for 5 pass bands ($F_{\text{BPF}} = 5$), the commutation quality factor of the capacitors should be more than 8000 for a 3-pole band-pass filter ($N = 3$) and more than $6 \cdot 10^4$ for a 5-pole band-pass filter ($N = 5$).

This means that a competitive ferroelectric varactor should have loss tangent on the level of 0.005–0.01 and tunability $n > 1.5$ at the frequencies of the interest. However, it should be noted that the commutation quality factor is not the only parameter representing the properties of the switchable microwave element. It is more a qualitative than a quantitative characteristic of the ferroelectric varactor. It tells us qualitatively whether the ferroelectric capacitor is potentially suitable for the design of a microwave tunable component with the required characteristics. Keeping the commutation quality factor constant it is possible to have an infinite number of values of the tunability and loss tangent (see Eq. (2.3)); it may be a combination of a high loss with a high tunability as well as that of a low loss with a low tunability. For almost all designs of microwave controllable components, the first case is preferable because the higher the tunability, the easier it is to combine the varactor with the passive microwave circuit to get a better performance. In addition, there are many methods to reduce the effective loss tangent and tunability of a varactor by loading it with various kinds of microwave passive elements. This also gives an advantage to the more tunable materials. Another issue is a rather high absolute value of the relative dielectric permittivity of the ferroelectrics, which is often not desirable for high frequency devices (too small dimensions are required). For this reason, for the same value of the commutation quality factor, materials with lower permittivity may be preferable. The typically high *TCE* of materials with high permittivity also suggests a preferable use of materials with lower permittivity. The main parameters of a tunable material (tunability, loss tangent, dielectric permittivity, and *TCE*) are dependent one on the other so that the material optimization has to be done with great care. The interplay among these parameters is the center of the rest of this paper.

## 3. Theory of Dielectric Response of Tunable Materials

### 3.1. Dielectric constant and tunability of ferroelectric materials

Ferroelectrics are known for their high dielectric constant and high tunability. These features of ferroelectrics are linked and the general trend "the higher the dielectric constant, the higher the tunability" generally holds in ferroelectric materials. It holds for an "ideal ferroelectric" where the dielectric response is controlled by the lattice dynamics of the material as well as for situations where the dielectric response is essentially influenced by the presence of defects or inclusions of a low-dielectric-constant non-tunable material. Below we will discuss these situations separately. The effects of dielectric anisotropy do not usually play an essential role in the performance of tunable materials. For this reason, these effects are not addressed here so that we will always consider the situation where the polarization and electric fields are collinear.

#### *3.1.1. Ideal ferroelectric*

*Dielectric permittivity*

The origin of the high dielectric permittivity of ferroelectrics in the paraelectric phase is a delicate compensation of various kinds of microscopic forces that maintains the material in a non-poled state in the absence of a microscopic electric field. Because of this compensation, the restoring force opposing the poling action of the applied field is relatively weak. This results in a high dielectric permittivity of the material. In the ferroelectric phase, the newly appeared spontaneous polarization is not very stable either. This makes the dielectric permittivity high in the ferroelectric phase. In addition to this, the permittivity may be further increased by contributions stemming from ferroelectric domains. Since the present paper is devoted to microwave materials, in the following discussion, we will address only the paraelectric phase of ferroelectrics (normal or incipient) of the displacive type; it is believed that only for this case will a material exhibit a high tunability and a relatively low dielectric loss at microwave frequencies, a combination of interest for microwave application. Order-disorder materials are inappropriate since they exhibit a much smaller permittivity at microwave frequencies[1], which implies a smaller tunability. The situation with the ferroelectric phase of the displacive ferroelectrics is not so clear. Though, in the ferroelectric phase, additional mechanisms (domain related, see e.g. [35], and quasi-Debye mechanisms, see Subsec 3.2.1) of the loss are active, their contributions might not be decisive in the case of real materials. Thus, the tunable materials in the ferroelectric phase should not be ruled out outright.

The most straightforward description of the dielectric response of ferroelectrics is given by the conventional Landau theory and is based upon an expansion of the Helmholtz free energy $F$ with respect to the vector macroscopic polarization $\mathbf{P}$. For the situation where the polarization is collinear with the macroscopic electric field $\mathbf{E}$ in the material, the first two terms of this expansion read:

$$F = \frac{\alpha}{2}P^2 + \frac{\beta}{4}P^4. \tag{3.1}$$

The equation of state $\partial F/\partial P = E$ then leads to a relation between the polarization and electric field:

$$E = \alpha P + \beta P^3. \tag{3.2}$$

Equation (3.2) enables us to present the relative dielectric permittivity of the material in the form:

$$\varepsilon = \frac{1}{\varepsilon_0}\frac{\partial P}{\partial E} = \frac{1}{\varepsilon_0}\frac{1}{\alpha + 3\beta P^2} = \varepsilon(0)\frac{1}{1 + 3\beta\varepsilon(0)\varepsilon_0 P^2} \tag{3.3}$$

where $\varepsilon(0) = (\varepsilon_0\alpha)^{-1}$ and $\varepsilon_0 = 8.85 \times 10^{-12}$F/m . This expression describes the dielectric permittivity both in the absence of a bias field and under it. In the first case, one sets $P = 0$ in Eq. (3.3) to obtain:

$$\varepsilon = \varepsilon(0) = \frac{1}{\varepsilon_0\alpha}, \tag{3.4}$$

whereas, for the situation under the bias, one sets $P = P_{dc}$ in Eq. (3.3) ($P_{dc}$ is the polarization induced by the bias field).

According to the Landau theory, the coefficient $\alpha$ is assumed to be a linear function of temperature and vanishes at the Curie-Weiss temperature $T_0$, the ferroelectric instability temperature of the material

$$\alpha = \alpha_L = \frac{1}{\varepsilon_0}\frac{T - T_0}{C}. \tag{3.5}$$

[1] The dielectric anomaly in order-disorder ferroelectrics suffers from very strong frequency dispersion already in the radio frequency range.

where $C$ is the Curie-Weiss constant. This assumption of the Landau theory[2] can be always justified for $|T - T_0| / T_0 << 1$. However, for displacive ferroelectrics, it has been shown [36] that the coefficient $\alpha$ can be taken with a good precision in the form given by Eq. (3.5) for a much wider temperature range: starting from the temperature of about a fraction of the Debye temperature $\theta$ of the material and up to its melting point. For real materials, this statement has been checked for the incipient ferroelectrics $SrTiO_3$ and $KTaO_3$, with the Debye temperatures of about 400 K, and it was shown that Eq. (3.5) applies with a reasonable accuracy from 50–80 K up to very high temperatures [12].

Equation (3.5) ceases to work at $T << \theta$ where the quantum statistics of the lattice vibration should be taken into account. In this regime, the temperature dependence of $\alpha$ strongly slows down. In terms of simple models, several explicit forms have been suggested for this dependence. The Barrett theory [37], which models the lattice dynamics of the ferroelectric with a system of identical oscillators, yields:

$$\alpha = \alpha_B = \frac{T_B}{\varepsilon_0 C}\left[\coth\left(\frac{T_B}{T}\right) - \frac{T_0}{T_B}\right]. \tag{3.6}$$

whereas according to Vendik [38] (who models the lattice dynamics of the ferroelectric with a system of spatially coupled oscillators) the following formula can be used for a description of the temperature dependence of $\alpha$:

$$\alpha = \alpha_V = \frac{T_V}{\varepsilon_0 C}\left[\sqrt{\frac{1}{16} + \left(\frac{T}{T_V}\right)^2} - \frac{T_0}{T_V}\right]. \tag{3.7}$$

Here $T_B$ and $T_V$ are parameters decreasing the slowing down of the temperature dependence of $\alpha$ on decreasing temperature. Calculations performed in a more precise model [36] give a temperature dependence of $\alpha$ similar to these given by Eqs. (3.6) and (3.7).

An essential feature of displacive ferroelectrics is the typical value of the Curie-Weiss constant $C$ of about $10^5$ K. Such a value of $C$ implies high values of the dielectric permittivity even far from the Curie-Weiss temperature $T_0$. Indeed, at $T = T_0 + 200$ K, Eqs. (3.4) and (3.5) implies $\varepsilon \cong 500$.

*Tunability*

The dependence of the dielectric permittivity on the applied dc bias electric field $E_0$ can be determined from Eq. (3.3). There are two parameters used for characterization of this dependences: tunability

$$n = \frac{\varepsilon(0)}{\varepsilon(E_0)}, \tag{3.8}$$

and relative tunability

$$n_r = \frac{\varepsilon(0) - \varepsilon(E_0)}{\varepsilon(0)} = \frac{n-1}{n}. \tag{3.9}$$

[2] In this article we do not discuss the critical phenomena which may impact the linear temperature dependence of $\alpha$ very close to $T_0$ since the available experimental data do not suggest that this effect is substantial in ferroelectrics.

The aforementioned trend—the higher permittivity $\varepsilon(0)$, the higher the tunability (for a given value of the bias field $E_0$) can be clearly traced. In the limit of weak nonlinearity, i.e. at $n_r << 1$, this trend is really strong. Indeed, in this case, according to (3.3), we have:

$$n = \frac{\varepsilon(0)}{\varepsilon(E_0)} = 1 + 3\beta\varepsilon(0)\varepsilon_0 P_{dc}^2 \approx 1 + 3\beta(\varepsilon(0)\varepsilon_0)^3 E_0^2. \qquad (3.10)$$

That means that the relative tunability is a very fast function of $\varepsilon$ (0), specifically. $n_r \propto \varepsilon(0)^3$. With increasing bias fields this dependence slows down. In the limit of ultrahigh fields where $n >> 1$, one readily finds $\varepsilon(E_0) \approx \beta^{-1/3} E_0^{-2/3}/(3\varepsilon_0)$. This implies $n \propto \varepsilon(0)$. It is instructive to give an expression for the field $E_n$ needed to achieve a given tunability $n$:

$$E_n = \frac{\sqrt{n-1}(2+n)}{\varepsilon_0\sqrt{27\beta\varepsilon_0}} \frac{1}{\varepsilon(0)^{3/2}}. \qquad (3.11)$$

Thus, it is seen that a high dielectric permittivity of the material is really essential for its tunable behavior.

For the description of the dielectric non-linearity at an arbitrary dc field, one can use Eq. (3.3) where the polarization $P$ is calculated from the solution of cubic equation (3.2). A convenient form for calculation of the dependence of the permittivity on the dc bias field has been suggested by Vendik [38, 39]:

$$\varepsilon(T, E_0) = \frac{\varepsilon_{00}}{\left[\left(\xi^2+\eta^3\right)^{1/2}+\xi\right]^{2/3} + \left[\left(\xi^2+\eta^3\right)^{1/2}-\xi\right]^{2/3} - \eta}; \qquad (3.12)$$

$$\varepsilon_{00} = \frac{C}{T_0}; \qquad \eta = \varepsilon_{00}\alpha\varepsilon_0; \qquad \xi = \xi_B = \frac{E_0}{E_N};$$
$$E_N = \frac{2}{\sqrt{27\beta\varepsilon_{00}^3\varepsilon_0^3}}. \qquad (3.13)$$

The field dependence of dielectric permittivity discussed above is illustrated in Fig. 3.1 for two values of $\varepsilon(0)$ and the value of $\beta = 8 \times 10^9 \mathrm{JC^{-4}m^{-5}}$ (typical for $SrTiO_3$). The regimes of small and high tunability are shown together with the asymptotic dependences of $\varepsilon(E)$ in these regimes.

### 3.1.2. Ferroelectric with random field defects

The thermodynamic model discussed in the previous subsection was extended by Vendik and Zubko [40] to the case of materials containing random built-in electric fields (or random built-in polarization). These fields can naturally come from the presence of dipolar or charged defects and should result in a suppression of local dielectric permittivity. The calculation of the dielectric response of a ferroelectric containing random electric field is a difficult task for an exact theoretical treatment. Vendik and Zubko treated the problem in an approximation equivalent to a model where the material is presented as a system of layers perpendicular to the applied field. The layers differ by the values of their built-in random electric fields, a Gaussian distribution of the latter

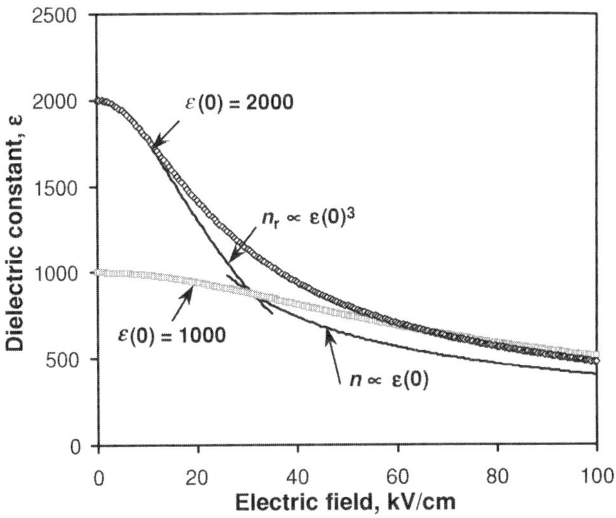


**Fig. 3.1.** Field dependence of the dielectric permittivity according to Eq. (3.12) plotted for $\beta = 8 \times 10^9 \text{JC}^{-4}\text{m}^5$ and for the values of $\varepsilon(0)$ 2000 and 1000. The regimes of small and high tunability are shown together with the asymptotic dependences of $\varepsilon(E)$ in these regions.

being assumed. It was shown that the results obtained this way for the temperature and field dependence of the dielectric permittivity can be reasonably well approximated by Eq. (3.12) where the parameter $\xi$, which controls the field effect, is set in the form:

$$\xi = \sqrt{\xi_B^2 + \xi_{St}^2}. \qquad (3.14)$$

In this framework, parameter $\xi_{St}$ controls the suppressing effect of the random fields on the dielectric permittivity of the ferroelectric material. The ratio $\xi_B/\xi_{St}$ has the physical meaning of the ratio between the applied field and the typical value of the random field.

This model was used for the description of the dielectric response in a number of materials of interest for tunable applications. For these maratials, the model parameters entering Eqs. (3.7) and (3.12–14) are listed in Table 3.1.

**Table 3.1.** Parameters of the Vendik's model for $KTaO_3$ and $SrTiO_3$ crystals and $(Ba_x Sr_{1-x})TiO_3$ ceramics [2, 41].

| | $SrTiO_3$ | $Ba_xSr_{1-x}TiO_3$ | $KTaO_3$ |
|---|---|---|---|
| $T_0$, K | 42 | $T_c(x) = 42 + 439.37 \cdot x - 95.95 \cdot x^2$ | 32.5 |
| $C \times 10^{-5}$ | 0.86 | $C(x) \times 10^{-5} = 0.86 + 1.1 \cdot x^2$ | 0.45 |
| $T_v$, K | 175 | 175 | 170 |
| $\varepsilon_{00}$ | 2080 | $\varepsilon_{00}(x) = C(x)/T_c(x)$ | 1390 |
| $E_N$, kV/cm | 19 | $E_N(x) = 8.4/\left(\varepsilon_0 (3\varepsilon_{00}(x))^{3/2}\right)$ | 15.6 |
| $\xi_s$ | 0.018 | 0.3 | 0.02 |

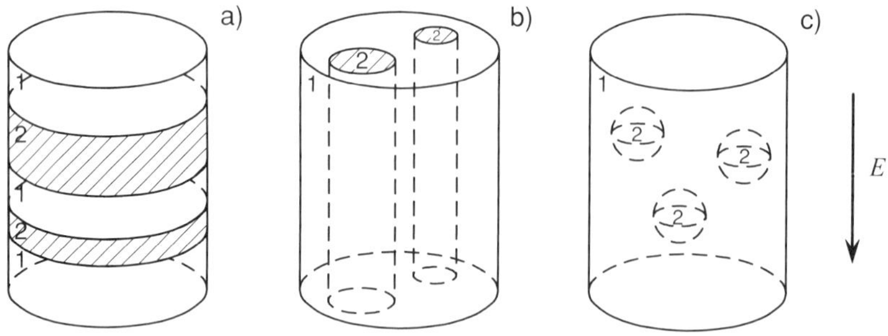


**Fig. 3.2.** Schematic of models of bi-component composite: a)—layered model, b)—columnar model, and c)—spherical inclusion model. 1—ferroelectric component, 2—dielectric component.

### *3.1.3. Ferroelectric containing inclusions of low-dielectric-constant non-tunable material*

There are some situations where the tunable material can be treated as a ferroelectric/dielectric composite. These situations occur due to the presence of near-by–electrode layers or secondary phases at grain boundaries as well as when a ferroelectric/dielectric composite is deliberately used as a tunable material. The dielectric permittivity and tunability of the materials are strongly affected by the presence of the "passive" component. Since no exact general theory of the dielectric constant and tunability of composites is available[3], we will overview some results obtained for three models of bi-component ferroelectric-dielectric composites. These models correspond to different structures illustrated in Fig. 3.2. In addition to the models shown in this figure, the so-called "brick-wall" model is often addressed in the literature (see e.g. [44]). It consists of a regular structure of cubes of one phase equally separated in three dimensions by thin layers of another phase. We will not specifically address this model since, in the case of the ferroelectric tunable composite (the cubes are of a ferroelectric and the layers are of a dielectric), this model is very close to the layered model (shown in Fig. 3.2a) with a 3 times reduced concentration of the dielectric.

We start from the layered model (Fig. 3.2a). This case can be described as the in-series connection of several capacitors corresponding to the different layers of the composite. Electrically, this system is equivalent to the in-series connection of only two capacitors which correspond to the two components of the mixture. The effective dielectric permittivity of such a composite can be found as:

$$\frac{1}{\varepsilon_{\mathrm{mix}}(q)} = \frac{1-q}{\varepsilon_{\mathrm{f}}} + \frac{q}{\varepsilon_{\mathrm{d}}} \tag{3.15}$$

[3] A comprehensive overview of the linear dielectric response of ferroelectric composites was offered by Hudak *et al.* [42]. The problem of the tunability of ferroelectric composite was theoretically approach by Liou and Chiou [43]. However, the analysis offered in this paper is based on many approximation and assumptions. For these reason in the present paper, we will not discuss the results of this work.

where $q$ is the volume concentration of the dielectric phase in the composite. $\varepsilon_{mix}$, $\varepsilon_f$, and $\varepsilon_d$ are the dielectric constants of the composite, ferroelectric and dielectric, respectively.

The field dependence of the composite dielectric constant originates from that of the dielectric constant $\varepsilon_f(E_f)$ of the ferroelectric component, where $E_f$ is the electric field in the ferroelectric. This dependence $\varepsilon_f(E_f)$ can be obtained by applying Eqs. (3.2)–(3.4) to the ferroelectric component of the composite to get

$$\varepsilon_f^{-1}(E_f) = \varepsilon_0\alpha + 3\beta\,\varepsilon_0\,P_f^2 \tag{3.16}$$

$$\beta\,P_f^3 + \alpha\,P_f - E_f = 0. \tag{3.17}$$

To find the dielectric constant of the composite material as a function of the bias field $E_0$ and concentration $q$ one appends Eqs. (3.15–17) with the condition of continuity of electrical displacement:

$$\varepsilon_0\varepsilon_b E_f + P_f = \varepsilon_0\varepsilon_d\,E_d \tag{3.18}$$

and the equation for the voltage drops across the composite components:

$$E_f\,(1-q) + E_d\,q = E_0. \tag{3.19}$$

Here $E_d$ is the electric field in the dielectric and $\varepsilon_b$ is the background dielectric constant[4] of the ferroelectric, which describes the contribution of the hard ("non-ferroelectric") optical phonon modes and the electronic degrees of freedom to the permittivity of the material. Using Eqs. (3.15–19) the solution of the problem can be presented in the form [45]:

$$\beta\,P_f^3 + \alpha^*\,P_f - E_f^* = 0, \tag{3.20}$$

$$\begin{cases} \alpha^* = \alpha + \dfrac{q}{\varepsilon_0\varepsilon_b q + \varepsilon_0\varepsilon_d\,(1-q)} \\ E_f^* = E_0\dfrac{\varepsilon_d}{\varepsilon_b q + \varepsilon_d\,(1-q)} \end{cases}, \tag{3.21}$$

$$\varepsilon_{mix}^{-1}(q, E_0) = \varepsilon_0\alpha^* + 3\beta\varepsilon_0\,P_f^2 + \frac{q^2\,(\varepsilon_b - \varepsilon_d)}{\varepsilon_d\,[\varepsilon_b q + \varepsilon_d\,(1-q)]} - q\varepsilon_0(\alpha + 3\beta\,P_f^2). \tag{3.22}$$

In the limit of small concentration of the dielectric material in the mixture ($q << 1$) and when $\varepsilon_f/\varepsilon_d >> 1$, the last two terms in the Eq. (3.22) can be neglected and the renormalized electric field $E^*$ in Eq. (3.20) becomes equal to the electric field $E_0$. Thus, Eqs. (3.20) and (3.22) become identical to the equations for the permittivity and polarization of the pure ferroelectric material, Eqs. (3.2) and (3.3), but with the coefficient $\alpha$ replaced by $\alpha^*$. One can easily see that, in this case, the behavior of the composite is identical to that of the pure ferroelectric whose coefficient $\alpha$ is increased, which corresponds to an effective decrease of the Curie-Weiss temperature of the material.

The second model, the columnar structure shown in Fig. 3.2b, is clearly much simpler to analyze [43]. This system is equivalent to the parallel connection of two

---

[4] In Sec. 3.1.1. the difference $\varepsilon_b - 1$ was neglected because for the situation treated in that Section this approximation leads only to a small (about $\varepsilon_b/\varepsilon_f << 1$) inaccuracy in the results. In Eqs. (3.15) and (3.23) the contribution of $\varepsilon_b$ is also neglected to within the same accuracy.

capacitors corresponding to the two components of the mixture. The effective dielectric permittivity of such composite can be presented as:

$$\varepsilon_{\text{mix}}(q) = \varepsilon_{\text{f}}(1-q) + \varepsilon_{\text{d}}\, q. \tag{3.23}$$

In the case of parallel connection of the composite components, the electric field is the same in all parts of the composite, being equal to the applied field $E_0$. This means that the field dependence of dielectric constant of the composite is simply given by Eq. (3.23) where $\varepsilon_{\text{f}}$ is given by Eqs. (3.2) and (3.3).

The spherical inclusion model shown in Fig. 3.2c consists of a random distribution of dielectric spheres in the ferroelectric host medium. The well-known electrostatic solution of the problem for a dielectric sphere embedded into a dielectric medium gives the electric field distribution in the sphere and in the medium. This solution also enables the calculation of the effective dielectric permittivity of a two-component material in the limit of low concentration of the spherical inclusions ($q << 1$) [46]. For this dielectric-ferroelectric bi-component system, the dielectric constant $\varepsilon_{mix}$ and the electric field in the ferroelectric matrix $\vec{E}_{\text{f}}$ read [46]:

$$\varepsilon_{\text{mix}}(q) = \varepsilon_{\text{f}} + 3q\,\varepsilon_{\text{f}}\frac{\varepsilon_{\text{d}} - \varepsilon_{\text{f}}}{\varepsilon_{\text{d}} + 2\varepsilon_{\text{f}}}, \tag{3.24}$$

$$\vec{E}_{\text{f}}(\vec{r}) = \vec{E} + \sum_{i}^{N} \frac{\varepsilon_{\text{d}} - \varepsilon_{\text{f}}}{\varepsilon_{\text{d}} + 2\varepsilon_{\text{f}}} \frac{R^3}{|\vec{r} - \vec{r}_i|^3} \overleftrightarrow{G}\vec{E}, \tag{3.25}$$

where $R$ is the radius of the dielectric inclusions, $N$ is the total number of the dielectric spheres distributed in the composite, and $\vec{r}_i$ is the radius-vectors pointing to the center of the $i^{\text{th}}$ spherical inclusion. The matrix $\overleftrightarrow{G}$ reads

$$\overleftrightarrow{G} = 3n_t n_s - \delta_{ts},$$

where $\vec{n} = (\vec{r} - \vec{r}_i)/|\vec{r} - \vec{r}_i|$ and $\delta_{ts}$ is the Kroneker symbol.

In the limit where $\varepsilon_{\text{d}} << \varepsilon_{\text{f}}$ these expressions can be reduced to:

$$\varepsilon_{\text{mix}}(q) = \varepsilon_{\text{f}}\left(1 - \frac{3}{2}q\right), \tag{3.24a}$$

$$\vec{E}_{\text{f}}(\vec{r}) = \vec{E} - \frac{1}{2}\sum_{i}^{N} \frac{R^3}{|\vec{r} - \vec{r}_i|^3} \overleftrightarrow{G}\vec{E}. \tag{3.25a}$$

This solution takes into account a non-uniform distribution of the electric field in the host material but does not take into account its dielectric nonlinearity. Once this nonlinearity is taken into account, the problem becomes much more complicated. A very rough solution of this problem, in the case of weak dc fields corresponding to a small relative tunability, $n_r << 1$ (a typical situation in bulk ceramics), is available in Ref. [45]. In this solution, when calculating the dielectric nonlinearity of the composite the second term in Eq. (3.25a) is neglected so that in the ferroelectric matrix one set $\vec{E}_{\text{f}} \approx \vec{E}$. Thus, in this approximation, to calculate the effective dielectric permittivity at a non-zero $E_0$ one still use Eq. (3.24a), where $\varepsilon_{\text{f}} = \varepsilon_{\text{f}}(E_0)$. This implies that, at small $q$, the concentration dependence of the relative tunability is very weak (the linear term in $q$ is absent)). Further calculations performed beyond this approximation (analytical

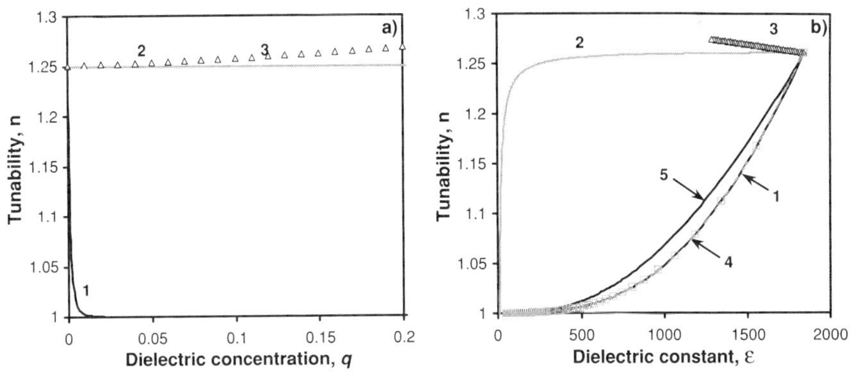


**Fig. 3.3.** Tunability $n$ of ferroelectric/dielectric composites as a function of dielectric concentration $q$ (a) and dielectric constant of the material $\varepsilon$ (b) (black solid line 1—layered model, grey solid line 2—columnar model, and triangles 3—spherical inclusions model). For comparison, the $\varepsilon$ dependence of tunability $n$ is also given for a pure ferroelectric (grey dashed line 4 with squares) and ferroelectric with random-field impurities (black solid line 5). Modeling was performed for material parameters of $SrTiO_3$ given in Tab.3.1. In the case of a pure ferroelectric, the permittivity is influenced by a temperature variation, in the case of a ferroelectric with random-field impurities it is influenced by a variation of the random-field intensity, and in the case of composites it is influenced by a variation of the concentration of dielectric inclusions.

and finite element analysis FEMLAB$^{TM}$) confirmed the weakness of this dependence, however, with a non-vanishing linear term in $q$. Specifically, it is obtained [47] for the relative tunability of the composite $n_{r,\text{mix}}$ that

$$n_{r,\text{mix}} = (1 + 0.2q)n_r. \tag{3.26}$$

where $n_r$ is the tunability of the ferroelectric at the same value of the applied field. This result is quite unexpected since it predicts an increase (though weak) of the tunability as of a ferroelectric with delusion it with a non-tunable dielectric. However, it agrees with the prediction that could be drawn from the theories of other authors on the optical [48] and dielectric [49] nonlinearity of composite.

The presented above analysis enables a comparison of the concentration dependences of the tunability for the three model for the ferroelectric/dielectric composite. These dependences are shown in Fig. 3.3a (the results for the spherical inclusion model being valid in the limit $n_r << 1$ and $q << 1$). Here it is of interest to note that the tunability calculated for the spherical inclusion model does not lay between the results obtained for the "in-series" and "parallel" models. This fact might be considered as unexpected, however, in contrast to the case of dielectric permittivity, there is no reason to consider the "in-series" and "parallel" models as those giving upper and lower bound for the tunability of the composite.

### *3.1.4. Tunability as a function of the permittivity*

When discussing the tunability of the ideal ferroelectric we have seen that the tunability is an increasing function of the permittivity of the material. This trend does not always hold for ferroelectric materials containing imperfections or inclusions of non-tunable dielectrics, substantially varying from one system to another. The knowledge of this dependence for a given system may be useful for various purposes. First, when working with a "non-ideal" ferroelectric, the comparison of the measured $n$ vs. $\varepsilon$ dependence with those simulated for various models may provide information on the character of the imperfection of the material. Second, depending on the application, a combination of a high tunability with larger $\varepsilon$ or smaller $\varepsilon$ is desirable. Thus, the knowledge of possible $n$ vs. $\varepsilon$ dependences may help in tailoring properties of the material.

In Fig. 3.3b, the results of simulation of the $n$ vs. $\varepsilon$ dependence for a pure ferroelectric, a ferroelectric with random field impurities, and various ferroelectric/dielectric composites are shown. It is seen that these dependences may be very different from each other (for material parameters used in the simulation see Table 3.4). A feature to mention is that the behavior of the layered composite is very close to that of pure ferroelectric in the whole interval of the permittivity. Remarkable is the behavior of the columnar composite (and the spherical model composite in the limit of small $q$ and $n_r$) where an essential reduction of the permittivity may take place without essential reduction of the tunability or even with a slight increase of the latter.

## 3.2. Dielectric loss in ferroelectrics

Materials used for microwave tunable applications are incipient or regular ferroelectrics (in the paraelectric phase) of displacive type. For the frequencies of interest both fundamental phonon loss mechanisms and extrinsic mechanisms (owing to the coupling of the microwave field with defects) are relevant. Below we discuss these contributions as well as the loss in ferroelectric/dielectric composites and the impact of dc bias electric field on the loss.

### *3.2.1. Intrinsic loss*

The origin of the fundamental loss is the interaction of the ac field with the phonons of the material. The theory of the loss stemming from this interaction has been developed for crystalline materials with a well defined phonon spectrum (see for a review [6]), i.e., for materials where the damping of phonons (average frequency of the inter-phonon collisions), $\Gamma$, is much smaller than their frequencies. According to this theory, in terms of quantum mechanics, the fundamental loss mainly corresponds to the absorption of the energy quantum of the electromagnetic field $\hbar\omega$ ($\omega$ is the ac field frequency) in collisions with the thermal phonons, which have much higher energies. This large difference in the energies makes it difficult to satisfy the conservation laws in these collisions. In such a complicated situation there exist three efficient schemes of absorption of the $\hbar\omega$-quanta, which correspond to the three main fundamental loss mechanisms: 1) three-quantum, 2) four-quantum, and 3) quasi-Debye.

*Three-quantum mechanism*

The three-quantum mechanism corresponds to a process involving a $\hbar\omega$-quantum and two phonons. The theory of this contribution has been developed by different authors [6, 50–53]. Owing to a very big difference between the energy quantum of the electromagnetic field $\hbar\omega$ and those of the thermal phonons, the three-quantum processes can take place only in the regions of the wave vector space ($\vec{k}$-space) where the difference between the frequencies of two different phonon branches is small, specifically, of the order of $\omega$ and/or of the phonon damping $\Gamma$. These regions are usually located in the vicinity of the degeneracy lines of the spectrum, i.e. the lines in $\vec{k}$-space where the frequencies of different branches are equal. In the materials of interest for tunable applications, the phonon spectrum contains low-lying (soft) optical modes controlling their high dielectric permittivity and the degeneracy lines formed with participation of these modes are of primary importance for the loss. Since the degeneracy of the spectrum is mainly controlled by the symmetry of the crystal, the explicit temperature dependence (which does not take into account the temperature dependence of the dielectric permittivity) and frequency dependence of the three-quantum loss are very sensitive to the symmetry of the crystal. Depending on crystalline symmetry, temperature interval, frequency range and some parameters of the phonon spectrum, the temperature and frequency dependence of the imaginary part of the dielectric permittivity can be described by power laws [6], i.e:

$$\varepsilon'' \propto \omega^n T^m \tag{3.27}$$

where $n = 1 - 5; m = 1 - 9$.

For microwave high-$\varepsilon$ and tunable materials, which are typically centrosymmetric, cubic or pseudocubic, the $T-$ , $\omega-$ , and $\varepsilon-$ dependence of the three-quantum loss can be evaluated in an analytical form. The calculations made for a model, in which the dielectric response is assumed to be controlled only by the lowest polar transverse optical (TO) mode ($\Omega_{\mathrm{TO}}$ is the frequency of this mode at $\vec{k} = 0$), give for $T \geq \hbar\Omega_{\mathrm{TO}}/k_{\mathrm{B}}$ [52, 54, 55]:

$$\varepsilon'' \cong A\omega\left[\ln\left(\frac{\Omega_{\mathrm{LO}}^2}{\omega^2 + 4\Gamma^2}\right) + \frac{\omega}{\Gamma}\tan^{-1}\left(\frac{\omega}{2\Gamma}\right)\right]; \quad A = \frac{\mu\Gamma\varepsilon^2}{12\Omega_{\mathrm{LO}}^2}\left(\frac{\varepsilon}{\varepsilon_\infty}\right)^{x-2}, \tag{3.28}$$

where $\Gamma$ is the damping of these phonons ($\Gamma \propto T$) and $\Omega_{\mathrm{LO}}$ is the typical frequency of longitudinal optical phonons; $\varepsilon_\infty$ is the optical permittivity; $\mu$ is the small anharmonicity parameter [6, 56], which can be evaluated as $k_B T/Mv^2$ ($M$ and $v$ are the average mass of the atoms and the average sound velocity in the crystal). The parameter $x$ has been calculated for two limiting cases: (i) for the case of a strong dispersion of the TO mode (the typical case for ferroelectric soft modes) and (ii) for the case of a negligible TO mode dispersion (the case close to the situation in non-ferroelectric crystals). In these cases, $x = 2.5$ and $x = 5$ have been found, predicting a fast growth of dielectric loss with increasing permittivity[5].

For high-$\varepsilon$ but non-ferroelectric materials, the theoretical result has been compared to the data of submillimeter frequency studies of dielectric loss in a system of complex perovskites $\mathrm{Ba(M^{(I)}_{1/2}\,M^{(II)}_{1/2})O_3}$ ($\mathrm{M^{(I)}}$ = Mg, In, Y, Gd, Nd and $\mathrm{M^{(II)}}$= W, Ta, Nb)

[5] Actually, the $\varepsilon$-dependence in Eq. (3.28) was derived only for the two aforementioned cases of strong and vanishing dispersion of the optical mode; for the case of an arbitrary dispersion, the dependence $A \propto \varepsilon^x$ with $2.5 < x < 5$ should be considered as an interpolation.

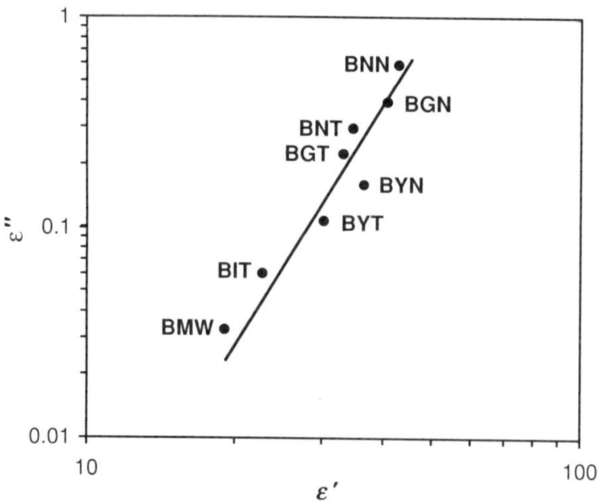


**Fig. 3.4.** Imaginary part of the dielectric permittivity plotted vs. its real part for $Ba(Mg_{1/2}W_{1/2})O_3$ (BMW), $Ba(In_{1/2}Ta_{1/2})O_3$(BIT), $Ba(Y_{1/2}Ta_{1/2})O_3$ (BYT), $Ba(Y_{1/2}Nb_{1/2})O_3$ (BYN), $Ba(Gd_{1/2}Ta_{1/2})O_3$ (BGT), $Ba(Nd_{1/2}Ta_{1/2})O_3$ (BNT), $Ba(Gd_{1/2}Nb_{1/2})O_3$ (BGN), and $Ba(Nd_{1/2}\ Nb_{1/2})O3$ (BNN) measured at $\omega = 300\ GHz$. The best fit $\varepsilon'' \propto \varepsilon^x$; $x = 4.1$ K [55,57].

[55, 57]. These are cubic non-ferroelectric (pseudo-cubic) centrosymmetric materials with the relative dielectric permittivity ranging from 20 to 40. In this frequency range the imaginary part of dielectric permittivity is expected to be controlled by the intrinsic contribution given by Eq. (3.28). A value of $x$ close to 5 is expected due to the non-ferroelectric nature of these materials. The comparison of the theory to the experiment is shown in Fig. 3.4. This figure shows that the experimental data on $\varepsilon''$ measured at 300 GHz do reveal a strong correlation between $\varepsilon''$ and $\varepsilon$, which can be approximated as a power law with the exponent close to 4. This value is close to the upper theoretical bound for this exponent. The difference between the upper bound for $x$ and its measured value is attributed to a small but appreciable dispersion of the optical modes in the studied materials.

For the case of a tunable material with the ferroelectric soft mode, the theory was tested [6, 54] by using the experimental data on microwave dielectric loss in $SrTiO_3$. In this case, Eq. (3.28) with $x = 2.5$ and $\omega \ll \Gamma$ leads to $\tan\delta \equiv \varepsilon''/\varepsilon \propto \omega T^2 \varepsilon^{1.5}$. As shown in Fig. 3.5, this expression is in good agreement with experimental data reported for crystalline $SrTiO_3$ by Buzin [58].

*Four-quantum mechanism*

The four-quantum mechanism corresponds to the field-quantum absorption processes involving three phonons [59–61]. The conservation (energy and quasi-momentum) laws do not impose strong restrictions on the type and energy of the phonons participating in these processes. For this reason, in contrast to the three-quantum processes, not only the degeneracy lines but also the whole thermally excited part of the $\vec{k}$-space nearly homogeneously contributes to the loss. Due to this fact, the explicit

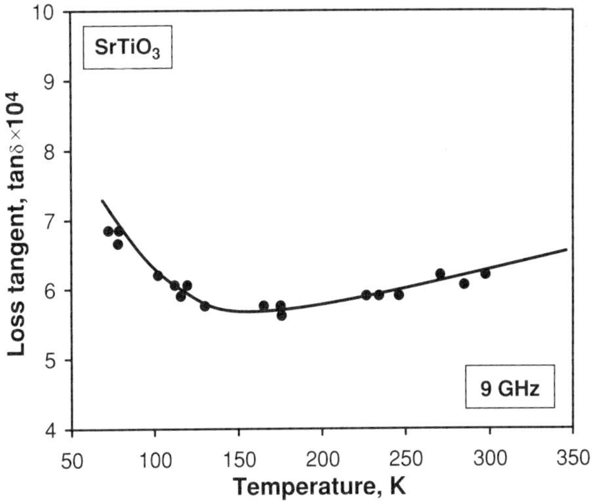


**Fig. 3.5.** Temperature dependence of the loss tangent in $SrTiO_3$ crystals at the ac electric field frequency 9 GHz: points—experimental data [58], curve—the prediction of the phonon theory [54] $\tan\delta \propto \omega T^2 \varepsilon^{1.5}$. The theory also correctly predicts the order of magnitude for the proportionality coefficient in this relation.

temperature and frequency dependence of the contribution of this mechanism to the imaginary part of the dielectric permittivity appears to be insensitive to the symmetry of the crystal. For $T \geq \hbar\Omega_{\mathrm{TO}}/k_{\mathrm{B}}$, this mechanism provides $\varepsilon'' \propto \omega T^2$; whereas at lower temperatures its contribution to the loss drastically decreases with decreasing temperature, e.g., for $T << \hbar\Omega_{\mathrm{TO}}/k_{\mathrm{B}}$, this mechanism provides [6] $\varepsilon'' \propto \omega T^9$.

This theory is quite applicable to the materials of interest for tunable applications. In this case, for the high temperature situation of practical importance where $T \geq \hbar\Omega_{\mathrm{TO}}/k_{\mathrm{B}}$, one finds [60, 61] $\tan\delta = \varepsilon''/\varepsilon \propto \omega T^2 \varepsilon^{1.5}$ for the four quantum contribution to the loss. This dependence is identical, to within a logarithmic factor, to result for the three-quantum contribution at $\omega \leq \Gamma$ (see Eq. (3.28) with $x = 2.5$). However, based on order-of-magnitude estimates [6, 54], one expects that, in these materials, the three-quantum contribution is still leading in the total balance of the intrinsic loss.

*Fundamental loss mechanism in centrosymmetric crystal*

The three- and four-quantum mechanisms are the only two mechanisms which control the intrinsic loss in centrosymmetric crystals [6]. In the case of practical interest of the low-microwave-loss materials, the theory of phonon transport, which has been used to obtain the results reviewed in the two previous sections, is applicable with quite good accuracy. This means that, possessing the complete information on the phonon spectrum and the constants of inter-phonon and ac field-phonon couplings, one can calculate the intrinsic dielectric loss with an accuracy equal to the relative damping of the typical phonons participating in the absorption of the ac field [6]. Thus, the result that the intrinsic loss in centrosymmetric crystals is given by the sum of the three- and four-quantum contribution can be considered as perfectly justified. However, in

**Table 3.2.** Functional dependence of the loss tangent $\tan\delta = \varepsilon''/\varepsilon$ of a cubic incipient ferroelectric according to different models used for the description of the intrinsic loss

| Frequency range | Phonon transport theory | Vendik model | Damped oscillator |
|---|---|---|---|
| $\omega \leq \Gamma$ | $\omega T^2 \varepsilon^{3/2}\,[1 + b \ln(\Omega_{TO}/\Gamma)]$ | $\omega T^2 \varepsilon^{3/2}$ | $\omega T \varepsilon$ |
| $\omega >> \Gamma$ | $\omega^2 T \varepsilon^{3/2}$ | – | $\omega T \varepsilon$ |

practice, when discussing the intrinsic microwave loss, two models differing from that discussed above are often used. These two models are the Vendik model [38, 61] and the so-called damped oscillator model [62]. Let us discuss the correspondence between the results of the phonon transport theory and of these models.

The Vendik model, as was mentioned in the original paper [61], is a version of the theory of the four-quantum mechanism. It gives correct dependences of the four-quantum-contribution to the loss tangent on frequency, temperature and dielectric permittivity for the temperature range of practical importance.

In the damped oscillator model, the loss is calculated by using the expression for the shape of the far-infrared absorption lines (associated with transverse optical branches) at the ac field frequency, which is much below the resonance frequencies. In this framework, the contribution to the complex dielectric permittivity from one branch (with frequency $\Omega_{TO}$) is presented in the form [62]:

$$\varepsilon^* \propto \frac{1}{\Omega_{TO}^2 - \omega^2 + 2i\omega\Gamma} \tag{3.29}$$

and used at $\omega << \Omega_{TO}$ for the description of the dielectric loss. The analysis in terms of the phonon transport theory shows that the use of this formula for calculating the dielectric loss at $\omega << \Omega_{TO}$ is not justified [6], except for crystals of symmetry $\frac{4}{m}$ and $\frac{6}{m}$. Thus, the damped oscillator model, in general, is not supported by the comprehensive theory. Table 3.2 compares the predictions of the two aforementioned models with these of the phonon transport theory. The comparison is performed for the case of a cubic incipient ferroelectric at $T \geq \hbar\Omega_0/k_B$ where $\Omega_0$ is the soft-mode frequency (like $SrTiO_3$ or $KTaO_3$ at $T \geq 20 - 30$ K). One concludes from this table that, for the frequency range $\omega \leq \Gamma$ (for $SrTiO_3$ or $KTaO_3$ that means $\omega \leq 100$ GHz), to within the logarithmic factors, the Vendik model gives correct functional dependences of the loss tangent on the frequency, temperature and the dielectric permittivity[6], whereas the prediction of damped oscillator model substantially differs from that of the comprehensive phonon theory.

*Quasi-Debye loss mechanism*

The origin of this mechanism is the relaxation of the phonon distribution function of the crystal [6, 51, 62, 63]. In non-centrosymmetric crystals, the phonon frequencies

[6] This does not contradict the expected leading role of the 3-quantum contribution in this frequency range. The point is that, in this range, the these functional dependences for the 3-quantum and 4-quantum contributions (including Vendik's model) are very close, though the 3-quantum one is expected to be numerically larger.

are linear functions of a small electric field applied to the crystal [51]. Thus, the oscillations of the ac field result in time modulation of the phonon frequencies; the latter in turn induces a deviation of the phonon distribution function from its equilibrium value. A relaxation of the phonon distribution functions gives rise to dielectric loss in a similar way as a relaxation of the distribution function of the dipoles gives rise to the loss in the Debye theory [64]. This analogy is expressed by the name "quasi-Debye". The frequency dependence of the quasi-Debye contribution to the loss factor is of the Debye type, the average relaxation time of the phonon distribution function playing the role of the Debye relaxation time. The contribution of the quasi-Debye mechanism to the tensor of the imaginary part of the relative dielectric permittivity of a crystal, $\varepsilon''_{\alpha\beta}$, can be expressed in terms of phonon dispersion curves, $\Omega_j(\vec{k})$ phonon damping, $\Gamma_j(\vec{k})$, and the so-called vector of electrophonon potential, $\vec{\Lambda}_j(\vec{k})$. Here the index $j$ designates the phonon branches and $\vec{k}$ is the phonon wave-vector. Calculated in the relaxation-time approximation $\varepsilon''_{\alpha\beta}$ can be presented in quadratures [6]:

$$\varepsilon''_{\alpha\beta} = \varepsilon_0^{-1} \sum_j \int \frac{d^3k}{(2\pi)^3} N_j \left(N_j + 1\right) \frac{\hbar^2 \Omega_j^2 \Lambda_j^{(\alpha)} \Lambda_j^{(\beta)}}{k_B T} \frac{2\Gamma_j \omega}{4\Gamma_j^2 + \omega^2} \tag{3.30}$$

where $\varepsilon_0$ and $N_j = [\exp(\hbar\Omega_j/k_B T) - 1]^{-1}$ are the permittivity of free space and the Planck distribution function; the integration volume is the first Brillouin zone. The electrophonon potential $\vec{\Lambda}_j(\vec{k})$ is defined as the relative change of the phonon frequencies per unit of the applied electric field [6]:

$$\Delta\Omega_j\left(\vec{k}\right) = \Omega_j\left(\vec{k}\right) \Lambda_j^{(\alpha)}\left(\vec{k}\right) E^{(\alpha)} \quad \text{or} \quad \Lambda_j^{(\alpha)}\left(\vec{k}\right) = \frac{1}{\Omega_j\left(\vec{k}\right)} \frac{\partial \Omega_j\left(\vec{k}\right)}{\partial E_\alpha}. \tag{3.31}$$

One can notice an analogy between Eq. (3.30) and the expression for the Debye loss contribution [64]:

$$\varepsilon'' \propto \frac{p^2}{k_B T} \frac{\tau\omega}{1 + (\tau\omega)^2} \tag{3.32}$$

($p$ is the dipole moment of the Debye dipoles) as well as between Eq. (3.31) and the expression for the change of energy of a dipole in an electrical field:

$$\Delta U = -p^{(\alpha)} E^{(\alpha)}, \tag{3.33}$$

with a correspondence $p \Leftrightarrow -\hbar\Lambda\Omega$ and $\tau \Leftrightarrow 1/2\Gamma$. Apparently, in non-centrosymmetric crystal, the phonons behave similarly to the Debye dipoles, as far as the dielectric loss is concerned [51].

An important feature of the quasi-Debye mechanism is that, once allowed by symmetry, it can be easily a few orders of magnitude greater than that of three- and four-quantum mechanisms. To give an idea about the strength of this mechanism, rough estimates for the three-, four-quantum and quasi-Debye contributions to the loss tangent $\tan\delta = \varepsilon''/\varepsilon'$ are presented in Fig. 3.6. Specifically, this figure schematically shows the contributions arising from the interaction of the ac field with the soft-mode phonons of a perovskite-type ferroelectric in the ferroelectric phase. It is seen that in this material

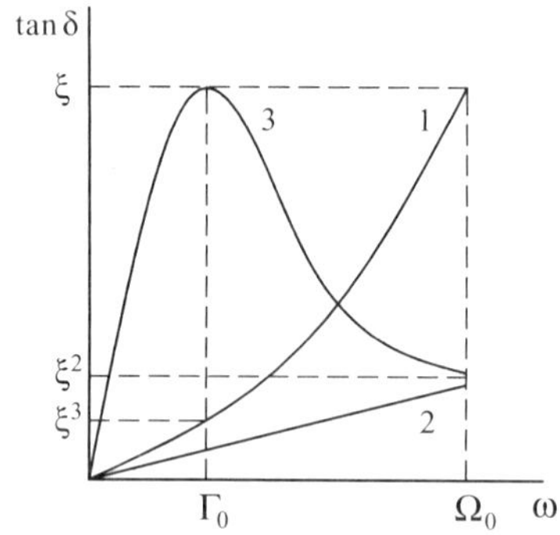


**Fig. 3.6.** Results of phonon transport theory. Schematic off-scale plot of the frequency dependence of the three-quantum (1), four-quantum (2), and quasi-Debye (3) contributions to the loss tangent due to interaction of the ac electric field with the phonons of the soft mode of a displacive type ferroelectric. $\Omega_0$ and $\Gamma_0$ are the soft-mode frequency and damping, respectively. $\xi = \Gamma_0/\Omega_0$. The range of applicability of the theory is $\Gamma_0 \ll \Omega_0$ and $\omega \ll \Omega_0$ (at $\omega$ tending to $\Omega_0$, the curves are only shown to guide the eye).

the quasi-Debye contribution dominates the intrinsic loss in the frequency range $\omega \leq \Gamma$(for $SrTiO_3$ and $KTaO_3$ at $T \geq 20-30$ K that means $\omega \leq 100$ GHz).

In microwave materials for tunable application, which are typically centrosymmetric, the quasi-Debye mechanism does not contribute to the loss in the absence of the tuning bias. However, under a dc bias field, $\vec{E}_0$, it becomes active due to the breaking of the central symmetry so that one is dealing with the *dc-field-induced quasi-Debye mechanism*. The contribution of this mechanism has been modeled for $SrTiO_3$, (Ba,Sr)$TiO_3$ and $KTaO_3$ using the above formulae and the available information on the material parameters and parameters of the phonon spectrum of these materials [65–67]. It was shown that for not too high frequencies ($\omega \leq \Gamma$), the contribution in question can be presented as:

$$\tan\delta_{QD}(E_0) = A\,\omega \cdot I(E_0) \cdot n_r \tag{3.34}$$

where the function $I(E_0) \rightarrow 1$ in the limit of small relative tunability $n_r = \frac{\varepsilon(0)-\varepsilon(E_0)}{\varepsilon(0)} \ll 1$. The theoretical estimates for the parameter $A$ are given in Table 3.3 together with the experimental data available in the literature [2, 7] (see also Sec. 5 of the present paper). The agreement between the theory and experiment can be considered as reasonable despite significant spread of the theoretical estimates, which is related, first of all, to the lack of reliable information on the damping of the phonons participating in the microwave absorption.

For dc fields corresponding to small relative tunability $n_r$, Eq. (3.34) suggests a quadratic field dependence of the dc field induced contribution to the loss tangent, $\tan\delta_{QD}(E_0)$. For higher fields, the $\tan\delta_{QD}$ vs. $E_0$ dependence deviates from this simple law and appears to be different for different materials. The result of simulations of this dependence for $SrTiO_3$ and $Ba_{0.6}Sr_{0.4}TiO_3$ crystals is shown in Fig. 3.7. In this simulation, only $\tan\delta_{QD}(E_0)$ was modeled whereas $\tan\delta(0)$ was taken from experimental

**Table 3.3.** Experimental and simulated data for $A$ in Eq. (3.34) for $KTaO_3$, $SrTiO_3$ and $Ba_{0.6}Sr_{0.4}TiO_3$

| | $A \times 10^{-3}$, GHz$^{-1}$ | | |
|---|---|---|---|
| | $KTaO_3$ ($T = 50$ K) | $SrTiO_3$ ($T = 80$ K) | $Ba_{0.6}Sr_{0.4}TiO_3$ ($T = 300$ K) |
| Experiment [2, 7] | 23 | 17 | — |
| Calculation | 8–27 | 5–19 | 0.1 |

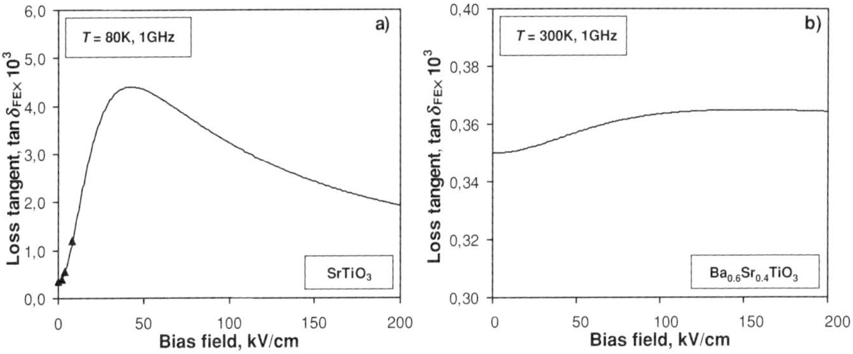


**Fig. 3.7.** Simulated field dependence of the loss tangent (solid lines) plotted together with experimental data [2] (black squares) for $SrTiO_3$ (a) and $Ba_{0.6}Sr_{0.4}TiO_3$ (b) crystals; $A_{SrTiO_3} = 17 \times 10^{-3}$GHz$^{-1}$, $A_{Ba_{0.6}Sr_{0.4}TiO_3} = 0.1 \times 10^{-3}$GHz$^{-1}$. The values of $\tan\delta(0)$ are taken from Refs [2] and [68], respectively.

data [2, 68]. For the case of $SrTiO_3$, the value of the $A$ parameter was set to fit the experimental $\tan\delta(E_0)$ dependence [2] at small $E_0$.

The striking difference between the field dependences of the loss in $SrTiO_3$ and $Ba_{0.6}Sr_{0.4}TiO_3$ shown in Fig. 3.7 is attributed to the difference in the strength of the flexoelectric coupling between the soft mode and acoustic branches in these materials. The non-monotonic field dependence of the loss tangent in $SrTiO_3$ is related to the contribution of the acoustic phonon to the quasi-Debye loss, which is activated by this flexoelectric coupling [65, 66]; in $Ba_{0.6}Sr_{0.4}TiO_3$, where the flexoelectric coupling is much smaller, this contribution is small and cannot be seen.

The above analysis implies that the contribution of the field-induced quasi-Debye loss can be a major contribution to the total loss in a material, especially in the case of thin films where high dc fields are readily achievable. For instance, for $SrTiO_3$ and $Ba_{0.6}Sr_{0.4}TiO_3$ @ $E_{dc} = 100$ kV/cm & $\omega = 25$ GHz, according to Fig. 3.7 and Eq. (3.34) one expects $\tan\delta_{QD}$ about 0.1 and 0.02 at $T = 80$ K and $T = 300$ K, respectively.

An important feature of the quasi-Debye loss is that as a function of frequency, similar to the Debye loss, its contribution to the loss tangent passes through a maximum at $\omega \cong \Gamma$ (see Fig. 3.5). For the aforementioned materials, where the phonon damping $\Gamma$ is of the order of $10^2$ GHz, this means a certain slowing down of the linear frequency dependence of $\tan\delta_{QD}$, Eq. (3.34), in the higher part of the microwave-frequency range.

### 3.2.2. Extrinsic loss

The role of the intrinsic mechanisms in the total balance of the dielectric loss of a material is strongly dependent on the dielectric permittivity of the material and the measuring frequency: typically, the higher the frequency and permittivity, the more important the intrinsic loss. In the case of tunable ferroelectric materials at microwave frequencies, the intrinsic and extrinsic contributions are comparable so that the dominating contribution to the loss may be extrinsic or intrinsic depending on the quality of the material. A kind of extrinsic/intrinsic crossover in loss may also take place under the action of a dc bias field, i.e. without the field, the extrinsic contribution dominates the loss, whereas under the field the intrinsic one does. Among the known extrinsic loss mechanisms those listed below are considered as significantly contributing to the loss in tunable microwave materials: (i) loss owing to charged defects, (ii) universal relaxation law mechanism, (iii) quasi-Debye contribution induced by random-field defects.

*Loss owing to charged defects*

Motion of charged defects caused by an ac electric field results in a generation of acoustic waves at the frequency of the applied field. This brings about an additional loss mechanism that was proposed by Schlöman [69], formulated for high-dielectric-constant materials by Vendik and Platonova [70], and developed by Garin [71]. The contribution of this mechanism to the loss tangent can be approximated as follows:

$$\tan \delta_{ch} = F\varepsilon \frac{n_d}{\rho v_t^3} \frac{Z^2}{4\pi} \omega \cdot \left[ 1 - \frac{1}{(1 + \omega^2/\omega_c^2)^2} \right] \tag{3.35}$$

where $Z$ and $n_d$ are the effective charge of the defects and their atomic concentration; $\rho$ and $v_t$ are the density and average transversal sound velocity of the material; $F$ is a material-dependent numerical constant of the order of unity; $\omega_c = v_t/r_c$ where $r_c$ is the correlation length of the charge distribution in the material. The physical meaning of $r_c$ is the minimal distance at which the elecroneutrality is maintained. For Schottky defects $r_c$ is of the order of the typical distance between the positively and negatively charged defects. This mechanism may play an essential role in thin film based tunable capacitors, where an elevated defect concentration compared to the bulk material is expected. In this case, the effect of semiconductor depletion of the carriers from the deep traps (caused by ferroelectric/electrode contact) [72, 73] will further increase the contribution of this mechanism via a strong reduction of $\omega_c = v_t/r_c$. This occurs in the depleted areas due to a strong increase of $r_C$ up to the depletion length. An essential feature of this mechanism is that its contribution to the loss tangent is proportional to the permittivity of the material. This implies that this contribution is inversely dependent on the applied dc field.

*Universal-relaxation-law mechanism*

For all of the loss mechanisms discussed above, a linear frequency dependence of the loss tangent is typical at least for microwave frequencies and below. In reality, this dependence is usually observed at the microwave and higher frequencies. For lower frequencies, a much weaker frequency dependence is usually observed, which is consistent with the so-called universal-relaxation law which in turn corresponds to the following expression for the complex dielectric permittivity $\varepsilon^*$ [74]:

$$\varepsilon^* = G(i\,\omega)^{n-1} = G(\cos(n\pi/2) - i\,\sin(n\pi/2))(\omega)^{n-1} \tag{3.36}$$

where $G$ is a frequency-independent constant and $0 < n < 1$. In perovskite thin films in both frequency and time domains, dielectric relaxation corresponding to this equation (for $n$ close to but smaller than unity) has been reported up to the microwave frequency range, e.g. in (Ba,Sr)$TiO_3$ [75]. The physical origin of this behavior is attributed to a variation in charge transport barriers, e.g. at the grain boundaries [76], or to creep of the boundary of the near-by-electrode depletion layer [77]. No information is available on the dependence of $G$ on the dielectric constant of the material.

*Impact of local polar regions*

The typically centrosymmetric tunable materials may have local polar regions induced by various defects and structural imperfections. For example, recent infrared reflectivity investigations of $SrTiO_3$ ceramics revealed a clear presence of grain-boundary-induced polar phase inclusions [78]. Random field defects can also be responsible for the appearance of local polar regions (see e.g. [4]). In all these polar inclusions, the quasi-Debye mechanisms is expected to be active. Though the volume fraction of the polar phase is typically small, this "defects-induced" quasi-Debye mechanism may be important due to its large (compared to other intrinsic loss mechanisms) contribution per unit of volume. Though the theory of the "defect-induced" quasi-Debye mechanism is not developed, using the results on the quasi-Debye loss in the ferroelectric phase [54] one can conclude that the contribution of the "defect-induced" quasi-Debye mechanism should be strongly $\varepsilon$-dependent. According to Ref. [54] in the ferroelectric phase of a displacive ferroelectric $\tan\delta_{QD} \propto \varepsilon^{3/2} P_S^2$ where $P_S$ is the spontaneous polarization. In the case of non-polar tunable material with polar inclusions one can substitute the volume average of the local polarization squared, $< P_{loc}^2 >$, for $P_S^2$ to evaluate the contribution of the "defect-induced" quasi-Debye mechanism as

$$\tan\,\delta_{dQD} \propto \varepsilon^{3/2} \cdot < P_{loc}^2 > . \tag{3.37}$$

We can also use the relation between the ferroelectric correlation length $\xi$ and $< P_{loc}^2 >$[78]: $< P_{loc}^2 > \propto \xi^{3-d}$ where $d$ is the dimension of the defect, i.e. $d = 2$ for a planar defects like grain boundaries, $d = 1$ for linear defects, and $d = 0$ for point defects. Finally, making allowance that $\xi \propto \sqrt{\varepsilon}$ we arrive at a rather strong $\varepsilon$-dependence of the contribution of this mechanism, namely:

$$\tan\,\delta_{dQD} \propto \varepsilon^{4.5-d}. \tag{3.38}$$

### 3.2.3. Loss in tunable composites

As we have mentioned above there are some situations where the tunable material can be treated as a ferroelectric/dielectric composite (see Sec. 3.2.2). Clearly, not only the dielectric permittivity and tunability of the materials are strongly affected by the presence of the "passive" component but the dielectric loss is affected as well. Since "normal" low-$\varepsilon$ dielectrics typically exhibit much lower loss tangent than ferroelectrics, a reduction of the loss in composites may be expected with increasing volume fraction of the dielectric. Recently performed simulation of the problem for a number of simple models of the composite showed that this trend does not necessarily take place [45]. In what follows we will review the results of these simulations which were performed for the layered, columnar and spherical inclusion models already discussed in Sec. 3.1.3 and illustrated in Fig. 3.2.

First of all an important remark should be made. Mixing a tunable ferroelectric with a linear dielectric modifies the electrical properties of the material due to following effects: (i) the effect of mutual doping of the composite components (inter-diffusion), (ii) effects of inter-component bondaries, (iii) effects of mixing-induced modification of microstructure of the ferroelectric component, and (iv) the effect of the electric field redistribution between the different components of the mixture. In the following we will address the last but not least effect, which can be treated in terms of electrostatic considerations. Thus, when discussing the composite effects on the electric parameters of a real material, the result presented below should be applied together with arguments related to the other aforementioned effects.

The calculation of the imaginary part of the permittivity of the composite, $\varepsilon''_{\text{mix}}$, can be performed based on the equation of the balance of the dissipated energy in the material[7]

$$\frac{\omega}{2}\varepsilon''_{\text{f}} < \tilde{E}_{\text{f}}^2 > (1-q) + \frac{\omega}{2}\varepsilon''_{\text{d}} < \tilde{E}_{\text{d}}^2 > q = \frac{\omega}{2}\varepsilon''_{\text{mix}}\tilde{E}^2 \tag{3.39}$$

where $\tilde{E}$ is the applied ac field amplitude whereas $< \tilde{E}_{\text{f}}^2 >$ and $< \tilde{E}_{\text{d}}^2 >$ stand for the averaged squared amplitudes of the ac field in the ferroelectric and dielectric, respectively; $q$ is the volume fraction of the dielectric.

In the case of columnar and layered composites, where the ac fields in the ferroelectric and dielectric remain homogeneous, the loss tangent of the mixture can be readily expressed (using Eqs. (3.39), (3.15), and (3.23)) in term of the volume fraction and permittivity of the components, namely:

$$\tan\delta_{\text{mix}}^{\text{col}} = \frac{\tan\delta_{\text{f}}\varepsilon_{\text{f}}(1-q) + \tan\delta_{\text{d}}\,\varepsilon_{\text{d}}\,q}{(1-q)\,\varepsilon_{\text{f}} + \varepsilon_{\text{d}}\,q} \tag{3.40}$$

for the columnar composite and

$$\tan\delta_{\text{mix}}^{\text{lay}} = \frac{\tan\delta_{\text{f}}\,\varepsilon_{\text{d}}(1-q) + \tan\delta_{\text{d}}\,\varepsilon_{\text{f}}\,q}{(1-q)\,\varepsilon_{\text{d}} + \varepsilon_{\text{f}}\,q} \tag{3.40a}$$

for the layered composite. Here $\tan\delta_{\text{f}}$ and $\tan\delta_{\text{d}}$ are the loss tangent of the ferroelectric and dielectric. These expressions can also be used for calculation of the loss tangent of the composites in the presence of a dc bias field $E_0$. In the case of the columnar composite under dc bias, the permittivity and loss tangent of the ferroelectric in Eq. (3.40) should be set equal to their values corresponding to dc field equal $E_0$, i.e. $\varepsilon_{\text{f}}(E_0)$ and $\tan\delta_{\text{f}}(E_0)$. In the case of the layered composite, these parameters should be set equal to $\varepsilon_{\text{f}}(E_{\text{f}})$ and $\tan\delta_{\text{f}}(E_{\text{f}})$ where the field seen by the ferroelectric $E_{\text{f}}$ is to be calculated by taking into account the field redistribution between the components

[7] This is not the only way to address this problem, for an alternative description see Ref. [42].

(see Sec. 3.1.3). Due to this, in the latter case, the fraction $q$ dependence of the loss tangent is more complicated than the explicit $q$-dependence of Eq. (3.40a).

In the case of the spherical inclusion model, the field in the ferroelectric matrix is not homogeneous (see Eqs. (3.25) and (3.25a)) so that the extra contribution to $< \tilde{E}_f^2 >$ arises due to the inhomogeneous component of the field. This essentially complicates the problem. Below we will present some results of calculations performed for the case of small volume concentration of the dielectric spherical inclusions (i.e. $q \to 0$) and small relative tunability of the composite. In these calculations, it was also assumed that the real and imaginary parts of permittivity are much greater in the ferroelectric matrix than in the dielectric inclusions (i.e. $\varepsilon_d << \varepsilon_f, \varepsilon''_d << \varepsilon''_f$), which is true to the real experimental situation. In this case, using Eqs. (3.39) and (3.25a) one finds for the imaginary part of the composite permittivity [45]:

$$\varepsilon''_{\mathrm{mix,sph}} \approx \varepsilon''_f \left(1 - \frac{3}{2}q\right), \tag{3.41}$$

This result essentially differ from those for columnar and layered models which give in the same limiting case ($\varepsilon_d << \varepsilon_f, \varepsilon''_d << \varepsilon''_f$, and $q \to 0$):

$$\varepsilon''_{\mathrm{mix,col}} \approx \varepsilon''_f (1 - q), \tag{3.42}$$

$$\varepsilon''_{\mathrm{mix,lay}} \approx \varepsilon''_f \left(1 - 2\frac{\varepsilon_f}{\varepsilon_d}q\right), \tag{3.43}$$

respectively. It is instructive to compare the above results for the imaginary part of permittivity with the predicted concentration dependences of the real part of permittivity

$$\varepsilon_{\mathrm{mix,col}} \approx \varepsilon_f (1 - q), \tag{3.44}$$

$$\varepsilon_{\mathrm{mix,lay}} \approx \varepsilon_f \left(1 - \frac{\varepsilon_f}{\varepsilon_d}q\right), \tag{3.45}$$

$$\varepsilon_{\mathrm{mix,sph}} \approx \varepsilon_f \left(1 - \frac{3}{2}q\right). \tag{3.46}$$

and loss tangent

$$\tan \delta_{\mathrm{mix}}^{\mathrm{col}} \approx \tan \delta_f, \tag{3.47}$$

$$\tan \delta_{\mathrm{mix}}^{\mathrm{lay}} \approx \tan \delta_f \left(1 - \frac{\varepsilon_f}{\varepsilon_d}q\right), \tag{3.48}$$

$$\tan \delta_{\mathrm{mix}}^{\mathrm{sph}} \approx \tan \delta_f. \tag{3.49}$$

which are evaluated in the same limiting case. A remarkable feature of these results is that the dilution effect on the loss tangent can be much less pronounced than for the real part of the permittivity as it is seen in the columnar and spherical inclusion models. That actually means a rather weak $\tan \delta$ vs. $\varepsilon$ dependence which is absolutely untypical for the loss mechanisms in "non-composite" materials discussed in the preceding sections. This result holds for the case of higher concentrations of the dielectric as it is seen from the modeling performed in a wide range of the permittivity variation. The results of this modeling are presented in Fig. 3.8 where the loss tangent of the composites is plotted

**Table 3.4.** Characteristics of the composite components

| | Dielectric constant $\varepsilon$: | | |
|---|---|---|---|
| Material: | $E=0$ | $E=7.3$ kV/cm | Loss tangent ($\tan\delta$) |
| Ferroelectric ($\beta = 7.37 \times 10^9 JC^{-4}m^5$) | 3300 | 2640 | $10^{-2}\frac{\varepsilon(E)}{\varepsilon(0)}$ |
| Linear dielectric | 8.4 | 8.4 | $10^{-4}$ |

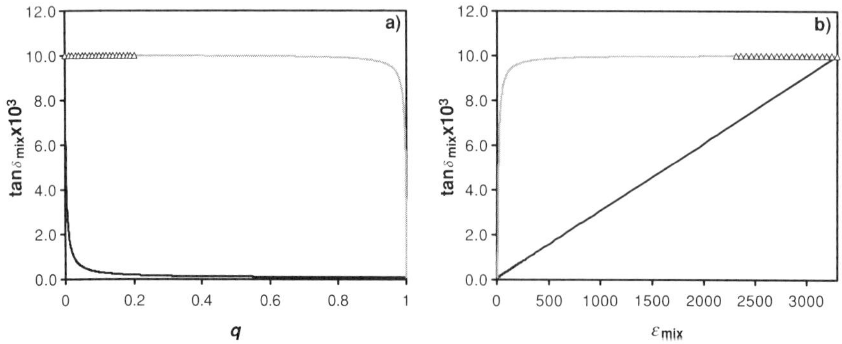


**Fig. 3.8.** Calculated loss tangent of ferroelectric/dielectric composites, $\tan\delta_{mix}$, as a function of volume concentration of dielectric $q$ (a) and dielectric constant of the composite material $\varepsilon_{mix}$ (b). (Black lines—layered model; grey lines—columnar model; triangles—spherical inclusion model) [45].

as a function of their permittivity and the concentration of the dielectric. The material parameters used in the modeling are listed in Table 3.4.

## 3.3. Commutation quality factor of tunable materials

As was discussed in Sec. 2 the Vendik's Commutation Quality Factor of the ferroelectric $K$ (introduced by Eq. (2.3) is a most suitable parameter for characterization of performance of tunable ferroelectric materials. It is instructive to consider the implication of the theoretical results reviewed above on the $K$-factor of tunable materials. Two issues will be addressed below: the limitation on the $K$-factor imposed by the dc field induced quasi-Debye mechanism and the expected impact on the $K$-factor of the dilution of ferroelectrics with a low-permittivity-low-loss dielectric.

The dc-field-induced increase of the loss tangent of a tunable material imposes intrinsic limitations on the $K$ factor. Since the field induced contribution to the loss tangent may be comparable to its typical values, this contribution ought to be taken into account when evaluating the intrinsic limitations on the $K$-factor of various materials. The intrinsic limit for the $K$-factor can be evaluated by using the result on tunability from Sec. 3.1.1, equation (3.34) for the contribution of the field induced quasi-Debye mechanism to the loss, and the data on the intrinsic component of loss tangent. The latter can be presented as $\tan\delta(0) = A_0\omega$ where $A_0$ is a frequency independent constant.

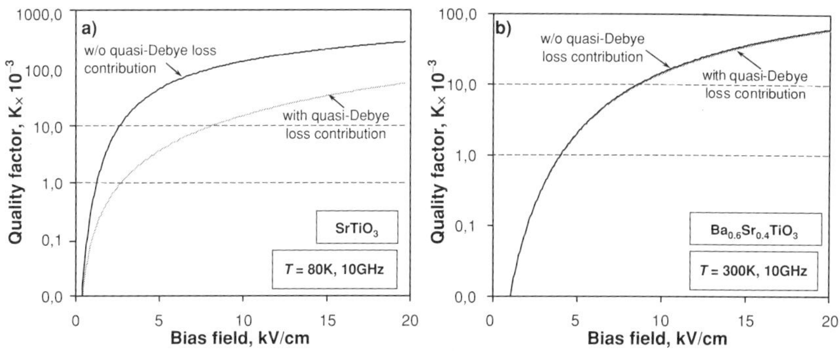


**Fig. 3.9.** Calculated field dependences of the quality factor $K$ with (grey line) and without (black line) taking into account the field induced quasi-Debye loss contribution for $SrTiO_3$ (a) and $Ba_{0.6}Sr_{0.4}TiO_3$ (b) ferroelectrics. The calculations were performed for the same material parameters as those used for plotting the graphs in Fig. 3.7 [67].

This gives

$$K = \frac{n-1}{A_0 A \omega^2 I(E_0)}. \tag{3.50}$$

The $K$-factor calculated this way for $SrTiO_3$ (a) and $Ba_{0.6}Sr_{0.4}TiO_3$ (b) [67] is presented in Fig. 3.9 as a function of the tuning bias dc field $E_0$.

The modeling of impact of the dilution of ferroelectrics with a low-permittivity-low-loss dielectric on the $K$-factor is motivated by the current activity on the tunable performance of ferroelectric/dielectric composites [68, 79, 80]. The calculations performed in Sections 3.1.3 and 3.2.3 enable such modeling. Figure 3.10 shows the results of

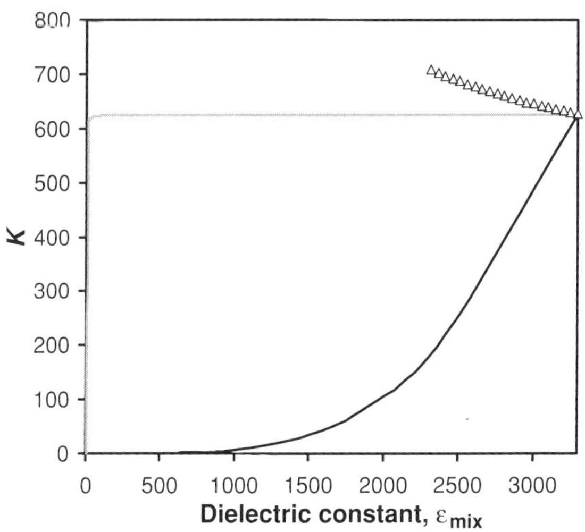


**Fig. 3.10.** Calculated quality factor of composite, $K$, as a function of its dielectric constant $\varepsilon_{\text{mix}}$. (Black lines – layered model; grey lines – columnar model; triangles – spherical inclusion model).

this modeling performed for a ferroelectric/dielectric composite whose parameters are listed in Table 3.4. The three types of the composite illustrated in Fig. 3.2 are addressed. The field dependence of the loss in the ferroelectric is set as $\tan \delta_f (E) = \tan \delta_f (0) \frac{\varepsilon_f(E)}{\varepsilon_f(0)}$, which is consistent with the trend known for the extrinsic contributions to the loss in real materials as will be discussed in Section 5.2.

Several features of the results shown in Fig. 3.10 are to be commented upon.

First, in contrast to the case of linear dielectric response, the result for the spherical inclusion model does not lay between those for the layered and columnar models. Second, the results for the spherical inclusion model shows that the composite effect may result in an increase (though slight) of the $K$-factor owing to dilution of a ferroelectric with non-tunable dielectric. Of interest is that this effect takes pace due to an increase of tunability rather than decrease of the loss tangent. Concerning these results one should also mentioned that these were obtained only for a very limited case of small relative tunability and small concentration of the dielectric. Thus, the problem of tunability of a ferroelectric composite remains open for theoretical studies. Third, considering the above results of modeling from the practical point of view, one can state that, in ferroelectric/dielectric composites, the dilution of the ferroelectric may lead to a substantial reduction of the composite permittivity having, at the same time, a minute effect on the $K$-factor.

The above analysis enables us to conclude that, at least due to the considered effect of the field redistribution, it is difficult to expect a substantial improvement of the tunable performance of a ferroelectric material by using a composite. At the same time, one cannot exclude an appreciable increase of the $K$–factor caused by the addition of a dielectric in the ferroelectric, which is due to a kind of doping effect, e.g. due to a doping induced reduction of the loss in the ferroelectric matrix. In addition, the composite effect may be advantageous for the tunable applications where a lower dielectric permittivity of the active material is desirable. This advantage is coming from the aforementioned possibility to have in composites a lower permittivity for the same value of the $K$-factor.

## 3.4. Dielectric response of tunable thin films

The theory reviewed in the preceding section is equally applicable to bulk and film forms of tunable materials. However, in the case of films, there exist additional effects that may substantially influence the tunable performance of the material. These effects are responsible for the difference between the tunable performance of the films and bulk samples having nominally the same chemical composition. Specifically, the films typically exhibit lower dielectric permittivity and higher loss than the bulk materials of similar composition (see Sec. 5 for a discussion of experimental results). In the case of films thinner than a few micrometers, which are commonly called *thin films* and which we are going to approach below, this difference is always significant, at least when the permittivity of the material exceeds 1000. The origin of this difference relates to two groups of factors. First, the processing conditions for bulk materials and thin films are different (e.g., a difference in the processing temperature, a possibility for a substantial loss of a volatile component, etc). Second, even if the bulk and film counterparts are chemically identical (the same compositions and defect populations), one can expect a difference in their properties due to several physical size effects. In the present section we give an overview the physics behind these effects.

### 3.4.1. Passive layer effects

Several reasons may cause a ferroelectric thin film to exhibit identical or similar dielectric response to that of a perfect ferroelectric film with a thin dielectric layer at its surfaces. The simplest possibility is the presence of a secondary phase at the surface of the films. This situation can be excluded by proper processing control. In addition, there exist two intrinsic reasons for this kind of dielectric behavior of the films. The first one is a near-by-surface variation of the polarization (field induced or spontaneous). This situation was theoretically addressed by Kretschmer and Binder [81] in a general manner and further analyzed by Vendik and Zubko [40, 82]. Second, if the dielectric response of a film is monitored by using a parallel plate capacitor, this response is affected by the fact that the free charges in the electrode form a finite thickness layer that behaves as a capacitor connected in-series with the material of the films [83, 84]. Treatment of the electron gas in the electrode in the Thomas-Fermi approximation [85] shows that this capacitor has the capacitance per unit area of $\varepsilon_0/l_s$ where $l_s$ is the Thomas-Fermi screening length. This evidentially leads to the following expression for the apparent out-of-plane dielectric permittivity in the film:

$$\varepsilon_{\text{eff}}^{-1} = \varepsilon^{-1} + \frac{2l_s}{h}. \tag{3.51}$$

The two models mentioned above have an essential common feature, namely, the in-plane and out-of-plane components of the dielectric permittivity are differently affected by the considered effects. In the case of the "electrode effect", only the out-of-plane component is affected since it is this component that is measured using the plate-capacitor set up. For the "near-by-surface variation effect", the out-of-plane component is more strongly affected than the in-plane one, however this is less obvious than in the case of the "electrode effect". This issue will be addressed below following the results from Refs [40, 81, 82]. However, these results will be corrected by the impact of the background dielectric permittivity of the ferroelectric, which was ignored in these papers.

The starting point of the model is the assumption that the surface value of the polarization in a ferroelectric is not affected by the electric field as strongly as that in the bulk. This is consistent with the microscopical arguments that the ferroelectric softness of the lattice is somehow suppressed near to the surface. The simplest modeling of this point is performed in the framework of the continuous Landau theory for a situation where the polarization at the two surfaces of the film is completely blocked. This corresponds to the equation for the polarization:

$$E = \alpha P - \kappa \frac{\partial^2 P}{\partial x^2} \tag{3.52}$$

with the boundary conditions

$$P(0) = 0 \text{ and } P(h) = 0, \tag{3.53}$$

the surfaces of the film being at $x = 0$ and $x = h$. In the following treatment we will also take into account that the relative dielectric permittivity of the material is always large, i.e. $\varepsilon = (\alpha\varepsilon_0)^{-1} \gg 1$.

In the case of the in-plane component of the effective permittivity of the system, which is actually monitored in the planar capacitor setup (shown in Fig. 2.3) having the gap $g$ much greater than the film thickness $h$, the expected inhomogeneity of polarization across the film does not create depolarizing field. Thus, the field $E$ entering Eq. (3.52) is equal[8] to the "applied field" $E_{\text{ext}} = U/g$, where $U$ is the voltage applied to the gap. In this case, the distribution of the polarization across the film (i.e. in the direction of the film normal) can be readily found in the form

$$P(x) = \frac{E_{ext}}{\alpha}\left(1 - \frac{\cosh\dfrac{x - h/2}{\xi}}{\cosh\dfrac{h}{2\xi}}\right) \tag{3.54}$$

where

$$\xi = \sqrt{\frac{\kappa}{\alpha}} \tag{3.55}$$

is the so-called correlation length having the meaning of the scale on which the polarization changes near the surfaces. Typically, $\xi$ rarely exceeds a few nm. For the case of practical interest where $h \gg \xi$, Eq. (3.54) leads to the following value of the average polarization in the films:

$$\bar{P} = \frac{1}{h}\int_0^h P(x)dx = \frac{E_{ext}}{\alpha}\left(1 - \frac{2\xi}{h}\right). \tag{3.56}$$

This corresponds to effective dielectric permittivity of the system:

$$\varepsilon_{eff} = \varepsilon\left(\frac{h - 2\xi}{h}\right). \tag{3.57}$$

This equation means that, effectively, two layers of thickness $\xi$ have the dielectric constant much smaller than $\varepsilon$ so that they, actually, do not contribute to the polarization response. Alternatively, these layers can be considered as connected in parallel with ferroelectric films of thickness $h - 2\xi$.

In the case of the out-of-plane component of the effective permittivity of the system, which corresponds to the situation in the parallel plate ("through thickness") capacitor, the polarization is normal to the plane of the films so that its variation *does create a depolarizing field.* This is a crucial difference compared to the preceding case. The relation between the "applied field" $E_{\text{ext}} = U/h$, where $U$ is the voltage applied across the films, and the field seen by the ferroelectric can be found from the Poisson equation. Taking into account the background contribution to the displacement $\varepsilon_0\varepsilon_b E$, where $\varepsilon_b \ll \varepsilon$ is the contribution to the electric permittivity of the non-ferroelectric lattice

[8] To within $h/g << 1$.

modes of the crystal, the Poisson equation reads in our case as $d(\varepsilon_0\varepsilon_b E + P)/dx = 0$ leading to the sought relation:

$$E = E_{ext} - \frac{1}{\varepsilon_0\varepsilon_b}(P - \bar{P}). \tag{3.58}$$

Using this relation, Eq. (3.52) can be rewritten as

$$E_{ext} = \alpha P - \kappa\frac{\partial^2 P}{\partial x^2} + \frac{1}{\varepsilon_0\varepsilon_b}(P - \bar{P}). \tag{3.59}$$

The solution of this equation that satisfies the boundary conditions Eq. (3.53) reads

$$P(x) = \frac{E_{ext}}{\alpha}\left(1 - \frac{\cosh\dfrac{x - h/2}{\xi_1}}{\cosh\dfrac{h}{2\xi_1}}\right)\frac{1}{1 + 2\dfrac{\varepsilon}{\varepsilon_b}\dfrac{\xi_1}{h}\tanh\dfrac{h}{2\xi_1}} \tag{3.60}$$

where $\xi_1 = \xi/\sqrt{1 + \varepsilon/\varepsilon_b} \approx \xi/\sqrt{\varepsilon/\varepsilon_b} = \sqrt{\varepsilon_b}\sqrt{\kappa\varepsilon_0}$ has the meaning of the scales on which, in this geometry, the polarization changes near the surfaces. Note that $\xi_1$ is yet smaller than $\xi$ so that, for any case of practical interest, $\xi_1/h << 1$. Under this conditions, Eq. (3.60) leads to the expression for the effective dielectric permittivity of the film:

$$\varepsilon_{\text{eff}}^{-1} = \varepsilon^{-1} + \frac{2\xi_1}{h}\varepsilon_b^{-1}. \tag{3.61}$$

This relation corresponds to the in-series connection of the ferroelectric film with two dielectric layers of thickness $\xi_1$ and dielectric permittivity$\varepsilon_d = 1$.

Thus we see that the surface region with partially suppressed dielectric response behaves as a passive layer, however in contrast to a real dielectric layer its thickness is essentially different for in-plane and out-of-plane geometries.

The above theoretical treatment of the problem has been presented for the situation where the polarization at the surfaces of the films is completely blocked. The more general situation where the blocking is not complete can be simulated by using the mixed boundary conditions

$$P_{(0)} - \lambda\left.\frac{\partial P}{\partial x}\right|_{x=0} = 0 \quad \text{and} \quad P(h) + \lambda\left.\frac{\partial P}{\partial x}\right|_{x=h} = 0 \tag{3.62}$$

instead of (3.53). These conditions interpolate the situation between the blocked (at $\lambda = 0$) and free (at $\lambda \to \infty$) polarization at the surfaces of the films. Under this modification the above results, Eqs. (3.57) and (3.61) hold however with the substitutions

$$\xi \Rightarrow \frac{\xi}{1 + \lambda/\xi} \quad \text{and} \quad \xi_1 \Rightarrow \frac{\xi_1}{1 + \lambda/\xi_1}. \tag{3.63}$$

Thus, as one can expect, a weakening of the surface blocking leads to a reduction of the effective passive layer thickness, the effect vanishing in the limit of free polarization at the film surface ($\lambda \to \infty$).

It is instructive to illustrate schematically the difference between the spatial distribution of the polarization for the "in-plane" and "out-of-plane" situations (Fig. 3.11). In the "in-plane" situation where there is no impact of the depolarizing effects, the

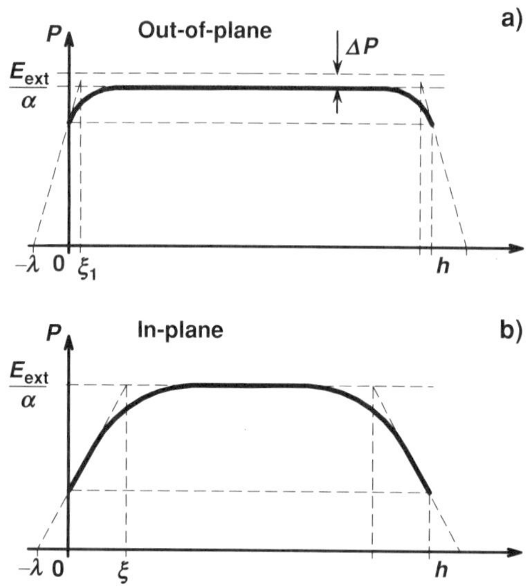


**Fig. 3.11.** Schematic distribution of polarization across a film of thickness $h$ for out-of-plane (a) and in-plane (b) cases. $\Delta P\alpha/E_{ext} = 2\varepsilon\xi_1/h$ where $\xi_1 = \xi\sqrt{\varepsilon_b/\varepsilon}$ (see Eq. (3.60)). Here, $\varepsilon$ is the dielectric constant of the material, $\varepsilon_b$ its background dielectric constant, $h$ is the thickness of the film, $\xi$ is the correlation length, and $E_{ext}/\alpha$ is the "bulk value" of the polarization.

polarization reaches exponentially fast the value that is expected in the bulk sample for the same value of the applied filed. On the other hand, in the "out-of-plane" situation, due to the depolarizing effects the polarization changes yet faster with the distance from the electrode, however finally reaching a value which is smaller than the polarization reached in the bulk situation. This figure illustrates the geometrical meaning of parameter $\lambda$ as the so-called extrapolation length.

All in all, the two models predict an increasing thickness dependence of the dielectric permittivity of the films, Eqs. (3.51), (3.57), and (3.61). Let us now address the applicability of these predictions to a real experimental situation in $(Ba,Sr)TiO_3$. To do so, some information on the correlation length $\xi$ in the material is needed in addition to its known dielectric parameters. This information may be extracted from data on the dispersion of the soft mode. Using the data from Refs [86, 87], the temperature independent parameter $\xi/\sqrt{\varepsilon} = \sqrt{\kappa\varepsilon_0}$ is evaluated[9] as 0.03 Å for $BaTiO_3$ and 0.08 Å for $SrTiO_3$. For $\varepsilon = 1600$, this leads to the estimates for the correlation length 1.2 Å for $BaTiO_3$ and 3.2 Å for $SrTiO_3$. This estimate enables us to make important conclusions:

(i) For situations of practical interest, the expected thickness dependence of the in-plane component of the measured permittivity is very weak. According to Eq. (3.57) the expected relative correction to the bulk permittivity is about $2\xi/h$.

[9] Actually the correlation length is anisotropic, in other words the parameter $\kappa$ (from Eq. (3.52)) is different for different orientations of the polarization and its gradient. Since here we are interested in an estimate, this anisotropy will be neglected.

For 200 nm thick films, this makes a 1% correction at most. However, this result may be taken only as qualitative since the theoretical situation corresponds to the limit of the range of applicability of the continuous Landau theory; the latter is applicable if the typical scale of the polarization variation ($\xi$ in this case) is essentially larger than the lattice constant of the material (4 Å in the case of (Ba,Sr)$TiO_3$).

(ii) For the out-of-plane component of the measured permittivity, a much stronger impact of the surface blocking of the polarization is found. According to Eq. (3.61) the expected relative correction to the bulk permittivity is about $2\frac{\xi_1}{h}\frac{\varepsilon}{\varepsilon_b}$, that is some $\sqrt{\varepsilon/\varepsilon_b}$ times stronger than in the case of the in-plane component. On the other hand, assuming $\varepsilon_b \cong 10$ the scale $\xi_1 = \sqrt{\varepsilon_b}\sqrt{\kappa\varepsilon_0}$ (i.e. 0.1–0.2 Å) is much below the value of the lattice constant of the material, so that the use of the theoretical prediction, even as an order-of-magnitude estimate, cannot be justified. However, the physics behind Eq. (3.61), namely the in-series connection of the surface passive layer with the "bulk" ferroelectric part of the film looks reasonable. In this context, Eq. (3.61) might be used as a semi-empirical relation, $\xi_1/\varepsilon_b$ being a fitting parameter. An analysis of the thickness dependence of the in-plane component of dielectric constant in terms of Eq. (3.61) performed by Vendik and Zubko [82] for (Ba,Sr)$TiO_3$ thin films yields the values of $\xi_1/\varepsilon_b$ in the range 0.2–2.5 Å.

One more useful qualitative conclusion can be done on the basis of the above modeling for the out-of-plane component of the permittivity. Namely, its thickness dependence may be sensitive to the electrode material since the effective thickness of the surface dead layer is a function of the boundary condition at the electrodes, which in turn may be dependent on the electrode material. Here two hypotheses for the relation between the boundary condition in question and the electrode material should be mentioned. First, it was suggested that the surface blocking of polarization is much less pronounced in the case of a similarity between the ferroelectric and electrode, specifically, when the electrode is oxide and the ferroelectric is an oxide perovskite [40]. The second hypothesis is that in the case of oxide electrode, the effective surface passive layer is short-circuited [88]. Thus, both hypotheses imply that ferroelectric capacitors with oxide electrodes are expected to exhibit a weaker thickness dependence of the measured dielectric constant compared to those with metallic electrodes, which according to Vendik and Zubko is compatible with the existing experimental data [40].

Thus, though not being quantitatively justified, the mechanism relating the size effect to the surface blocking of the polarization looks relevant to the experimental situation in ferroelectric thin films. On the other hand, the "electrode" mechanism should not be discarded. Its prediction Eq. (3.51), which is strictly justified mathematically, is identical to that of the "surface blocking" model where $\xi_1$ is replaced by screening length $l_s$. The experimental data [82] for (Ba,Sr)$TiO_3$ thin films correspond to $l_s$ in the range 0.2–2.5 Å, which is comparable to the typical value of Thomas-Fermi screening length, $\approx$0.5 Å.

The presented analysis is readily extended to the description of the tunability of a ferroelectric capacitor containing the aforementioned effective or real passive layers. In the case of a real dielectric layer or the "electrode" effect, the film will obviously exhibit the tunable behavior of a layer composite, which was discussed in Sec. 3.1.3. In the case of the "surface polarization blocking" scenario, the situation with the out-of-plane component of the permittivity is also of this type of tunable behavior. The physical reason[10] for that is that the spatial scale $\xi_1$ weakly depends on $\varepsilon$ which is the only field dependent parameter of the system. Thus, in all these cases, the field dependence of the measured permittivity of the films is controlled by the simple "in-series-capacitor" formula:

$$\varepsilon_{\text{eff}}^{-1}(E) = \varepsilon^{-1}(E_f) + \frac{l}{h} \tag{3.64}$$

where $l$ is a constant that is independent of $E$ and $h$ (c.f. Eqs. (3.15) and (3.21)). As was mentioned in Sec. 3.1.3, the impact of the passive layer on the polarization response (permittivity and tunability) is identical to that of a lowering of the Curie-Weiss temperature of the material. Thus, all the passive layer models addressed above predict the polarization response of the films identical to that of an ideal film with a modified value of the Curie-Weiss temperature.

Concluding the discussion of the passive layer effects it is instructive to give a table for the thickness of a dielectric layer with $\varepsilon_d = 1$, which influence the dielectric permittivity of the film identically to a given passive layer effect.

### *3.4.2. Depletion effect*

Electrochemical interaction of conductive electrodes with electronic carriers of the ferroelelectric may result in the so-called depletion effect i.e. the removal of the carriers from narrow near-by-electrode regions so that charged regions of bare impurities are formed near the electrodes. In the case of a wide gap and heavily compensated semiconductor, which is the case of typical ferroelectrics used for tunable applications, this built-in charge is related to deep trapping centers and oxygen vacancies [72]. It can be considered as immobile during electrical measurements at room temperatures and homogeneous inside the depletion layers [73]. The depletion phenomenon is characterized by the parameter $W$ called *depletion layer thickness*, which is a function of the electrochemical parameters of the electrode-ferroelectric interface. If the film thickness $h$ is larger than $2W$, the film contains two space-charge layers at the two electrodes and a neutral region in the middle. This is the case of *partial depletion.* If $h << 2W$, the film is homogeneously filled with space charge. This is the case of the so-called *full depletion.* For the two case discussed above, the charge density distribution $\rho_d(x)$ across the films is shown in Figs. 3.12 (a) and (b).

This space charge can influence the ferroelectric and dielectric properties of the films [89, 90]. The impact of the depletion effect on the effective dielectric permittivity of a parallel plate capacitor containing a ferroelectric which is subjected to the depletion effect of the electrodes can be elucidated using the following simple arguments. The built-in depletion charge results in a certain built-in electric field which in turn creates the built-in $x$-dependent polarization $P_b(x)$, $x$ being the coordinate normal to the plane of the film. For the dielectric response of the film, this implies a local $x$-dependent

[10] This can also be shown by considering Eq. (3.59) with the non-linear term $\beta P^3$.

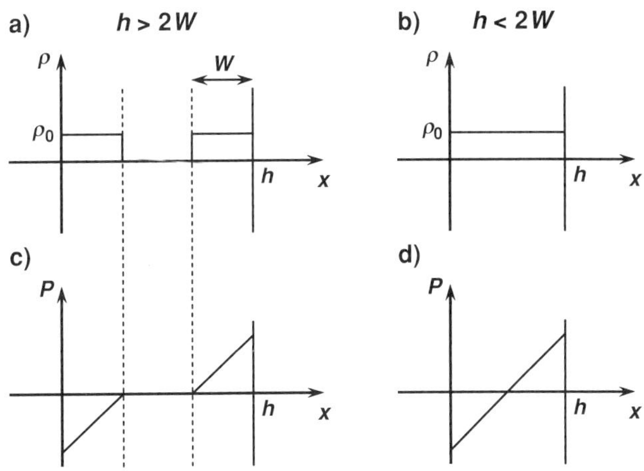


**Fig. 3.12.** Space charge distributions of partially (a) and fully (b) depleted films. The distributions of the build-in polarization (c) and (d), in these cases, respectively.

permittivity of the films whose out-of-plane component according to Eq. (3.3) reads:

$$\varepsilon_{loc}(x) = \frac{1}{\varepsilon_0} \frac{1}{\alpha + 3\beta P_b^2(x)}. \tag{3.65}$$

Taking into account the "in-series" geometry of the problem the effective dielectric constant of the system can be found by averaging $1/\varepsilon_{loc}(x)$ across the film thickness, namely

$$\frac{1}{\varepsilon_{eff}} = \frac{1}{h} \int_0^h \frac{dx}{\varepsilon_{loc}(x)} = \frac{1}{\varepsilon} + \frac{3\beta\varepsilon_0}{h} \int_0^h P_b^2(x)dx. \tag{3.66}$$

The exact value of $P_b(x)$ can be obtained from a rather cumbersome solution of a set of equations: the Poisson equation

$$d(\varepsilon_0\varepsilon_b E + P_b(x))/dx = \rho(x), \tag{3.67}$$

where $\rho(x)$ is the density of the built-in charge and the equation of state for the ferroelectric, Eq. (3.2). This solution should also obey the condition

$$\int_0^h E dx = 0. \tag{3.68}$$

A good approximation for the built-in polarization can be readily obtained if, in Eq. (3.67), one neglects the term $\varepsilon_0\varepsilon_b E$. This approximation is justified by a large value of the dielectric permittivity of the ferroelectric (i.e. $\varepsilon >> \varepsilon_b$). Then, Eqs. (3.67) and (3.68) yield

$$P_b(x) = (x - h/2)\rho_0 \tag{3.69}$$

**Table 3.5.** Thickness of a dielectric layer with $\varepsilon_d = 1$, which impacts the dielectric permittivity of the film identically to a given passive layer effect or depletion effect. h—film thickness; d—thickness of the layer; $\xi$—correlation length; $\varepsilon$ and $\varepsilon_b$—the permittivity and background permittivity of the ferroelectric; $\lambda$—extrapolation length for the polarization boundary conditions; $l_s$—the Thomas-Fermi screening length; $\beta$—coefficient of the dielectric non-linearity, $\rho_0$—the space charge density in the depletion layer; W—depletion layer width.

| | in-plane permittivity | out-of-plane permittivity |
|---|---|---|
| Real layer with $\varepsilon_d = 1$ | $d$ | $d$ |
| Full surface blocking of polarization | $\xi$ | $\xi_1 = \frac{\xi}{\sqrt{\varepsilon\varepsilon_b}}$ |
| Partial surface blocking of polarization | $\frac{\xi}{1+\lambda/\xi}$ | $\frac{\xi_1}{1+\lambda/\xi_1}$ |
| Electrode effect | – | $l_s$ |
| Partial depletion effect, $h > 2W$ | – | $h_d = 2\varepsilon_0\beta\rho_0 W^3$ |
| Full depletion effects $h < 2W$ | – | $h_d\left(\frac{h}{2W}\right)^3$ |

for the full-depletion case and

$$
\begin{aligned}
P_b(x) &= (x - W)\rho_0 & & 0 < x < W \\
P_b(x) &= 0 & \text{for} \quad & W < x < h - W \\
P_b(x) &= (x - h + W)\rho_0 & & h - W < x < h
\end{aligned}
\tag{3.70}
$$

for the case of partial depletion. Here $\rho_0$ is the space charge density in the depleted regions. The profiles given by Eqs. (3.69) and (3.70) are illustrated in Fig.3.12 (c) and (d).

Using this result for the built-in polarization and Eq. (3.66) we find that the depletion leads to a reduction of the effective dielectric permittivity of the film. In the cases of the partial and full depletion, one finds

$$
\frac{1}{\varepsilon_{eff}} = \frac{1}{\varepsilon} + \frac{2\varepsilon_0\beta\rho_0^2 W^3}{h}
\tag{3.71}
$$

and

$$
\frac{1}{\varepsilon_{eff}} = \frac{1}{\varepsilon} + \frac{\varepsilon_0\beta\rho_0^2 h^2}{4},
\tag{3.72}
$$

respectively. Equations (3.71) and (3.72) imply that, for the effective dielectric permittivity of the film, the depletion effect is equivalent to the in-series connection of the ferroelectric with a linear capacitor. In the case of the partial depletion, this capacitance is independent of the film thickness so that the film behaves like a film with a real passive layer. However, in the case of the full depletion, according to Eq. (3.72) this series capacitance goes up with decreasing thickness as $h^3$. Comparison of the "passive layer" performance of the depletion effects with that of other passive layer effects is presented in Table 3.5.

It is instructive to evaluate the possible impact of the depletion effect on the permittivity. Numerically, it is not strong but, it is, however, comparable with that of the effective passive layers discussed in the previous section. For values of $W$ and $\rho_0$ compatible with the data on ferroelectric perovskite with metallic electrodes, $W = 0.2$ ìm and $\rho_0 \cong 1.6 \times 10^{-19} \times 10^{18}$ Ccm$^{-3}$ $= 0.16$ Ccm$^{-3}$ [72, 89, 91] one finds that, in the case of partial depletion (where $h > 2W$), a single depletion layer works as a dielectric layer with $\varepsilon_d = 1$ and thickness $h_d = \varepsilon_0 \beta \rho_0^2 W^3 \approx 0.1$ Å. In the case of full depletion, the depletion effect on the permittivity can be readily identified by a drastic reduction of the effect with decreasing film thickness (see Eq. (3.72)). It is also worth mentioning that, as for any "in-series model", the impact of the depletion on the dielectric constant of a ferroelectric material in paraelectric phase is formally equivalent to a lowering of its Curie-Weiss temperature [90].

The modification of tunability by the depletion effect has not been addressed theoretically so that, at present, it is not clear whether under a dc bias electric field a depleted film still behaves as in-series connection of ferroelectric and a dielectric layers.

There is another obvious effect of the depletion: the built-in polarization will induce the quasi-Debye loss mechanism. However, theoretical treatment which can provide reliable estimates for the magnitude of the dielectric loss induced by this effect is not available in the literature.

### 3.4.3. Strain effects

Ferroelectric films deposited onto dissimilar substrates are subjected to in-plane deformation. This deformation modifies the dielectric permittivity of the films as well as its temperature dependence due to the electrostrictive coupling. This effect can be readily described in terms of the Landau expansion for the free energy of the material. Let us address it for the case of a (001) film of a cubic ferroelectric perovskite in the paraelectric phase. To do that we will generalize Eq. (3.1) by making allowance for the electrostrictive coupling and crystalline anisotropy of the material:

$$\begin{aligned} F = {} & \frac{\alpha}{2}\left(P_1^2 + P_2^2 + P_3^2\right) - q_{11}\left(u_{11}P_1^2 + u_{22}P_2^2 + u_{33}P_3^2\right) \\ & - q_{12}\left[u_{33}\left(P_1^2 + P_2^2\right) + u_{22}\left(P_3^2 + P_1^2\right) + u_{11}\left(P_2^2 + P_3^2\right)\right] \\ & - 4q_{44}\left(u_{12}P_1P_2 + u_{23}P_3P_2 + u_{13}P_1P_3\right) + \ldots \end{aligned} \tag{3.73}$$

where $u_{ij}$ are the components of the strain tensor. Typically, the films are isotropically strained laterally so that $u_{11} = u_{22} \equiv u_m$ and $u_{12} = 0$ whereas in other directions they are mechanically free, the latter implying $u_{13} = u_{23} = 0$ and $u_{33} = -u_m\eta/(1-\eta)$ where $\eta$ is the Poisson ratio of the ferroelectric. In this case, on the lines of Section 3.3 one readily finds:

$$\varepsilon_{11}^{-1} = \varepsilon_{22}^{-1} = \varepsilon^{-1} - 2\varepsilon_0 u_m\left[q_{11} + q_{12} - \frac{2\eta}{1-\eta}q_{12}\right] \tag{3.74}$$

for the in-plane and

$$\varepsilon_{33}^{-1} = \varepsilon^{-1} - 2\varepsilon_0 u_m\left[2q_{12} - \frac{2\eta}{1-\eta}q_{11}\right] \tag{3.75}$$

for the out-of-plane components of the dielectric permittivity of the film, where $\varepsilon = (\alpha\varepsilon_0)^{-1}$ is the dielectric permittivity of the non-strained ferroelectric. The in-plane deformation of the film, usually called misfit strain, is mainly controlled by two contributions: the strain that appears during the film crystallization and the thermal strain due to the difference between the thermal expansion coefficients of the substrate and ferroelectric[11]. In the simplest case of liner thermal expansion of the substrate and ferroelectric, the temperature dependence of $u_m$ reads

$$u_m = u_{mR} + (\alpha_s - \alpha_f)(T - T_R) \tag{3.76}$$

where $\alpha_s$ and $\alpha_f$ are the thermal expansion coefficients of the substrate and ferroelectric, $u_{mR}$ being the misfit strain at a certain temperature $T_R$. When the dielectric permittivity of the ferroelectric obeys the Curie-Weiss law $\varepsilon = C/(T - T_0)$, relations (3.74)–(3.76) imply a renormalization of the Curie-Weiss constant and Curie-Weiss temperature of the films [93, 94]. Clearly, the renormalized values of $C$ and $T_0$ will be different for the in-plane ($C_{in}$, $T_{in}$) and out-of-plane ($C_{out}$, $T_{out}$) components of the permittivity:

$$C_{in}^{-1} = C^{-1} - 2\varepsilon_0\left[q_{11} + q_{12} - \frac{2\eta}{1-\eta}q_{12}\right](\alpha_s - \alpha_f), \tag{3.77}$$

$$C_{out}^{-1} = C^{-1} - 2\varepsilon_0\left[2q_{12} - \frac{2\eta}{1-\eta}q_{11}\right](\alpha_s - \alpha_f), \tag{3.78}$$

$$T_{in} = T_0\frac{C_{in}}{C}\left[1 + \frac{2C\varepsilon_0}{T_0}(q_{11} + q_{12} - \frac{2\eta}{1-\eta}q_{12})\left[u_{mR} + (\alpha_s - \alpha_f)T_R\right]\right], \tag{3.79}$$

$$T_{out} = T_0\frac{C_{out}}{C}\left[1 + \frac{2C\varepsilon_0}{T_0}(2q_{12} - \frac{2\eta}{1-\eta}q_{11})\left[u_{mR} + (\alpha_s - \alpha_f)T_R\right]\right]. \tag{3.80}$$

The strength of the impact of the mechanical effect on the dielectric response of the ferroelectric films can be assessed from the relations for the value of misfit strain $u_R$ that is required to shift the Curie-Weiss temperature to a given temperature, namely

$$u_R = \frac{T_R - T_0}{2C\varepsilon_0\left[q_{11} + q_{12} - \frac{2\eta}{1-\eta}q_{12}\right]} \tag{3.81}$$

for the in-plane and

$$u_R = \frac{T_R - T_0}{2C\varepsilon_0\left[2q_{12} - \frac{2\eta}{1-\eta}q_{11}\right]} \tag{3.82}$$

for the out-of-plane polarization instabilities.

Let us evaluate the strength of the effects discussed above using the numerical values of the parameters for $SrTiO_3$ ($T_0 = 42$ K, $C = 86\ 000$ K, $q_{11} = 2.2 \times 10^{10}$ m/$F$, $q_{12} = 0.2 \times 10^{10}$ m/$F$, $\alpha_f = 11 \times 10^{-6}$ K, $\eta = 0.24$ [95]) and typical substrates (MgO $\alpha_s = 13 \times 10^{-6}$ K; $Al_2O_3$ $\alpha_s = 6 \times 10^{-6}$ K; Si $\alpha_s = 4 \times 10^{-6}$ K).

---

[11] The misfit strain may be also be affected by the formation of the misfit dislocation in the already crystallized film (see e.g. the work by Speck and Pompe [92]. We will not address this issue in the paper, since this effect does not seem to be of practical importance for the tunable thin films.

From Eqs. (3.81) and (3.82) we find that to shift the in-plane ferroelectric instability temperature up to room temperature, the in-plane tensile deformation of some 0.7 % is required whereas, for the out-of-plane instability, it is the out-of-plane compressive deformation of some 1.5 %. Note that the in-plane deformation results in opposite shifts the Curie-Weiss temperatures of the in-plane and out-of-plane components of the permittivity. This seems to be a general feature of perovskite thin films (see also Refs. [96, 97]).

The impact of the mechanical effects on the Curie-Weiss constant depends on the relation between $\alpha_s$ and $\alpha_f$. If $\alpha_s > \alpha_f$, the in-plane Curie-Weiss constant is larger whereas the out-of-plane Curie-Weiss constant is smaller than the original value of the Curie-Weiss constant. This is the case of MgO substrate where $(C_{in} - C)/C_{in} \approx 0.07$ and $(C_{out} - C)/C_{out} \approx -0.03$. If $\alpha_s < \alpha_f$, the opposite relation takes place. This is the case of $Al_2O_3$ and Si substrates, for which $(C_{in} - C)/C_{in} \approx -0.17$ and $(C_{out} - C)/C_{out} \approx 0.08$ and $(C_{in} - C)/C_{in} \approx -0.24$ and $(C_{out} - C)/C_{out} \approx 0.11$, respectively.

All these estimates show that the mechanical effects in tunable thin films are large enough to be measured experimentally and to influence the behavior of film based tunable devices.

Concluding this section, important reservations are to be made:

(i) The above consideration was aiming at the elucidation of the basic trend related to the impact of the misfit strain effects on the dielectric permittivity of the ferroelectric thin films. On the other hand, the real parameter of interest of thin films is their tunability. To obtain information on the latter, an analysis of the impact of the mechanical effects on the coefficient of the dielectric non-linearity is required. Such analysis is available in recent papers on the subject [96–98]. It predicts a certain renormalization of these coefficients which, for tunable materials of interest like (Ba, Sr)$TiO_3$ does not seem to be significant. Thus, the impact of mechanical effects on the tunability of the films is actually reduced to a change of the initial permittivity. This corresponds to the replacement of $\alpha$ in the equation of state Eq. (3.2) with the $1/(\varepsilon_0\varepsilon_{cor})$ where $\varepsilon_{cor}$ is the permittivity of the films in the absence of a dc field corrected by the mechanical effects, i.e. $\varepsilon_{11}$ or $\varepsilon_{33}$ in Eqs. (3.74) and (3.75).

(ii) The relations derived above are quantitatively valid when the permittivity of the ferroelectric obeys the Curie-Weiss law; the validity of Eqs. (3.77)–(3.80) also requires the linearity of the thermal expansion in the ferroelectric and the film. If it is not the case, e.g. at low temperatures, Eqs. (3.74) and (3.75) still can be used, however using the real temperature dependence of $\varepsilon$ (e.g. from Eq. (3.7)) and values of $u_m$ evaluated taking into account the temperature dependence of $\alpha_s$ and $\alpha_f$.

(iii) The mechanical effects may also influence other structural instabilities of the ferroelectric which in turn can modify the dielectric response of the system. This is the case for $SrTiO_3$ where an antiferrodistorsive phase transition at 105 K takes place. A theoretical treatment of the resulting two-instability system has been offered in Ref. [98]. Though, a focused theoretical treatment of the tunability in this situation is still missing, the analysis of the result obtained in this paper suggests that the additional instability does not affect the tunable behavior of the material in a crucial manner.

# 4. Measurement Techniques

In this section we address the methods and techniques that enable the measurements of the complex dielectric permittivity and tunability of different forms of ferroelectric materials.

All measurement techniques require two steps. First, the electrical characteristic of a ferroelectric containing device is monitored. Second, the dielectric permittivity and loss tangent of the material are evaluated from the obtained data. Thus, the problem of determination of the complex dielectric permittivity of the material can be divided into two: measurement problems and evaluation problems.

All measurement methods can be divided into three groups:

- *direct methods* where either the capacitance and loss tangent of a capacitor containing the investigated material are measured using an impedance analyzer, or else the scattering matrix [99] of the ferroelectric capacitor is measured by a network analyzer,
- *waveguide methods* in which the scattering matrix of the ferroelectric containing waveguide is measured using a network analyzer.
- *resonance methods* where the characteristics of a resonator which contains the investigated material are measured.

The choice of the measurement method and correspondingly the type of the measurement setup strongly depends on the frequency range and on the form of the ferroelectric material (thin film, thick film or bulk sample). For frequencies below some tens or hundreds of megahertz (radio frequencies), the tunable capacitor which is made of any form of ferroelectric material is regarded as a lumped element since its dimensions are much less than the wavelength of the electromagnetic signal. The capacitance and loss tangent of this capacitor can be measured directly by a standard impedance analyzer.

At higher frequencies, direct measurements of the capacitances are less applicable because, at higher frequencies, the dimensions of the capacitors become comparable with the length of the electromagnetic wave and they cannot be considered any more as lumped elements. Moreover, the impedance of the capacitors becomes very small in comparison to the resolution of the impedance analyzer. Thus, these methods are to be replaced by filled-waveguide characterization techniques and resonance methods.

## 4.1. Lumped capacitance measurement methods

Capacitors used for the characterization of tunable materials can be fabricated by depositing electrodes onto the parallel sides of a ferroelectric bulk sample (Fig. 4.1a) or by making a set of small top electrodes on a ferroelectric film with a common bottom electrode (Fig. 4.1b). Capacitors with lower capacitance values, which are more suitable for high frequency measurements, can be made from thick- or thin-ferroelectric films by forming an interdigital electrode pattern (Fig. 4.1c) or rectangular electrodes with an air gap in between (Fig. 4.1d). In the two last types of capacitors, often called "*planar capacitors*", only the top side of the film is subjected to metallization.

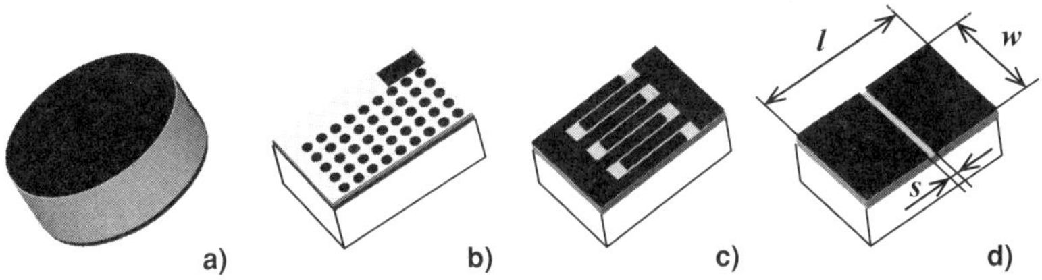


**Fig. 4.1.** Different types of ferroelectric capacitors: a) parallel plate capacitor, b) array of thin/thick film parallel plate capacitors, c) interdigital capacitor, d) air gapped planar capacitor (in all pictures black color represents conducting electrodes, grey—ferroelectric material and white color-substrate).

*Parallel plate capacitors*

At low frequencies the ferroelectric parallel plate capacitor (Fig. 4.1a,b) is described by a simple electrical equivalent circuit consisting of a capacitance ($C_p$), a frequency dependent resistance ($R_p$) connected in parallel, and a series resistance ($R_s$) corresponding to the conducting losses in the electrodes [100–102]. Measurements up to a few megahertz can be done by an impedance analyzer, directly giving the values of the capacitance and loss tangent. Higher frequencies require the use of a network analyzer to measure the complex reflection coefficient ($S_{11}$). In this case, the loss tangent and capacitance of the ferroelectric capacitor are calculated using the following equation:

$$\tan\delta \cdot \omega C + j\omega C = \frac{1}{Z_0}\left(\frac{1 - S_{11}}{1 + S_{11}}\right) \tag{4.1}$$

where $\omega$ is the angular frequency of the a.c. field, $C$ and $\tan\delta$ are the measured capacitance and loss tangent, respectively; $Z_0 = 50\ \Omega$.

The loss in the electrodes (represented by $R_s$), which is often not negligibly small, is to be taken into account. This can be done by subtracting the conducting loss using the following equation:

$$\frac{\varepsilon''}{\varepsilon} = \tan\delta_e = \tan\delta_{meas} - \omega R_s C_p \tag{4.2}$$

where $\tan\delta_e$ and $\tan\delta_{meas}$ are the loss tangent of the ferroelectric material and measured loss tangent of the capacitor, respectively. The parameter $R_s$ can be evaluated analytically [100, 101] or determined experimentally [102]. For parallel plate capacitors, the relative dielectric permittivity of the ferroelectric material can be extracted from the measured capacitance using the simple formula:

$$\varepsilon = \frac{Ct}{\varepsilon_0 S}, \tag{4.3}$$

where $C$ is the measured capacitance, $t$ is the thickness of the sample and $S$ is the electrode area.

The so-called quasi-lumped approach to the characterization of a capacitor made of a ferroelectric bulk ceramics was proposed by Li [103]. This approach allows expanding the frequency range of the measurements up to 2GHz, while the conventional lumped approach does not work at so high frequencies because the dimensions of the sample are comparable to the wavelength in the ferroelectric material. In this approach, the measured sample is modeled as an equivalent electrical circuit consisting of two parts: the first part of this circuit is a lossy transmission line of a length equal to the thickness of the sample. This line is terminated by a lumped capacitor. Measuring the reflection coefficient from the sample, one can calculate its complex dielectric permittivity.

*Planar capacitors*

Modeling of the planar capacitor structures (Fig. 4.1c,d) is more complicated and should take into account the electric field distribution between the ferroelectric layer, substrate and air. Based on the conformal mapping technique, mathematical models for the capacitance of the interdigital structure [104] and air-gapped electrodes structures [105] were developed. For the latter case (Fig. 4.1d), the capacitance can be calculated as [105]:

$$C = \varepsilon_0 w \left[ \frac{2}{\pi} \ln\left(\frac{4l}{s}\right) + \frac{\varepsilon_3 - 1}{\pi} \ln\left(16\frac{h_3 + h_2}{\pi s}\right) + \frac{\varepsilon_2 - \varepsilon_3}{\frac{s}{h_2} + \frac{4}{\pi}\ln 2} \right], \tag{4.4}$$

where $\varepsilon_2$ and $\varepsilon_3$ are the relative dielectric permittivities of the ferroelectric film and the substrate respectively, $h_2$ and $h_3$ are the thicknesses of the ferroelectric film and the substrate, respectively, $l$ is the length of the capacitor, $s$ is the gap between the electrodes, and $w$ is the capacitor width (see Fig. 4.1.(d)). This equation was derived for a situation typical for ferroelectric capacitors. According to Vendik et al. [105] it provides some 1–2% accuracy on condition that $h_2 < s \leq 10h_2$, $s \leq 0.25l$, $s \leq 0.5h_3$, $\frac{\varepsilon_2}{\varepsilon_3} > 100$. However, a numerical simulation of the problem performed by Deleniv [106] showed that this formula still can be used with an accuracy better than 5% on a relaxed condition $s \leq 100\, h_2$.

## 4.2. Waveguide and transmission line methods

*Filled waveguide*

The waveguide method is one of the first developed methods for characterization of dielectric materials at microwave frequencies. This method, described in detail by Altschuler in Ref. [107], involves the measurements of the scattering matrix of a waveguide filled with the investigated dielectric. This method can be readily used for the characterization of materials with relatively low dielectric permittivity but it is less suitable in the case of very high dielectric permittivity materials, because of the problem of impedance matching.

*Transmission line*

The transmission line method is a special case of the waveguide method. It consists of the measurement of the scattering matrix of a transmission line which is patterned onto the ferroelectric film. The coplanar transmission line structure shown in Fig. 4.2a is often used for the characterization of both thick and thin ferroelectric films. The microstrip line structure with the bottom electrode shown in Fig. 4.2b can be used only for the characterization of thick ferroelectric films (thicker than 50μm). Only in

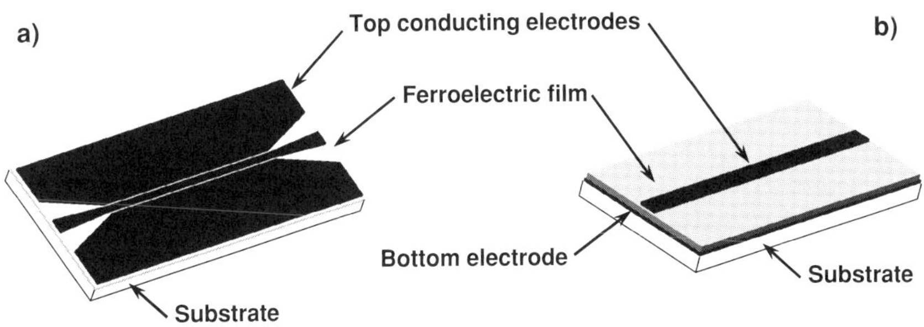


**Fig. 4.2.** Coplanar line a) and microstrip line b) structures patterned on the ferroelectric films (black color represents conducting layers).

this case can the microstrip line have a value of the wave impedance comparable to 50 Ω.

The full wave analysis of such structures [108] or models developed by the conformal mapping technique [109] enable the extraction of the parameters of the ferroelectric material from the measured scattering matrix. To minimize the error in the determination of the loss tangent of the ferroelectric, the conduction loss of the electrodes is to be taken into account when calculating the parameters of the material from the measured frequency characteristics [110, 111].

## 4.3. Resonance methods

*Hakki-Coleman method and its modifications*

The last group of measurement techniques is represented by various resonance methods. These methods are well developed for both bulk samples and films (thick and thin).

The classical Hakki-Coleman method [112] consists of the measurement of the resonance frequency and quality factor of the $TE_{01n}$ mode, excited in a dielectric rod which is placed between two metal discs ($L = h$ in Fig. 4.3a). To reduce the influence of loss in the metal discs, Krupka [113] proposed to keep the distance between the discs much greater than the sample thickness ($L \gg h$ in Fig. 4.3a).

Another frequently used modification of this method is shown in Fig. 4.3b. It employs a metal cavity with dimensions 2–3 times greater than those of the dielectric sample. The cavity has two coupling loops to feed the microwave power to the resonator. It was shown that in this structure the resonance frequency and the quality factor of the system are identical to those of the dielectric resonator in a free space [114].

Measuring the parameters of the resonator, the dielectric permittivity ($\varepsilon$) and loss tangent ($\tan\delta$) of the sample can be evaluated [112, 113].

As in the case of the waveguide method, the dielectric resonator method is very difficult to use for ferroelectrics because their high dielectric constant and relatively high loss impede effective excitation of the resonator.

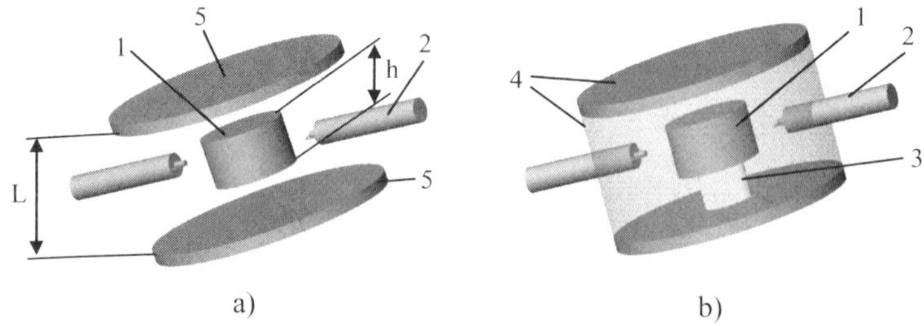


**Fig. 4.3.** Sketch of resonance measurement setups for characterization of dielectric bulk materials: Hakki-Coleman setup (Fig. 4.3a, $L = h$), modified Hakki-Coleman setup (Fig. 4.3a, $L >> h$), and metal cavity with the dielectric resonator (b). 1—dielectric resonator, 2—coaxial cables terminated with coupling loops, 3—dielectric support, 4—metal cavity, 5— metal plates with the diameter much bigger than that of the sample [113].

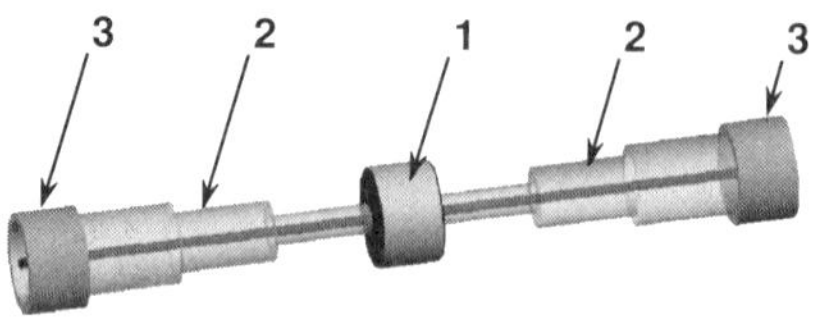


**Fig. 4.4.** Sketch of the coaxial line based measurement setup (1—both sides metallized dielectric resonator, 2—set of quarter-wavelength coaxial impedance transformers, 3—microwave connectors).

*Characterization techniques based on the excitation of the transverse magnetic field (TM) resonance modes*

Another resonance technique was used by Vendik *et al.* [115]. It consists of excitation and characterization of the $TM_{mn0}$ resonances in a cylindrical sample placed in a break of the central conductor of a coaxial line. To match the impedance of this line with that of the network analyzer a set of quarter-wavelength coaxial lines with different wave impedances is used (Fig. 4.4). The dielectric permittivity and loss tangent can be simply evaluated from the resonance frequency and $Q$-value of a $TM_{mn0}$ mode of the resonator using the following equation [115–117]:

$$\varepsilon = \left(\frac{k_{on}c}{2\pi r f_0}\right)^2, \quad \tan\delta = \frac{1}{Q_0} - \frac{R_{sur}c}{120\pi^2 h f_0}, \tag{4.5}$$

where $k_{0n}$ is the $n$-th root of the Bessel function, $c = 3 \cdot 10^8$ m/s, $r$ is the radius of the cylinder, $R_{sur}$ is the surface resistance of the electrode, $h$ is the thickness of the resonator, $f_0$ is the resonance frequency, and $Q_0$ is the unloaded quality factor.

The unloaded quality factor ($Q_0$) can be calculated from the loaded (or measured) one either analytically or numerically. An alternative simple and fast numerical method of determination of the unloaded quality factor has been proposed by Kaifez [118]. This method enables the evaluation of the loaded and unloaded quality factors as well as the coupling coefficient only from the measured scattering matrix of the resonator, not involving an analysis of the resonator itself.

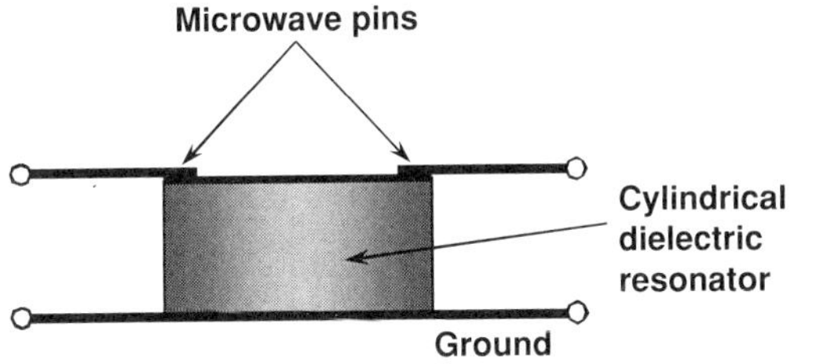


**Fig. 4.5.** Measurement setup for excitation of resonator by using microwave pins [119].

The excitation of the lowest $TM_{mn0}$ modes in the cylindrical resonator can also be done through galvanic couplings where the pins of the microwave connectors are in direct contact with the top conductive electrode [119] while the bottom electrode is grounded (Fig. 4.5). This method, potentially, can be applied to the characterization of thick films deposited on conductive substrates.

## 4.4. Resonance perturbation techniques

Another group of resonance measurement techniques employs the resonance perturbation approach. In these methods the sample is introduced into a high-$Q$ resonance cavity or a high-$Q$ resonator is loaded with the sample. The sample can be either a piece of bulk dielectric or a thin film deposited onto the dielectric substrate with known dielectric characteristics. The presence of the sample in the resonator leads to a shift of the resonance frequency ($\Delta f_0$) and to a decrease of the quality factor ($\Delta Q_0$). The complex dielectric permittivity of the material can be evaluated from these two parameters and the sample dimensions.

*Rectangular cavity*

The cavity perturbation technique was described by Altschuler in [107] and adopted for high dielectric permittivity materials by Dube et al. [120] [121] for the setup shown in Fig. 4.6.

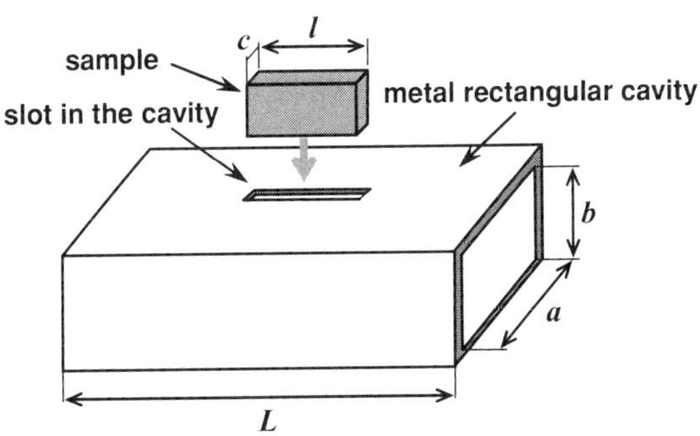


**Fig. 4.6.** Rectangular cavity with a dielectric sample to be inserted into it through the slot [120, 121].

For this setup, the real and imaginary parts of the material dielectric permittivity can be evaluated by using the following relations [121]:

$$\begin{aligned} \varepsilon &= \frac{f_1 - f_2}{f_2} \frac{2V_c}{b\left[1 + \frac{L}{n\pi}\sin\left(\frac{n\pi l}{L}\right)\right]\left[c + \frac{a}{\pi}\sin\left(\frac{\pi c}{a}\right)\right]} + 1 \\ \varepsilon'' &= \frac{Q_1 - Q_2}{Q_1 Q_2} \frac{V_c}{b\left[1 + \frac{L}{n\pi}\sin\left(\frac{n\pi l}{L}\right)\right]\left[c + \frac{a}{\pi}\sin\left(\frac{\pi c}{a}\right)\right]} \end{aligned}, \tag{4.6}$$

where $f_1$ and $Q_1$, $f_2$ and $Q_2$ are the resonance frequencies and quality factors of the cavity without and with the measured sample; $a$, $b$, and $L$ are the dimensions of the cavity; $V_c$ is its volume and $c$ and $l$ are the dimensions of the sample having the shape of a parallelepiped.

*Microwave scanning microscopy*

Recently, a new method of characterization of dielectric materials based on a near-field imaging technique, the so-called microwave scanning microscopy, has been developed. Conceptually this method is the cavity perturbation methods. A microwave scanning microscope consists of a coaxial resonator with a very sharp central conductor at the end, which is used as a microwave probe tip. The dielectric sample under test plays the role of a load for the coaxial resonator (Fig. 4.7.) [122, 123]. Comparing the resonance frequency and $Q$ values of the unloaded coaxial resonator with those of the resonator loaded with the ferroelectric, one can evaluate the dielectric permittivity and loss tangent of the investigated material. An important feature of this method is that it provides information of the local values of the dielectric parameter of the material, a lateral resolution better than 1 μm being readily achieved.

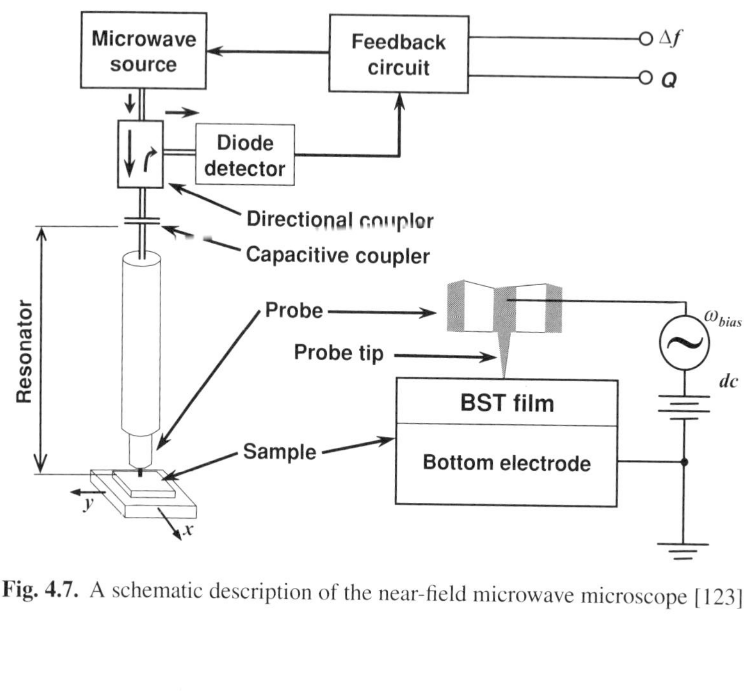


**Fig. 4.7.** A schematic description of the near-field microwave microscope [123].

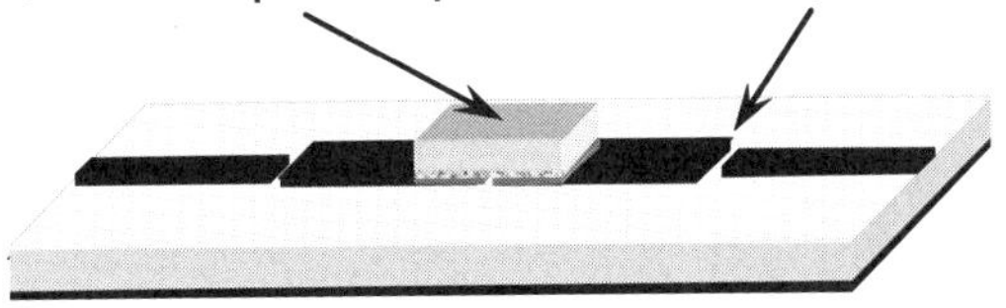


**Fig. 4.8.** Microstrip half wave resonator with the ferroelectric planar capacitor (see Fig. 4.1d) mounted as a flip-chip element with the gap centered between two parts of the resonator [125].

*Transmission-line resonator methods*

The most often used resonance method for the dielectric characterization of ferroelectric films is the so-called transmission-line resonator method. This method originates from the cavity perturbation method and employs transmission line resonators to characterize ferroelectric film based capacitors. Different transmission lines (microstrip lines, lines on the suspended substrate, slot lines) can be used in this method.

One practical realization of this method is a half-wavelength microstrip resonator interrupted at its center by a narrow air gap filled with the ferroelectric thin film [124] or by a ferroelectric planar capacitor mounted as a flip-chip element [125] (Fig. 4.8). The capacitance of the ferroelectric capacitor and loss tangent of the ferroelectric material are calculated from the measured resonance frequency and $Q$-value of the system by using the following formulas [125]:

$$C = \left[2Z_0\omega \cot\left(\frac{\omega\sqrt{\varepsilon_{eff}}l_{eff}}{c}\right)\right]^{-1}, \quad \tan\delta = \left(\frac{1}{Q_m} - \frac{1}{Q_{ref}}\right)\cdot\zeta,$$
$$\zeta = \left[\frac{1}{2} + \frac{Z_0}{X_c}\left(1 + \frac{X_c^2}{4Z_0^2}\right)\left(n\pi - \operatorname{arccot}\left(\frac{X_c}{2Z_0}\right)\right)\right]^{-1}, \quad X_c = \frac{1}{\omega C}, \tag{4.7}$$

where $Z_0$ is the wave impedance of the microstrip resonator, $\varepsilon_{eff}$ and $l_{eff}$ are the effective dielectric permittivity and length of the resonator respectively, $c = 3 \cdot 10^8$ m/s, $Q_m$ and $Q_{ref}$ are the measured and reference unloaded quality factors (the latter can be determined by measuring the $Q$-factor of the resonator loaded with a low loss capacitor with known parameters), $\zeta$ is the inclusion coefficient of the ferroelectric capacitor into the resonator, $n$ is the number of the harmonic of the resonator. The dielectric permittivity of the ferroelectric film is calculated from the obtained capacitance using Eq. (4.4).

## 4.5. About the precision and limits of measurement techniques

All sources of the errors appearing in different characterization techniques can be divided into measuring errors and evaluation errors.

As described above, it is either the capacitance of the ferroelectric capacitor (at radio frequencies) or the elements of the scattering matrix of the ferroelectric containing device (at higher frequencies) that are measured by the corresponding equipment. The relative instrument error of these measurements does not exceed 1–2% and depends on the quality of calibration and calibrating standards. Table 4.1 contains the measurement precisions and limitation for the different measuring techniques.

**Table 4.1.** Basic measurement errors and limits of the different characterization techniques. $\omega C$ [1/ Ω] is the reactance of the ferroelectric capacitor. [1] The measurement errors and limits were taken and calculated from the accuracy of the HP4284 LCR Meter and HP8722 Network Analyzer, [2] The precision of the waveguide methods is basically determined by the quality of the calibrating standards and by the possibility to separate conducting, radiative and dielectric losses; the absolute errors for the phase and transmission coefficient are to be taken into account when calculating the parameters of the ferroelectric materials using the measured transmission characteristics of the waveguide or line. [3] The limits depend on the impedance and quality factor of the non perturbed resonator

| Methods | | Instrumental errors[1] | Limits[1] |
|---|---|---|---|
| Direct capacitor measurements | capacitance<br>loss tangent | 0.05–0.1%<br>0.0005–0.001% | $10^{-17} < C[F] < 10$<br>$10^{-6} < tan\delta < 10$<br>$f[Hz] < 10^6 - 10^8$ |
| Characterization of the capacitor through the measurements of the reflection coefficient | capacitance<br>loss tangent | 2–5% (HP8722)<br>5–6% | $0.005 < \omega C < 0.05$<br>$0.0005 < \omega C\tan\delta < 0.02$ |
| Waveguide and transmission line methods[2] | absolute error for the phase<br>absolute error for the transmission coefficient | <5 degrees<br><0.02 | $f[GHz] > 0.5$<br>length of the line:<br>$l\,[m] > \dfrac{2}{f\,[GHz]\sqrt{\varepsilon}}$ |
| Resonance methods[3] | capacitance or dielectric permittivity<br>loss tangent | <1%<br>~5% | $10^{-4} - 10^{-2} < \tan\delta < 0.1$ |

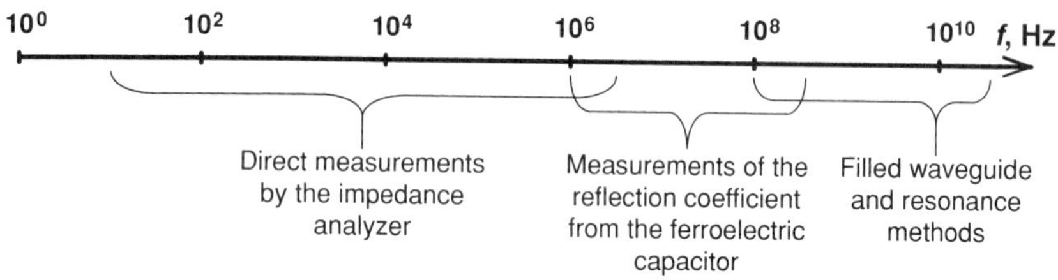


**Fig. 4.9.** Frequency range of applicability of the main types of techniques used for dielectric characterization of materials.

The second source of errors is related to the calculation of the complex dielectric permittivity of the ferroelectric from the measured capacitance and loss of the ferroelectric capacitor or from the elements of the scattering matrix. Usually the analytical formulas used for these calculations have their own errors, which are in the range 1–2%. The source of errors is due to the inaccuracy of the measurements of the linear dimensions of the sample, which are needed for computations. This error can reach 5–6% depending on the sample size and used equipment.

The measurement techniques described in this chapter cover a large frequency range, from some tens of hertz to some tens of gigahertz. They have different limitations and precisions and their choice depends on many parameters. Figure 4.9 gives an idea of the frequency ranges of applicability of the main types of these measurement techniques.

## 4.6. Measurements of dielectric permittivity and loss as a function of dc electric field

To measure the complex dielectric permittivity of a ferroelectric under a dc bias electric field, the experimental setup including the sample itself should satisfy the three following conditions:

(i) The sample has to be coated with electrodes (the mechanical contact between the non-metallized dielectric and the metal plates of the sample fixture is not sufficient because of an air gap which always exists between two surfaces in mechanical contact).
(ii) The setup should have a possibility to apply a dc voltage to the sample without subjecting the source of the ac signal to danger and short-circuiting the ac signal with the dc voltage source.
(iii) The distribution of the dc electric field in the sample is to be homogeneous and/or well modeled; otherwise it would not be possible to evaluate the electric field acting in the material.

The first two conditions cannot be satisfied in the filled waveguide methods and in some resonance methods where the metal plates should not be in the contact with the sample (Fig. 4.3a, $L \gg h$ and Fig. 4.3b). Moreover, the second condition requires the use of an external blocking circuit consisting of low frequency LC or RC filters in order to isolate the dc voltage source from the ac signal and to protect the measuring equipment from high voltage.

The last condition can be satisfied, first, for the parallel plate electrode geometry of the active area of the sample (like shown in Fig. 4.1a and b, Fig. 4.2b, Fig. 4.3a (for $h = L$), Fig. 4.4, Fig. 4.5). Second, in the planar electrode configurations (Fig. 4.1c and d, Fig. 4.2a) in the case where the thickness of the sample (actually film) is much less than the distance between the electrodes. In such structures, the electric field can be considered as homogeneous and evaluated directly from the known values of the dc voltage applied to the sample and the distance between the electrodes. In the case, where the field in the varactor is strongly inhomogeneous (like Fig. 4.1c and d for the film thickness larger or comparable to the distance between the electrodes), the evaluation of the tunability requires a solution of the electrostatic problem with a field dependent permittivity, which is a very difficult task.

The precision of all the methods of determination of the tunability is close to that of the dielectric permittivity. This holds up to the microwave frequency range. By contrast, the situation with the field dependence of the loss tangent is very different. In the RF range, where the contribution of the conductive loss in electrodes is not essential; the precision of the determination of the loss tangent of a material with and without a dc bias field is roughly the same. However, at higher frequencies, where the contribution of the conductive loss in electrodes becomes appreciable, the measurement of the field dependence of the loss tangent of the material often becomes a much more difficult task than that of the tunability.

## 5. Materials for Tunable Applications

The appropriate level of tunability for tunable applications ($n \geq 1.5$ i.e. $n_r \geq 0.3$ see Sec. 2) is practically achievable in materials with relatively high dielectric permittivity at microwave frequencies. For non-composite bulk materials, a value of the dielectric constant exceeding some 1000 is required whereas in thin films (where the application of higher fields is possible) the dielectric constant should be lager than some 300. This limitation together with the requirement of relatively low values of the dissipation factor (below a few percent) narrows the scope of the materials of interest to the displacive type ferroelectrics (incipient or regular in the paraelectric phase). Up to now, many materials have been addressed as possible candidates for tunable applications, the most attention has being paid to $SrTiO_3$ and its solid solution with $BaTiO_3$: $(Ba,Sr)TiO_3$. In the present section we will address the properties of some these materials. The properties of bulk materials and thin films will be discussed separately.

### 5.1. Bulk materials

*$KTaO_3$ and $SrTiO_3$ crystals*

High quality single crystals of $SrTiO_3$ (STO) exhibit a high dielectric permittivity (up to tens of thousands) and a low loss factor at microwave frequencies (below $10^{-3}$). This enables their use in tunable microwave components. As incipient ferroelectric, STO stays paraelectric down to 0 K and its dielectric constant strongly increases on cooling (See Fig. 5.1). STO is a cubic ($m\bar{3}m$) perovskite at $T > T_{St} = 105$ K, and undergoes at $T = T_{St}$ an antiferrodistorsive phase transition to a centrosymmetric tetragonal phase ($4/mm$). In its tetragonal phase, the crystal becomes dielectrically anisotropic. However, the anisotropy is apparent only at rather low temperatures, being about 3% at 80 K and

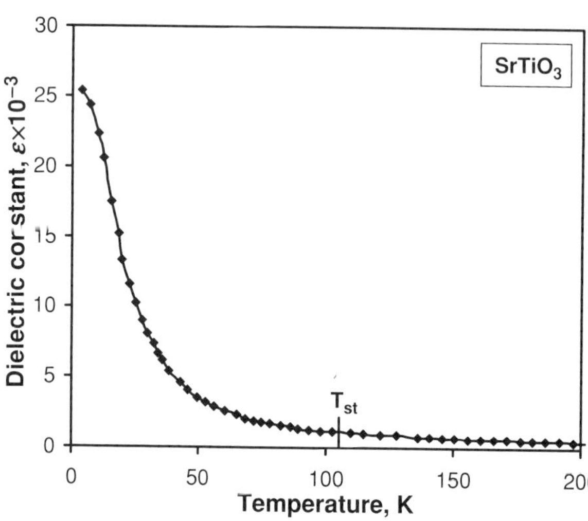


**Fig. 5.1.** Temperature dependence of the dielectric constant of $SrTiO_3$ single crystal measured at 10kHz in the direction perpendicular to the tetragonal axis of the 4/mm phase [126].

more than 100% at 10 K [126]. When cooled below $T_{St}$ without special precautions [126], STO crystals typically break into twins, so that their dielectric constant represents an average of its values for the different crystallographic directions. For this reason, the dielectric constant of different STO crystals measured below 30 K may substantially differ due to differences in their twin structures.

Another reason for this spread is a possible difference in the defect population in differently processed crystals; as clear from simple thermodynamic arguments, the higher the permittivity of the material, the stronger the impact of crystal imperfections on it. The temperature dependence of the dielectric permittivity of the material reasonably follows the predictions of the Landau theory [126]. For temperatures above some 50 K, it can be with good accuracy approximated by Eq. (3.12) with the parameters coming from Tab.3.1 or, alternatively, by Eqs. (3.4) and (3.6). However, there are publications reporting features of the dielectric response of STO crystals, which cannot be covered by the conventional thermodynamic approach that takes into account the polarization and the order parameter of the antiferrodistorsive phase transition. Namely, double [7, 127] and single [128] *P-E* hysteresis loops were reported for STO crystal. Despite some hypotheses put forward, the nature of these phenomena is still unclear.

The bulk form for STO (single crystals or ceramics) exhibits a high enough tunability only at rather low temperatures whereas at room temperatures the required electric field is unrealistically large . This can be seen from the following relation for the bias field that is required to reach a given value of tunability $n$:

$$E_n = 0.8 \cdot 10^6 \frac{\sqrt{n-1}(2+n)}{\varepsilon^{3/2}} \quad \left[\frac{\mathrm{kV}}{\mathrm{cm}}\right]. \tag{5.1}$$

This relation readily follows from Eq. (3.11) with $\beta = 8 \times 10^9 \mathrm{JC^{-4}m^{-5}}$. Using this relation one finds that the tunability of interest ($n > 1.5$) requires a field > 30 kV/cm at 77 K ($\varepsilon \approx 1500$) and > 400 kV/cm at RT ($\varepsilon \approx 300$).

In single crystals of STO the loss tangent at 10 GHz is typically about $10^{-3}$ [2] and, as was mentioned in Sec. 3, the data for the best quality crystals are consistent with the theory of the intrinsic phonon loss mechanisms (three and four quantum contributions in the absence of a bias electric field). Under an electric field, the theory predicts a steep rise of the intrinsic dielectric loss with increasing field (the so called field-induced quasi-Debye mechanism, see Sec. 3). An essential prediction is that the field corresponding to small values of the relative tunability may lead to a many-time increase of the loss tangent. This behavior was experimentally documented by Vendik *et al.* [115]. Figure 5.2a shows a strong decrease of the quality factor of an STO containing resonator with superconductor electrodes, which is due to the quasi-Debye mechanism.

The properties of $KTaO_3$ (KTO) crystals are very close to those of STO. This material is also a cubic incipient ferroelectric whose extrapolated instability temperature $T_0$ and Curie-Weiss constant $C$ are somewhat lower than those of STO (see Tab. 3.1), whereas the coefficient of the dielectric non-linearity[12] is slightly larger [129]. Another

[12] This statement holds for the case where both ac and dc fields are parallel to a cubic axis of the material, in the general situation it may not be the case. Detailed information on the anisotropy of the dielectric non-linearity can be found in Ref. [129].

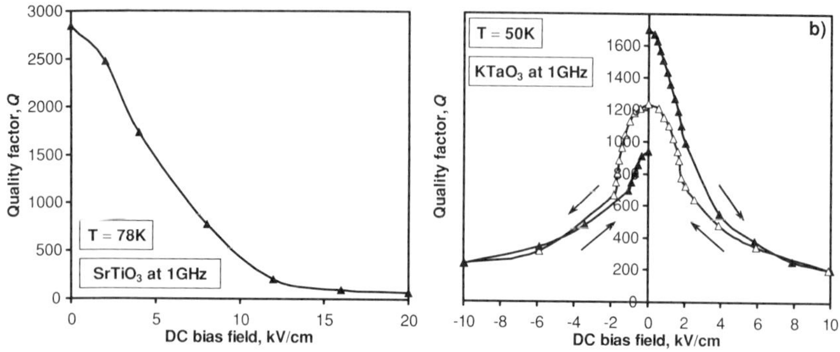


**Fig. 5.2.** Field dependences of the quality factors of the $SrTiO_3$ (a) and $KTaO_3$ (b) based bulk resonators with lossless superconducting $YBa_2Cu_3O_7$ electrodes (see Refs. [115] and [7], respectively). The directions of measurements are shown in the figure b) by the arrows. The dc field applied to the samples resulted in a small (<20%) variation of the dielectric permittivity of material.

difference between these two crystals is that KTO remains cubic down to 0 K. Though the set of material parameters of KTO and STO are rather close, the tunablility of KTOis substantially inferior to that of STO. As it follows from the Tab. 3.1 and Eq. (3.11), for the same values of tunability $n$ at $T = 77$ K, the tuning field in KTO should be three time larger than for STO. On the other hand the loss level exhibited in these materials is comparable. According to Gevorgian *et al.* [7], at $T = 77$ K @ $\omega = 1$ GHz, the crystal of KTO yields $\tan\delta = 0.5 \cdot 10^{-3}$. Under the action of a dc filed, KTO, similar to STO, exhibits an increasing field dependence of the loss consistent with the quasi-Debye contribution. The data [7] on the dc field dependence of the quality factor of a disc resonator containing $KTaO_3$, which illustrate this phenomenon, are shown in Fig. 5.2b.

An important issue to discuss is the level of the extrinsic contribution to the loss at microwave frequencies in KTO and STO crystals. Though any systematic study of the problem is absent, some conclusions on STO crystals at cryogenic temperatures can be done using the available data. The analysis [54] of the data collected by Buzin [58] strongly suggests that, in STO at $T = 77$ K, the intrinsic loss corresponds to $\tan\delta \propto \omega$ with $\tan\delta/\omega = 0.7 \cdot 10^{-4}$ GHz$^{-1}$ in the frequency range 1–30 GHz. This is to be compared to the data reported for STO crystals by other authors: at $T = 77$ K @ $\omega = 1 - 2$ GHz, it was reported that $\tan\delta/\omega = 3.3 \cdot 10^{-4}$ GHz$^{-1}$ [7], $\tan\delta/\omega = 1.2 \cdot 10^{-4}$ GHz$^{-1}$ [130], and $\tan\delta/\omega = 4.2 \cdot 10^{-4}$ GHz$^{-1}$ [115]. The spread of the reported values, clearly exceeding the accuracy of the measurements, can be attributed to extrinsic loss contributions. Thus, $\tan\delta/\omega \cong 10^{-4}$ seems to be a reasonable estimate for these contributions at $T = 77$ K @ $\omega = 1 - 2$ GHz.

The presented parameters of KTO and STO crystals correspond to an excellent tunable performance of these materials at cryogenic temperatures ($T = 77$ K). Taking the tunability $n = 1.5$ and $\tan\delta = 5 \cdot 10^{-3}$ we arrive at the Commutation Quality Factor $K \approx (n-1)^2/(n \cdot \tan^2\delta) \cong 7000$, which suffices for many application (see Sec. 2).

*(Ba,Sr)TiO$_3$ crystals and ceramics*

The solid solution $Ba_xSr_{1-x}TiO_3$ (BST) is a very suitable material for tunable applications. Depending on the barium concentration this material exhibits the permittivity

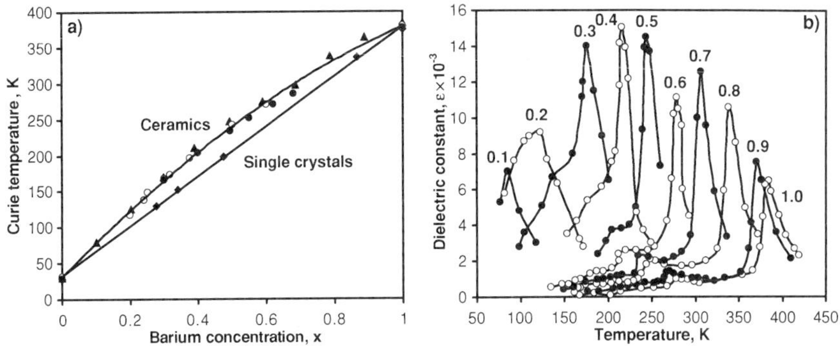


**Fig. 5.3.** a) - Curie temperature of $Ba_xSr_{1-x}TiO_3$ ceramic and single crystal materials as a function of Ba concentration $x$ [82]. b) - Temperature dependences of dielectric constant of $Ba_xSr_{1-x}TiO_3$ ceramic materials for different Ba concentration $x$ [14].

maxima within the temperature range 0–390 K, behavior of single crystals and ceramics being very close. The composition and temperature dependence of the permittivity of BST single crystals and bulk ceramics is illustrated in Fig. 5.3. As is clear from Fig. 5.3b, for a given temperature, a high enough value of the dielectric permittivity can be obtained by choosing a proper barium concentration. The coefficient of the dielectric non-linearity of BST $\beta_{BST}$ is expected to be smaller than that of STO because of a smaller value of $\beta$ in $BaTiO_3$ (BTO) [131], i.e. $\beta_{STO} \gg \beta_{BTO}$. Approximating the coefficient of the dielectric non-linearity of BST as $\beta_{BST} = (1-x)\beta_{STO} + x\beta_{BTO} \approx (1-x)\beta_{STO}$, from (3.11) and (5.1), one obtains an estimate for the bias field required for reaching the given tunability $n$ in $Ba_xSr_{1-x}TiO_3$:

$$E_n \approx \frac{0.8 \cdot 10^6}{\sqrt{1-x}} \frac{\sqrt{n-1}(2+n)}{\varepsilon^{3/2}} \left[\frac{\text{kV}}{\text{cm}}\right]; \quad x < 0.7. \tag{5.2}$$

This relation implies that, for the same permittivity, the tuning ability of $Ba_xSr_{1-x}TiO_3$ decreases with increasing barium content.

The parameter of BST single crystals important for tunable applications are given in the Tab. 5.1 offered by Vendik and coworkers [2].

This table illustrates several essential issues. First, the operational temperature range does not correspond to a close vicinity to $T_0$ where the permittivity and tunability are

**Table 5.1.** Parameters for BST crystal (after Vendik et al. [2])

| Barium concentration | Ferroelectric instability temperature ($T_0$, K) | Average value of tan $\delta$ at 10 GHz in operational temperature range | Operational temperature range (K) |
|---|---|---|---|
| 0.00 | 42 | 0.001 | 4–100 |
| 0.27 | 125 | 0.005 | 150–200 |
| 0.34 | 155 | 0.007 | 175–250 |
| 0.47 | 203 | 0.01 | 240–300 |
| 0.65 | 260 | 0.02 | 300–350 |
| 0.87 | 338 | 0.05 | — |
| 1.0 | 387 | 0.1 | — |

maximal. The reason for this[13] is the low temperature stability of the permittivity at $T$ close to $T_0$. The upper limit of the operational temperature range is roughly determined by the condition $\varepsilon \approx 1000$. Second, the trend "the higher the barium concentration, the higher the microwave loss" is clearly seen. This trend is consistent with the prediction of the phonon transport theory for the dielectric loss (see Sec. 3). It is related to a very big difference between the damping of soft-mode phonons in the end members of the BST solid solution. This damping essentially controls the level of the intrinsic dielectric loss at microwave frequencies. In STO the soft mode is under-damped, e.g. at $T = 77$ K, $\Gamma/\Omega_{TO} \cong 0.02$ [132], whereas in BTO it is overdamped, e.g. at $T = 300$ K, $\Gamma/\Omega_{TO} \cong 2.3$ [133]. On the other hand, the intrinsic loss in cubic centrosymmetric perovskites like STO follows the estimate (3.28) that yields $\tan\delta \propto \Gamma$ as long as the soft-mode stays underdamped. Thus, for barium concentrations corresponding to this situation, the intrinsic loss in $Ba_xSr_{1-x}TiO_3$ may be expected to grow steeply with increasing "$x$", which is the trend clearly seen in the above table. However, for higher barium concentrations where the soft mode is overdamped, the intrinsic loss is not expected to grow with increasing $\Gamma$(this can be checked by using a simple damped oscillator model for the loss). Thus, the origin of the concentration dependence of the loss at the BTO end of the composition needs to be further clarified.

The above discussion may have important implications for the microwave applications of BST. The intrinsic loss, which steeply increases with increasing barium concentration "$x$", is expected to be the major part of the total loss at high barium concentrations. This actually implies that the higher the "$x$", the less sensitive its microwave loss to the optimization of the material. At this point one should mention one more prediction of the phonon theory. Namely, as was discussed in Sec. 3, in $Ba_{0.5}Sr_{0.5}TiO_3$ the field-induced quasi-Debye contribution to the loss is substantially smaller than in pure STO where it was experimentally documented. This means that, in BST with intermediate concentrations of barium, which is of interest for practical applications, one does not expect a substantial dc field induced increase of the loss like those reported for STO and KTO crystals.

*Doping and composite effects in (Ba,Sr)$TiO_3$ ceramics*

Modification of BST by adding non-ferroelectric oxides in the range up to very high concentrations has been attempted by many workers (see e.g. [68, 134–137]). However the full set of parameters needed to evaluate the tunable performance of the material at microwave frequencies is rarely available in the literature. A rather complete set of experimental data of BST-MgO system was reported by Sengupta and Sengupta [137], namely the permittivity, tunability (for a dc field of 20 kV/cm), and loss tangent at 10 GHz were reported for 0.45/0.55, 0.5/0.5, 0.55/0.45, and 0.6/0.4 compositions of BST with the MgO content from 0.25 to 60 Wt.% (0.4 -70 vol.%). These data are presented in Figs. 5.4–5.6. Despite a substantial spread of the data and the absence of

[13] Elevated values of the loss in the vicinity of $T_0$ does not look to be a limiting factor for applications of the material in tunable devices. One can show that, in this temperature range, elevated values of the tunability will probably compensate the loss increase.

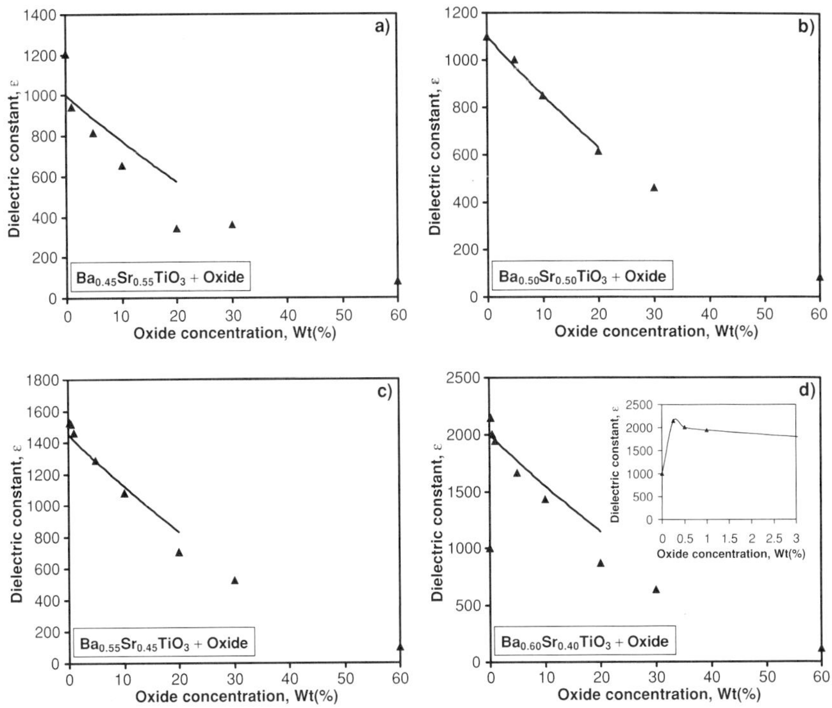


**Fig. 5.4.** Dielectric constant of ferroelectric-dielectric composite material as a function of weight concentration of the dielectric inclusions in the mixture for a)—$Ba_{0.45}Sr_{0.55}TiO_3$/oxide, b)—$Ba_{0.50}Sr_{0.50}TiO_3$/oxide, c)–$Ba_{0.55}Sr_{0.45}TiO_3$/oxide, d)—$Ba_{0.60}Sr_{0.40}TiO_3$/oxide material compositions. The black triangles are the experimental data, and the solid curves correspond to Eq.(3.24a) $\varepsilon_{\text{mix}}(q) = \varepsilon_f\left(1 - \frac{3}{2}q\right)$ with recalculated values of the concentration of dielectric inclusions $q$ from the volume values into the percents by weight Wt(%). RT data.

the information on their accuracy, these data can be discussed in terms of the available knowledge in the field.

As was discussed in Subsection 3.2.3, mixing a tunable ferroelectric with a linear dielectric may modify the electrical properties of the material due to many effects. Two of these whose physics is rather clear are (i) "doping effect",—effect of doping of the ferroelectric lattice via the substitution of the ions of the host material and (ii) "composite effect"—effects of redistribution of the electric field in the material due to the precipitation of the non-ferroelectric phase at the grain boundaries or in the bulk of the material. In addition, two more effects may be active especially in the case of the dielectric loss: (iii) effects of inter-components boundaries, and (iv) effects of mixing-induced modification of microstructure of the ferroelectric component. In a real material and possessing a limited set of experimental data, the distinction among the contributions of these effects is a difficult task. In the following discussion, we will mainly address the first two effects.

The doping effect is limited by the solubility of the additive. The main impact of this effect is the shift of the Curie-Weiss temperature $T_0$ of the material. In the case

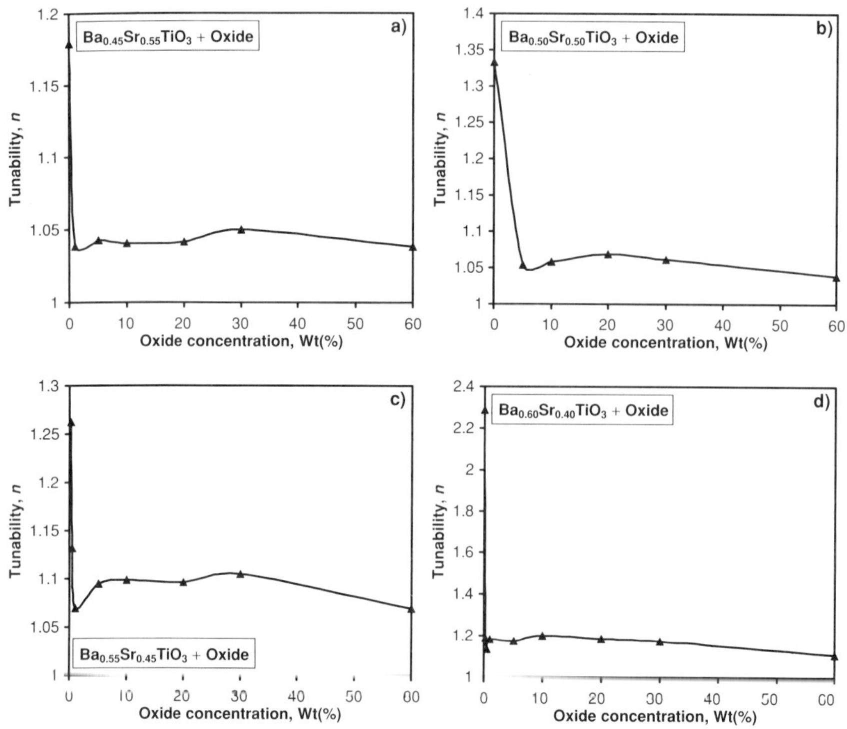


**Fig. 5.5.** Tunability of ferroelectric-dielectric composite material as a function of weight concentration of the dielectric inclusions in the mixture for a)—$Ba_{0.45}Sr_{0.55}TiO_3$/oxide, b)—$Ba_{0.50}Sr_{0.50}TiO_3$/oxide, c)—$Ba_{0.55}Sr_{0.45}TiO_3$/oxide, d)- $Ba_{0.60}Sr_{0.40}TiO_3$/oxide material compositions. RT data.

of the BST-MgO system, the doping effect is limited at least to the MgO concentration below 2%. This is illustrated in Fig. 5.7 where it is seen that, in $Ba_{0.8}Sr_{0.2}TiO_3$ with increasing MgO content, a shift of $T_0$ takes place for dopant concentrations below 2 mol% (0.4 Wt%) [138]. Typically, the non-ferroelectric additives shift $T_0$ down as shown in this figure. If the non-doped material is in its ferroelectric phase, the progressive doping may lead to a non monotonic variation of the permittivity. Such a behavior is seen in the insert in Fig. 5.4d. The abrupt variation of the tunability at small MgO concentrations, which is seen in Fig. 5.5, can also be attributed to the doping effect. The behavior of the permittivity and tunability at higher MgO content is consistent with manifestations of the composite effect. As seen from the result of modeling for various ferroelectric/dielectric composites, the impact of the dielectric component on the permittivity and tunability of the material should be very sensitive to the geometry of its distribution in the host material. Namely, layers of the dielectric perpendicular to the direction of the electric fields impact both the permittivity and tunability much stronger than sphere-type inclusions or columns parallel to the field (see Fig. 3.3a). Comparison of the results of modeling with the experimental data on BST-MgO system (see Figs. 5.4 and 5.5) suggests that the experimental situation is close to the model of the spherical

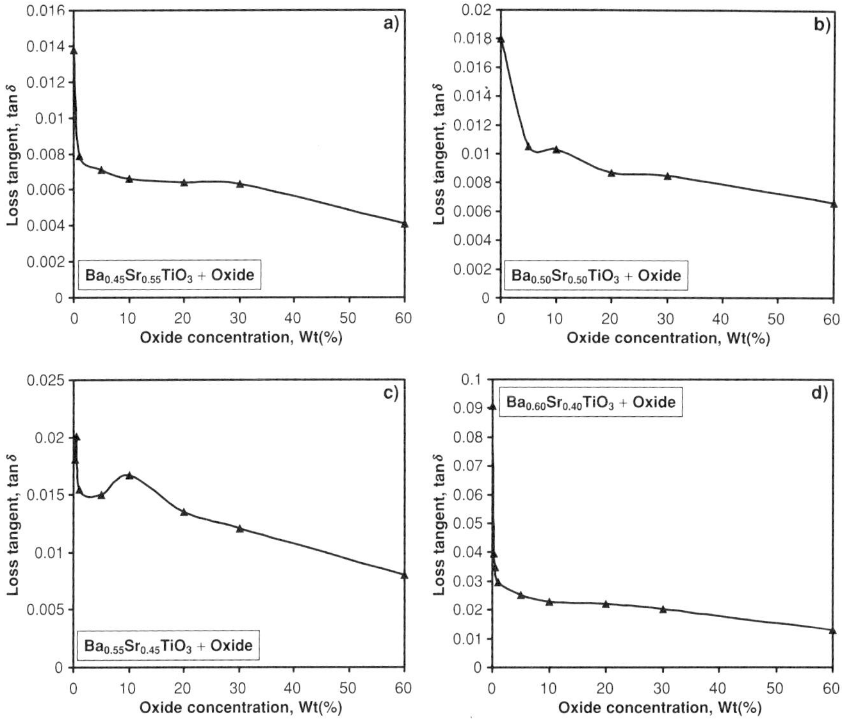


**Fig. 5.6.** Loss tangent of ferroelectric-dielectric composite material as a function of weight concentration of the dielectric inclusions in the mixture for a)—$Ba_{0.45}Sr_{0.55}TiO_3$/oxide, b)—$Ba_{0.50}Sr_{0.50}TiO_3$/oxide, c)—$Ba_{0.55}Sr_{0.45}TiO_3$/oxide, d)—$Ba_{0.60}Sr_{0.40}TiO_3$/oxide material compositions. RT data.

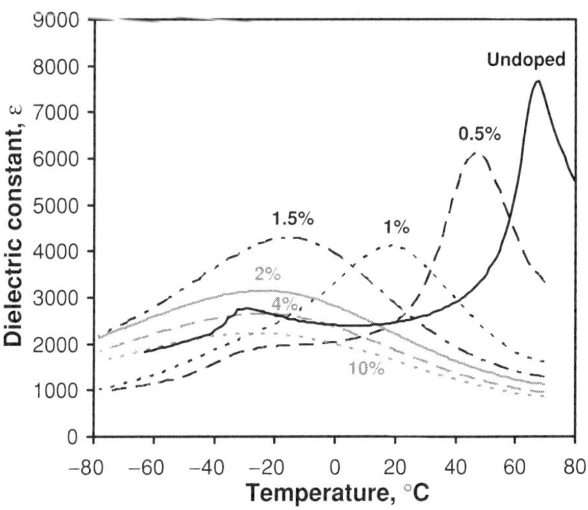


**Fig. 5.7.** Temperature dependences of the dielectric constant of $Ba_{0.8}Sr_{0.2}TiO_3$ ceramics with different levels of MgO content (indicated in mol %) [138].

inclusion. The supporting arguments are as follows. First, the concentration dependence of the permittivity in the range of 1–30 wt% is close to that given by Eq. (3.46). Second, for not too small concentrations, the tunability is weakly concentration dependent in accordance with the prediction of the model.

The data on the concentration dependence of the loss shown in Fig. 5.6 cannot be fully interpreted in terms of the doping-induced shift of $T_0$ and composition effect. The fast drop of the loss tangent at small MgO concentrations may be at least partially attributed to the doping-induced shift of $T_0$ as well as to a reduction of the loss related to the transition from the ferroelectric to paraelectric phase in the case of 0.6/0.4 BST composition[14]. However, the further reduction of the loss with increasing MgO concentration (for higher concentrations) does not look to be consistent with the flat concentration dependence of the tunability, as is clear from the comparison of Figs. 5.5 and 5.6 with the results of modeling the composite effect (Figs. 3.3a and 3.8a). The origin of this disagreement is not clear for the moment. First, the cited theory for the spherical inclusion model is strictly applicable only to the case of small oxide concentrations whereas the experimental data correspond to rather high oxide concentrations. Thus, more advance theory might be in better agreement with the experiment. Alternatively the loss reduction with increasing doping can be attributed to some non-composite mechanisms. These mechanisms may be related to a variation of the population of the point defects of the material with increasing MgO content and to the impact of the additive on the microstructure of the ceramics (e.g. grain size). The loss tangent can be sensitive to these factors. These mechanisms can also contribute to the drop of the loss tangent at the small concentrations.

The overall tunable behavior on the BST/MgO systems under consideration (data from Ref. [137]), as given by the Vendik's Commutation Quality Factor K is illustrated in Fig. 5.8. The presented $K - \varepsilon$ dependences of four different compositions of BST typically reveal two features. First, the initial drop of $\varepsilon$ is accompanied by a substantial drop of the K factor, which can be associated with the shift of $T_0$. Second, the K factor increases on the further reduction of $\varepsilon$. This effect is too strong to be explained by the purely composite effect at least for simple models (c.f. Fig. 3.10). This suggests that the change in $K$ with increasing content of the dielectric in BST relates to other effects than the composite effect. As in the case of the concentration dependence of the loss, it may be related to a variation of the population of the point defects of the material with increasing MgO concentration and to the impact of the additive on the microstructure of the ceramics.

The data presented in Fig. 5.8 correspond to $K$ factors $\leq 300$. These values are not very far from 900 that is required from a material to be of real interest for applications in phase-shifters (Sec 2). Since $K \propto 1/\tan^2\delta$, for the bulk BST ceramic materials discussed above, a two–time reduction of the loss level is actually sufficient to reach the required level of the $K$ factor.

---

[14] Judging from the non-monotonic concentration dependence of the permittivity shown in Fig. 5.4d, at RT, the samples of pure BST 60/40 studied in this work were in the ferroelectric phase whereas some Wt.1% mixing with the oxide shifts the transition temperature below RT. This explains a high loss level in pure material and a strong reduction of the loss with doping, which are shown in Fig. 5.6d.

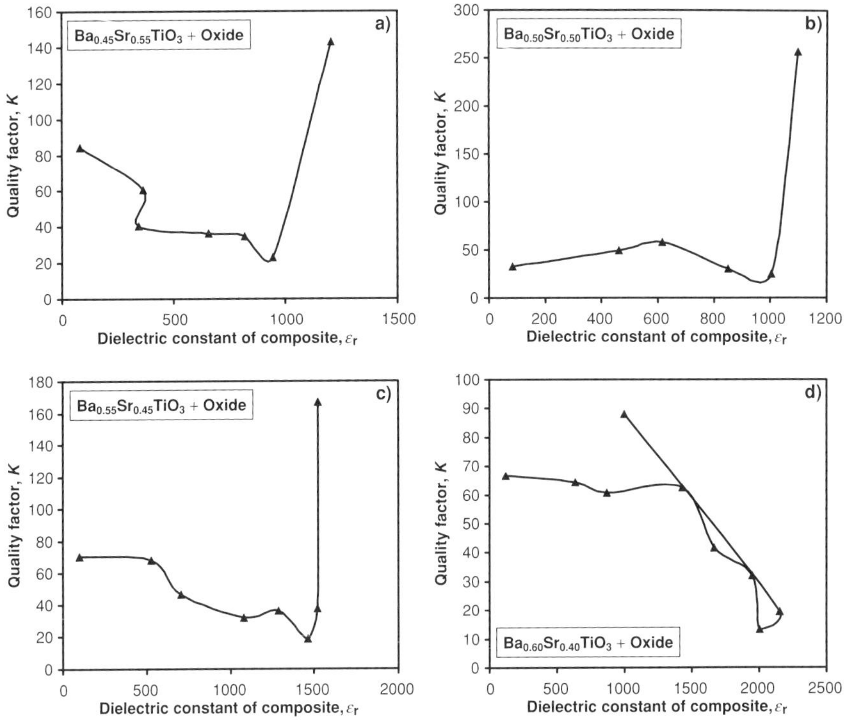


**Fig. 5.8.** Quality factor of ferroelectric-dielectric composite material as a function of dielectric constant of composite for a)—$Ba_{0.45}Sr_{0.55}TiO_3$/oxide, b)—$Ba_{0.50}Sr_{0.50}TiO_3$/oxide, c)—$Ba_{0.55}Sr_{0.45}TiO_3$/oxide, d)—$Ba_{0.60}Sr_{0.40}TiO_3$/oxide material compositions. RT data.

*Other candidates for tunable applications*

Several materials with high dielectric permittivity (>1000) were investigated as possible candidates for the tunable applications.

The compounds $Cd_2Nb_2O_7$ [139] and $(Cd_xPb_{1-x})_2Nb_2O_7$ .[140] with the pyrochlore structure have been recently explored. The material exhibits a very high tunability below a diffuse dielectric anomaly at about 180 K (see Fig. 5.9). However, the lack of information on the dielectric loss at microwave frequencies impedes the assessment of this material as a candidate for tunable applications. Actually, the required low level of the microwave loss may not be achieved due to the relaxor nature of the main dielectric anomaly in this material.

Rather high tunability was reported for double substituted BTO: $(Ba_{(1-x)}Ln_{(x)}Ti_{(1-y)}M_{(y)}O_3$ with donor doping at A site ($Ln = La^{3+}$, $Sm^{3+}$, $Gd^{3+}$, $Dy^{3+}$) and acceptor doping at B-site ($M = Al^{3+}$, $Fe^{3+}$, $Cr^{3+}$) [141]. It was found that, with increasing the $LnMO_3$ content $T_0$ shifts downwards. With 4% substitution of $LnMO_3$ almost all the compositions show $T_0$ near room temperature, all have values of tunability comparable to those of $Ba_{0.6}Sr_{0.4}TiO_3$. The lack of information on the loss at microwave frequencies impedes the assessments of the Commutation Quality factor of these materials.

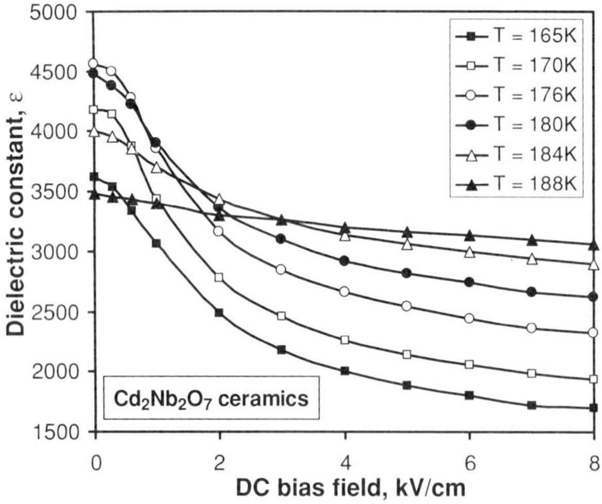


**Fig. 5.9.** dc electric field dependence of dielectric constant of $Cd_2Nb_2O_7$ ceramics measured at 5kHz at temperatures around 180K, the paraelectric-ferroelectric transition temperature [139].

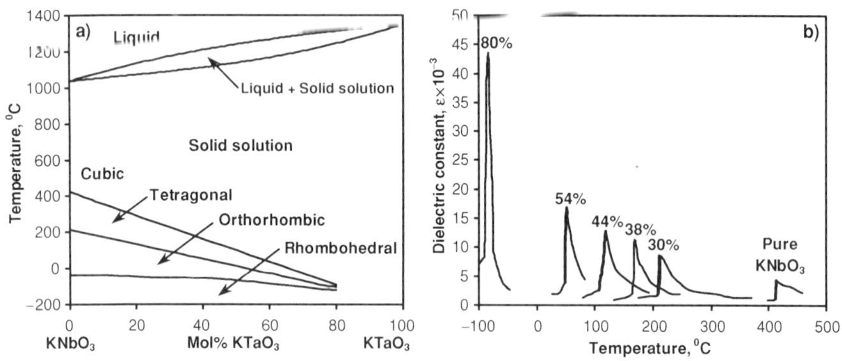


**Fig. 5.10.** a)—Phase diagram of $KNbO_3$—$KTaO_3$ solid solution [142]. b)—Dielectric constant vs. temperature for solid solution single crystals of $KTaO_3$ and $KNbO_3$. The indicated percentage are in moles of $KTaO_3$, Measuring ac field is ~5 V/cm and 10 kHz [142].

As a possible candidate for microwave tunable applications the solid solution $K(Ta,Nb)O_3$ (KTN) is worth mentioning. Being a solid solution of an incipient ferroelectric KTO (very similar to STO) and a ferroelectric $KNbO_3$(with a sequence of phase transitions identical to that in BTO) this material represents a "direct" analog of BST. It exhibits solubility in the full range of concentrations having $T_0$'s in the range for 0 to 700 K [142] (see Fig. 5.10). KTN single crystals undergo sharp first order phase transitions with dielectric permittivity as high as a few thousands in a wide temperature range (see Fig. 5.10). KTN ceramics exhibits a smeared dielectric anomaly with a substantial relaxor-type frequency dispersion [143–145]. The dielectric constant of the ceramics is smaller than in single crystals but still high (see Fig. 5.11), the tunability

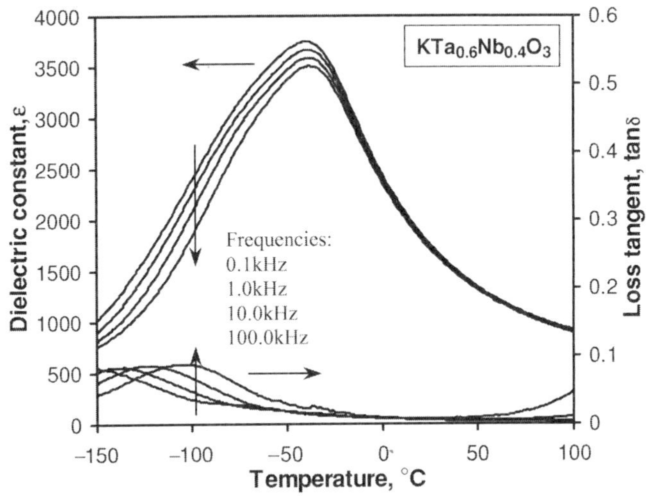


**Fig. 5.11.** Temperature dependences of the relative dielectric permittivity and loss tangent of $KTa_{0.6}Nb_{0.4}O_3$ ceramic at different frequencies [145]. The vertical arrows in the figure indicate the direction of the frequency changing (from small to high frequency).

being even larger than in BST ceramics (for the same value of the permittivity and applied field). For example, $K(Ta_{0.6}Nb_{0.4})O_3$ ceramics at RT and $\varepsilon_{RT} = 1700$ shows $n = 1.2$ at a dc bias field of 11 kV/cm [145]. On the other hand, in $Ba_{0.65}Sr_{0.45}TiO_3$ ceramics according to Eq. (5.2), the same values of the tunability and permittivity correspond to a dc bias field of 28 kV/cm. The matter of concern with KTN ceramics is the relaxor-like nature of the material, which is seemingly related to difficulty of processing high-volatility potassium-containing material. This feature of KTN ceramics may result in a dispersion of the tunability and high dielectric loss at microwave frequencies.

## 5.2. Ferroelectric thin films

The problem of the high bias voltage required for efficient tuning of the bulk ferroelectric based microwave components and the interest in a further miniaturization of the ferroelectric based microwave devices inspired wide investigations of the ferroelectric thin film materials since the beginning of the 1980s. The potential to produce in one technological cycle microwave integrated circuits based on ferroelectric thin films, thus reducing manufacturing costs, and the possibility to combine the microwave circuits with well developed semiconductor integrated circuits have additionally motivated an activity in the field of ferroelectric thin films[15].

The discovery of high-temperature superconductivity (HTSC) in complex metal oxides in 1987 has stimulated a special interest in strontium titanate and potassium tantalate thin film ferroelectrics. The structural and chemical compatibility of the superconductive oxides (in particular, $YBa_2Cu_3O_7$) with STO and KTO ferroelectrics

[15] Here and further the term "*thin film*" means the film which is thinner than 2–3 μm.

assures a high-quality interface between HTSC electrodes and ferroelectric thin films. This improves the quality of the films from the point of view of the dielectric loss and offers a substantial scope of the implementation of cryogenic microwave circuits using the unique properties of nonlinear dielectrics and superconducting materials.

On the other hand, the complexity of achieving low temperatures and the cost of the cryogenic plants necessitated the development of ferroelectric thin film materials with operating temperatures close to room temperature. This made BST films attractive. These films were found to exhibit an excellent tunable behavior even at room temperature (in the case of 40–60% concentration of barium), however, the properties of the bulk BST and BST thin films were found to be rather different. The present section addresses the dielectric properties of ferroelectric thin films with the main focus on STO and BST thin films.

The quality of the ferroelectric thin films also strongly depends on the substrates used for film deposition. The typical substrates for STO and BST thin films are single crystal sapphire, lanthanum aluminate, and magnesium oxide. These substrates have similar values of the lattice parameters with those of the aforementioned ferroelectrics, which provide the epitaxial growth of the films on the substrates, and high values of the thermal conductivity coefficient, which increase the power handling capability of the high power microwave devices based on these films [146].

*Dielectric permittivity of ferroelectric thin films*

The dielectric properties of ferroelectric thin films deposited onto single crystal substrate usually differ from those of bulk ferroelectric materials in relative dielectric permittivity and in tunability.

An example of direct comparison of the properties of the BST thin films deposited by RF magnetron sputtering technique and bulk ceramic samples was offered by Outzourhit *et al.* [147]. The films were deposited onto $LaAlO_3$ (100) substrates at 550 °C and showed (100)-preferred growth orientation. It was found that the values of the dielectric constants and electrical tunability were more than an order of magnitude lower than those of bulk samples with the same composition (see Fig. 5.12).

One should also note that the temperature corresponding to the maximum of dielectric constant of the film was 70 K which was about 15 K smaller than that of bulk ceramics samples with the same concentration of barium (see Fig. 5.3b).

The thickness dependences of properties of ferroelectric thin films were investigated by several groups. Li *et al.* [148] and the group of Kingon [93, 149] investigated the influence of the film thickness on its dielectric properties. It was found that the dielectric permittivity of the films can substantially change with the thickness of the film (see Fig. 5.13).

The detailed analysis of the dielectric response of BST thin films of different thicknesses and different compositions has been performed by Streiffer *et al.* [93], Basceri *et al.* [150], Vendik *et al.*[82], and others. As has already been discussed (see Section 3.4) and in those works, one of the possible causes for the suppression of the effective dielectric constant of the ferroelectric film is the so-called passive layer effect (see Section 3.4.1). This effect corresponds to a useful formula for the inverse capacitance density, A/C of a ferroelectric film based parallel plate capacitor containing (real or

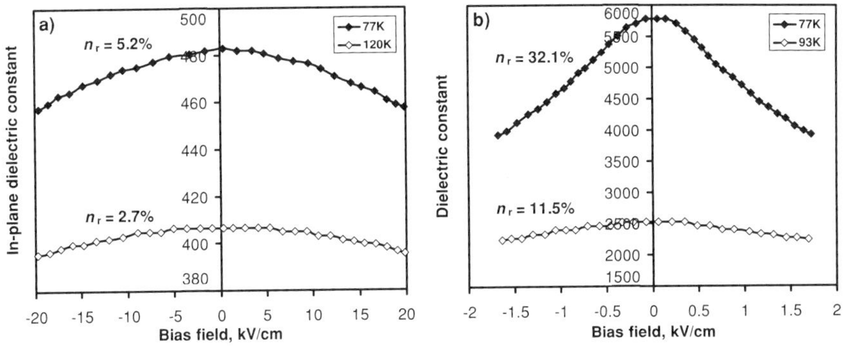


**Fig. 5.12.** Electric field dependences of the in-plane dielectric constant of (a) $Ba_{0.1}Sr_{0.9}TiO_3$ film 0.7 μm thick deposited by RF magnetron sputtering onto $LaAlO_3$ substrate; and of (b) bulk polycrystalline $Ba_{0.1}Sr_{0.9}TiO_3$ sample [147]. Measurement temperatures and maximum relative tunability achieved are indicated in the graphs.

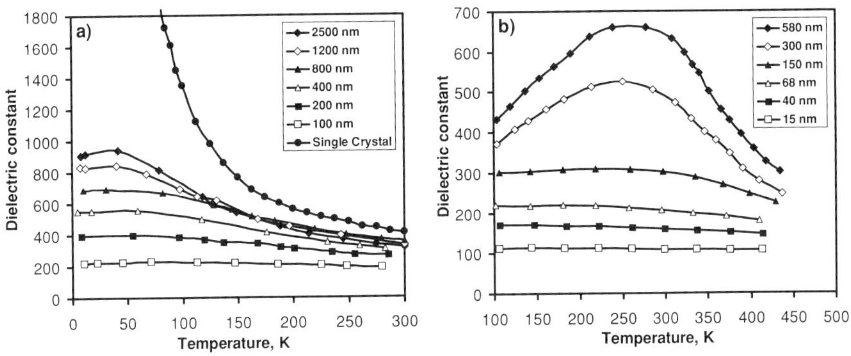


**Fig. 5.13.** (a) Out-of-plane dielectric constant as a function of temperature for $SrTiO_3$ films of difference thickness[148] grown on $SrRuO_3/LaAlO_3$ by laser pulse deposition technique and bulk dielectric constant of [100] $SrTiO_3$ single crystal sample [127]; (b) Out-of-plane dielectric constant as a function of temperature for $Ba_{0.7}Sr_{0.3}TiO_3$ films of different thickness deposited by metalorganic chemical vapour deposition technique onto $Pt/SiO_2/Si$[149] (in bulk $Ba_{0.7}Sr_{0.3}TiO_3$ material $T_c = 305$K [14]).

effective) passive layers near to the electrodes:

$$\frac{A}{C} = \frac{h}{\varepsilon_0 \varepsilon_f} + \frac{2\delta}{\varepsilon_0 \varepsilon_{pass}},$$

where $C$ and $A$ are the total capacitance and area of the capacitor, $h$ and $\varepsilon_f$ are the thickness and dielectric constant of the ferroelectric thin film, $\delta$ and $\varepsilon_{pass}$ are the thickness and dielectric constant of the passive layers; the formula is obtained from Eq. (3.64).

It is seen that by measuring the capacitance of the parallel plate capacitor based on ferroelectric thin film and plotting the inverse capacitance density $A/C$ as a function of film thickness $h$, one can obtain the thickness to dielectric constant ration $\delta/\varepsilon_{pass}$ of the passive layer, by interpolating the $A/C$ vs. $h$ curve to zero film thickness. Figure

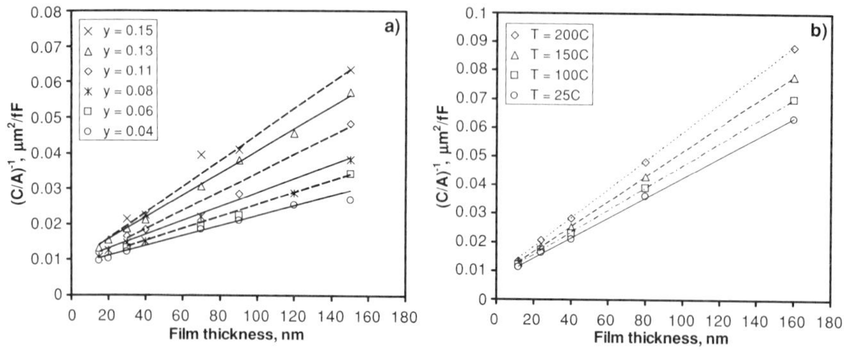


**Fig. 5.14.** (a) Film thickness dependences of the inverse zero-bias capacitance density of $(Ba_{0.7}Sr_{0.3})Ti_{1+y}O_{3+z}$ film based plate capacitors at different Ti content $y$ [93]; (b) The inverse of the zero-bias capacitance density of $Ba_{0.7}Sr_{0.3}TiO_3$ film based parallel plate capacitors as a function of film thickness at different temperatures [150].

5.14 shows the measured inverse capacitance densities of BST film based parallel plate capacitors as a function of film thickness for different Ti contents (Fig. 5.14a) and different temperatures (Fig. 5.14b). From these figures one readily finds that in both cases the ratio $\delta/\varepsilon_{\text{pass}}$ is about 0.5 Å. This value is compatible with the result of the analysis of the dielectric properties of BST films offered by Vendik (see Ref. [82]). However, based on these data, one cannot decide which of the mechanisms for this kind of size effect (surface blocking model, "electrode" mechanism, or depletion effect, see Sec. 3.4.1) operates in this material since all these mechanisms deal with the values of $\delta/\varepsilon_{\text{pass}}$ of this order.

The dielectric properties of the ferroelectric thin films become somehow closer to those of bulk materials when the thickness of the films is higher than 2 micrometers (see Fig. 5.13). Bearing in mind microwave tunable applications of the films, an important remark should be made. These applications usually imply the planar structure of varactors (see Figs. 4.1e,d and 4.2a), where the in-plane component of the permittivity matters. As was pointed out in Sec. 3, this component of the permittivity is much less sensitive to the passive layer effects than the out-of-plane one. Thus, the size effect on the out-of-plane component of the permittivity discussed above may not be relevant to the real situation in the microwave devices. To evaluate the strength of the size effect on the in-plane component of the permittivity, experimental information on the thickness dependence of this component is required. The relevant experimental information available in the literature is very scarce. Data on the thickness dependence of the in-plane tunability in BST films ($h = 22 - 1150$ nm) were recently reported by Bellotti *et al.* [151]. The authors found a significant reduction of the tunability with decreasing film thickness, which was also accompanied by an increase of the absolute value of the in-plane strain disregarding to its sign. The physical origin of this size effect remains unclear, since none of the presently available theories can predict such kind of behavior. Clearly, the passive layer scenario might be summoned however this scenario would require the existence of an unrealistically thick passive layer.

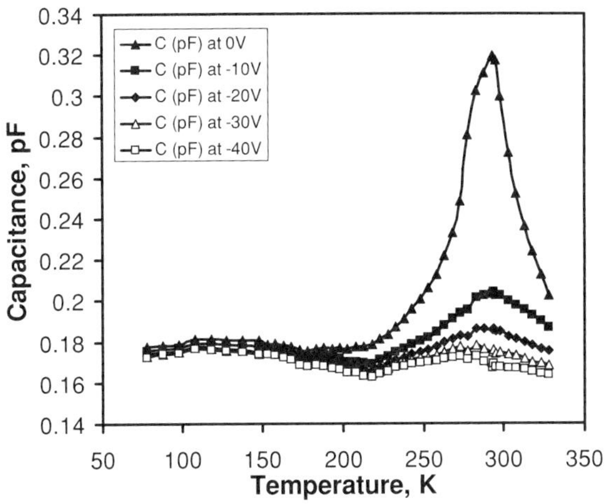


**Fig. 5.15.** Capacitance of the planar inter-digital capacitor (capacitive gap 12 μm) based on $SrTiO_3$ thin film deposited by MBE on (110) $DyScO_3$ substrate as a function of temperature under different dc bias applied [152].

Though the situation discussed above where dielectric permittivity of the films is reduced compared to its bulk counterpart is rather typical, it is not the only one possible. An example of an alternative behavior was recently reported by Schlom *et al.* who investigated the dielectric properties of STO thin films deposited by the molecular beam epitaxy technique onto (110) $DyScO_3$ substrates [152]. They observed a substantial shift of the phase transition temperature in the film (up to 300 K, see Fig. 5.15) and consequently a high tunability of planar capacitors based on this film even at room temperature ($C(0)/C(U_{max}) \approx 2$ at $T = 300$ K). These films were reported to be strongly strained exhibiting at RT an in-plane lattice constant of $a = b = 3.92$ Å (the RT bulk value of STO lattice constant is 3.905 Å), which corresponds to an in-plane misfit tensile strain of about 0.4%. This value is comparable with the in-plane misfit tensile strain of 0.7% needed to shift the transition temperature in STO films up to room temperature (see Sec. 3.4.3.). Thus, the effect of mechanical clamping of the film by the substrate seems to play an essential role in the change of the transition temperature and dielectric permittivity of the film compared to the bulk material.

At present, the effect of the misfit strain is often considered as the main reason for the difference between the dielectric properties of the ferroelectric films grown on different substrates and/or objected to different annealing procedures. For example, Chang *et al.* investigated the effect of annealing on dielectric properties of $Ba_{0.5}Sr_{0.5}TiO_3$ thin films deposited by the pulsed laser deposition technique onto (100) MgO and (100) $LaAlO_3$ substrates [153], the results are listed in Table 5.2.

One can see that the effect of annealing is different in the case of different substrates. For the same capacitor geometry, the as-deposited BST film on MgO shows a higher capacitance than as-deposited BST film on $LaAlO_3$, which means a higher dielectric constant. After annealing, the capacitance $C$ and tunability of BST film on MgO decrease whereas the opposite effect is observed for films deposited onto $LaAlO_3$.

**Table 5.2.** 10 GHz dielectric properties of as-deposited and low temperature ($\leq 1000°C$) annealed $Ba_{0.5}Sr_{0.5}TiO_3$ films on MgO and $LaAlO_3$ substrates. (The capacitors have an interdigitated structure with eight fingers with 6 μm gap, 80 μm length, and 10 μm width) [153]

| | $Ba_{0.5}Sr_{0.5}TiO_3$ on MgO | | $Ba_{0.5}Sr_{0.5}TiO_3$ on $LaAlO_3$ | |
|---|---|---|---|---|
| | as-deposited | annealed | as-deposited | annealed |
| $C_0$, pF | 0.775 | 0.387 | 0.440 | 1.281 |
| $n$ ($E = 67$ kV/cm) | 1.85 | 1.53 | 1.24 | 2.16 |
| tan $\delta$ ($E = 0$) | 0.14 | 0.07 | 0.05 | 0.17 |
| tan $\delta$ ($E = 67$ kV/cm) | 0.06 | 0.03 | 0.02 | 0.05 |

**Table 5.3.** Low-frequency (1 MHz) dielectric constants ($\varepsilon$) at 300 K and at the peak temperature ($T_{max}$) for various $Ba_{0.4}Sr_{0.6}TiO_3$ thin films [154]

| Film | $\varepsilon$(300 K ) | $\varepsilon(T_{max})$ | $T_{max}$, K |
|---|---|---|---|
| As-dep. BST/MgO | 2620 | 4130 | 242 |
| Annealed BST/MgO | 2290 | 6020 | 227 |
| As-dep. BST/LAO | 2960 | 3320 | 267 |
| Annealed BST/LAO | 4660 | 5170 | 277 |

The authors attributed the large difference in the dielectric behavior for the BST thin films deposited onto MgO and $LaAlO_3$ substrates to the difference in the stress in the films, which in turn is due to a difference in lattice constants and thermal expansion coefficients of the substrates.

Similar behavior of the ferroelectric film dielectric properties after post-deposition annealing has been also observed by Carlson et al. in $Ba_{0.4}Sr_{0.6}TiO_3$ thin films deposited by laser ablation on MgO and $LaAlO_3$ substrates (see Table 5.2) [154].

Though the explanation of the impact of annealing and the substrate material on the lines of the aforementioned "mechanical" scenario is possible, without information on the in-plane lattice constants of the films this scenario cannot be justified. A recent comprehensive study [155] of similar systems actually questions the importance of the mechanical effects in this situation. In this study, the influence of annealing procedure on the dielectric properties of $SrTiO_3$ thin films deposited by laser ablation onto (100) MgO and (100) $LaAlO_3$ substrates was addressed. An example of the temperature dependences of the dielectric permittivity in these films is presented in Fig. 5.16.

One can see that STO thin films deposited under the same conditions onto different substrates show completely different behavior. In the case of $LaAlO_3$ substrate we can see an evident maximum in the temperature dependence of the permittivity, which suggests the appearance of ferroelectricity in STO thin films, whereas in the case of MgO substrate the STO thin films exhibit a monotonic temperature dependence of the permittivity.

As in the work of Chang et al. [153], an impact of the annealing procedure on film properties has been observed for $SrTiO_3$ thin films [155], which was qualitatively the same as that reported by Chang et al. for BST films. A post-deposition annealing at

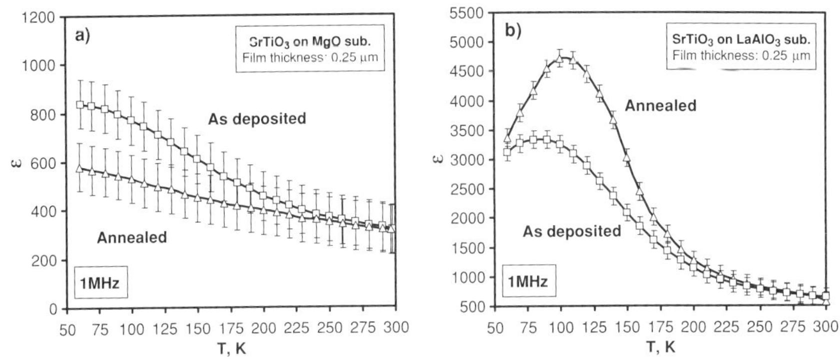


**Fig. 5.16.** Relative in-plane dielectric constant of the $SrTiO_3$ thin film deposited on MgO (a) and $LaAlO_3$ (b) substrates as a function of temperature measured at 1MHz [155].

**Table 5.4.** The crystal cell lattice parameters for $SrTiO_3$ thin films on different substrates [155].

| Crystal cell lattice parameters (Å): | $SrTiO_3$ on $LaAlO_3$ Before annealing | $SrTiO_3$ on $LaAlO_3$ After annealing | $SrTiO_3$ on MgO Before annealing | $SrTiO_3$ on MgO After annealing |
|---|---|---|---|---|
| *a* | 3.903 | 3.903 | 3.894 | 3.899 |
| *b* | 3.912 | 3.909 | 3.896 | 3.902 |
| *c* | 3.905 | 3.905 | 3.909 | 3.909 |

1100 °C in the oxygen atmosphere during 4 hours results in a shift of the temperature peak up in the case of $LaAlO_3$ substrate (Fig. 5.16b), which means increase of the film dielectric constant at certain temperature, and a decrease of the film dielectric constant in the case of MgO substrate (Fig. 5.16a).

To clarify the origin of the annealing-induced variation of the permittivity, in addition to dielectric measurements, the crystal cell lattice parameters of these STO thin films was measured by grazing-angle X-ray diffraction techniques [156] (see Table 5.4).

One can see that the annealing procedure results in an increase of the in-plane lattice constant in the case of MgO substrate[16] and in a decrease in the case of $LaAlO_3$. As it follows from the thermodynamic theory for the influence of misfit strain in the film on its dielectric properties and phase transition temperature [98, 157](see Sec. 3.3.3), the increase of the in-plane lattice constant should lead to an increase of the in-plane Curie-Weiss temperature and correspondingly, the decrease of the in-plane lattice constant should lead to its decrease[17]. It is clearly seen that the behavior of the dielectric permittivity in these systems, as illustrated in Fig. 5.16, is just opposite to this trend. This suggests that in this case the misfit strain does not play the principal role

[16] The origin of the anisotropy of the in-plane lattice constants ($a \neq b$) of the films deposited onto cubic MgO substitutes remains unclear.

[17] This can be readily checked using Eq. (3.74) and the materials parameters of STO listed in Sec. 3.4.3.

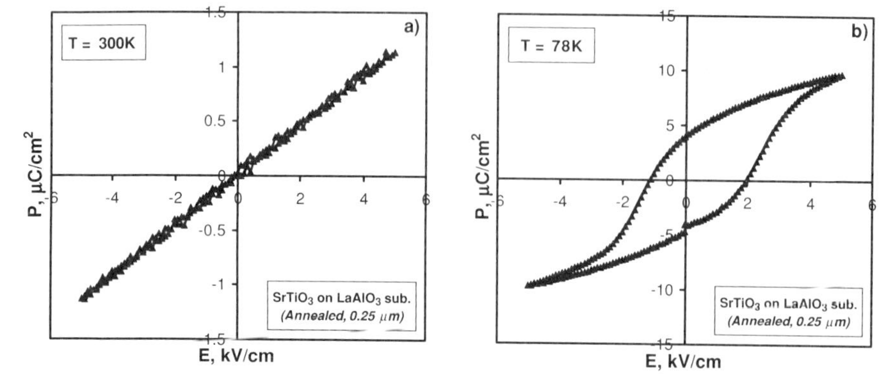


**Fig. 5.17.** Hysteresis loops for $SrTiO_3$ thin film based planar capacitors at room (a) and liquid nitrogen (b) temperatures [155].

in the shift of the Curie-Weiss temperature of the films. A possible explanation for the impact of annealing and the substrate material on the dielectric properties of these films is non-stoichiometry-driven shifts of the Curie-Weiss temperature, which are sensitive to the annealing and the substrate material.

The maxima in the temperature dependence of the permittivity mentioned in the discussion of the dielectric properties of STO films [152, 155] are strongly believed to be a manifestation of a ferroelectric phase transition. For instance, to prove the presence of ferroelectricity in the $SrTiO_3$ thin film on $LaAlO_3$ substrate [155], the hysteresis loops at room and at liquid nitrogen temperatures (the temperatures which are higher and lower than the supposed phase transition temperature) were experimentally tested (see Fig. 5.17). In Fig. 5.17 one can readily see the presence of the remanent polarization at liquid nitrogen temperature whereas at room temperature the polarization response is linear. Also, the observed value of the remanent polarization in the film is in a good agreement with the theoretical estimates which can be obtained from the Landau theory of STO films [155]. ($P_R \approx 5.6$ ìC/cm$^2$ for $(T - T_c) \approx 20$ K). It is worth mentioning that the appearance of the ferroelectricity in STO films is not rare. For example, the presence of the ferroelectric phase in $SrTiO_3$ thin films grown by RF magnetron sputtering technique on (001)-oriented $YBa_2Cu_3O_{7-x}$ bottom electrode deposited onto an STO substrate has been observed by D. Fuchs *et al.* [158]. They reported a maximum in the temperature dependence of dielectric constant around 100 K and non-zero remanent polarization in hysteresis loops at temperatures below 120 K.

Summarizing the discussion of the present section one may identify three phenomena influencing the dielectric response of tunable thin films: (i) effects manifesting themselves in a reduction of the permittivity similar to that caused by a thin dielectric layer at the film surface (surface blocking model, "electrode" mechanism, and depletion effect), (ii) misfit-strain-driven variation of the Curie-Weiss temperature, and (iii) non-stoichiometry-driven shifts of the Curie-Weiss temperature. This list should be completed with a defect-induced suppression of the dielectric permittivity in ferroelectrics (Sec. 3.1.2). Being also active in bulk materials, according to the number of authors [4, 159, 160] this effect seems to be of importance for thin ferroelectric films,

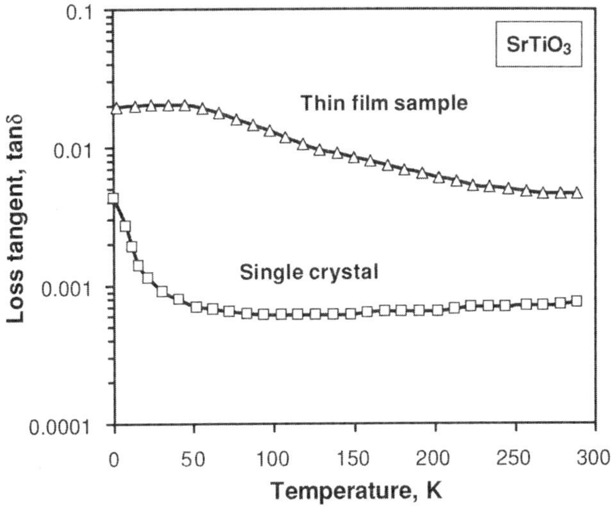


**Fig. 5.18.** Estimated loss tangent of the $SrTiO_3$ single crystal and thin film as a function of temperature at frequency $f = 10$ GHz and bias field $E = 0$ [2].

especially when the suppression of permittivity is large (like in Fig. 5.13a at very low temperatures).

*Dielectric loss in ferroelectric thin films*

High values of the dielectric loss in thin films in comparison with single-crystals and bulk ceramics make a problem which sets limits to the practical application of ferroelectric thin films in microwave devices. According to Vendik et al. [2] and other experimental works [115, 161–163] the dielectric loss in the ferroelectric thin films are roughly one order of magnitude higher than those in the bulk samples (see Fig. 5.18).

Lemaitre *et al.* have investigated at microwave frequencies (2 GHz) the dc field dependences of the dielectric constant and loss tangent of the $SrTiO_3$ thin film deposited by RF magnetron sputtering onto $LaAlO_3$ substrates [163]. Plotting the loss tangent as a function of the film dielectric constant taken at the same value of the electric field, one can find that the loss tangent increases almost linearly with the dielectric constant (Fig. 5.19a). A similar behavior has been observed at radio frequencies (10 kHz) by Moeckly [30] in $SrTiO_3$ thin films deposited by laser ablation technique (see Fig. 5.19b). The data from Ref. [153] given in Table 5.1 also confirm the trend: the higher the permittivity, the higher the loss tangent.

The level of the microwave loss in thin films and its dependence on the permittivity is compatible with the existing knowledge on the main mechanisms of dielectric loss in ferroelectrics reviewed in the previous parts (Chapters 3.2 and 3.4). The excessive loss in the films compared to single crystals clearly attests to their mainly extrinsic origin. The observed increasing permittivity dependence of the loss tangent corresponds to the theoretical predictions for all loss mechanisms except the universal-relaxation–law, which does not look decisive at microwave frequencies. The increasing permittivity dependence of the loss tangent taken into account along with the decreasing dc field

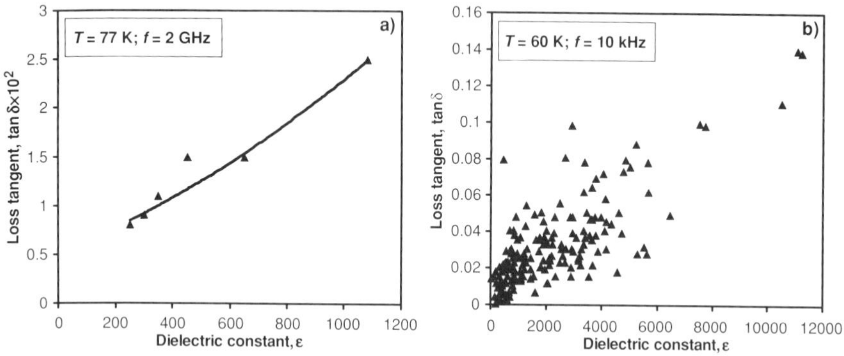


**Fig. 5.19.** (a) Loss tangent of the $SrTiO_3$ thin film at microwave frequencies (2 GHz) as a function of the in-plane dielectric constant of the film (the dielectric constant was changing by applying dc bias field to the samples) [163]. (b) Loss tangent of the $SrTiO_3$ thin film at radio frequencies (10 kHz) as a function of the film dielectric constant (the each points correspond to the different samples with different dielectric constant) [30].

dependence of the permittivity readily suggest a decreasing dc field dependence of the loss tangent. This trend is typically observed experimentally (see e.g. [163]). However, in high quality films, it is not always observed. This situation was reported by K. Astafiev *et al.* for the loss tangent at microwave frequencies (8 GHz) in STO thin films deposited by laser pulse deposition technique onto MgO and $LaAlO_3$ substrates [164]. It has been found that when the initial value of the loss tangent (at zero bias field) is high (more than 0.01) the dielectric loss decreases with the applied field (Fig. 5.20b), at the same time, when the loss tangent is small in the absence of a bias field, one can observe field independence or an increasing field dependence of dielectric loss (Fig. 5.20a). This behavior can be attributed to a manifestation of the dc-field-induced quasi-Debye loss (see Sec. 3.2.1) documented for single crystals of STO and KTO (see Sec. 5.1): in better quality (lower loss) films the contribution of this mechanism, which rapidly increases with the increasing dc field, can be seen on the background of the extrinsic loss whereas it is impossible to detect it in the films with a higher (extrinsic) loss level.

Results on frequency dependence of the dielectric loss in thin films were offered by Baniecki *et al.* [75]. They found that the loss tangent of $Ba_{0.7}Sr_{0.3}TiO_3$ thin film does not depend on the frequency in the range of radio frequencies (1 Hz–100 MHz). For microwave frequencies (1–20 GHz) their data suggest an increasing frequency dependence of the loss tangent[18] (Fig. 5.21a). A similar behavior of the loss tangent at microwave frequencies was observed by Razumov *et al.* [165] in $Ba_{0.3}Sr_{0.7}TiO_3$ thin film deposited by RF sputtering onto ceramic alumina and single crystal sapphire substrates (Fig. 5.21b).

The frequency dependence of the loss in thin films discussed above can be rationalized in terms of the existing knowledge on the loss mechanisms in ferroelectrics (see Sec.3.2). As it has already been shown in Sec. 3.2, for all of the loss mechanisms (intrinsic and extrinsic) the dielectric loss tangent is almost a linear function of the

[18] The interpretation of these data is not clear. According to the authors of the paper the measured increasing frequency dependence of the loss tangent may be an artifact of the measurement setup.

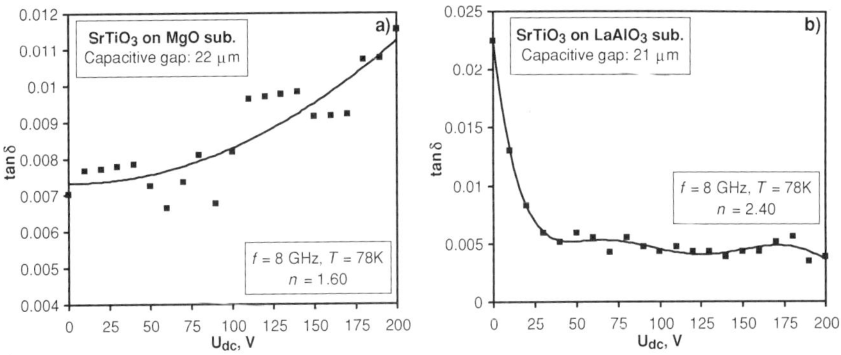


**Fig. 5.20.** Field dependences of the loss tangent of $SrTiO_3$ thin films deposited by laser ablation on to MgO (a) and $LaAlO_3$ (b) substrates [164].

frequency, except the so-called universal-relaxation-law loss mechanism. The data shown in Fig. 5. 21a may be interpreted as a crossover (around $f = 1$ GHz) between the almost flat area, which probably corresponds to universal-relaxation-law loss mechanism, and an increasing frequency dependences of the loss tangent, which in turn corresponds to others loss mechanisms. In Fig. 5.21b one can see a linear frequency dependence of the loss tangent of BST thin film in the range of 1–30 GHz. It suggests that the aforementioned universal-relaxation-law loss mechanism does not contribute much to the total balance of the dielectric loss at microwave frequencies.

*Dependence of the ferroelectric thin film properties on the film preparation conditions and ways to improve the quality of the grown films*

Obviously, many properties of ferroelectric films strongly depend on the conditions of the film deposition. The properties relevant to tunable applications are in particularly sensitive in this respect. Razumov *et al.*, for example, showed that the dielectric properties of the BST thin films could be substantially changed only by changing the deposition temperature (see Fig. 5.22) [165]. It is seen in Fig. 5.22 that, in the case of $Ba_{0.5}Sr_{0.5}TiO_3$ thin film, a change of the deposition temperature by 20% leads to a

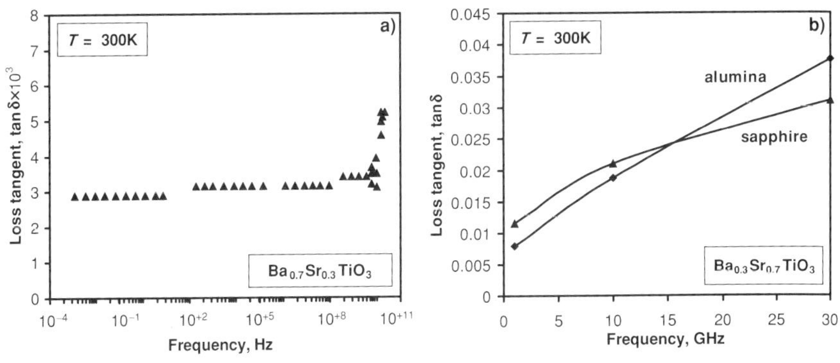


**Fig. 5.21.** Frequency dependences of the loss tangents of (a) $Ba_{0.7}Sr_{0.3}TiO_3$ thin films [75]; and (b) $Ba_{0.3}Sr_{0.7}TiO_3$ thin films deposited onto alumina and single crystal sapphire substrates [165].

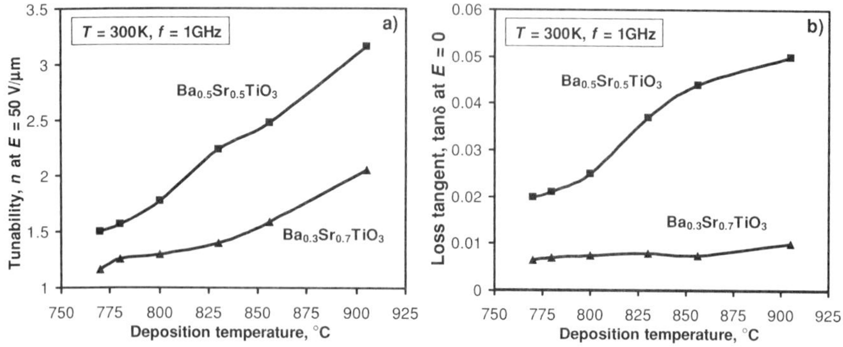


**Fig. 5.22.** Dependence of the electrical tunability (a) and loss tangent (b) of the BST thin films based planar capacitors on deposition temperature [165].

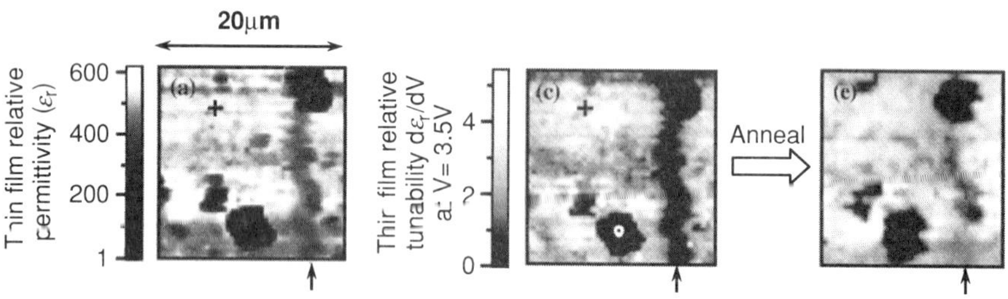


**Fig. 5.23.** Distribution of the dielectric permittivity before annealing (a) and tunability before annealing (c) and after annealing (e) of the $Ba_{0.6}Sr_{0.4}TiO_3$ thin film along the surface of substrate measured at frequency 7.2GHz [123].

more than 2 times increase of the tunability, however, the loss tangent also increases more than 2 times. This behavior may be attributed to the influence of the deposition temperature on the film stoichiometry and defect population in it, which in turn affect the Curie-Weiss temperature of the material. In the case of the data presented in the Fig. 5.22 an increase of the deposition temperature should result in a shift of the Curie-Weiss temperature up, which according to the existing theories (see Sec.3) will lead to an increase of the permittivity, loss, and tunability.

Along with the dependence of the film properties on deposition conditions there exist problems with the homogeneity of grown ferroelectric films. The dielectric response of the films can significantly differ over the area of the substrate. Steinhauer *et al.* investigated by scanning microwave microscopy described in Sec. 4.4 the distribution of the dielectric permittivity and electrical tunability of 370 nm thick $Ba_{0.6}Sr_{0.4}TiO_3$ films across the $LaAlO_3$ substrate, where the films were deposited by pulsed laser deposition technique (see Fig. 5.23) [123]. They found that there exist regions in the film with a reduced dielectric permittivity and tunability. This was attributed to a higher density of defects or to smaller thickness of the film in these regions. It was also shown [123] that the post-deposition annealing procedure (e.g. 650 °C, in the air) lead to an improvement of the homogeneity of film properties and to a decrease of the "bad" regions with low dielectric permittivity and tunability (see Fig. 5.23).

**Table 5.5.** Effect of Mg doping on the dielectric properties of $Ba_{0.6}Sr_{0.4}TiO_3$ thin films. (Film thickness: ~0.26 μm, room temperature measurements at 100 KHz)

| Mg (mol %) | $\varepsilon$ | $\tan\delta$ | $n$ ($E = 20$ V/μm) | $K$ |
|---|---|---|---|---|
| 0 | 450 | 0.013 | 1.39 | 650 |
| 5 | 386 | 0.007 | 1.21 | 730 |
| 20 | 205 | 0.009 | 1.09 | 84 |

**Table 5.6.** Dielectric properties of pure and multilayer MgO:BST thin films at 1 MHz and room temperature [80]

| Film: | $\varepsilon$ | $\tan\delta$ | $n$ | $T_{max}$, K | $K$ |
|---|---|---|---|---|---|
| Pure BST | 2714 | 0.0215 | 2.08 | 267 | 1213 |
| MgO:BST layered | 1729 | 0.0049 | 1.34 | 258 | 3593 |

Another possible way to improve the quality of ferroelectric thin films from the point of view of its dielectric properties is doping of the film by different dopants like magnesium or manganese. Joshi *et al.* have reported the microstructural and dielectric properties of Mg-doped $Ba_{0.6}Sr_{0.4}TiO_3$ thin films prepared by the metalorganic solution deposition technique [166]. They observed an increase of the quality factor at 5 mol % content of Mg in the mixture, and a decrease of that factor with further increasing the Mg content in the mixture up to 20 mol % (see Table 5.5).

The behavior substantially differ from that of the bulk ceramics of the same composition discussed in Section 5.1 and it can not be interpreted in terms of simple doping and composite effects.

A different behavior of a multicomponent ferroelectric film was recently reported by Jain *et al.* [80] (see Table 5.6) for a layered structure of sol-gel-derived $Ba_{0.5}Sr_{0.5}TiO_3$ thin films and MgO thin layers (four layers of BST each 70 nm thick and three layers of MgO, 40 nm each). A planar capacitor sturucture was used for the characterization so that, electrically, the system was equivalent to the columnar (parallel) composite model (see Secs. 3.1.3 and 3.2.3). The composite film yielded a three times higher value of the quality factor $K$ than an identically processed pure BST film. At the same time, a shift of phase transition temperature was not practically observed. This behavior by no means can be attributed to a manifestation of the composite effect which predicts for the parallel composite a weakly decreasing $\varepsilon$-dependence of the $K$-factor. Specifically, the observed reduction of the loss is too large to be explained by the composite effect. Two mechanisms can be invoked to rationalize the reported behavior of the $K$ factor and loss: a small mutual doping effect at the borders of layers and an improvement of the film morphology (e.g, grain size). Indeed, Jain *et al.* indicated a better densification of the composite films.

*New technologies and alternative thin film ferroelectric materials for tunable microwave applications*

The desire to combine the advantages of the thin films with the high quality of single-crystals has inspired the development of a completely new technique of thin

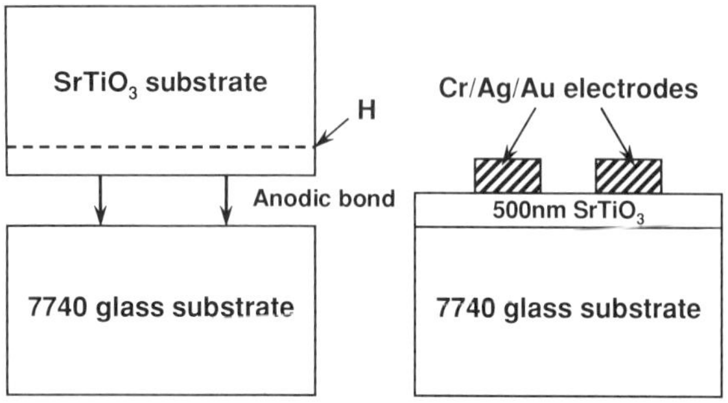


**Fig. 5.24.** Interdigitated finger capacitor fabrication approach using separation by hydrogen ion implantation [168].

ferroelectric layer production: a single-crystal separation by hydrogen implantation or so-called "Smart-Cut" process [167]. The typical "Smart-Cut" process procedure consists of a few main steps:

- Ion implantation of hydrogen ions in a single-crystal STO wafer;
- A bonding at room temperature of the STO wafer to another wafer, for example to an insulating glass substrate;
- Two-step heat treatment of the two bonded wafers: During the first step (400–600 °C), the implanted STO wafer splits into two parts: a thin layer of monocrystalline strontium titanate remaining bonded to glass substrate and the rest of STO wafer. The second high temperature treatment phase (>1000 °C) aims to strengthen the chemical bonds;
- Polishing: after splitting, the thin monocrystalline layers exhibit a microroughness which makes chemo-mechanical polishing of the surface necessary.

Recently, Kub *et al.* characterized interdigitated planar capacitors based on a 500 nm thick monocrystalline strontium titanate layers produced by this technique [168]. Planar interdigitated finger capacitors were fabricated on the as-hydrogen split surface with no further cleaning or processing steps (see Fig. 5.24). The metal electrodes for the interdigitated finger capacitors consisted of a 20 nm thick chromium adhesion layer, 1.5 μm thick silver layer, and a 50 nm gold cap layer to ensure a good probe contact and to prevent tarnishing. A capacitor contained six finger pairs with 10 μm finger width, 6 μm finger gaps, and 80 μm overlap length.

The properties of the obtained thin ferroelectric layer based planar capacitor are presented in Fig. 5.25. The reported rather low tunability of the planar capacitor ($n_{cr}$ $\approx$ 0.3%) (Fig. 5.25a) is compatible with the dielectric properties of STO bulk crystals at RT. On the other hand, the dielectric loss is much higher than that in STO bulk crystals and the expected increase in dielectric loss with bias and frequency (like in the bulk material) is not observed (Fig.5.25b). The origin of the elevated loss level in the "smart cut" STO films remains unclear: it may be an effect of the hydrogen splitting or a manifestation of the loss induced by the electrode-ferroelectric interface.

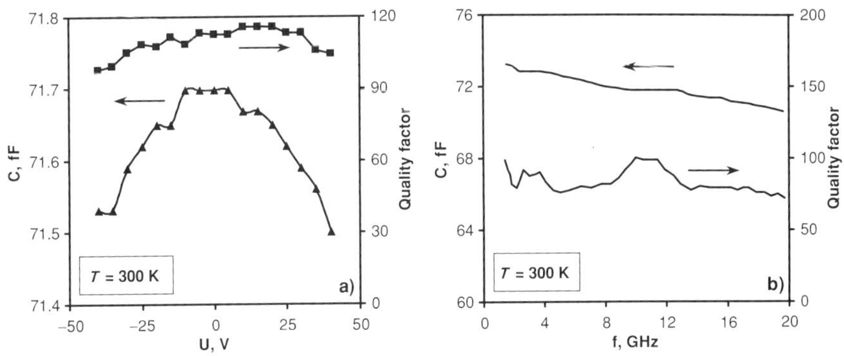


**Fig. 5.25.** Field (a) and frequency (b) dependences of the capacitance and quality factor of the interdigitated planar capacitors based on 500 nm thick monocrystalline $SrTiO_3$ layer [168].

A possible application of materials other than BST and STO as a part of microwave tunable devices has been also pursued in the last decade. Shareef *et al.* have reported on the effect of dc field on the dielectric properties of $Pb(Zr,Ti)O_3$ and BTO thin film materials grown on $Al_2O_3$ and $LaAlO_3$ substrates [169]. Even though the data reported on tunability is promising (up to 50%), large hysteresis has been observed in the dielectric constant versus electric field characteristics, which is disadvantageous for the tunable applications. The loss tangent of the film at high frequencies (>1GHz) were not presented so that one can not assess the suitability of these materials for microwave tunable components.

Another material recently reported is a series of $Ag(Ta,Nb)O_3$ (ATN) solid solutions. All these materials show a series of phase transitions from high temperature cubic phase (>800k) down to low temperature tetragonal, orthorhombic, rhombohedral, and three different types of monoclinic phases. The room temperature phase is monoclinic. One of the interesting properties exhibited by these compounds is their relatively low dielectric loss ($\tan\delta < 0.002$ at 1MHz) in the ferroelectric phase [170]. A recent report on ATN films grown on MgO and $LaAlO_3$ substrates reveals that tunability measured at 1 kHz is around 6% and the loss tangent varies between 0.0025 and 0.0034 at 1 MHz [171]. However, the suitability of these materials for microwave tunable components still cannot be assessed because of the lack of data on microwave dielectric loss.

## 6. Conclusions

This chapter reviewed and analyzed the properties of ferroelectric materials that are relevant to tunable devices at microwave (MW) frequencies. We centered on effects that can be analyzed in terms of physical mechanisms related to ferroelectricity and in terms of interaction of the material with electric field. Chemical effects such as the inter-diffusion between different components of the material, or the mechanisms which relate the processing conditions to the microstructure were out of the scope of this chapter and are discussed in the previous chapter (by Auciello et al.). Experimental results on 'imperfect' materials both from the literature and from authors' laboratory were analyzed in light of the theory and often showed a reasonable agreement with it. However, one should recognize that the descriptive ability of the existing theory is limited. The theory for the tunability available in the literature is adequate for the description of real materials, except for the tunability of composites where only limited amount of results is reported. The theory of the dielectric loss is properly developed for the case of intrinsic phonon loss whereas the available models for the extrinsic microwave loss do not seem to cover the spectrum of the loss mechanisms operating in real materials. More detailed conclusions that can be drawn from the critical analysis offered in this chapter can be summarized as follows.

*Which materials are suitable for tunable MW applications?*

The most interesting materials for tunable ferroelectric applications are the paraelectric phases of displacive ferroelectrics (or incipient ferroelectrics). The high Curie-Weiss constant of displacive ferroelectrics implies high values of dielectric permittivity and tunability even far above the Curie temperature. This also reduces the sensitivity of the permittivity to temperature changes. The absence of ferroelectric domains helps in keeping the loss level low.

*The relation between tunability and permittivity*

In both the ferroelectric and the paraelectric phases, the tunability strongly depends on the permittivity: At low fields the tunability varies cubically with the permittivity, and at higher fields this is reduced to a linear dependence (Fig. 3.1).

Real ferroelectric ceramics can be modeled as composites containing a minor dielectric (non ferroelectric) phase. In the case of a layered composite, e.g. a passive layer at the ferroelectric-electrode interface, the permittivity and the tunability can be strongly suppressed. This is particularly important for thin films. In fact, the behavior of a layered ferroelectric composite is very close to that of a pure ferroelectric with a decreased Curie temperature. Two more cases are of interest. The columnar composite, e.g. a columnar microstructure of thick films with a passive layer at the grain boundary (in the case of out-of-plane component of the permittivity) and a composite of a ferroelectric material with dielectric spherical inclusions, e.g. a parasitic phase at the triple points. A remarkable feature of both cases is that while the permittivity can be strongly reduced, the tunability is much less affected (Fig. 3.3), though for the spherical inclusion model this statement should be taken with a reservation for the roughness of the approximation used in the calculations.

*Intrinsic and extrinsic dielectric loss in 'ideal' and in 'real' ferroelectrics*

At microwave frequencies, in contrast to lower frequencies, the intrinsic and extrinsic contributions to the losses in tunable ferroelectrics can be of the same order of magnitude.

The intrinsic dielectric loss of ferroelectrics in the paraelectric phase increases with the dielectric permittivity. This loss originates from the interaction of the ac field with the phonons in the material through 3-quantum and 4-quantum mechanisms. The contribution to losses by the 3-quantum mechanism is expected to be of a higher importance. In the so-called high-K dielectrics (low loss paraelectrics with relative permittivity in the 20-100 range), it has been shown both theoretically and experimentally that $\tan\delta$ due to the 3-quantum mechanism increases by power of 3–4 with the dielectric permittivity. In $SrTiO_3$ this power law has an exponent of 1.5.

Electric field induces an additional loss mechanism in paraelectrics, the quasi-Debye loss, due to the breaking of the center of symmetry. At the frequencies of interest (<100 GHz) this loss mechanism can be substantially higher than 3- and 4-quantum mechanisms. It varies with the electric field in a complex manner—having a quadratic dependence on the field at low fields, and flattening at higher fields. At the higher MW frequency range, a slowing down of the loss dependence of frequency is expected. As an order-of-magnitude estimate, an intrinsic loss of 0.02 was calculated for BST (60:40) at 25 GHz and 100 kV/cm (see Sec. 3.2.1).

Extrinsic losses emanating from charged defects (which are expected to be stronger in thin films) are directly dependent on the permittivity and inversely dependent on applied field. Local polar regions induced by defects will also contribute to the losses, with a strong dependence on the permittivity and on the value of the local polarization.

Altogether, for the materials of interest, the quasi-Debye mechanism is the only mechanism that leads to a higher loss under field, while the other mechanisms, intrinsic and extrinsic alike are more severe in the zero field situation where the permittivity is at maximum. In this way, the quasi-Debye mechanism imposes an upper limit on the tunable performance as a function of dc field (Fig. 3.9). However, it should be kept in mind that the importance of Quasi-Debye loss varies from material to material, depending on the phonon damping and the strength of the flexoelectric coupling. Because of this, the losses are less field dependent in BST than in pure STO.

The effect of the addition of a second phase of a lower permittivity and loss was analyzed in terms of various composite configurations, layered structure, columnar structure, and spherical inclusions of the dielectric phase into the ferroelectric matrix. The electric field redistribution due to the introduction of the second phase was considered, while possible chemical modifications such as inter-diffusion were ignored. While the layered composite undergoes a strong reduction in the loss upon addition of a small fraction of dielectric material, the loss in the columnar and the spherical inclusion configurations are much less sensitive to the second phase addition.

*Can the commutation quality factor be increased by the addition of a dielectric phase?*

Combining the composite effects on the permittivity and the dielectric loss it was shown that, the Vendik's commutation quality factor, $K$, cannot be appreciably increased by the introduction of a dielectric phase. On the other hand, it was found that the columnar configuration and spherical inclusions lead to a substantial reduction of the permittivity, without essential impact on the $K$ factor (for spherical inclusion model this statement should be taken with a reservation for the limitations of the approximations used in the calculations). This may be advantageous in applications where a lower permittivity is desired. An important conclusion of the analysis is that if an improvement in the $K$ factor is gained by adding a dielectric to the ferroelectric, the reason for that should probably be sought in chemical or microstructure effects rather than in the composite effect. To be more specific concerning this issue, one is to possess a better theory on the dielectric non-linearity of ferroelectric composites.

*Dielectric response of tunable ferroelectric thin films*

The dielectric response of tunable ferroelectric thin films differs from that of bulk materials: their permittivity is often lower and the loss is higher. Not all the differences are accounted for so far.

Size effects are responsible for some differences. The ferroelectric lattice is harder near the surface in comparison to the bulk. The ferroelectric film can be therefore considered as a composite of a ferroelectric layer sandwiched between two dielectric layers of a very small thickness. The thickness of this effective dielectric layer varies depending on the measurement configuration (in-plane or out-of-plane), and it may also be sensitive to the electrode material. Depletion effects are also important in thin films. Depletion effects can be described too in terms of a layer composite model, though the in-plane and out-of-plane dielectric response will not be the same as in the previous situations. The depletion effect has not been studied yet in detail. Another competing scenario for the explanation of the size effect in films is related to the finite screening length in the electrodes ("electrode effect"). The dielectric-layer, depletion and "electrode" effects all of them can explain the thickness dependence of the out-of-plane component of dielectric permittivity of thin films, however, the reported experimental results do not suffice for the discrimination between these three possible scenarios.

Misfit strain may substantially shift the Curie temperature and modify the permittivity. This has been proven theoretically and experimentally. The effect can be as strong as to make the normally incipient ferroelectric STO be ferroelectric at room temperature. In addition, the Curie temperature for the in-plane and out-of-plane components of the permittivity should remarkably differ. Another important conclusion is that the influence of strain on losses and on the tunability is expected to be limited to that obtained indirectly through the modification in the dielectric permittivity.

The dielectric loss in thin films is roughly one order of magnitude higher than in bulk materials. This is believed to be of extrinsic origin. A more detailed investigation revealed that, in well-prepared films (that have low loss at zero dc field), an increase of loss under dc field is observed, compatible with the Debye-loss mechanism.

*Additional practical conclusions and needed work*

The last but not least interesting question is how to proceed in order to evaluate the potential of ferroelectrics that have not been studied so far. For displacive type ferroelectrics the permittivity and tunability at low frequencies are indicative of these properties at MW frequencies—this eases the characterization problem. However, this is not correct for relaxor ferroelectrics which exhibit a strong frequency dispersion.

In general, the losses measured at rf are not indicative for the loss in high frequencies since the mechanisms involved are different. More information about MW loss in various ferroelectrics is needed. Thickness dependence of dielectric properties are needed for both in-plane and out-of-plane configurations. Finally very little is reported about the field dependence of MW loss – these experiments are complex to carry but are valuable for further development of the subject.

## 7. Acknowledgements

The Swiss National Science Foundation is acknowledged for financial support. Part of the work has been done within the European project MELODY - The Swiss Office of Science and Education is acknowledged for the financial support of the authors' participation in this project.

The authors are indebted to Dr. Stephen K. Streiffer for his numerous suggestions on the improvement of the manuscript.

## References

[1] M. D. Domenico, D. A. Johnson, and R. H. Pantell, *Journal of Applied Physics*, **33**, 1697, (1962).

[2] O. G. Vendik, E. K. Hollmann, A. B. Kozyrev, and A. M. Prudan, *Journal of Superconductivity*, **12**, 325, (1999).

[3] M. J. Lancaster, J. Powell, and A. Porch, *Supercund. Sci. Technol.*, **11**, 1323 (1998).

[4] X. X. Xi, H. C. Li, W. D. Si, A. A. Sirenko, I. A. Akimov, J. R. Fox, A. M. Clark, and J. H. Hao, *Journal of Electroceramics*, **4**, 393, (2000).

[5] F. A. Miranda, F. W. Van Keuls, R. R. Romanofsky, C. H. Mueller, S. Alterovitz, and G. Subramanyam, *Integrated Ferroelectrics*, **42**, 131, (2001).

[6] V. L. Gurevich and A. K. Tagantsev, *Adv. Phys.*, **40**, 719, (1991).

[7] S. Gevorgian, E. Carlsson, E. Wikborg, and E. Kollberg, *Intergrated Ferroelectrics*, **22**, 245, (1998).

[8] S. A. Wolf and D. Treger, *Integrated Ferroelectrics*, **42**, 39, (2001).

[9] O. G. Vendik, *Ferroelectrics at microwaves (in Russian)* (Sovyetskoye Radio, Moscow, 1979).

[10] N. M. Alford, S. J. Penn, A. Templeton, X. Wang, J. C. Gallop, N. Klein, C. Zuccaro, and P. Filhol, *IEE Colloquium on Electro-technical ceramics—processing, properties and applications*, 9/1, (1997).

[11] S. J. Penn, N. McNAlford, A. Templeton, N. Klein, J. C. Gallop, P. Filhol, and X. Wang, *IEE Colloquium on Advances in Passive Microwave Components*, 6/1, (1997).

[12] G. Rupprecht and R. O. Bell, *Physical Review*, **135**, A748, (1964).

[13] A. Linz, *Physical Review*, **91**, 753, (1953).

[14] G. A. Smolenskii and V. A. Isupov, *Zhurnal Tekhnicheskoi Fiziki*, **24**, 1375, (1954).

[15] G. S. Khizha, I. B. Vendik, and E. A. Serebryakova, *Microwave Phase Shifters Based on p-i-n Diodes ( in Russian)* (Radio i Svyas, Moscow, 1984).

[16] I. Vendik, O. Vendik, and E. Kollberg, *IEEE Trans. Microwave Theory and Techniques*, **48**, 802, (2000).

[17] A. Deleniv, A. Eriksson, and S. Gevorgian, *2002 IEEE MTT-S digest*, 197, (2002).

[18] J. Rao, D. Patel, and V. Krichevsky, *IEEE Trans. Antennas and Propagation*, **47**, 458, (1999).

[19] F. D. Flaviis, N. G. Alexopoulos, and O. M. Stafsudd, *IEEE Trans. Microwave Theory Tech.*, **45**, 963, (1997).

[20] O. G. Vendik, L. T. Ter-Martirosyan, A. I. Dedyk, S. F. Karmanenko, and R. A. Chakalov, *Ferroelectrics*, **144**, 33, (1993).

[21] A. Kozyrev, V. Osadchy, A. Pavlov, and L. Sengupta, *IEEE MTT-S Digest*, 1355, (2000).

[22] B. Acikel, T. R. Taylor, P. F. Hansen, J. S. Speck, and R. A. York, *IEEE Microwave and Wireless Components Letters*, **12**, 237, (2002).

[23] A. Kozyrev, A. Ivanov, A. Prudan, O. Soldatenkov, E. Hollmann, V. Loginov, D. S. Ginley, and T. Rivkin, *Integrated Ferroelectrics*, **24**, 287, (1999).

[24] V. Sherman, K. Astafiev, N. Setter, A. Tagantsev, O. Vendik, I. Vendik, S. Hoffmann-Eifert, U. Bottger, and R. Waser, *IEEE Microwave and Wireless Components Letters*, **11**, 407, (2001).

[25] O. G. Vendik, I. B. Vendik, and V. O. Sherman, *Integrated Ferroelectrics*, **43**, 81, (2002).

[26] I. Vendik, O. Vendik, E. Kollberg, and V. Sherman, *IEEE Transactions on Microwave Theory and Techniques*, **47**, 1553, (1999).

[27] F. A. Miranda, G. Subramanyam, F. W. V. Keuls, R. R. Romanofsky, J. D. Warner, and C. H. Mueller, *IEEE Trans. on Microwave Theory And Techniques*, **48**, 1181, (2000).

[28] A. Kozyrev, A. Ivanov, V. Keis, M. Khazov, V. Osadchy, T. Samoilova, A. Pavlov, G. Koepf, C. Mueller, D. Galt, and T. Rivkin, *IEEE MTT-S Digest*, **2**, 985, (1998).

[29] V. Keis, A. Kozyrev, M. Khazov, J. Sok, and J. Lee, *Electronics Letters*, **34**, 1107, (1998).

[30] B. H. Moeckly and Y. Zhang, *IEEE Transaction on Applied Superconductivity*, **11**, 450, (2001).

[31] I. Vendik, O. Vendik, V. Pleskachev, A. Svishchev, and R. Woerdenweber, *IEEE MTT-S Digest*, **3**, 1461, (2001).

[32] T. B. Samoilova, K. F. Astafiev, T. Rivkin, and D. S. Ginley, *Journal of Applied Physics*, **90**, 5703, (2001).

[33] J. G. Colom, R. A. Rodrigues-Solis, J. Almodovar, and M. Castaneda, *Integrated Ferroelectrics*, **42**, 313, (2002).

[34] C. Weil, P. Wang, H. Downar, J. Wenger, and R. Jakoby, *Frequenz*, **54**, 250, (2000).

[35] J. O. Gentner, P. Gerthsen, N. A. Schmidt, and R. E. Send, *J. Appl. Phys.*, **49**, 4585, (1978).

[36] V. G. Vaks, *Introduction to the Microscopic Theory of Ferroelectrics* (Nauka, Moscow, 1973).

[37] J. H. Barrett, *Phys. Rev.*, **86**, 118, (1952).

[38] O. G. Vendik, L. T. Ter-Martirosyan, and S. P. Zubko, *J. Appl. Phys.*, **84**, 993, (1998).

[39] O. G. Vendik, *Sov. Phys. Solid state*, **14**, 849, (1972).

[40] O. G. Vendik and S. P. Zubko, *J. Appl. Phys.*, **82**, 4475, (1997).

[41] O. G. Vendik, S. P. Zubko, and M. A. Nikol'ski, *J. Appl. Phys.*, **92**, 7448 (2002).

[42] O. Hudak, I. Rychetsky, and J. Petzelt, *Ferroelectrics*, **208**, 429, (1998).

[43] J. W. Liou and B. S. Chiou, *Journal of Physics-Condensed Matter*, **10**, 2773, (1998).

[44] M. Vollman and R. Waser, *J. Am. Ceram. Soc.*, **77**, 235 (1994).

[45] K. F. Astafiev, V. O. Sherman, A. K. Tagantsev, and N. Setter, *Journal of the European Ceramic Society*, **23**, 2381 (2003).

[46] L. D. Landau, E. M. Lifshitz, and L. P. Pitaevskii, *Electrodynamics of Continuous Media*, (Butterworth-Heinemann, 1995).

[47] V. Sherman, A. Tagantsev, K. Astafiev, and N. Setter, *to be published*, (2004).

[48] J. E. Sipe and R. W. Boyd, *Physical Review A*, **46**, 1614, (1992).

[49] K. W. Yu, P. M. Hui, and D. Stroud, *Physical Review B*, **47**, 14150, (1993).

[50] V. S. Vinogradov, *Fiz. Trerd. Tela*, **4**, 712, (1962).

[51] V. L. Gurevich, *Fiz. Tverd. Tela*, **21**, 3453, (1979).

[52] A. K. Tagantsev, *Sov. Phys. JETP*, **53**, 555, (1981).

[53] K. A. Subbaswamy and D. L. Mills, *Phys. Rev. B*, **33**, 4213, (1986).

[54] A. K. Tagantsev, *Sov. Phys. JETP*, **59**, 1290, (1984).

[55] A. K. Tagantsev, J. Petzelt, and N. Setter, *Solid State Commun*, **87**, 1117, (1993).

[56] V. L. Gurevich, *Transport in Phonon Systems* (North-Holland, Amsterdam, 1986).

[57] R. Zurmulen, J. Petzelt, S. Kamba, G. Kozlov, A. Volkov, B. Gorshunov, D. Dube, A. Tagantsev, and N. Setter, *J. Appl. Phys.*, **77**, 5351, (1995).

[58] I. M. Buzin, *Vestn. Mosk. Univ. Fiz. Astron.*, **18**, 70, (1977).

[59] R. Stolen and K. Dransfeld, *Phys. Rev.*, **139**, 1295, (1965).
[60] B. Y. Balagurov, V. G. Vaks, and B. I. Shklovskii, *Fiz. Tverd. Tela*, **12**, 89, (1970).
[61] O. G. Vendik, *Sov. Phys. Solid State*, **17**, 1096, (1975).
[62] G. J. Cooms and R. A. Cowley, *J. Phys. C*, **6**, 121, (1973).
[63] A. K. Tagantsev, *Sov. Phys. JETP*, **50**, 948, (1979).
[64] C. Kittel, *Introduction to Solid State Physics* (John Wiley & Sons, Inc., New York, London, 1971).
[65] A. Tagantsev, *Appl. Phys. Lett.*, **76**, 1182, (2000).
[66] A. K. Tagantsev and K. F. Astafiev, *Integrated Ferroelectrics*, **39**, 251, (2001).
[67] K. F. Astafiev, A. K. Tagantsev, and N. Setter, J. Appl. Phys., 97, (2005), (in press).
[68] L. C. Sengupta, S. Stowell, E. Ngo, M. E. O'Day, and R. Lancto, *Integrated Ferroelectrics*, **8**, 77, (1995).
[69] E. Schlöman, *Phys. Rev.*, **135**, A413, (1964).
[70] O. G. Vendik and L. M. Platonova, *Sov. Phys. Solid State*, **13**, 1353, (1971).
[71] B. M. Garin, *Sov. Phys. Solid state*, **32**, 1917, (1990).
[72] C. J. Brennan, *Integrated Ferroelectrics*, **7**, 93, (1995).
[73] J. G. Simmons, *J. Phys. Chem. Solids*, **32**, 2581, (1971).
[74] A. K. Jonscher, *Universal relaxation law* (Chelsea Dielectrics Press, London, 1996).
[75] J. D. Baniecki, R. B. Laibowitz, T. M. Shaw, P. R. Duncombe, D. A. Neumayer, D. E. Kotecki, H. Shen, and Q. Y. Ma, *Appl. Phys. Lett.*, **72**, 498, (1998).
[76] R. Waser, in *Science and Technology of Electroceramic Thin Films*, vol. 284, *NATO ASI; Series E: Applied Science*, O. Auciello and R. Waser, Eds., 1995, pp. 223.
[77] Y. Fukuda, K. Numata, K. Aoki, and A. Nishimura, *Jpn. J. Appl. Phys.*, **35**, 5178, (1996).
[78] J. Petzelt, T. Ostapchuk, I. Gregora, I. Rychetsk, S. Hoffmann-Eifert, A. V. Pronin, Y. Yuzyuk, B. P. Gorshunov, S. Kamba, V. Bovtun, J. Pokorn, M. Savinov, V. Porokhonskyy, D. Rafaja, P. Vanek, A. Almeida, M. R. Chaves, A. A. Volkov, M. Dressel, and R. Waser, *Phys. Rev. B*, **64**, 184111, (2001).
[79] L. C. Sengupta, S. Stowell, E. Ngo, and S. Sengupta, *Integrated Ferroelectrics*, **13**, 203, (1996).
[80] M. Jain, S. B. Majumder, R. S. Katiyar, D. C. Agrawal, and A. S. Bhalla, *Applied Physics Letters*, **81**, 3212, (2002).
[81] R. Kretschmer and K. Binder, *Phys. Rev. B*, **20**, 1065, (1979).
[82] O. G. Vendik and S. P. Zubko, *J. Appl. Phys.*, **88**, 5343, (2000).
[83] I. P. Batra and B. D. Silverman, *Solid State Communications*, **11**, 291, (1972).
[84] R. D. Tilley and B. Zeks, *Ferroelectrics*, **134**, 313, (1992).
[85] J. M. Ziman, *Principles of the theory of solids* (Cambridge University Press, Cambridge, 1972), p. 435.
[86] A. K. Tagantsev, E. Courtens, and L. Arzel, *Phys. Rev. B*, **64**, 224107, (2001).
[87] Y. Yamada, G. Shirane, and A. Linz, *Phys. Rev.*, **177**, 848, (1969).
[88] A. K. Tagantsev and I. A. Stolichnov, *Appl. Phys. Lett.*, **74**, 1326, (1999).
[89] A. K. Tagantsev, C. Pawlaczyk, K. Brooks, and N. Setter, *Integrated Ferroelectrics*, **4**, 1, (1994).
[90] A. M. Bratkovsky and A. P. Levanyuk, *Phys. Rev. B*, **61**, 15042, (2000).
[91] R. Waser and M. Klee, *Integrated Ferroelectrics*, **2**, 23, (1992).
[92] J. S. Speck and W. Pompe, *J. Appl. Phys.*, **76**, 466, (1994).
[93] S. K. Streiffer, C. Basceri, C. B. Parker, S. E. Lash, and A. I. Kingon, *J. Appl. Phys.*, **86**, 4565, (1999).
[94] N. A. Pertsev, A. G. Zembilgotov, S.Hoffman, R. Waser, and A. K. Tagantsev, *J. Appl. Phys.*, **85**, 1698 (1999).

[95] Landolt-Bornstein, *Numerical data and functional relationships in science and technology* (Springer, New York, 1981), vol. New Series Vol. III/29a,b.

[96] N. A. Pertsev, A. G. Zembilgotov, and A. K. Tagantsev, *Phys. Rev. Lett.*, **80**, 1988, (1998).

[97] A. K. Tagantsev, N. A. Pertsev, P. Muralt, and N. Setter, *Phys. Rev. B*, **65**, 012104, (2002).

[98] N. A. Pertsev, A. K. Tagantsev, and N. Setter, *Phys. Rev. B*, **61**, R825, (2000).

[99] R. E. Collin., *Fondations for Microwave Engineering* (McGraw-Hill, New York, 1992), p. 924.

[100] D. C. Dube, J. Baborowski, P. Muralt, and N. Setter, *Applied physics letters*, **74**, 3546, (1999).

[101] A. Tombak, J. P. Maria, F. Ayguavives, Z. Jin, G. T. Stauf, A. I. Kingon, and A. Mortazawi, *IEEE Microwave and Wireless Components Letters*, **12**, 3, (2002).

[102] T. Ayguavives, A. Tombak, J. P. Maria, G. T. Stauf, C. Ragaglia, J. Roeder, A. Mortazawi, and A. Kingon, *Proc. 12th ISAF*, **1**, 365, (2000).

[103] S. Li, J. Sheen, Q. M. Zhang, S.-J. Jang, A. S. Bhalla, and L. E. Cross, *Proc. 8th ISAF*, 480, (1992).

[104] S. S. Gevorgian, T. Martinsson, P. L. J. Linner, and E. L. Kollberg, *IEEE Trans. on Microwave Theory And Techniques*, **44**, 896, (1996).

[105] O. Vendik, S. Zubko, and M. Nikolski, *Technical Physics*, **44**, 349, (1999).

[106] A. N. Deleniv, *Technical Physics*, **44**, 356, (1999).

[107] M. Sucher and J. Fox, *Handbook of Microwave Measurements* (Interscience, New York, 1963), vol. 2.

[108] C. Krowne, S. Kirchoefer, and J. Pond, *IEEE MTT-S Digest*, 1193, (2000).

[109] E. Carlsson and S. Gevorgian, *IEEE Transactions on Microwave Theory and Techniques*, **47**, 1544, (1999).

[110] A. T. Findikoglu, Q. X. Jia, C. Kwon, B. J. Gibbons, K. O. Rasmussen, Y. Fan, D. W. Reagor, and A. R. Bishop, *Materials Research Sosiety Symposium Proceedings*, **603**, 27, (2000).

[111] A. T. Findikoglu, D. W. Reagor, K. O. Rasmussen, A. R. Bishop, N. Gronbech-Jensen, Q. X. Jia, Y. Fan, C. Kwon, and L. A. Ostrovsky, *Journal of Applied Physics*, **86**, 1558, (1999).

[112] B. W. Hakki and P. D. Coleman, *IRE Trans. on Microwave Theory and Technique*, 402, (1960).

[113] J. Krupka, *5th International Conference on Dielectric Materials, Measurements and Applications*, 322, (1988).

[114] J. Delaballe, P. Guillon, and Y. Garault, *AUE, Electronics and Communication*, **35**, 80, (1981).

[115] O. G. Vendik, E. Kollberg, S. S. Gevorgian, A. B. Kozyrev, and O. I. Soldatenkov, *Electronics Letters*, **31**, 654, (1995).

[116] A. Eriksson, P. Linner, and S. Gevorgian, *IEE Proc.-Microw. Antennas and Propag.*, **148**, 51, (2001).

[117] J. Watkins, *Electronics Letters*, **5**, 524, (1969).

[118] D. Kajfez, *IEEE Trans. on Microwave Theory And Techniques*, **42**, 1149, (1994).

[119] S. Gevorgian, E. Carlsson, P. Linner, E. Kollberg, O. Vendik, and E. Wikborg, *IEEE Trans. on microwave Theory And Techniques*, **44**, 1738, (1996).

[120] R. Thomas and D. C. Dube, *Electronics Letters*, **33**, 218, (1997).

[121] D. C. Dube, *Ferroelectrics*, **225**, 141, (1999).

[122] Y. G. Wang, M. E. Reeves, W. Chang, H. J.S., and W. Kim, *Materials Research Sosiety Symposium Proceedings*, **603**, 289, (2000).

[123] D. E. Steinhauer, C. P. Vlahacos, F. C. Wellstood, S. M. Anlage, C. Canedy, R. Ramesh, A. Stanishevsky, and J. Melngailis, *Appl. Phys. Lett.*, **75**, 3180, (1999).

[124] D. Galt, J. Price, J. Beall, and T. Harvey, *IEEE Trans. on Applied Superconductivity*, **5**, 2575, (1995).
[125] A. B. Kozyrev, V. N. Keis, G. Koepf, R. Yandrofski, O. I. Soldatenkov, K. A. Dudin, and D. P. Dovgan, *Microelectronic engineering*, **29**, 257, (1995).
[126] T. Sakudo and H. Unoki, *Phys. Rev. Lett.*, **26**, 851, (1971).
[127] M. A. Saifi and L. E. Cross, *Phys. Rev. B*, **2**, 677, (1970).
[128] J. Hemberger, P. Lunkenheimer, R. Viana, R. Bohmer, and A. Loidl, *Phys. Rev. B*, **52**, 13159, (1995).
[129] G. V. Belokopytov, *Ferroelectrics*, **168**, 69, (1995).
[130] J. Krupka, R. G. Geyer, M. Kuhn, and J. H. Hinken, *IEEE Transactions on Microwave Theory and Techniques*, **42**, 1886 (1994).
[131] F. Jona and G. Shirane, *Feroelectric crystals* (Macmillan, New York, 1962).
[132] L. Arzel, PhD thesis, University of Montpellier II, Montpellier, 2001.
[133] J. Harada, J. Axe, and G. Shirane, *Phys. Rev. B*, **4**, 155, (1971).
[134] E. Ngo, P. C. Joshi, M. W. Cole, and C.W. Hubbard, *Appl. Phys. Lett.*, **79**, 248, (2001).
[135] L. Wu, S. Wu, F.-C. Chang, Y.-T. Shen, and Y.-C. Chen, *J. of Materials science*, **35**, 5945, (2000).
[136] D. M. Potrepka, S.C. Tidrow, and A. Tauber, *Integrated ferroelectrics*, **42**, 97, (2002).
[137] L. C. Sengupta and S. Sengupta, *Mat. Res. Innovat.*, **2**, 278 (1999).
[138] B. Su, J. H. Holmes, and T. W. Button, *J. Am. Ceram. Soc.*, submitted, (2003).
[139] C. Ang, A. S. Bhalla, R. Guo, and L. E. Cross, *J. Appl. Phys.*, **90**, 2465 (2001).
[140] M. Daglish, M. Presland, and S. Batbedat, *Integrated Ferroelectrics*, **39**, 339, (2001).
[141] D. Li and M. A. Subramanian, *Soild state Science*, **2**, 507, (2000).
[142] S. Triebwasser, *Phys. Rev.*, **114**, 63, (1959).
[143] B. Cristopher, C. B. DiAntonio, and S. M. Pilgrim, *J. Am. Ceram. Soc.*, **84**, 2547, (2001).
[144] P. Debely, P. Gunter, and H. Arend, *Am. Ceram. Soc. Bull.*, **58**, 606, (1979).
[145] J. Venkatesh et al., (unpublished).
[146] A. Kozyrev, A. Ivanov, T. Samoilova, O. Soldatenkov, K. Astafiev, and L. C. Sengupta, *Journal of Applied Physics*, **88**, 5334, (2000).
[147] A. Outzourhit, J. U. Trefny, T. Kito, B. Yarar, A. Naziripour, and A. M. Hermann, *Thin Solid Films*, **259**, 218, (1995).
[148] H.-C. Li, W. Si, A. D. West, and X. X. Xi, *Appl. Phys. Lett.*, **73**, 464 (1998).
[149] C. B. Parker, J. P. Maria, and A. I. Kingon, *Applied Physics Letters*, **81**, 340, (2002).
[150] C. Basceri, S. K. Streiffer, A. I. Kingon, and R. Waser, *Journal of Applied Physics*, **82**, 2497, (1997).
[151] J. Bellotti, E. K. Akdogan, A. Safari, W. Chang, and S. Kirchoefer, *Integrated Ferroelectrics*, **49**, 113, (2002).
[152] D. Schlom, *unpublished*, (2003).
[153] W. Chang, J. S. Horwitz, A. C. Carter, J. M. Pond, S. W. Kirchoefer, C. M. Gilmore, and D. B. Chrisey, *Appl. Phys. Lett.*, **74**, 1033, (1999).
[154] C. M. Carlson, T. V. Rivkin, P. A. Parilla, J. D. Perkins, D. S. Ginley, A. B. Kozyrev, V. N. Oshadchy, and A. S. Pavlov, *Appl. Phys. Lett.*, **76**, 1920, (2000).
[155] K. F. Astafiev, V. O. Sherman, A. K. Tagantsev, N. Setter, P. K. Petrov, T. Kaydanova, D. S. Ginley, S. Hoffmann-Eifert, U. Bottger, and R. Waser, *Integrated Ferroelectrics*, (Submitted), (2003).
[156] P. K. Petrov, Z. G. Ivanov, and S. S. Gevorgyan, *Materials Science and Engineering a-Structural Materials Properties Microstructure and Processing*, **288**, 231, (2000).
[157] Z. G. Ban and S. P. Alpay, *Journal of Applied Physics*, **91**, 9288, (2002).
[158] D. Fuchs, C. W. Schneider, R. Schneider, and H. Rietschel, *J. Appl. Phys.*, **85**, 7362, (1999).

[159] O. G. Vendik and L. T. Ter-Martirosyan, *J. Appl. Phys.*, **87**, 1435, (2000).
[160] J. H. Chen, C. L. Lia, K. Urban, and C. L. Chen, *Applied Physics Letters*, **81**, 1291, (2002).
[161] G. Rupprecht and R. O. Bell, *Physical Review*, **125**, 1915, (1962).
[162] M. J. Dalberth, R. E. Stauber, J. C. Price, C. T. Rogers, and D. Galt, *Appl. Phys. Lett.*, **72**, 507, (1998).
[163] Y. Lemaitre, B. Marcilhac, D. Mansart, J. Siejka, and J. C. Mage, *Physica C*, **372**, 667, (2002).
[164] K. F. Astafiev et al., *unpublished*, (2003).
[165] S. Razumov, A. Tumarkin, O. Buslov, M. Gaidukov, A. Gagarin, A. Ivanov, A. Kozyrev, Y. W. Song, and C. S. Park, *Integrated Ferroelectrics*, **39**, 1317, (2001).
[166] P. C. Joshi and M. W. Cole, *Appl. Phys. Lett.*, **77**, 289, (2000).
[167] M. Bruel, *Electronics Letters*, **31**, 1201, (1995).
[168] F. J. Kub, K. D. Hobart, J. M. Pond, and S. W. Kirchoefer, *Electronics Letters*, **35**, 477, (1999).
[169] H. N. Al-Shareef, D. Dimos, M. V. Raymond, and R. W. Schwartz, *Journal of Electroceramics*, **1**, 145, (1997).
[170] M. Hafid, G. E. Kugel, A. Kania, K. Roleder, and M. D. Fontana, *Journal of Physics-Condensed Matter*, **4**, 2333, (1992).
[171] J. H. Koh and A. Grishin, *Applied Physics Letters*, **79**, 2234, (2001).

# Chapter 13

## Microfabrication of Piezoelectric MEMS

J. Baborowski[1]
*Ceramics Laboratory*
*Faculty of Engineering*
*Swiss Federal Institute of Technology EPFL*
*1015 LAUSANNE*
*Switzerland*
[1] *Present affiliation:* **CSEM**
*Swiss Center of Electronics and Microtechnology,*
*RF & Piezo Components Group, 2007 Neuchâtel,*
*SWITZERLAND*

In this chapter an overview of processes for fabrication of MEMS using piezoelectric thin films as active layer in planar structures is presented. These structures are used in cantilever-like and membrane configurations for sensing and actuation. Key issues consist in findings of a compatible dry etching sequence for piezoelectric layer, electrodes and silicon. The method of compensation of the gradient of mechanical stresses in the thin films in order to obtain flat multilayer structures is demonstrated. Membrane thickness definition and liberation are obtained by DRIE of silicon on insulator (SOI) substrates or by surface micromachining. The platinum bottom electrode turned out to be a useful mask for deep silicon etching of narrow grooves. The complete process has successfully been used to fabricate flat cantilevers, ultrasonic transducers and pressure sensors. Excellent permittivity and transverse piezoelectric coefficient of PZT and AlN have been obtained with complete devices. Several examples of piezoelectric devices (ultrasonic transducers, acoustic sensor, damping control, RF switches) presented in this paper have been developed during the last years. It has been demonstrated that the microfabrication of piezoelectric MEMS based on PZT or AlN thin films is the versatile and reliable technology. Most of the developments are based on PZT sol-gel textured films, however some efforts have been done for integration of AlN for piezoelectric MEMS. In this context, the advantage of AlN is the low thermal budget processing and the possibility of post-CMOS above-IC integration. Other examples of application like ferroelectric memories, nanopatterning and local growth of PZT are presented as well.

# 1. Introduction

Piezoelectric thin film are useful in various actuation and sensing devices requiring large output signals, low noise, or high frequency operation (see [1], [2] for a review). An important advantage lies in the fact that a planar structure is able to give excursion and strain detection in the out-of-plane direction, which is very useful for instance in scanning probe techniques [3]. Today, there is a growing interest in the field of micro-electro-mechanical systems (MEMS) for the integration of smart materials with good actuation and/or sensing capabilities. In particular, $Pb(Zr_xTi_{1-x})O_3$ (PZT) and AlN thin films are of primary interest [4].

Piezoelectric Micro-Electro-Mechanical Systems (pMEMS) contain at least two elements: bulk silicon frame and built on it a piezoelectric deflection element with its electrodes. Micromachining of silicon, which refers to the fashioning of microscopic mechanical parts out of a silicon substrate, has emerged as an extension of IC's fabrication technology [5], [6]. During the last 25 years, it has been successfully employed to produce a variety of mechanical microstructures of a great diversity (beams, diaphragms, grooves, orifices, sealed cavities, pyramids and needles) and has driven the rapid progress of MEMS. However, the silicon substrate represents often only the **structural** element defining the mechanical properties of the device and, usually, a specific functionality needs to be added. It is the role of **functional materials** such as piezoelectric thin films to provide a direct transformation between a driving signal or a read out signal and a sensor or an actuator parameter. The basics of silicon MEMS microfabrication technologies such as photolithography, pattern transfer with dry and wet etching techniques, and common thin films deposition ($SiO_2$, $Si_3N_4$, poly-Si, Al, Cr, Ni, Au) have been described extensively in the literature [7]. The microfabrication of piezoelectric MEMS suffers from inadequate micromachining processes and very few references have been published on the subject [8], [9], [10]. The integration of piezoelectric thin films like PZT or AlN, and "exotic materials" used as diffusion barrier ($TiO_2$), electrodes (Pt) or seed layers ($PbTiO_3$) requires new and specific micromachining processes that are not available in standard IC's processing.

The critical tasks, the dry etching of PZT, AlN and electrode thin films are, not trivial as the volatility of their reactive by-products is low or very different for each compound (e.g. the boiling point of: $PbCl_2$ : 950 °C, $PbF_2$ : 1290 °C, $TiCl_4$ : 136 °C, $TiF_4$: 284 °C, $ZrCl_4$: 334 °C, $ZrF_4$: 600 °C [11]). There is thus a need for new patterning processes to afford micro scale resolution at reasonable etching rates and to find a process flow that avoids any degradation of the piezoelectric film. Another critical issue is the above-IC integration of piezoelectric MEMS. This can be solved by using low temperature (<450 °C) post-CMOS process for deposition and patterning of the piezoelectric layer (e.g. AlN).

The aims of this paper are as follow:

- To provide an overview of micromachining tools developed to produce piezoelectric MEMS with microscale features,
- To highlight key issues such as stress compensation, process flow, advanced SOI substrates,
- To demonstrate the fabrication and some properties of PZT/Si microdevices.

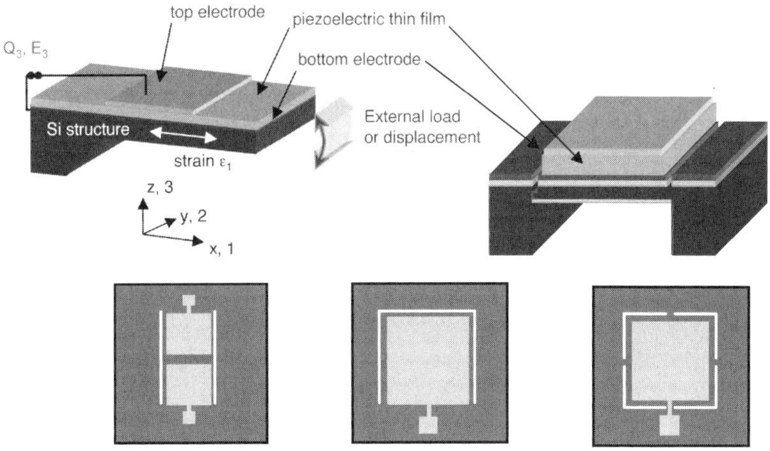


**Fig. 1.** Laminated piezoelectric layer—silicon deflecting structures used in MEMS: bridge, cantilever and suspended membrane.

## 2. Materials and Substrates for pMEMS

Most piezoelectric devices are based on laminated PZT/Si planar structures, which are defined as a flexible supporting silicon structure (5–20 μm thick) coated with a piezoelectric thin film (see Figure 1). Depending on applications, various geometry and clamping conditions can be considered: from the very simple cantilever to a fully clamped membrane or any partially clamped diaphragms (for instance, membrane fixed to the frame by few bridges). The electromechanical coupling is provided by the transverse piezoelectric coefficient $e_{31,f}$.

To succeed in the microfabrication of such 3D device, a number of important issues have to be addressed:

- the deposition of uniform, high quality PZT or AlN and electrode thin films directly on a wafer has to be solved,
- specific micromachining methods and process flow have to be developed,
- the mechanical stress through the structure need to be compensated to avoid any residual bending,
- the thickness of the supporting Si structural membrane should be uniform and the border conditions defined precisely.

Two different techniques have been used to fabricate devices based on planar PZT/Si structures. First, surface micromachining has been mainly employed to make simple accelerometers [12] (ZnO), [13] (PZT). One advantage of this method is that e.g. the poly-silicon layer can be used to define precisely the thickness of the supporting structure. However, the etching of the sacrificial layer (PSG) in HF solution often destroys the PZT thin film. Even with a silicon nitride encapsulation, pinholes occurred and the film was damaged. It has been also observed that the roughness of the polysilicon layer disturbs strongly the growth of the PZT film resulting in low piezoelectric

**Table 1.** Typical materials used in piezoelectrically actuated Si based deflecting structures

| Materials | Growth methods | Thickness | Typical stress (MPa) | Patterning methods |
|---|---|---|---|---|
| Silicon wafer/ SOI | Czochralski/bonding and polishing | 390 micron; 1 to 50 μm of device Si on burried $SiO_2$ | | DRIE; wet etching (KOH, TMAH) |
| Amorphous $SiO_2$ | Thermal wet oxidation | max. 2000 nm | $-300 \pm 5$ | ICP plasma etching; wet etching (BHF) |
| $Si_3N_4$ | LPCVD | max. 200 nm | 0 to +700 | ICP plasma etching |
| Pt bottom and top electrode | PVD sputtering | 100 to 300 nm | +550 | ICP plasma etching |
| PZT 53/47 {100} | PVD Sputtering or CSD sol-gel | 500 to 4000 nm | +110 (*unpoled*) +180 (*poled*) | ECR/RF plasma etching; wet etching (HCl/HF) |
| AlN | PVD Sputtering | 500 to 2000 nm | −200 to +220 | ICP plasma etching; wet etching ($H_3PO_4$) |
| Au/Cr top electrode | PVD evaporation | 100 nm/10 nm | +280 | lift-off |

coefficients. Finally, the geometries and size of surface micromachined piezoelectric devices are limited by the rate of the lateral etching of the sacrificial layer. Beams larger than 50 μm are difficult to release. The second technique is based on classical silicon bulk micromachining. Among reported devices, one can point out micromotors [14], [15] (PZT), [16] (ZnO), accelerometers, [17], [18] (PZT), audio microphone and microspeakers [19] (ZnO), 2-D scanners [20] (PZT), AFM tips [21], [3] (PZT) and ultrasonic transducers [22], [23] (PZT). Here, the limiting factor is a uniformity of the silicon structure obtained with standard wafers. In order to obtain the uniform thickness of a thin Si structure, silicon on insulator wafers (SOI) need to be used. Membranes in thickness range from 0.5 to 50 microns with tolerances better than 1% can be fabricated by DRIE using the buried oxide as etch stop layer. The inconvenience in this type of substrate is the unknown value of the stress in the buried $SiO_2$.

Films in laminated structures exhibit a mechanical stress (tensile or compressive), which is function of materials and deposition methods. The stress gradient in the thin film induces bending to the freestanding structure or buckling to the clamped structures. If the overall residual stress is non zero, a residual bending occurs that can reduce or even destroy the device sensitivity. The stress in piezoelectric PZT thin films depends on composition, texture, thickness and poling [10]. To achieve precise stress compensation, the effect of poling has to be considered as well. Due to the re-orientation of c-domains along the poling field (z-axis), the tensile stress increases further. Table 1 shows that the film stresses of a laminated PZT/Si structure are all tensile except the thermal $SiO_2$, which is compressive (−300 MPa). Stress compensation is achieved by adjusting the thickness of thermal $SiO_2$ to about 1000 to 1200 nm, depending on thickness of PZT (Figure 2).

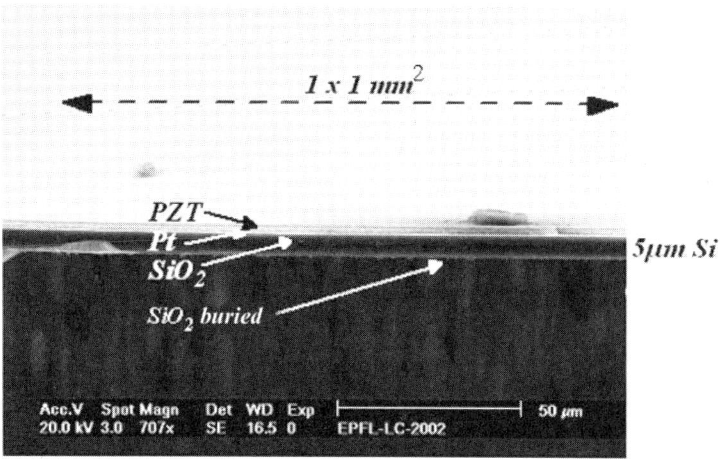


**Fig. 2.** SEM cross-section of flat PZT/Pt/$SiO_2$/Si square membrane (7 micron thick and 1 × 1 mm$^2$ large).

## 3. Processes and Tools for Patterning of pMEMS

### 3.1. Process flow for the microfabrication of PZT/Si planar structures

Like any IC's classic process flow, the microfabrication of planar PZT/Si deflecting structure is based on several iterations of film deposition—photolithography—etching [7]. Figure 3 shows the generic process flow for the microfabrication of planar cantilever structures based on SOI substrate and AlN as active layer. In general, starting with double side polished and oxidized silicon based substrate (Si wafer, SOI, polySi on $SiO_2$/Si), an adhesion layer (e.g. Ti/$TiO_2$), a bottom electrode (e.g. Pt, Ir, $IrO_2$, $RuO_2$) and eventually a seed layer (e.g. $PbTiO_3$, $TiO_2$) are deposited by PVD sputtering [24, 25]. On top of electrode or of seed layer, the piezoelectric thin film is deposited by sol-gel CSD [10] or by sputtering [26]. Top electrode is deposited and patterned either by lift-off (Au/Cr) or by plasma dry etching (Pt, Al). To give access to the bottom electrode, vias are opened through the piezoelectric film by wet or dry etching. The frontside shape of the structure (the grooves or gap surrounding the membrane) is then patterned through the piezoelectric layer/electrode/$SiO_2$ stack and through Si defining the depth of grooves. Bulk silicon micromachining from the wafer's backside is then performed to define the thickness of the structure (usually between 5 and 20 μm). The process is either dry (deep silicon etching in a plasma reactor) or wet in KOH solution (see [6] for review). The definition of precise and uniform silicon thickness (for example: $10 \pm 1$ μm thick silicon cantilever out of a 390 μm thick wafer) is of primary importance as the device sensitivity is usually a function of 1/thickness$^2$. In order to increase the precision of membrane thickness definition several solutions are proposed:

- accurate etching time measurement supposing constant process conditions,
- etch-stop obtained on n-doped epi-layer [27]
- Silicon-On-Insulator (SOI) used as substrate [23].

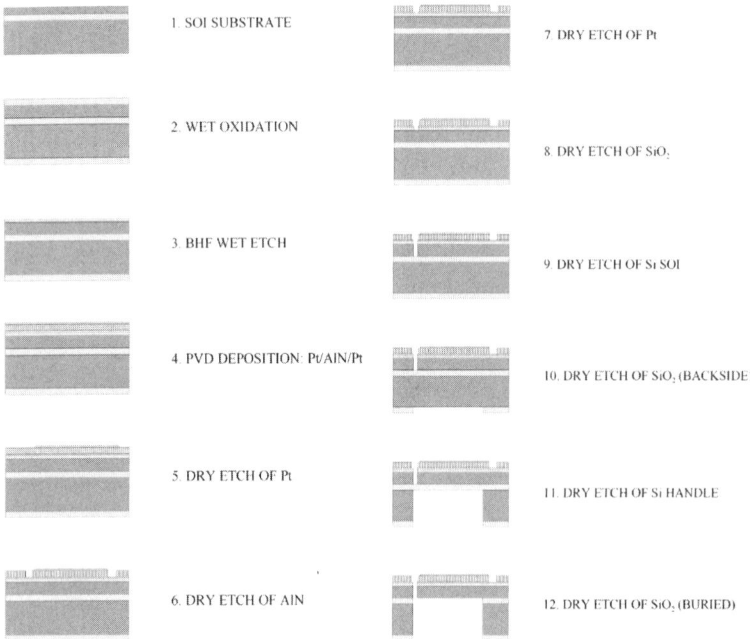


**Fig. 3.** Generic process flow for the microfabrication of piezoelectric actuated Si-based cantilever (cross section).

The micromachining steps of this process flow use many individual deposition and patterning processes. Details on general issues such as photolithography or silicon thermal oxidation can be found in [7] or in the process book of the Center of Microtechnology (CMI)—EPFL [28]. However, some of the processes are specific for microfabrication of piezoelectric MEMS and are described below.

### 3.2. Basics of patterning

Dry etching using reactive plasma has spread widely throughout the VLSI processing field because of its ability to fabricate fine-line semiconductor devices. However, the resolution of conventional RIE when decreasing the size of patterns is limited, because the operation pressure of RIE is too high to etch these patterns anisotropically. Moreover it is difficult to obtain a high etch rate because the plasma density of conventional RIE is limited. In case of piezoelectric MEMS the additional difficulty is related to the patterning of noble metals or refractory oxides, both exhibiting very low volatility. To overcome at least partially limitations of patterning such structures, a new process using high density plasma with low ion energy (to avoid damage of the piezoelectric layer and of the photoresist) and operating at lower pressure (lower backscattering, no fencing) is required. The most popular are actually inductively coupled plasma (ICP) sources (see [29] for review). However, other type of reactors like Helicon Wave Plasma (HWP), planar Electron Cyclotron Resonance

(ECR) or Multifrequency Reactors (ECR/RF, RF/HF) have been extensively studied for laboratory and industrial applications [30], [31], [32]. Recently the possibility of high anisotropic etching of submicron Ir/PZT/Ir ferroelectric capacitors by using a Dual Frequency High Density reactor operating at 380 °C has been reported [33], [34]. PZT piezoelectric MEMS microfabrication requires many different thin film patterning processes as well as silicon bulk micromachining. In following sections, selected topics that are specific to the microfabrication of piezoelectric devices will be presented:

- Top electrodes patterning,
- Wet and dry methods for patterning of thin (PZT, AlN) and thick (PMN-PT) films,
- Dry etching of bottom electrodes (platinum, $RuO_2$, iridium),
- Silicon bulk and surface micromachining.

### 3.3. Patterning of top electrodes

To collect the electrical charges on the PZT film different types of electrodes need to be deposited and patterned (Au/Cr, Pt).

Compared to platinum, Au/Cr top electrodes have two advantages:

- the patterning of Au/Cr top electrodes does not require plasma methods, as lift-off process is very versatile and offers sufficient resolution (micron scale) for MEMS applications,
- Au/Cr top electrodes do not need post-deposition annealing.

Au/Cr top electrodes are deposited by Joule effect (Au) and e-beam source (Cr) in common evaporator and patterned by lift-off using a negative photoresist. Figure 4 illustrates the 100 nm thick Au/Cr top electrode obtained by lift-off on top of the 50 μm × 1000 μm cantilever. The lateral resolution is only limited by the resolution of the mask and the thickness can reach few hundreds of nanometers.

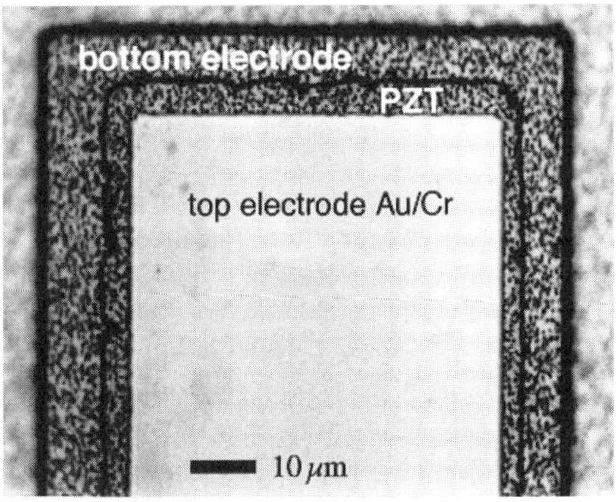


**Fig. 4.** Optical image of top view of the tip of a 100 × 500 $\mu m^2$ PZT/Si cantilever.

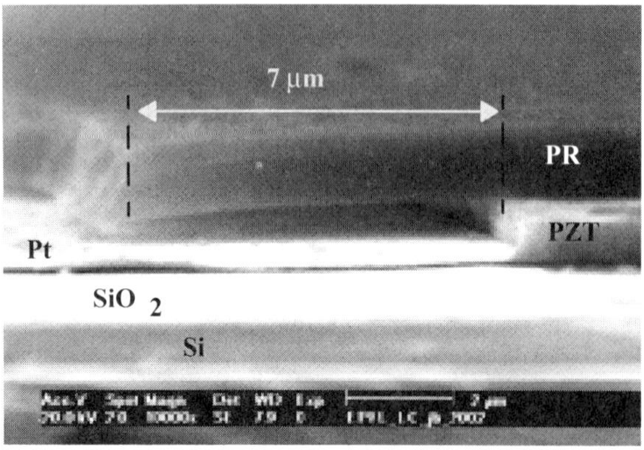


**Fig. 5.** SEM image cross-section of photoresist/PZT after wet etching of PZT (lateral underetch is 7 μm for 1 μm thick PZT layer).

As for platinum top electrode, it is obtained in three steps:

- PVD deposition of 50 to 200 nm Pt at room temperature,
- High temperature annealing to increase the adhesion, to restore the interface Pt/PZT and to recrystallize Pt
- Photolithography and dry etching to shape the top electrode (effective size of device).

With PVD deposited platinum, the PZT surface is damaged by the ions bombardment as well during deposition and etching. Thermal annealing is then required to restore the interface [35]. The maximum thickness is limited by the selectivity between platinum and photoresist mask during the dry etching.

### 3.4. Wet etching of piezoelectric thin films

The wet patterning of PZT thin films is a critical point in piezoelectric MEMS microfabrication. Despite the lack of etch rate control and photoresist undercut, the wet etching of PZT films remains a very versatile, fast and cheap method to open large areas (> 50 × 50 μm$^2$) like vias, with rather poor lateral resolution control. Because of high concentrations of hydrofluoric acid (HF), most of etch recipes are not selective to $SiO_2$ and to the $TiO_2$ (adhesion layer of the platinum bottom electrode). Delaminating of the structure can thus occur. In terms of etching rate, HCl solutions are very effective and selective etchants [36]. In our works, the standard solution of 30 ml of concentrated HCl (37%), 70 ml of water and 0.2 ml of HF has been used. At 56 °C, the etching rate is about 50–100 nm/s. Figure 4 shows the top view of a 500 × 100 μm$^2$ micro cantilever where the 1 μm PZT film has been patterned by wet etching. Underetching and profile irregularities as large as 5–10 μm can be observed which give a masks design rule (geometry tolerance) of about 10 times the PZT thickness (Figure 5). Under these conditions wet etching of narrow patterns (e.g. grooves of width 3–10 μm, capacitors below 25 μm$^2$) is thus impossible and dry etching techniques should be employed.

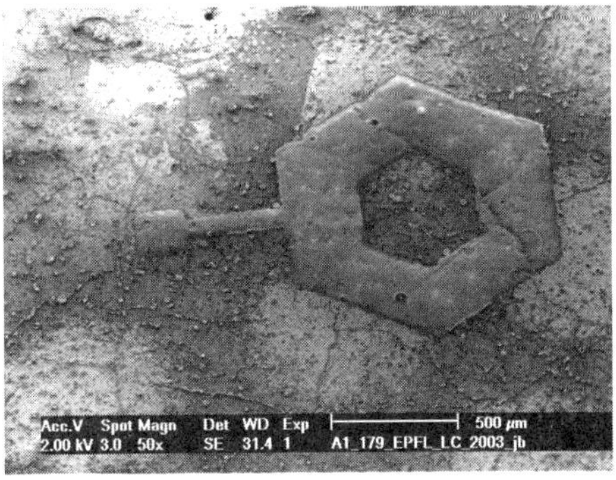


**(a)**

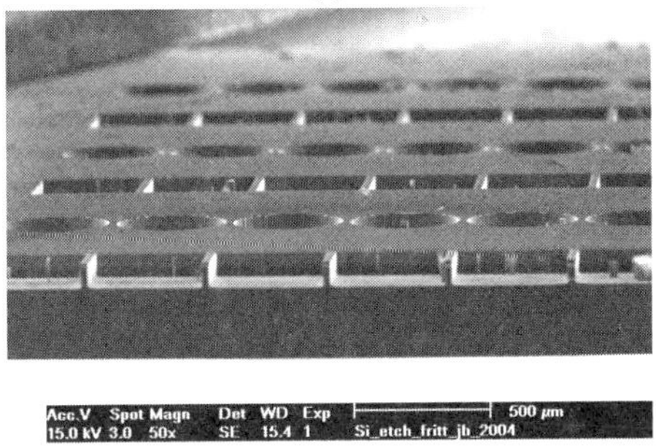


**(b)**

**Fig. 6.** (a) SEM top view of wet etched 10 micron PMN-PT thick film on Si substrate. (b) SEM view of backside of partially etched Si from wafer after the firing of PMN-PT film.

AlN sputtered thin films can be patterned in pure $H_3PO_4$ acid at 80 to 100 °C. The method is not isotropic and the etching rate depends on crystallographic orientation. Typically the etch rate is between 30 and 100 nm/sec.

Recently the integration of dense PMN-PT thick films on Si substrate has been attempted. The device fabrication requires patterning of up to 20 μm thick PMN-PT layer. Standard photoresist (Shipley 1818, 3.6 μm thick) could be applied because of the limited porosity in the sintered layer. Wet etching with a HF/HCl solution at 55 °C enables the patterning of relatively large devices (larger than $100 \times 100\ \mu m^2$, Figure 6). The anisotropy of patterns is satisfying and almost vertical sidewalls were observed. Some residues (Figure 7) visible at the pattern sidewalls and on the substrate can act as the etch stop layer preserving from larger underetch.

### 3.5. Dry etching of piezoelectric thin films

As the size of MEMS critical features shrank below 10 μm, the dry etching of PZT or AlN becomes mandatory to obtain a high degree of anisotropy, smoothness and preserved shape of profiles. The principal requirements for a dry etching process of piezoelectric films are:

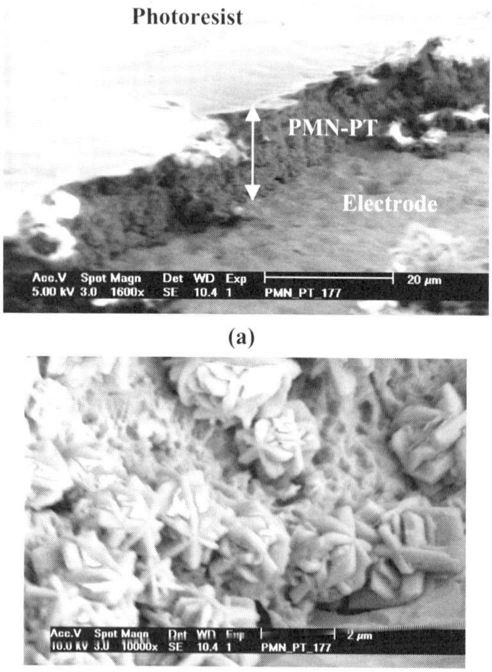


**Fig. 7.** (a) SEM side view image of of wet etched 10 micron PMN-PT thick film, (b) details of the sidewall of etched film.

- Sufficient selectivity to the photoresist (higher than 0.5) and with respect to the bottom electrode (Pt),
- High anisotropy to define precise micron-scale geometries (e.g. grooves, capacitors),
- High etching rate,
- No damage to the piezoelectric films and no patterning residue.

The dry etching of PZT films is not obvious as there is no common halogenous gas that forms volatile compounds with all three elements (Pb, Zr, Ti) to guarantee the residue free removal of film. The volatility of the reactive etch by-products of Pb, Zr and Ti is variable and limited [37] and energetic ion bombardment is often required to obtain the uniform removal of PZT layer. First developed processes used pure argon ion milling or Reactive Ion Etching (RIE) in simple parallel plate or inductively coupled plasma (ICP) reactor. Zeto [38] demonstrated etch rate up to 100–250 nm/min with argon ion milling. However, sloped walls (35°–70°), low selectivity to the platinum bottom electrode and photoresist mask as well as material redeposition on the sidewalls were the limiting factors. To overcome these drawbacks, Reactive Ion Etching (RIE) with halogenous gases should be employed. Chlorinated plasmas were generally employed because of the low melting point of metal chlorides with respect to the equivalent metal fluorides (e.g. $PbCl_3$, $PbF_4$). However, fluorine ion bombardment ($CF_4$, $SF_6$, $C_2F_4$) was shown to be more selective with respect to the platinum electrodes.

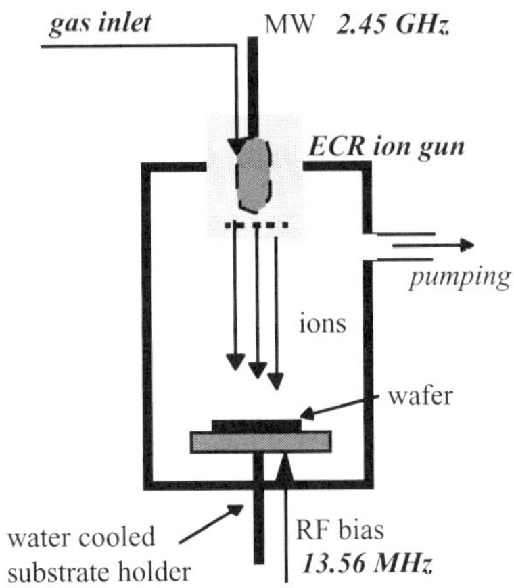


**Fig. 8.** Schema of dual frequency ECR/RF RIBE reactor [44, 45].

Etch rate of PZT has been increased up to 100 nm/min with monochlorotetrafluoroethane ($HC_2ClF_4$) as etch gas and 500 W RF power in a standard parallel plate reactor [38], but in this case, the stability of the photoresist was the limiting factor. For films thicker than 250 nm, the RF power has to be dramatically reduced down to 150 W to keep the photoresist mask removable. At such power level, the etching rate is only 13 nm/min. This example is representative of the problems occurring in RIE of PZT films. It is very difficult to find a trade-off between: reasonable PZT etch rate, photoresist dimensional stability and removability. Other studies gives similar results in term of etching rates (10 to 40 nm/min) in $HC_2ClF_4$ [39], in $C_2F_6/Cl_2$ [40] or in $CF_4/Cl_2$ [41] gases. Recently, Chung et al. [42] reported the etching process of PZT using ICP reactor and HBr/Ar gas mixture. An etch rate of 90 nm/min and a steep etch profile of 70° has been achieved.

Reactive Ion Beam Etching (RIBE) in a dual frequency ECR/RF reactor [43], [41], [44], [45] allows much lower working pressure than the RIE/ICP process (1 to 10 mPa compared to 1–10 Pa). The risk of material redeposition is reduced and the verticality of the sidewalls is improved thanks to the directionality of the bombarding ion beam. The bombardment of a reactive ion beam with gases such as $SF_6$, $CF_4$ or $CCl_4$ is very favorable to reduce the level of energy needed for the etching process and thus to obtain better trade-off between photoresist stability and etching rate.

The homemade reactor we used was based on a commercial ECR ion gun [44] as illustrate in Figure 8. The capacitive coupling of the substrate holder to a 13.56 MHz RF power supply contributes to the acceleration of the ions bombarding the substrate (negative bias voltage).

For RIBE of PZT films, a process with low ion energy ions and low operating pressure (see Table 2) has been developed. Best results were obtained by mixing both chlorine and fluorine gases and light Ar ion bombardment [45]. Photoresist stability was improved by hard baking at 150 °C in air and by keeping the moderate ion bombardment ($<0.75$ W/cm$^2$) during the process.

Dense, up to 2 μm thick PZT films have been etched with anisotropy close to 90°, without residue (Figure 9) and without degradation of PZT structure. Higher

**Table 2.** Process characteristics for etching of PZT and Pt with ECR/RF ion gun [30, 45]

| | PZT Etch Process | Platinum Etch Process |
|---|---|---|
| Gases | 40% $CCl_4$ / 40% $CF_4$ / 20% Ar | 85% $CCl_4$ / 15% Ar |
| Pressure | 10 mPa | 5 mPa |
| RF bias | 60 W (max. 0.75 W/cm$^2$) | 50 W ( max. 0.65 W/cm$^2$) |
| Etch rate | max. 70 nm/min | max. 45 nm/min |
| Selectivity | 0.5 to photoresist, 1.5 to Pt | 0.75 to photoresist, 2.5 to PZT |
| Mask | Shipley 1818, 2.3 μm, hard bake at 150 °C, 30 min, air oven | Shipley 1805, 1.5 μm, hard bake at 150 °C, 30 min, air oven |

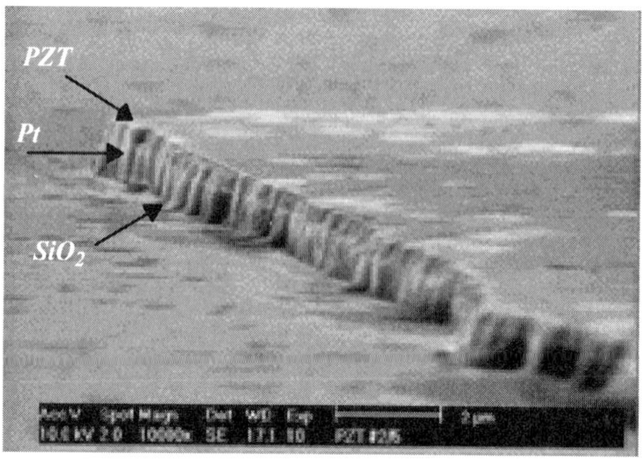


**Fig. 9.** SEM side view image of patterned PZT thin film, working pressure = 0.01 Pa, 60 W RF bias, optimized etching chemistry. Photoresist mask removed [45].

working pressure (> 0.5 Pa) or higher RF bias (> 100W) lead to destroying of PZT film (Figure 10).

As demonstrated for patterning of PZT for FERAMs cells the ferroelectric properties could be preserved during patterning processes. The impact of patterning process (plasma etching in ECR/RF RIBE) on properties of ferroelectric capacitors has been investigated. Properties of Pt/PZT/$IrO_2$-Ir capacitors have been monitored after all possible process steps:

- after top electrode etching (Figure 11a),
- after PZT dry etch (Figure 11b),
- after $IrO_2$/Ir/TiN dry etch (Figure 11c).

In contrast to most of literature data related to patterning of ferroelectric capacitors, here no degradation of PZT ferroelectric properties at different steps of patterning sequences was observed. Remanent polarization of 13 μC/cm$^2$ and maximum electric field of 400 kV/cm have been measured during all fabrication process and before final annealing in oxygen. This fact can be attributed to the patterning process (ECR RIBE) operating at very low working pressure (10 mPa) and using low energy ion bombardment (< 25V) in comparison to major ICP plasma reactors [46], [47].

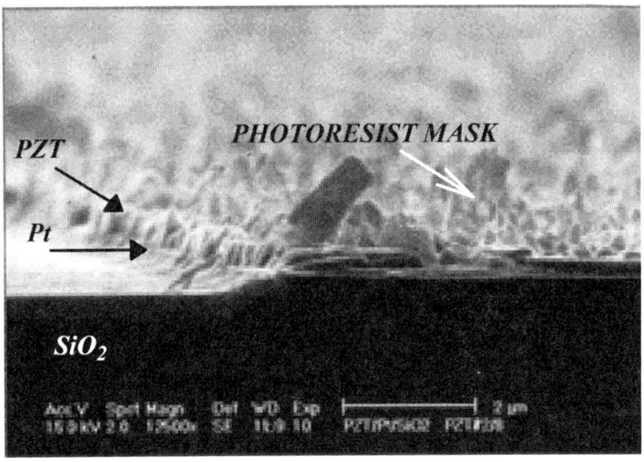


**Fig. 10.** SEM side view image of patterned PZT thin film, working pressure = 0.5 Pa, 120 W RF bias, optimized etching chemistry.

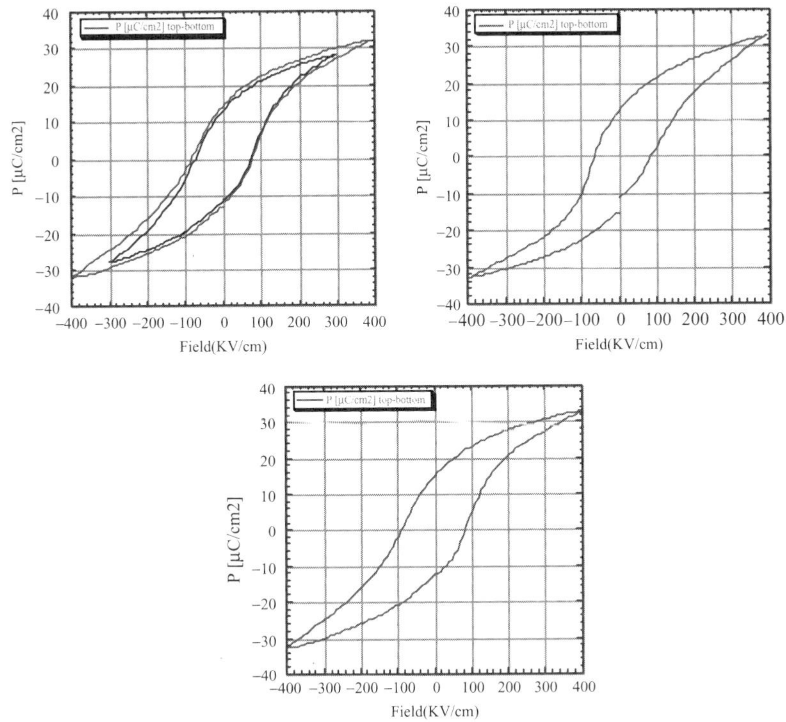


**Fig. 11.** Ferroelectric loops measured on $Pt/PZT/IrO_2/Ir$ stack: (a) after Pt top electrode etching, (b) after PZT dry etching, (c) after $IrO_2/Ir$ dry etching.

Dry etching of AlN is based on ICP process. Typically $Ar/Cl_2$ flux ratio is 1/3, the process pressure 8 mTorr and bias RF Power of 150W is applied. The etching rate yields 200 to 330 nm/min but it strongly depends on the morphology of AlN layer. Selectivity to Pt is 1:10 and to photoresist mask 1:0.5. When using a standard photolithography process, the AlN used to be attacked by basic solution of the developer. This increases the roughness of AlN and leads to some alteration of the etching rate, micromasking and residues formation.

## 3.6. Dry etching of bottom electrodes

### *3.6.1. Platinum*

Ferroelectric thin films for MEMS or memory applications require noble metal (platinum, iridium) or conductive metal oxide ($RuO_2$, $IrO_2$) electrodes. Due to their chemical inertness, the patterning of these materials is a difficult task. The platinum thin films can be patterned by wet etching: aqua regia [48] or electrochemical wet etching in concentrated HCl [49] are the most known methods. However, these recipes do not work with very thin structures (below 10 μm), are very sensitive to the platinum surface contamination, are often subject to underetching of the adhesion layer and, the selectivity to PZT is very low.

For piezoelectric MEMS, HDP dry etching techniques have to be chosen. There is only a small number of volatile compounds that can be formed with platinum (e.g. $PtCl_3$ decomposes at 370 °C in $Cl_2$, $PtI_2$ decomposes at 270 °C). In fact, the etching mechanisms are much more physical (ion bombardment) than chemical (formation of volatile reaction products). Most of recent developed methods have used RIE/ICP (Reactive Ion Etching with Inductive Coupled Plasma reactors), simple Ar ion milling or ECR/RIBE. In all cases [50], [51] several common limitations have been observed:

- Selectivity problems with photoresist
- Hard masks (TiN, $SiO_2$) masks are not common to use in presence of PZT,
- Metal residues on the sidewalls lead to fencing,
- Poor patterns anisotropy with ion milling,
- Due to the strong ion bombardment photoresist mask gets harder and after stripping the organic residues remain.

In our studies, chlorine based RIBE process with dual-frequency ECR/RF plasma [30], [31] (see Table 2). Figure 12 shows a SEM image cross-section of the edge of 100 nm thick Pt film deposited on $Si_3N_4/SiO_2$ membrane. The anisotropy of the pattern is close to 90° and the selectivity to $Si_3N_4$ is excellent (no overetch into $Si_3N_4$). With a RIE/ICP reactor from STS, best results have been obtained with $Cl_2$ flow rate of 20 sccm, Ar flow rate of 70 sccm, working pressure of 1 Pa, RF plasma power of 800 W and RF substrate bias power of 150 W. The etching rate was lower (32 nm/min) than the one obtained by RIBE and the selectivity was 0.35 and 0.5 with respect to standard photoresist (Shipley 1805) and $SiO_2$ respectively. Figure 13 shows the profile of $Pt/SiO_2$ after ICP etching at 1 Pa. The curved shape of the photoresist and the thin film is typical of highly physical dry etching process. However, any redeposition of the metal on the mask sidewalls has not been observed. Both processes for Pt etching (RIBE/ECR and RIE/ICP) offer a good selectivity to PZT ($> 2$) and have been successfully applied for the patterning of platinum top electrodes. The photoresist mask is striped in Shipley Microposit Remover 1165 at 70 °C followed by low power ashing in plasma oxygen [10].

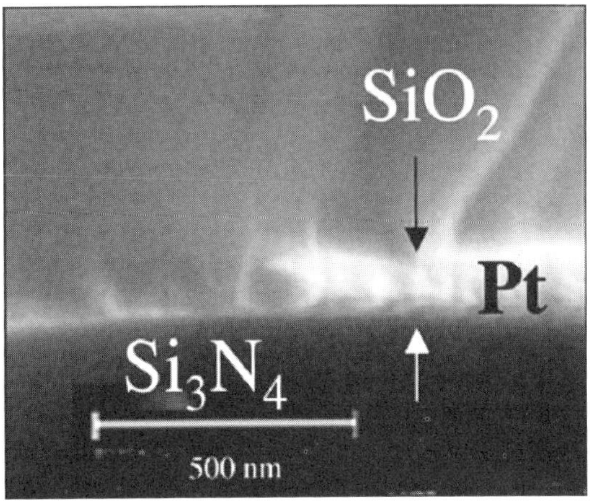


**Fig. 12.** SEM image cross-section of ECR dry etched Pt on $Si_3N_4$ membrane.

**Fig. 13.** SEM image cross-section of ICP dry etched Pt film.

### *3.6.2. Iridium*

The patterning of Ir electrode is mostly related to fabrication of FeRAMs capacitors [52], [53]. RIE/ICP etching of iridium has been related by several groups [54], [55], [56], [57]. Chung et al. has used a hard mask ($TiO_2$) [54] or photoresist mask [55] testing chlorine, bromide and fluorine chemistries ($Cl_2/O_2/Ar$; $HBr/O_2/Ar$; $C_2F_6/O_2/Ar$, $C_2F_6/Cl_2/Ar$). Etch rates 40 to 60 nm/min has been obtained with chlorine. Clean profiles with $Cl_2$ and HBr when using a hard mask and with fluorine when using the photoresist were observed. Residue-free patterns, 50° sidewalls with fluorine-chlorine gases and Ar addition have been finally obtained. Chongying Xu et al. [56] have established the process for etching of Ir electrical contacts. By using $XeF_2$, in presence of Si source a good selectivity to $IrO_2$ and Pt has been achieved and Ir/PZT/Ir capacitors remained non-affected. Chiang et al. [57] has employed the $Ar/O_2/BCl_3$ gas mixture and TiN hard mask. The selectivity Ir/TiN up to 10, the Ir etch rate up to 66 nm/min (60%Ar/20%$O_2$/20%$BCl_3$, 1500W bias), fence free patterns and vertical ($> 70°$) sidewalls were observed. Patterning of Ir/$IrO_2$/PZT submicron, ferroelectric stack has been

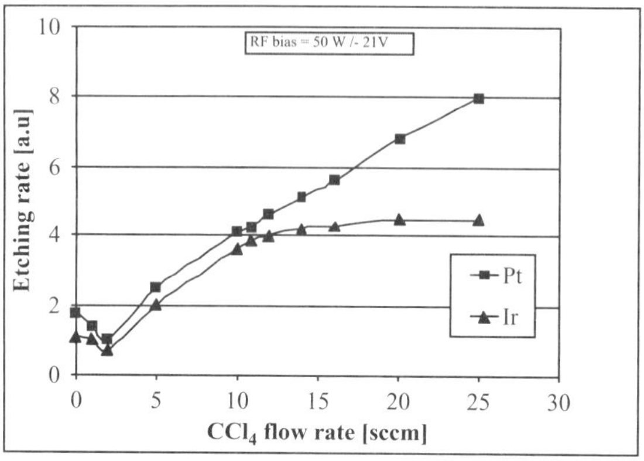


**Fig. 14.** Comparison of etching rates of Pt and Ir as function of $CCl_4$ percentage in $Ar/CCl_4$ mixture when using ECR/RF RIBE [10].

demonstrated and commercialized by Tegal [33]. Using the double frequency HDPe reactor operating at 380 °C the etching rate of Ir up to 100 nm/min with the sidewall profile better the 85° have been obtained.

Etching of $Ir/IrO_2$ stack for FeRAM applications was done with ECR/RF Ion Gun [10]. The 100 nm iridium bottom electrode was etched in $80\%CCl_4/20\%Ar$ mixture. Increasing of $CCl_4$ rate flow induces some enhancement of etching, however, a saturation of removal rate occurs when limited by formation of $IrCl_4$ product (Figure 14). Despite, the etch rate of Ir can be stimulated by RF substrate bombardment, the maximum etch rate 15 to 20 nm/min was obtained. Micron-scale resolution patterns without any impact on properties of ferroelectric capacitor or transistors have been performed (Figure 15).

### *3.6.3. Ruthenium oxide*

For $RuO_2$ it has been reported that the etching in oxygen plasma is effectuated by the formation of volatile $RuO_4$ [58] and this process can be catalyzed by small addition

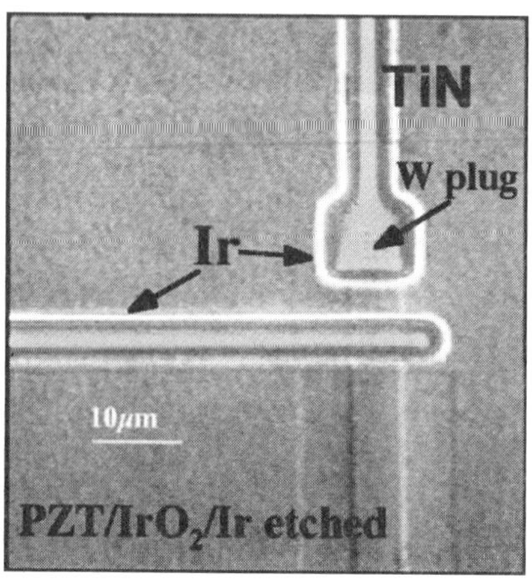


**Fig. 15.** Top view of $PZT/IrO_2/Ir$ stack etched with ECR/RF RIBE.

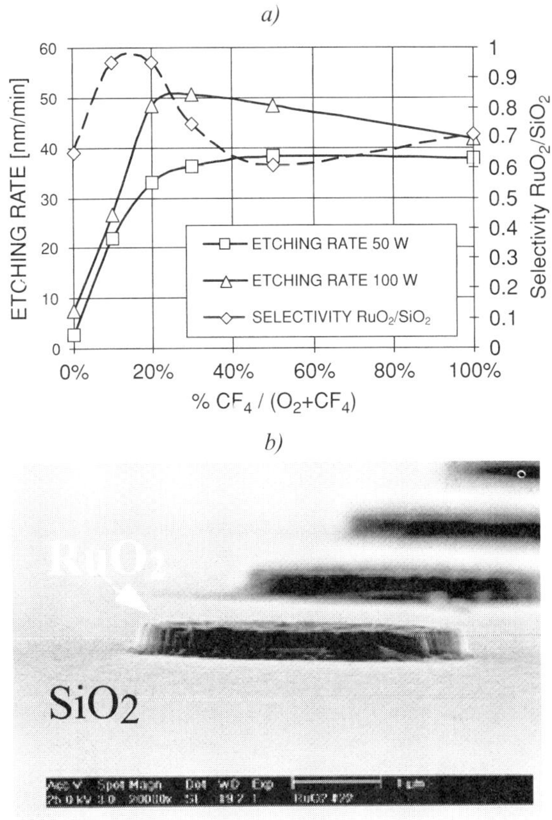


**Fig. 16.** Etching of $RuO_2$: a) effect of $CF_4$ concentration on etching rate of $RuO_2$ and the selectivity $SiO_2/RuO_2$, b) $3 \times 3$ micron$^2$ patterns obtained with ECR/RF RIBE [31].

of Cl- or F-based gas to $O_2$ plasma [59]. Most of these experiments were carried out in RIE/ICP systems where independent control of the energy and flux of the reaction species is not easy. Moreover, the working pressure range is usually about several Pa reducing the anisotropy and creating the polymer like residues. Unlike Lee [58], a very low etch rate with Cl was obtained even if $O_2$ was added. Only fluorine-containing gases provides reasonable etch rates [31]. As suggested by Desu [59] a small addition of fluorine gas to oxygen increases significantly the etching rate. When the $CF_4/O_2$ is used, the addition of 25% of $CF_4$ induces the increase of etching rate from 10 to 50 nm/min at 100W RF bias and from 7 to 35 nm/min at 50W RF bias. An interesting correlation with the selectivity to $SiO_2$ occurs—for low quantity of $CF_4$ the selectivity is poor because of low etching rate of $RuO_2$. On the other hand, the high (above 50%) proportion of $CF_4$ induces higher etch rate of $SiO_2$ which reduces the selectivity again. Finally, an optimal proportion of 20% $CF_4/O_2$ allows to achieve both high etching rate and good selectivity to $SiO_2$ underlayer. As shown in Figure 16 the $2 \times 2$ μm$^2$ patterns through 300 nm thick $RuO_2$ can be obtained with anisotropy higher than 80°. Smooth $SiO_2$ substrate surface is visible and no polymer-like residue due to decomposition of

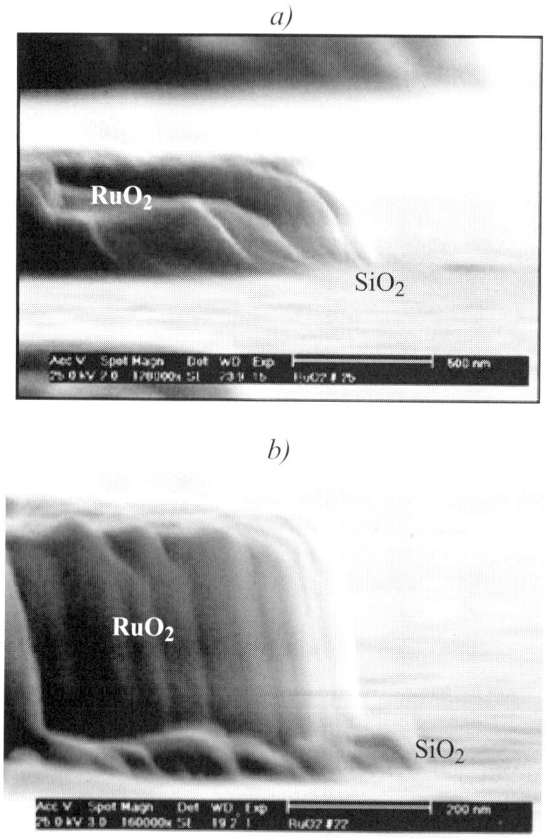


**Fig. 17.** Profiles of $RuO_2$ patterns obtained with a) $SF_6/O_2$, b) $CF_4/O_2$ mixture.

$CF_4$ [58] was observed; this fact is due principally to a low working pressure (typically 10 mPa) in ECR/RF system. Figure 17 shows the difference in etched profiles of $RuO_2$ when using a) $SF_6/O_2$ and b) $CF_4/O_2$ mixture; for the same process conditions the etch rates were similar, but etching in $SF_6/O_2$ results in lower anisotropy of patterns and lower selectivity to $SiO_2$.

### 3.7. Silicon bulk micromachining

To produce thin silicon supporting structures like beams or diaphragms, bulk silicon has to be etched in a uniform way, down to the desired thickness (usually thinner than 20 μm). As the sensitivity of a sensor based on a cantilever is proportional to $1/(\text{silicon thickness})^2$, precise control of the silicon thickness has a direct influence on the production of uniform and homogenous devices. To control accurately the thickness of the silicon structures, many wet-etching methods (KOH, TMAH) using an etch-stop on a very well defined n-doped epi [27] or highly p-implanted silicon layers have been developed [6]. However, these methods suffer from difficult and complicated process implementation and stress non-uniformity. Despite a very low running cost, bulk silicon micromachining in KOH has important drawbacks:

- the shapes of the structures are restricted by the etching anisotropy (for example, V-grooves of 10 μm wide cannot achieve a depth larger than 7 μm due to the lower etching rate of the (111) planes),
- the etch rate in KOH depends not only on solution concentration and temperature, but also on impurity presence in Si wafer,
- the KOH is very aggressive against the PZT, AlN, and electrodes,
- the uniformity is limited to about ± 10–15 μm after etching 380 μm of silicon.

The dry and anisotropic plasma etching process offers much more flexibility in terms of design and is the solution of choice to define accurately small structure like very narrow gaps, grooves and even large surface in case of membrane patterning. The most commonly used process for deep silicon etching is the so-called "Bosch process" patented by Lärmer and Schilp [60]. By sequentially alternating etching steps (with $SF_6$ chemistry) and passivation steps (polymerization of the sidewall with $C_4F_8$), a deep anisotropic etching can be obtained on feature sizes ranging from submicron to several millimeters [61]. High etching rate of 7–10 μm/min enables a rapid fabrication of deep trenches and membranes. Figure 18 a) shows a SEM cross-section of a 2 × 30 μm

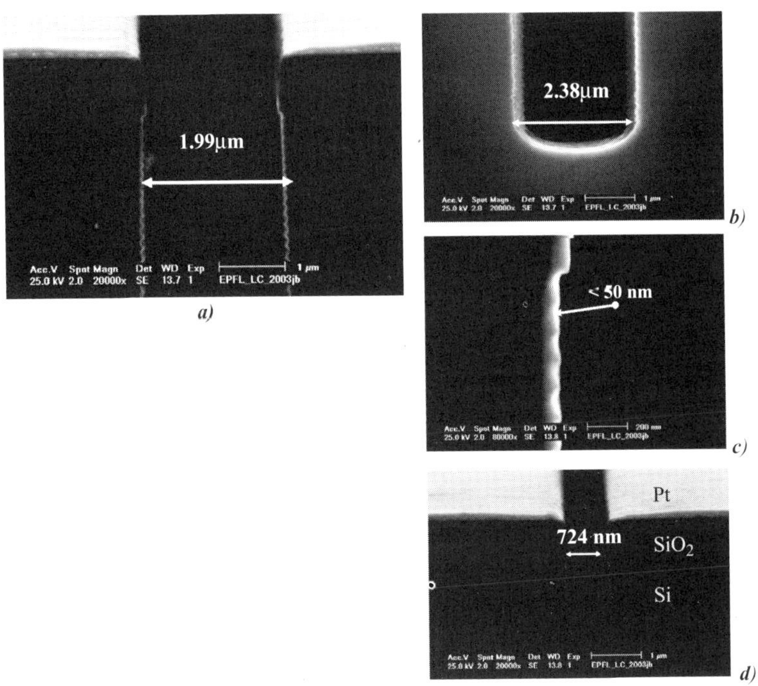


**Fig. 18.** SEM images cross section of 2 × 30 μm (AR = 15) groove etched through $Pt/Ti/SiO_2/Si$, and details of bottom and sidewalls of the slit (b,c). Pt has been used as a hard mask [10]. Submicron slit obtained by this method (d).

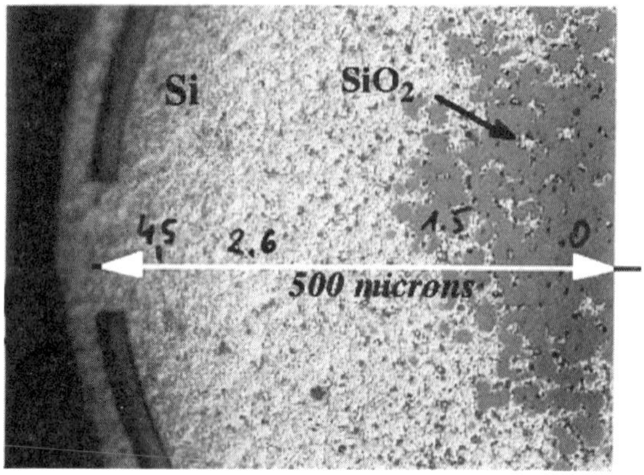


**Fig. 19.** Non uniformity of Si thickness at the bottom of 1 mm$^2$ large silicon membrane.

trench. Vertical sidewalls with aspect ratio of 1:15 have been obtained when using Pt bottom electrode as a hard mask even for submicron trenches (Figure 18 b, d) [10]. However, defining precise and uniform membrane thickness all over a wafer by deep silicon etching is a difficult task. It can be observed on Figure 18 c) that the bottom of the etched cavity is neither flat nor perpendicular to the sidewalls. This phenomenon can be explained by the ion scattering effect at the sidewalls and creates a non-uniformity of the membrane thickness (thicker near the clamping point) that can amount to several microns. As Figure 19 shows, the non uniformity of Si thickness at the bottom of 1 mm large membrane [9] is important. The Si thickness increases from 10 μm at the membrane outskirts to zero in the membrane centre, where the underlying $SiO_2$ layer is clearly visible. This difficulty in removing of Si near the clamping point or in the corners, where the ion bombardment is reduced, is critical issue, even if SOI substrates are used (Figure 20). For presented here $1000 \times 1000 \times 10$ μm bridge the nominal thickness of 10 μm could be achieved but the complete release requires longer etching time and precise wafer inspection. Furthermore, on the wafer level, a depth uniformity of about 3% (±5 μm in 380 μm deep cavities or 10 μm thick membranes) is typical even with well-tuned process. This leads to the quite large disparity of membrane properties and performances across the same wafer. To overcome, at least partially, these limitations, an etch-stop on a buried silicon oxide layer may offer an interesting solution.

The combination of short time isotropic and anisotropic plasma etching of Si using photoresist/$SiO_2$ mask is demonstrated in Figure 21. These methods are basically used for sharpening of AFM tips, micro-needles arrays or for surface micromachining dry release.

### 3.8. Integration of PZT films on SOI and polysilicon substrates

Silicon-on-Insulator (SOI) or polysilicon coated $SiO_2$/Si substrates can be used to improve the uniformity of the structure thickness on the wafer scale. These two substrates use mono- or polycrystalline silicon layer deposited on insulating and amorphous $SiO_2$ thin film; the latter acts as an etch-stop during the bulk silicon micromachining (DRIE) or can be used as sacrificial layer for surface micromachining.

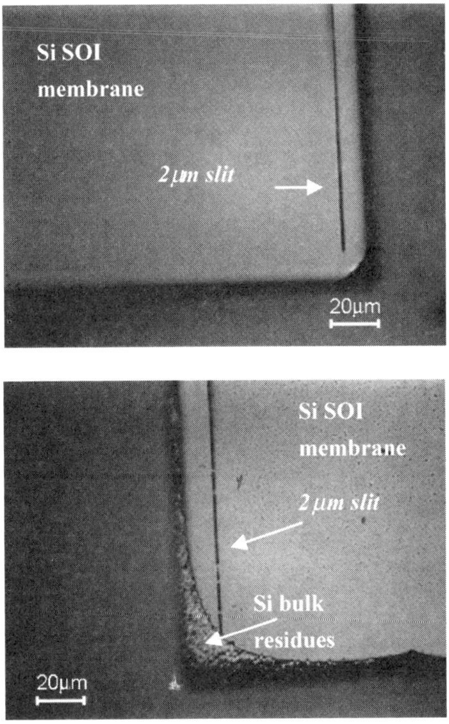


**Fig. 20.** Uniformly etched Si 5 µm thick SOI membrane (backside view).

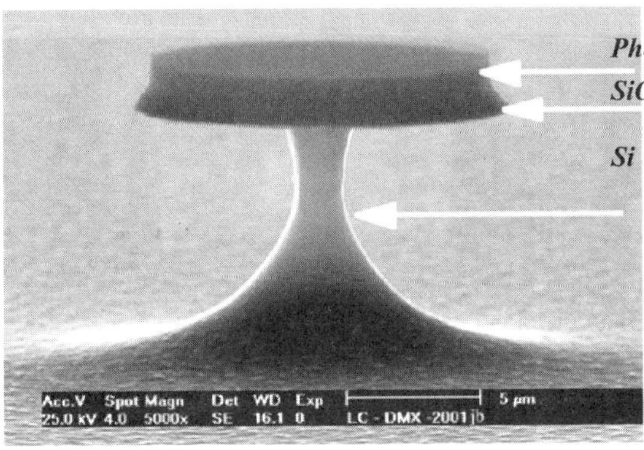


**Fig. 21.** SEM view of feature obtained by combined anisotropic and isotropic DRIE of silicon.

In SOI wafers, the device silicon thickness (top silicon layer) is usually ranging from 0.2 µm to several hundreds of microns and the deposition of PZT films and process flow do not differ from processing on classical silicon substrates. Figure 22 illustrates example of so fabricated PZT/SOI 5 µm thick membrane surrounded by the 20 × 4 µm grooves [9]. When using thick polysilicon layer (up to 4 microns typically), additional processing steps are required as the roughness of the film is high (~300 nm RMS)

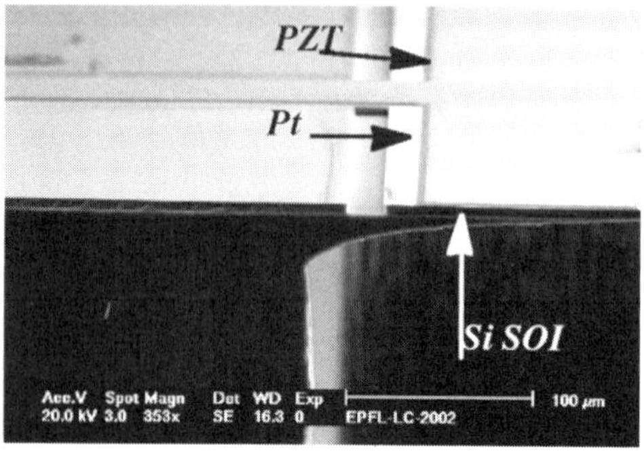


**Fig. 22.** SEM view of grooved membrane near the clamping point (2 μm PZT on 5 μm SOI substrate) [9].

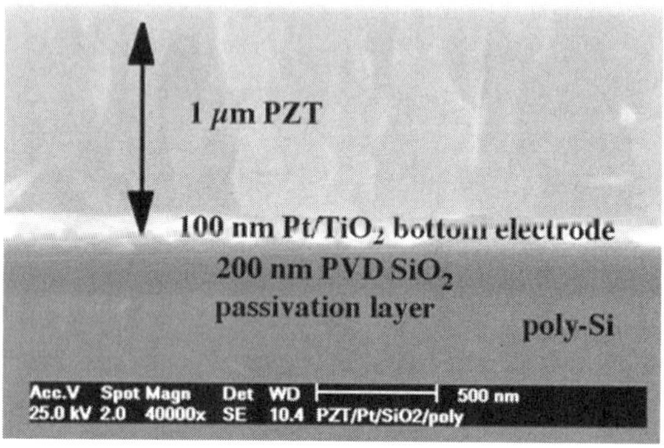


**Fig. 23.** SEM images cross-section of 1 μm PZT 53/47 {100}-textured thin film deposited on CMP-polished 4 μm poly-silicon layer.

after deposition by LPCVD. This roughness introduces random crystallization of PZT. To recover the surface state of polished silicon wafer, Chemical-Mechanical-Polishing (CMP) [62] of polysilicon has been used and 200 nm $SiO_2$ passivation layer have been then deposited by PVD.

Figure 23 shows SEM image of the cross section of 1 μm PZT 53/47 {100}-textured deposited on $PbTiO_3/Pt/TiO_2/SiO_2$/poly-Si/$SiO_2$/Si substrate. No cracks and no delaminating have been observed showing the excellent stability of the PVD $SiO_2$/poly-silicon structure during the processing of PZT and excellent functional properties have been also obtained: $e_{31,f} = -11\ C/m^2$, $\varepsilon = 1000$, $\tan\delta = 0.05$. Integration of PZT thin films on polysilicon substrate has been demonstrated and offers a valuable alternative to SOI wafers.

### 3.9. Surface Micromachining for pMEMS

Bulk micromachining means that 3-D features are etched into the bulk of crystalline and noncrystalline materials. In contrast, surface micromachined features are built up, layer by layer, on the surface of the substrate (e.g. a single crystal silicon wafer). Dry etching defines the surface features in the x, y plane and wet etching releases them from the plane by undercutting. The overview of traditional processes, materials and developed devices can be found in literature (e.g. [5], [7] for overview).

Recently, the surface micromachining has been used for fabrication of capacitive Micromachined Ultrasonic Transducers (cMUTs) [63], [64]. The feasibility of surface micromachining of Al/AlN/Pt TFBARs has been demonstrated by using PSG as sacrificial layer and Silox as removal agent.

However, there are still a lot of limitations concerning the compatibility of processes used for etching of the sacrificial layers with piezoelectric stack. Percin [65] has demonstrated the fabrication of piezoelectric (ZnO) ultrasonic transducer arrays by using the surface micromachining with low temperature oxide (LTO) as sacrificial layer. An interesting alternative to wet etching methods can be the new silicon sacrificial layer dry etching (SSLDE) for free-standing RF MEMS architectures [66]. This technique uses sputtered amorphous or LPCVD polycrystalline silicon as sacrificial layers and a dry fluorine-based ($SF_6$) plasma chemistry as releasing process. The process is capable of various applications in surface micromachining process, and can be applied in fabricating RF MEMS switches, tunable capacitors, high-Q suspended inductors and suspended-gate MOSFETs. It has been demonstrated that the SSLDE process can release metal suspended beams and membranes with excellent performance in terms of etch rate (up to 15 μm/min), $Si{:}SiO_2$ selectivity and is fully compatible with standard MEMS processing equipment and CMOS post-processing.

## 4. Examples of Fabricated Piezoelectric MEMS

### 4.1. Clamped-free and clamped-clamped cantilevers

Arrays of clamped-free micro cantilevers (10 to 100 μm in width, 50 to 1000 μm in length, nominal thicknesses of 5 and 10 microns) were fabricated using PZT on SOI processing (Figure 24). Clamped-clamped cantilevers with 2 micron of PZT built on 50 microns SOI were fabricated with different configuration of electrodes for simultaneous actuation and sensing (Figure 26). Electro-mechanical properties of piezoelectric micro cantilevers have been studied at the resonance in air and in vacuum (Figure 25) and under quasistatic excitations. FEM simulations have shown excellent agreements with respect to measured resonance frequencies and coupling factor (about few %) [67]. Whatever the geometry used (test samples on thick Si or micro cantilevers), highly reproducible $e_{31,f}$ values of $-10$ to $-12$ C/m$^2$ have been obtained with PZT 53/47 {100}-textured thin films [68]. Perfect accordance between the theoretical (calculated) and measured resonance frequency, as well very good uniphormity have been obtained for clamped-clamped cantilever. For the first resonance mode at 13.8 kHz, the deviation is lower than 1% (Figure 26).

### 4.2. Piezoelectric Ultrasonic Transducers (pMUTs)

The fabrication of piezoelectric micromachined ultrasonic transducers (pMUTs) with SOI wafers makes possible the formation of thin (2 to 50 μm) and uniform silicon diaphragm as well for single devices, 1D and 2D arrays. Suspended membranes down

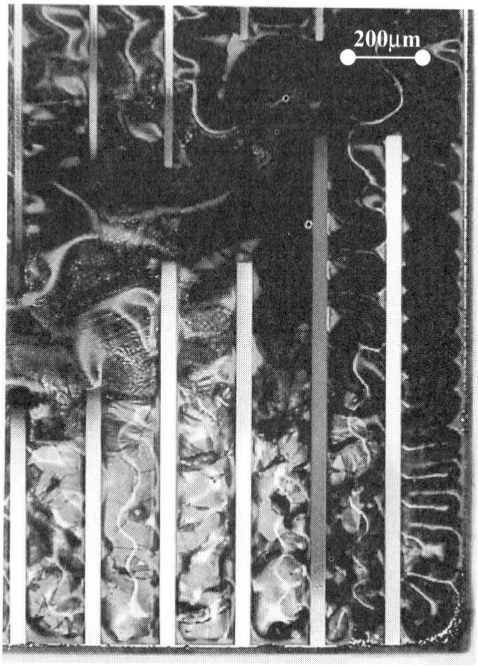


**Fig. 24.** Array of PZT/SOI microcantilevers. Backside view before the final release—etching of the buried $SiO_2$.

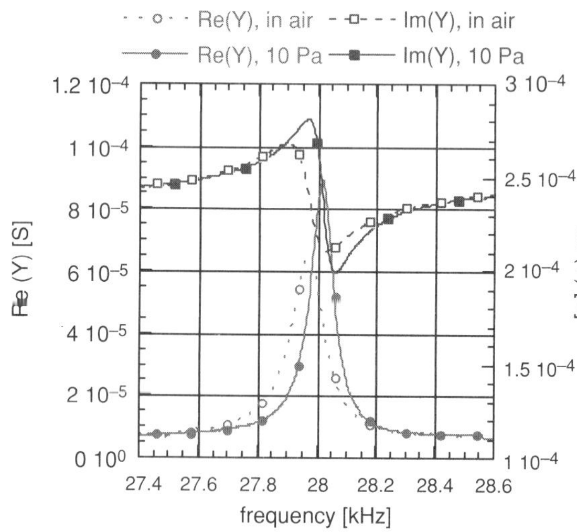


**Fig. 25.** Admittance curves of 1 mm × 100 μm large and 12 μm thick cantilever measured at atmospheric pressure and at 10 Pa.

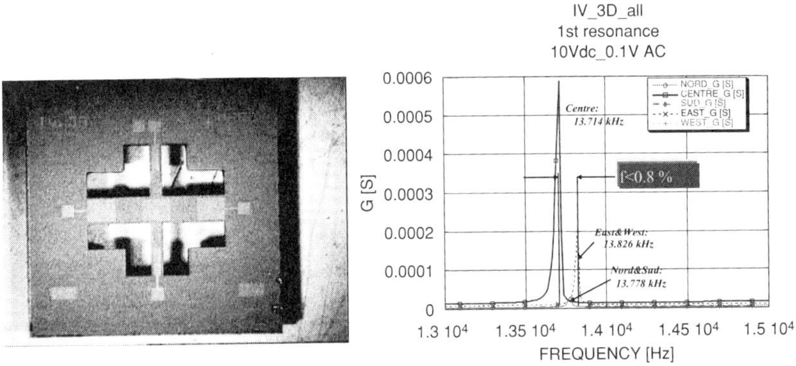


**Fig. 26.** Example of clamped-clamped cantilevers with multielectrode acting-sensing system. Admittance curves of 50 μm thick cantilever.

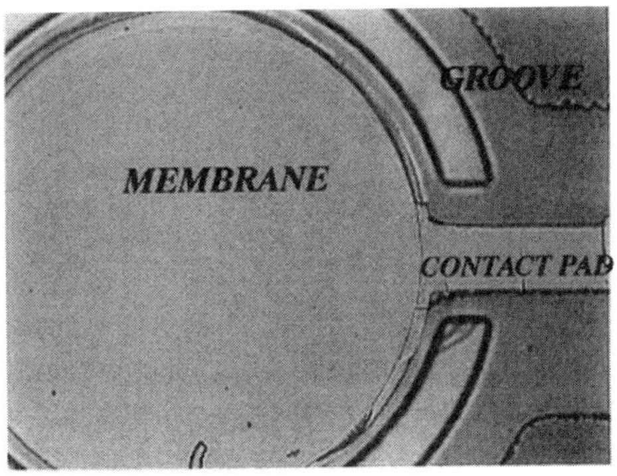


**Fig. 27.** Top view of 300 μm suspended membrane operating at 750 kHz [9].

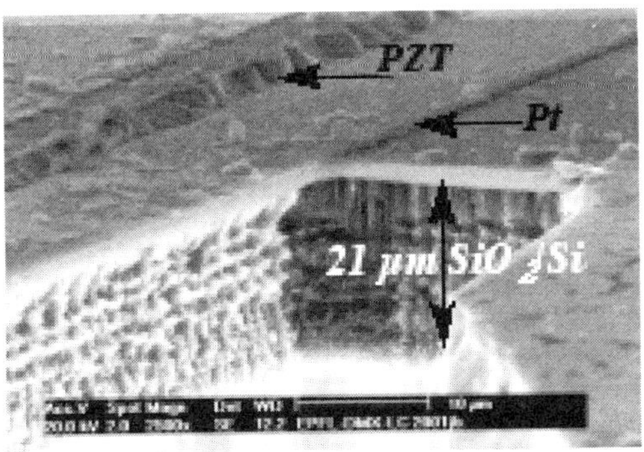


**Fig. 28.** SEM close-view of bridge and membrane of MUT operating at 100 kHz (2 μm PZT on 20 μm Si) [10].

to 300 μm pitch have been fabricated to achieve resonance frequencies in the 0.05 to 1 MHz range (Figure 27 and Figure 28). Values of the coupling factors $k^2$ as high as 6% have been measured with dc bias larger than 75 kV/cm for 300 μm large membrane at 750 kHz [9].

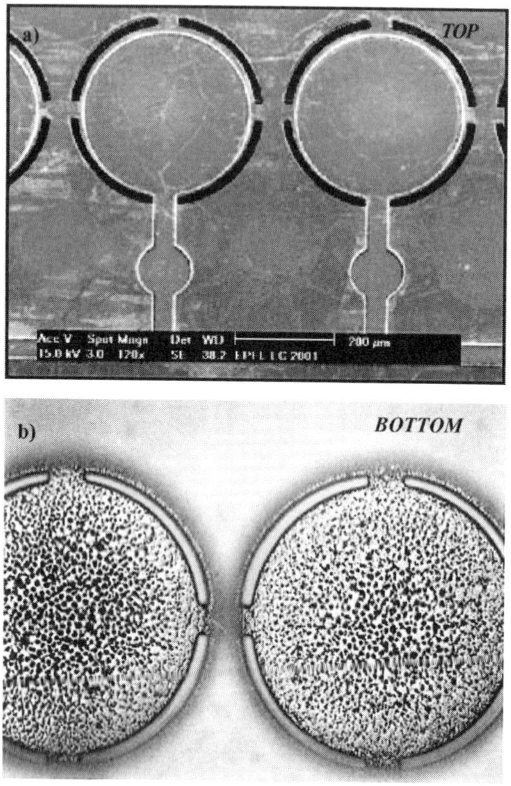


**Fig. 29.** Frontside and backside view of 450 micron pitch 1D array operating at 950 kHz (fully suspended membrane).

For applications as 1D (Figure 29a [9]) or 2D arrays it is crucial to have all the frequencies in a narrow range. Ideally, the scattering of frequencies should not be larger than the peak width of the resonances enabling the tuning. For the devices with the same geometry and boundary conditions the resonance frequency is defined by the thickness of Si membrane. Thus, the distribution of resonance frequencies is determined by the uniformity of etching of membranes (Figure 29 b). This was almost met with fabricated pMUT arrays. Figure 30 gives the frequency scattering within a 16 element linear array showing deviations within +/−2.5 %.

### 4.3. Pressure sensor and microphone

Piezoelectric cantilever and bridge acoustic sensors have been fabricated using the micromachining processes presented above. Two types of devices have been fabricated: cantilever and bridge structures (Figure 31 and 33). The thickness of membrane was 10 micron (+/−2 microns when standard Si substrate is used). To address the air leakage at low frequency, very narrow slits (3–5 μm) with very smooth and residue free sidewalls have been patterned around the structures through Pt/$SiO_2$/Si (Figure 32,

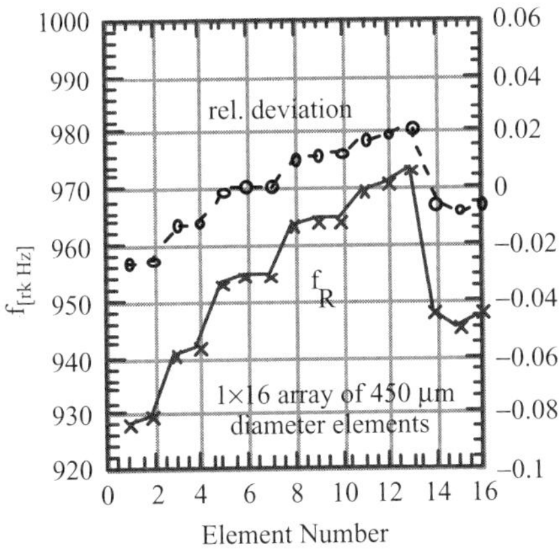


**Fig. 30.** Resonance frequency distribution of the 16 elements of a 1x16 array (450 μm diameter) measured in air.

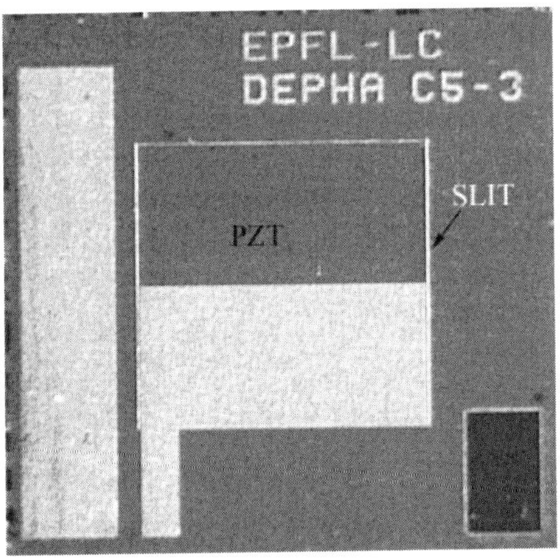


**Fig. 31.** Cantilever acoustic sensor, 1 micron of PZT, slit 5 μm [10].

34 [10]). It was thus possible to take advantage of the high sensitivity of cantilever structures to design acoustic sensors working at low frequencies (10 to 30 Hz). It has been found also that the slit conductance depends on many experimental factors such as its exact geometry, the surface roughness within of the walls or the presence of microfabrication residues. The integration of PZT/Si cantilever microphone into the photoacoustic gas detector has been demonstrated. Concentration of $CO_2$ down to 330 ppm (air) has been detected using miniaturized photoacoustic gas detector.

Recently, the acoustic sensor has been fabricated using sputtered Pt/AlN/Pt/Ti stack on thin SOI membranes (5 and 10 micron thick). Stress gradient compensation has been obtained for 2 micron thick AlN deposited on 1000 × 1000 × 10 μm SOI membrane. For thinner, 5 μm thick SOI membranes the distribution of tensile and compressive stresses is not uniform, inducing strong deflection and buckling of structures.

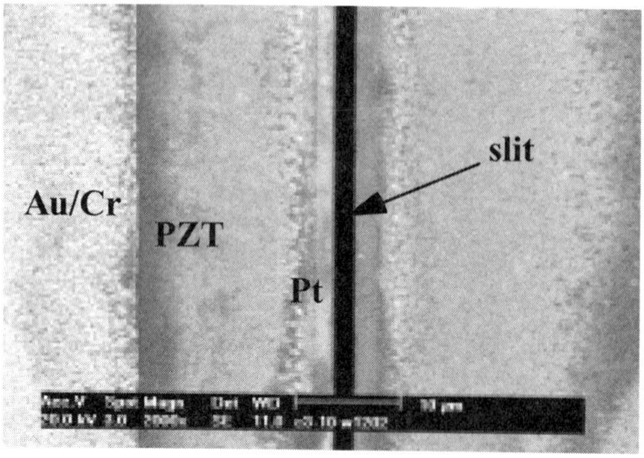


**Fig. 32.** SEM top view image of a patterned PZT/Pt/$SiO_2$/Si. Slit width = 2 µm [9,10].

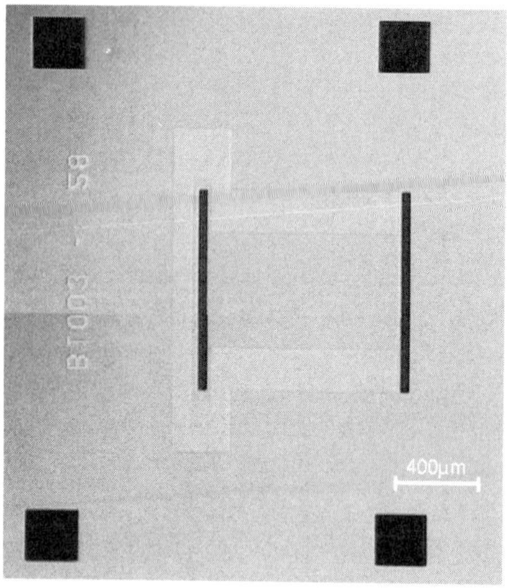


**Fig. 33.** Bridge acoustic sensor, based on 1.5 micron of AlN with surrounding slit of 1,2 or 3 µm

The acoustical characterization reveals the sensitivities of 40 to 70 µV/Pa for frequencies from 10Hz to 200Hz (the targeted range of application). The 10 µm deep and 2 µm wide slits (Figure 34), guaranties low air conductance and high sensitivity. All the process steps were conducted under 400 °C guarantying the post-CMOS compatibility.

### 4.4. RF Switches

MEMS switch have demonstrated useful performance at microwave frequencies using cantilever, rotary and membrane structures, because of their low insertion loss, high isolation, and ultra high linearity [69]. In general terms, cantilevered and multiple-supported surface micromachined electrostatic switches or relays are by far the most

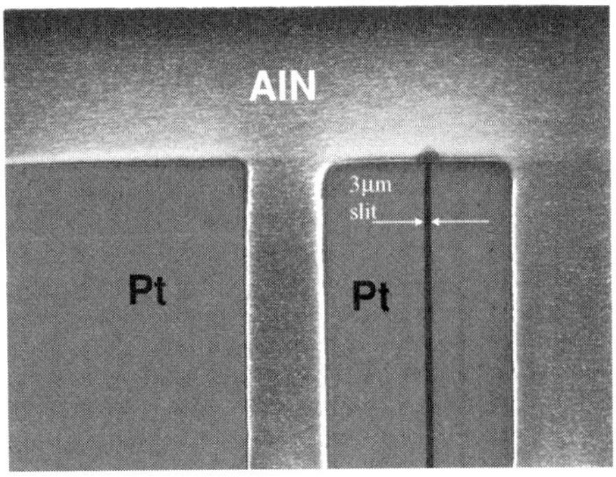


**Fig. 34.** SEM top view of patterned $AlN/Pt/SiO_2/Si$. Slit width = 3 μm.

widely studied devices to date. The need for low actuation voltage in MEMS switches (<15 V) has often caused excessive design and fabrication complexity as well as increase in the size of the device. The possible solution is to use the piezoelectric actuation for contact or shunt switches. There are two examples of recently developed switches using a piezoelectric actuation:

- capacitive switch developed by LG [70]; it uses a 4-arms piezoelectrically (PZT) activated movable membrane (LPCVD low stress SiN); in this configuration the required operating voltage is only 5 V,
- metal contact switches—one example uses the piezoelectric PZT actuation and is still under development at Penn State University [71]. A PZT unimorph cantilever actuator, fabricated on a sacrificial polysilicon layer and released using a xenon di-fluoride dry etch, was used to realize the switch. The PZT thin film was poled and driven with interdigitated electrodes to exploit the $d_{33}$ coefficient for switching actuation. The actuation voltage is about 40 V for 1 micron deflection. Accelerometer using the same type of actuation/detection (IDT and $d_{33}$ mode) has been developed recently by the same group [72].

## 4.5. Other devices (non mechanical-devices)

### *4.5.1. FeRams*

High density—low power memories, which are to be used in RF-ID cards, still represent a challenge for the semiconductor industry. The integration of ferroelectric thin films in CMOS technology represents one of the most important steps for the realization of such memories. We have investigated the complete integration of ferroelectric thin films (PZT) on W-plug in a standard 0.5 μm CMOS technology [73]. A special process flow compatible between ferroelectrics and CMOS integrations has been designed. Sputtered $IrO_2/Ir/TiN$ electrode has been used as well as barrier and seeding layer for growth of textured PZT. As barrier layer, it protects the W-plugs contacts against oxidation during the PZT deposition process up to 650 °C. Another important task was to observe the impact of patterning process (plasma etching in ECR/RF RIBE) on properties of ferroelectric capacitors. As shown in Figure 11, no alteration of ferroelectric properties has been measured [74].

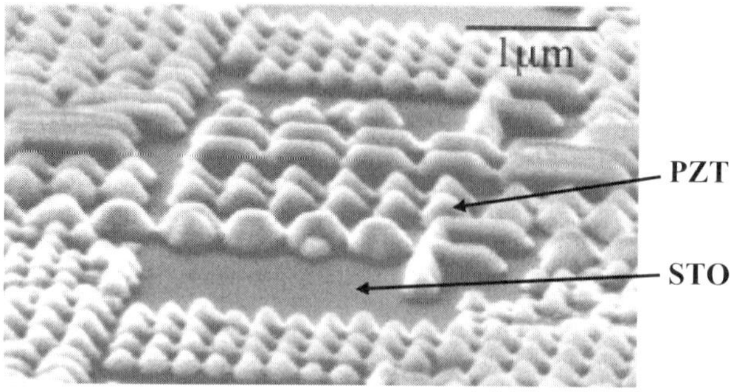


**Fig. 35.** SEM view of nanopatterned PZT features on STO [74].

## 4.6. Nanopatterning of PZT

Downscaling is an important step to achieve very high-density memories. Bühlmann et al., [75] have reported on nanopatterning of an epitaxial 200 nm thick film of $Pb(Zr_{0.40}Ti_{0.60})O_3$ (PZT) grown on conductive Nb-doped $SrTiO_3$ (100). Patterning of PZT has been done by e-beam lithography using PMMA (poly-methylmethacrylate) resist, Cr as hard and sacrificial mask and ECR/RF RIBE etching. For lithography, a 150 nm thick PMMA film was deposited. Patterns with lateral dimension between 50 and 200 nm were written by an e-beam. After the development, noncontact atomic force microscopy scans on the PMMA showed that holes down to 50 nm were obtained. A 75 nm thin Cr masking layer was evaporated onto the developed PMMA and the lift off was performed in acetone. Scanning electron microscopy observations on the Cr patterns showed that the patterning worked for all features with lateral dimensions larger than 50 nm. The PZT film and Cr sacrificial mask were then etched in a dual frequency ECR/RF reactor using a $CF_4$, $CCl_4$ and Are gas mixture at a low pressure of 10 mPa. This process assures a good balance between chemical and physical etching. Ion energies have been kept small. A small rf self-bias of only 25 V was applied to the substrate. The ions leaving the ECR ion gun had energies in the range of 200 to 300 eV at maximum.

The etching rate was found to be 10 nm/min for PZT and 3 nm/min for Cr. The remaining Cr was removed in an aqueous solution of ceric ammonium nitrate and perchloric acid. In contrast to Ar ion milling [76], this etching process did not leave any redeposited PZT. The dry etching process leads to rounded PZT patterns caused by a continuous reduction of the thinner border of the Cr-mask (Figure 35). The smallest items achieved had a lateral dimension of 100 nm. Piezoelectric sensitive scanning force microscopy in the contact mode revealed a strong increase of the piezoelectric response for feature sizes with lateral dimensions below 200 nm.

## 4.7. Local growth and crystallization of piezoelectric films

Piezoelectric and pyroelectric films of perovskite materials need high processing temperatures of more than 600 °C. This may damage the rest of the device and is especially critical for monolithic integration together with the read-out or driving circuit.

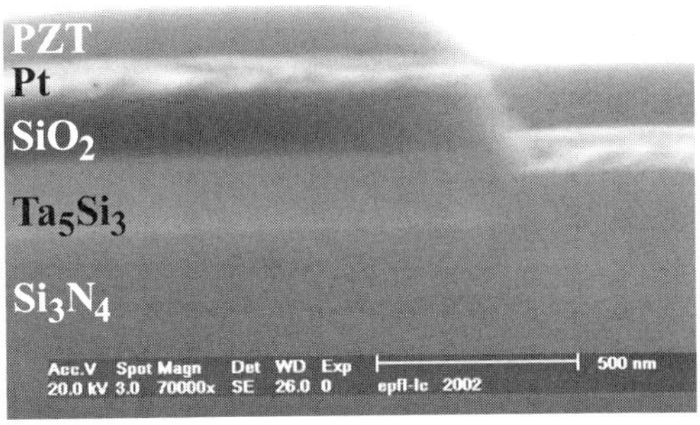


**Fig. 36.** SEM cross section view of pyrolized PZT grown locally on TaSi microhotplate [76].

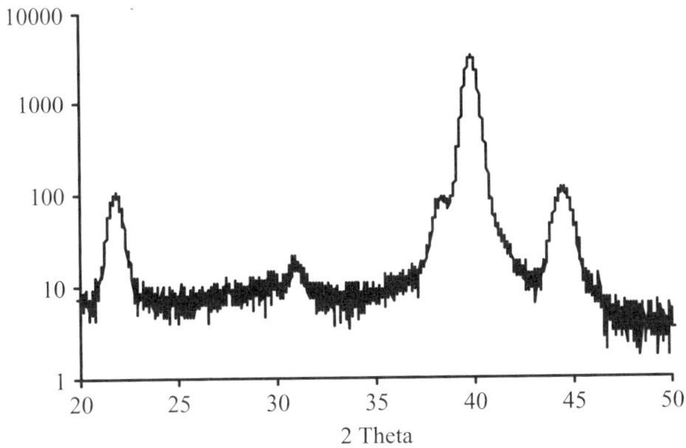


**Fig. 37.** XRD spectra showing the (100), (110), (111) and (200) orientation of PZT grown on the TaSi/SiN microhotplate.

In our studies, PZT ($PbZr_xTi_{1-x}O_3$) thin films have been locally grown by means of sol-gel deposition and local anneals on micro hotplate integrated in the substrate [77]. The micro hotplate was based on a tantalum silicide filament ($Ta_5Si_3$ [78]) formed on stress compensated SiN/$SiO_2$ membrane. On this filament, a passivation layer of $SiO_2$, a layer of Ti/$TiO_2$ and a bottom electrode of Pt were deposited by sputtering. After deposition of PZT sol-gel solution, the film is pyrolised and crystallized by applying the electrical power to the hot-plate (Figure 36). Due to the low heat conductivity of the membrane, crystallization of the PZT film occurs exactly ($+/-1$ μm) on the resistor filament. The formation of crystalline, i.e. perovskite PZT was verified by means of X-ray diffraction (Figure 37). A random PZT texture has been observed.

## 5. Final Remarks

In this chapter, critical issues for microfabrication of piezoelectric MEMS have been discussed. To overcome resolution problems in complex device structures, wet etching of PZT, AlN films and platinum bottom electrodes have been replaced by high density plasma dry etching methods. Micron scale patterns have been obtained using high density plasma (HDP ICP) or reactive ion beam etching (ECR/RF RIBE) processes. The developed processes exhibited reasonable etching rates and better results in term of anisotropy, photoresist stability and residues.

These new processes have been successfully integrated into a standard microfabrication sequence that fully preserves the functionality of piezoelectric thin films. For example for PZT based devices, at the end of the micromachining sequence, transverse piezoelectric coefficient $e_{31,f}$ of $-12$ C/m$^2$, dielectric permittivity $\varepsilon$ of 1100 and loss factor tan $\delta$ of 0.03 (at 1 kHz) have been measured on test samples with 1 μm PZT 53/47 $\{100\}$-textured thin film.

In-plane laminated PZT/Si and AlN/Si structures have been obtained by compensating the tensile stress of electrodes and piezoelectric films with compressive $SiO_2$ thermal oxide.

Finally, the surface micromachining for integration of PZT or AlN thin films on thin suspended membranes is under development. In this case SOI/$SiO_2$ or polysilicon/amorphous Si as combination of structural/sacrificial layers are widely investigated.

Several examples of piezoelectric devices (ultrasonic transducers, acoustic sensor, damping control, RF switches) presented in this paper have been developed during the last years. It has been demonstrated that the microfabrication of piezoelectric MEMS based on PZT or AlN thin films is the versatile and reliable technology. Most of the developments are based on PZT sol-gel textured films, however some efforts have been done for integration of AlN for piezoelectric MEMS. In this context, the main advantage of AlN is the low thermal budget processing and the possibility of post-CMOS above-IC integration.

## References


[1] Muralt, P., *Ferroelectric thin films for micro-sensors and actuators: a review.* J. Micromech. Microeng., 2000. 10: p. 136–146.

[2] Piezoelectric Materials and Devices, Edited by N. Setter, Lausanne 2002.

[3] Miyahara, Y., *PZT cantilever for non-contact atomic force microscopy*. Appl. Surf. Sci., 1999. 140: p. 428–431.

[4] P. Muralt, *PZT thin films for microsensors and actuators*: *Where do we stand*? IEEE Transactions on Ultrasonics, Ferroelectrics, and Frequency Control, 2000. 47(4):903–915.

[5] M. Elwenspoek, H. Jansen, *Silicon micromachining*, C.U. Press, 1998.

[6] W. Lang, *Silicon micromachining technology*, Materials Science and Engineering, R17 (1996) 1–55.

[7] Madou, M., *Fundamentals of Microfabrication*. 1997, Boca Raton: CRC Press.

[8] Gross, S.J., et al., *Reliable integration of piezoelectric lead zirconate titanate with MEMS fabrication process*. Proceedings of the SPIE, 2001. 4558: p. 72.

[9] J. Baborowski, N. Ledermann and P. Muralt, *Micromachining of Piezoelectric MEMS*, TRANSDUCERS '01 EUROSENSORS XV, The 11th International Conference on Solid-State Sensors and Actuators, Munich, 2001, p. 596–600.

[10] P.Muralt et al., *Integration of Piezoelectric* $Pb(Zr_xTi_{1-x})O_3$ *(PZT) thin films into micromachined sensors and actuators*, in *Materials and Process Integration for MEMS*, P.F.E.H. Tay, Editor. 2002, Kluwer Academic Publishers, The Netherlands, p. 3–27.

[11] Handbook of Chemistry and Physics, CRC, D.R.Lide Ed., CRC Press, 1995.

[12] DeVoe, D.L. and A.P. Pisano. *A Fully Surface-Micromachined Piezoelectric Accelerometer.* in *Transducers '97, International Conference on Solid-State Sensors and Actuators*. 1997. Chicago.
[13] Kim, J.H., et al., *Fabrication Process of PZT Piezoelectric cantilever unimorphs using surface micromachining*. Integrated Ferroelectrics, 1997. 15: p. 325–332.
[14] Flynn, A.M., et al., *Piezoelectric micromotors for microrobots*. J. Microelectromechanical Sytems, 1992. 1: p. 44–51.
[15] Dubois, M.-A. and P. Muralt, *PZT Thin Film Actuated Elastic Fin Micromotor*. IEEE Transactions on UFFC 1998. 45(5), p. 1169–1177.
[16] Racine, G.-A., P. Muralt, and M.-A. Dubois, *Flexural-standing-wave elastic force motor using ZnO and PZT thin film on micromachined silicon membranes for wristwatch applications*. Smart Mater. Struct., 1998. **7**: p. 404.
[17] Baborowski, J., et al., *Fabrication and Characterization of Micromachined Accelerometers based on PZT thin films*. Ferroelectrics, 1999. 224: p. 283–290.
[18] Kunz, K., P. Enoksson, and G. Stemme, *Highly sensitive triaxial silicon accelerometer with integrated PZT thin film detectors*. Sensors and Actuators A, 2001. 92: p. 156–160.
[19] Lee, S.S. and R.M. White, *Piezoelectric cantilever acoustic transducer*. J. Micromech. Microeng., 1998. 8: p. 230–238.
[20] Schroth, A., et al., *Application of sol-gel deposited thin PZT film for actuation of 1D and 2D scanners*. Sensors and Actuators A, 1999. 73: p. 144–152.
[21] Lee, C., T. Itoh, and T. Suga, *Self-excited piezoelectric PZT microcantilevers for dynamic SFM—with inherent sensing and actuating capabilities*. Sensors and Actuators A, 1999. 72: p. 179–188.
[22] Bernstein, J.J., et al., *Micromachined high frequency ferroelectric sonar transducers*. IEEE Tran. UFFC, 1997. 44(5): p. 960–969.
[23] Baborowski J., Ledermann N., Muralt P., *Piezoelectric Micromachined Transducers (pMUT'p) based on PZT thin films*, in Proceedings of IEEE Ultrasonic Symposium, Münich, Germany, 2002.
[24] Maeder, T., L. Sagalowicz, and P. Muralt, *Stabilized Platinum Electrodes for Ferroelectric Film Deposition using Ti, Ta and Zr Adhesion Layers*. Jpn. J. Appl. Phys., 1998. 37, p. 2007–2012.
[25] Muralt, P., et al., *Texture control of $PbTiO_3$ and $Pb(Zr,Ti)O_3$ thin films with $TiO_2$ seeding*. J. Appl. Phys., 1998. 83(7): p. 3835–3841.
[26] Maeder, T., et al., *$Pb(Zr,Ti)O_3$ Thin Films by In-situ Reactive Sputtering on Micromachined Membranes for Micromechanical Applications*. British Ceram. Proc., 1995. 54: p. 206–218.
[27] Kloeck, B., et al., *Study of electrochemical etch-stop for high-precision thickness control of silicon membranes*. IEEE Transactions on Electron Devices, 1989. 36(4): p. 663–669.
[28] Fluckiger, P., et al., *Standard processes at the Center of Microtechnology—EPFL*. 2002, http://microtechnique.epfl.ch/cmi/.
[29] J.Hopwood, *Review of inductively coupled plasmas for plasma processing*, Plasma Sources Science and Technology, 1, 109–116.
[30] Baborowski, J., P. Muralt, and N. Ledermann, *Etching of platinum thin films with dual frequency ECR/RF reactor*. Integrated Ferroelectrics, 1999. 27: p. 243–256.
[31] Baborowski, J., et al., *Etching of $RuO_2$ and Pt thin films with ECR/RF reactor*. Vacuum, 2000. 56: p. 51–56.
[32] M.Sugawara, *Plasma Etching Fundamentals and Applications*, Oxford Science Publications, 1998.
[33] S. Marks , L. G. Jerde, G. Celii, S. A. Aggarwal and J. S. Martin, *High Temperature Plasma Etching Characteristics of Sub-Micron Ir/PZT/Ir Ferroelectric Capacitor Stacks*, Proceedings of ISAF 2002, Nara.

[34] http://www.tegal.com/news_events/nr/nr020718.html

[35] Lee, E., et al., *Effects of top electrode annealing on the PZT thin film*. Integrated Ferroelectrics, 1997. 16: p. 165–174.

[36] Troiler-McKinstry, S., et al. *Etched piezoelectric structures*. in *International Symposium of Applied Ferroelectrics*, ISAF. 1986. Bethlehem.

[37] H.Macé, H.Achard, L.Peccoud, *Reactive ion etching of Pt/PZT/Pt ferroelectric thin film capacitors in high density DECR plasma*, Microelectronic Engineering 29 (1995) 45–48.

[38] Zeto, R., et al., *Dry etching of sol-gel PZT*. Mat. Res. Soc. Symp. Proc., 1999. 546: p. 159–164.

[39] Pan, W., et al., *Reactive ion etching of PZT and $RuO_2$* films by environmentally safe gases. J. Mater. Res., 1994. 9(11): p. 2976–2980.

[40] Chung, C.W., *Reactive ion etching of PZT thin films in inductively coupled plasma*. J. Vac. Sci. Technol. B, 1998. 16(4): p. 1894–1900.

[41] Jung, J.-K. and W.-J. Lee, *Dry etching of PZT films in $CF_4$ and $Cl_2/CF_4$ inductively coupled plasmas*. Jpn. J. Appl. Phys., 2001. 40: p. 1408–1419.

[42] Chee Won Chung, Yo Han Byun, Hye In Kim, *Inductively coupled plasma etching of $Pb(Zr_xTi_{1-x})O_3$ thin film in a HBr/Ar plasma*, Microelectronic Engineering 63 (2002) 353–361.

[43] Williams, K., et al. *Reactive ion beam etching of ferroelectric materials using a RF inductively coupled ion beam source*. in IEEE International Symposium on Application of Ferroelectrics (ISAF'96). 1996.

[44] Boukari, F., et al., *A very compact electron cyclotron resonance ion source*. Rev. Sci. Instrum., 1994. 65(4): p. 1097–1099.

[45] Baborowski, J., et al., *Mechanisms of $Pb(Zr_{0.53}Ti_{0.47})O_3$ Thin Film Etching with ECR/RF Reactor*. Integrated Ferroelectrics, 2000. **31**: p. 261–271.

[46] Myoung-Gu Kang, Kyoung-Tae Kim, Chang-Il Kim, *Recovery of plasma-induced damage in PZT thin film with $O_2$ gas annealing*, Thin Solid Films 398–399 (2001), 448–453.

[47] Myoung-Gu Kang, Kyoung-Tae Kim, Chang-Il Kim, *Plasma Induced damage in PZT thin films etched by inductively coupled plasma*, Thin Solid Films 435 (2003), 222–226.

[48] Vossen, J.L. and W. Kern, *Thin Film Process*, ed. 1. Vol. V-1. 1978, New-York: Academic Press.

[49] R.P. Frankenthal, and D.H. Eaton, J. Electrochem. Soc.: Solid-state science and technology, 1976. 123(5): p. 703–706.

[50] C.Farrell, K.Milkove, C.Wang and D.Kotecki, Integrated Ferroelectrics, Vol. 16, 109–138 (1997).

[51] K.L. Milkove, J.A. Coffin and C.Dziobkowski, J.Vac.Sci.Technol. A16(3), 1483 (1998).

[52] Celli et al., Texas Instruments, Integrated Ferroelectrics, vol. 27, 1999, 227.

[53] Sakoda T. et al., Japanese Journ. Appl. Phys. pI, 40, 2001, 2911.

[54] Chung et al., Integrated Ferroelectrics, vol. 39, 2001, 119.

[55] Chung et al., Journal of Vac. Sc&Technol. A, 19, 2001, 2400.

[56] Chongying Xu et al., Ferroelectric Thin Films VII, MRS, 1999, 129

[57] Chiang MC et al., Electrochemical Society Proceedings, Vol. 99, 2000, 114.

[58] Eunk-Jik Lee, Jin-Woong Kim, Won-Jong Lee, Jpn. J. Appl. Phys. Vol. 37 (1998), pp. 2634.

[59] Wei Pan, S.B.Desu, Phys.Stat.Sol., (a) 161, 201, (1997).

[60] F. Lärmer, and A. Schilp, Patent *US6284148: Method for anisotropic etching of silicon*. 2001: USA.

[61] Hynes, A.M., et al., *Recent advances in silicon etching for MEMS using the ASE™* process. Sensors and Actuators A, 1999. 74: p. 13–17.

[62] Yasseen, A.A., N.J. Mourlas, and M. Mehregany, *Chemical-Mechanical Polishing for polysilicon Surface Micromachining*. J. Electrochem. Soc., 1997. 144(1): p. 237–242.

[63] I. Ladabaum, X. Jin, B.T. Khuri-Yakub, *Miniature drumheads: microfabricated ultrasonic transducers*, Ultrasonic 36 (1998) 25–29.

[64] E. Cianci, V. Foglietti, G. Caliano, M. Pappalardo, *Micromachined capacitive ultrasonic transducers fabricated using silicon on insulator wafers*, Microelectronic Engineering 61–62 (2002) 1025–1029.

[65] G. Percin, B. T. Khuri-Yakub, *Piezoelectrically actuated extensional micromachined ultrasound transducers*, Ultrasonics 40 (2002) 441–448.

[66] S. Frédérico, C. Hibert, R. Fritschi, Ph. Flückiger, Ph. Renaud, and A.M. Ionescu, *Silicon sacrificial layer dry etching (SSLDE) for free-standing RF MEMS architectures*, Proceedings of the 16th IEEE International Conference on Micro Electro Mechanical Systems, Kyoto, Japan, pp. 570–573, 2003.

[67] J.Baborowski, N.Ledermann, P.Muralt, D.Schmitt, *Simulation and Characterization of Piezoelectric Micromachined Transducers Based on PZT/SOI Membranes*, International Journal of Computational Engineering Science (IJCES), Vol. 4, 3, 2003, 471–475.

[68] N. Ledermann, P. Muralt, J. Baborowski, S. Gentil, K. Mukati, M. Cantoni, A. Seifert, N.Setter, *{100} Textured, piezoelectric $Pb(Zr_x,Ti_{1-x})O_3$ thin films for MEMS: integration, deposition and properties*, Sensors and Actuators A 105 (2003) 162–170.

[69] Z. Jamie Yao, Shea Chen, Susan Eshelman, David Denniston, Chuck Goldsmish, *Micromachined low-loss microwave switches*, IEEE J.Microelectromech. Syst. 8 (2) (1999) 129–134.

[70] J.Y. Park, Y.J. Yee, H.J.Nam, J.U.Bu, *Micromachined RF MEMS tunable capacitors for RF MEMS application*, in IEEE MTTT-S International Microwave Symposium Digest, May 2001, 2111–2114.

[71] S. J. Gross, S. Tadigadapa, T. N. Jackson, S. Trolier-McKinstry, Q. Q. Zhang, *Lead-zirconate-titanate-based piezoelectric micromachined switch*, APPLIED PHYSICS LETTERS, Vol. 83, 1, 7 July 2003.

[72] Han Geun Yu, L. Zou, K. Deng, R. Wolf, S. Tadigadapa, S. Trolier-McKinstry *Lead zirconate titanate MEMS accelerometer using interdigitated electrodes*, Sensors and Actuators A 107 (2003) 26–35.

[73] L.Trupina, J.Baborowski, P.Muralt, V.Meyer, D.Bouvet, P.Fazan and M.Lobet, *Tungsten Based Electrodes for Stacked Capacitor Ferroelectric Memories*, Jpn. J. Appl. Phys., Vol. 41, Part 1, No. 11B, 30 November 2002 6862–6866.

[74] J.Baborowski et al., *Damageless patterning of ferroelectric devices*, Proceedings of ISIF 2004, Integrated Ferroelectrics, in press 2004.

[75] S. Bühlmann, B. Dwir, J. Baborowski, P. Muralt, *Size effect in mesoscopic epitaxial ferroelectric structures: Increase of piezoelectric response with decreasing feature size*; Applied Physics Letters, Vol. 80, 17, 2002, 3195.

[76] M. Hiratani, C. Okazaki, H. Hasegawa, N. Sugii, Y. Tarutani, and K.Takagi, Jpn. J. Appl. Phys., Part 1, 36, 5219, 1997.

[77] F. Calame, J. Baborowski, N. Ledermann, P. Muralt, *Local Growth of Sol-Gel Films by Means of Microhotplates*, International Journal of Computational Engineering Science (IJCES), Vol. 4, 3, 2003, 563–568.

[78] Pollien, N. Ledermann, J. Baborowski, P. Muralt, *High Temperature MicroHotplate*, Patent, WO 02/080620, 10.10.2002.

# Chapter 14

# Non-Conventional Micro- and Nanopatterning Techniques for Electroceramics

M. Alexe [1], C. Harnagea [1], and D. Hesse [1,2]

[1] *Max-Planck-Institut für Mikrostrukturphysik, Weinberg 2, D-06120 Halle, Germany*

[2] *hesse@mpi-halle.de*

An overview of non-conventional methods being used for micro- and nanopatterning of electroceramics is given, including various top-down and bottom-up approaches. Within the top-down approach, focussed ion beam patterning, electron-beam direct writing, nanoimprint lithography, and other next-generation lithography techniques are considered. Fabrication of single nanosize structures as well as ordered arrays of ferroelectric oxide structures with lateral sizes ranging from several microns down to 70 nm is described in detail. Major structural and electrical properties along with effects related to patterning itself, such as ferroelectric imprint, are also given and discussed in terms of intrinsic and extrinsic size effects.

Since below 70 nm lateral size the lithography methods are either at the physical limits or extremely expensive, several physical and chemical self-patterning routes are in parallel developed as bottom-up approaches. The bottom-up methods will in principle allow the inexpensive fabrication of structures with sizes of 10–20 nm in a parallel approach, but at the expense of a random positioning of the obtained nanostructures rather than in an ordered pattern. In any case, this is of very high scientific interest since in this size range the superparaelectric phase might develop. A short review on physical bottom-up approaches based on island (Volmer-Weber) or layer-than-island (Stranski-Krastanov) growth modes or microstructural instabilities of ultra-thin films as well as on chemistry-based bottom-up methods such as microemulsion route is given.

Also included is a chapter on ferroelectric testing of nanopatterned electroceramics, with emphasis laid on possibilities and limitations of piezoresponse scanning force microscopy. Correlation of the PFM signal to the local polarization considering the tensorial nature of the piezoelectric effect as well as most common sources of error in the PFM measurements and analysis are briefly described. Optimum measurement regimes for ferroelectric domain imaging and quantitative measurements (PFM hysteresis loops) are also discussed.

## 1. Introduction

Many applications of electroceramics require micro- or even nanopatterning of a thin film. For example, piezoelectric micro-sensors and micro-actuators involve piezoelectric thin films patterned into structures of micrometer-range sizes. For high-density memory applications in integrated microelectronics, ferroelectric thin films patterned into nanometer-range sizes are required [1]. For example, in a prospective 10 Gbit non-volatile ferroelectric random access memory (NV-FeRAM), the lateral area of the whole memory cell, consisting of a capacitor and a transistor, should not exceed 100 nm, implying ferroelectric capacitors having lateral dimensions well below 100 nm.

Fundamental questions arise in this respect, e.g., whether electroceramic cells of 100 nm lateral sizes are still piezo- or ferroelectric, considering the well-known "collective" or correlated character of ferroelectricity [2]. Other questions are, whether and how the piezoelectric or ferroelectric properties of the thin films are influenced by the patterning process itself, and whether the relatively large proportion of surfaces and interfaces leads to domain pinning or to other surface- or size-related problems, see, e.g. [3–5]. For an experimental study of these questions, micro- and nanopatterning techniques for electroceramic materials are fundamental prerequisites. In the following, an overview of non-conventional methods for micro- and nanopatterning of electroceramics is given, together with a description of testing techniques required for investigations of micro- and nanopatterned electroceramics.

## 2. Top-Down Approaches

### 2.1. Focussed Ion Beam Patterning

Focussed ion beam (FIB) patterning is nowadays used on rather large scale in the microelectronics industry to pattern structures with lateral sizes down to several tens of nanometers. The FIB equipment is conceptually similar to a scanning electron microscope (SEM), but instead of a beam of electrons, a highly focussed beam of gallium ions is used to scan the sample surface. This ion beam can be used for two purposes: imaging, and micromachining: Scanning a low-current beam over an area of the sample surface and simultaneously collecting the induced secondary electrons, one can produce a high-resolution image of the sample, whereas at high beam current, high-precision local sputtering and milling can be easily performed. FIB equipment also allows the fabrication of small structures by direct deposition of metals (such as tungsten, gold, or platinum) or insulators (like silicon dioxide) with a minimum feature size down to 40 nm and a precision of 10 nm, allowing studies on quantum transport mechanisms [6]. Various materials can be selectively etched in reactive gas atmospheres achieving aspect ratios up to 30 with a minimum feature size below 25 nm. Having the unique nanopatterning ability to add or remove features with a resolution of 20 nm or better, the FIB method is currently used to modify integrated circuits and masks, or to fabricate cross section transmission electron microscopy specimens.

In the attempts of preparing switchable ferroelectric structures as small as possible, FIB patterning was among one of the first methods to nanofabricate ferroelectric capacitors with sizes down to sub-100 nm lateral size. Ganpule et al. fabricated ferroelectric capacitors made of $Pb(Zr,Ti)O_3$ (PZT) and $SrBi_2Ta_2O_9$ (SBT) by cutting them out from a complex film structure [7,8]. For the PZT capacitors the starting complex heterostructure consisted of a sequence of Pt–LSCO–PNZT–LSCO–Pt layers deposited onto silicon. The polycrystalline $Pb(Nb_xZr_yTi_z)O_3$ (PNZT) layer of 160–210 nm thickness was prepared by sol-gel deposition, and a $La_{0.5}Sr_{0.5}CoO_3$ (LSCO) conducting barrier layer of about 100 nm thickness was deposited onto a bottom Pt layer by PLD. For the SBT-based capacitors the starting structure was $Pt/SrBi_2Ta_2O_9/Pt$.

Milling was performed using a Micrion FIB-2500 machine equipped with a liquid gallium ion source, operating at 50 kV accelerating voltage, 2–1500 pA beam current, and a corresponding beam diameter of about 6–100 nm. Scanning the beam within the chosen area was performed by a digital serpentine scan with a pixel dwell time of 1–50 ms and a distance of 10 nm between pixels, to provide sufficient overlap of the beam. The film heterostructures were milled down to the bottom conductive layer (which required the removal of up to 400–450 nm of material) in a designed pattern, *viz.* an island with the top Pt/LSCO layer forming one electrode (Fig. 1).

The advantage of this technique is the possibility of using an established method to produce high-quality thin films, including epitaxial films, such as PLD or MOCVD. This allows a comprehensive macroscopic characterisation of the film before nanopatterning, avoiding the potential compositional or structural deviations which might be

**Fig. 1.** Scanning ion-beam images of ferroelectric test structures fabricated by focussed ion-beam milling in the size range from 1 µm × 1 µm to 0.1 µm × 0.1 µm. The contrast from the various layers in the heterostructure is clearly visible. The inset is a schematic of the shape and dimensions of the 0.1 µm test structure. (After Ref. [7].)

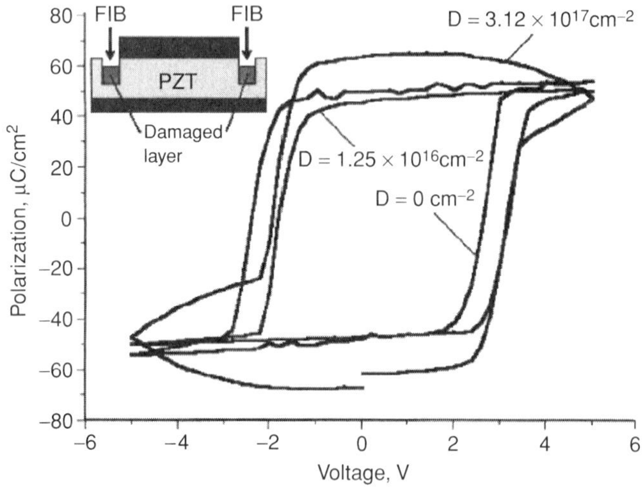


**Fig. 2.** Effect of the ion dose on the polarization hysteresis loops of FIB-patterned ferroelectric capacitors. (After Ref. [10].)

involved in other nanofabrication techniques based on bottom-up approaches. FIB patterning has actually opened the possibility of direct comparisons between the properties of a patterned structure and the original film, establishing the large influence of in-plane clamping, induced by the substrate, onto the electromechanical properties of a ferroelectric film [9].

The major drawback of the method is associated with the high damage that occurs during milling and imaging, in particular, if a dual-beam FIB machine (where imaging is performed by a primary *electron* beam) is not available. While most of the structural defects can be healed by a high-temperature thermal annealing after milling, a gallium doping is basically unavoidable and in special cases might be relatively harmful to the final electrical properties. Detailed studies on radiation damage during the FIB process and its recovery were performed by Stanishevsky et al. [10]. They discovered that the exposure of a PZT film to a gallium ion beam of 50 kV dramatically changes the composition of a near-surface layer and leads to a more or less pronounced modification of the ferroelectric properties, depending on the radiation dose (Fig. 2). The modified near-surface layer is rich in gallium and depleted in oxygen. The gallium concentration is lower on the side walls due mostly to sputtering and shallower implantation. Gallium has a strong tendency to segregate into nanosize droplets during the milling and additionally might precipitate and diffuse along the grain boundaries of the film, creating local conducting channels and increasing the leakage current.

A thermal annealing of the cut-out structures recovers the ferroelectric properties, but in general a damaged layer up to 10 nm thick still remains, in which the ferroelectric properties are altered. Obvously, this would affect only structures with lateral sizes less that 100 nm. Despite this major drawback, the nanopatterning via FIB is an extremely versatile method that allows fabrication of diverse nanoscale ferroelectric systems starting from their film counterpart. Whereas FIB patterning is well suited to prepare single nanostructures, the preparation of cell arrays requires methods that are less time-consuming and have more of a parallel character. Among them are electron beam direct writing, nanoimprint lithography, and other next generation lithographies.

## 2.2. Electron beam direct writing

At the end of the nineties, electron beam direct writing (EBDW)—prior applied to other materials [11,12]—was first used to prepare ferroelectric arrays. Arrays from $Bi_4Ti_3O_{12}$ [13], $SrBi_2Ta_2O_9$ [14] and PZT [15], with individual cells of 150 nm [13] and 100 nm lateral size [14,15], respectively, were prepared and shown to display ferroelectric properties. An electron-beam lithography approach using a PMMA film as resist, a chromium mask, and a dry etching process has recently been applied to this end as well, *cf.* [5].

As an example for EBDW, Fig. 3 shows scanning electron microscope (SEM) images of PZT arrays. Starting from a solution containing metalorganic compounds or metal colloids, a substrate is spin-coated with a thin precursor film that contains

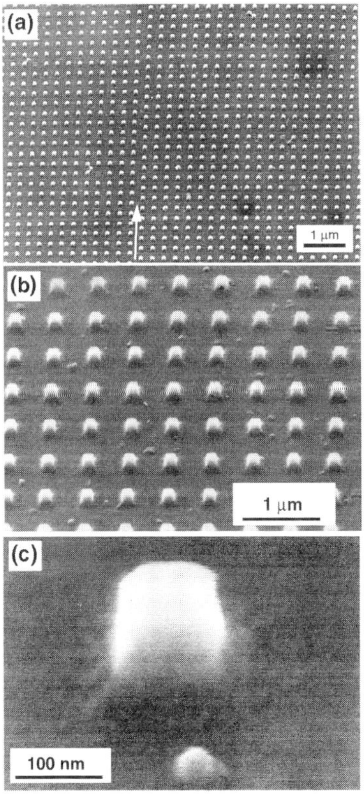


**Fig. 3.** (a) Scanning electron micrograph of an array of PZT cells with 100 nm lateral size prepared by EBDW. The array contains a purposely introduced antiphase boundary (see arrow). (b) and (c) Differently magnified images of this array.

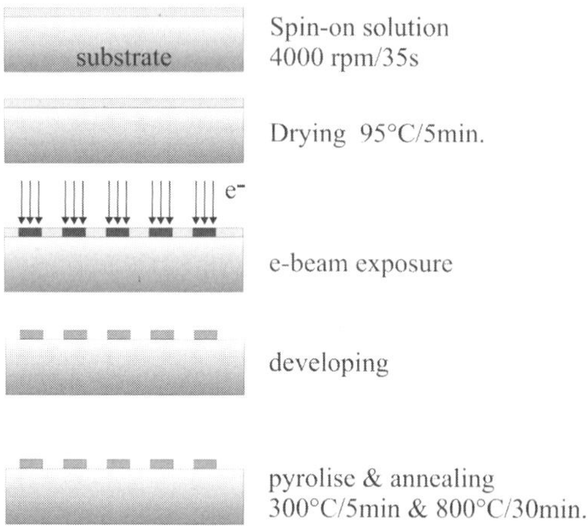


**Fig. 4.** Process flow chart of electron-beam direct writing of ferroelectric cell arrays.

the required elements (various metals and oxygen) in a proportion corresponding to that of the later ferroelectric phase (Fig. 4). For example, precursor films for SBT and PZT structures can be prepared from Sr-, Bi-, and Pb-ethylhexanoate, Ti- and Zr-isopropylene, and Ta-methoxide solutions, using xylene and 2-methoxyethanole, respectively, as solvents. The obtained precursor film is then patterned by scanning an electron beam of, e.g., 3 nm diameter over selected areas of the film ("pattern exposure"), and immersing the exposed sample for 1 minute in toluene ("pattern development"), followed by dry-blowing with nitrogen. In our case, EBDW was performed using a commercial electron beam lithography system (ELPHY Plus, Raith Co.) adapted to a JEOL JSM SEM with $LaB_6$ cathode, working at 40 kV acceleration voltage. The electron doses used during exposure were from 600 to 1200 $\mu C/cm^2$ for SBT, and 1500 to 6000 $\mu C/cm^2$ for PZT, depending on film thickness and cell size. After pattern development, the still metalorganic mesas are transformed into an amorphous mixed oxide phase by low-temperature annealing in air, and subsequently the required ferroelectric phase is obtained by a high-temperature anneal in air. For example, the low-temperature anneal is performed at 300 °C for 5 min., and the subsequent high-temperature anneal at 600 to 850° for one hour. The two annealing processes lead to quite a shrinkage of the original mesas in size, of the order of 40 to 50 % overall, and they may result in more or less pronounced shape changes, depending on the material. PZT mesas were observed to almost fully maintain their original shape, although shrinking in size by 50%.

The resulting crystalline patterns (*cf.* Fig. 3) consist of individual cells that are polycrystalline, implying very small grains with sizes of the order of 10 to 20 nm. As it was shown by piezoresponse scanning force microscopy (PFM; for a detailed description, see Section 4.2), the cells respond to a d.c. field with a piezoelectric hysteresis loop, which clearly shows that they are ferroelectric [16]. Figure 5 presents such hysteresis loops, one for a cell of 1 μm lateral size, the other for one of 100 nm lateral size. From investigations of this kind it became clear that PZT cells $100 \times 100 \times 100$ nm$^3$ in size still display ferroelectric properties. As the interaction between the tip of the

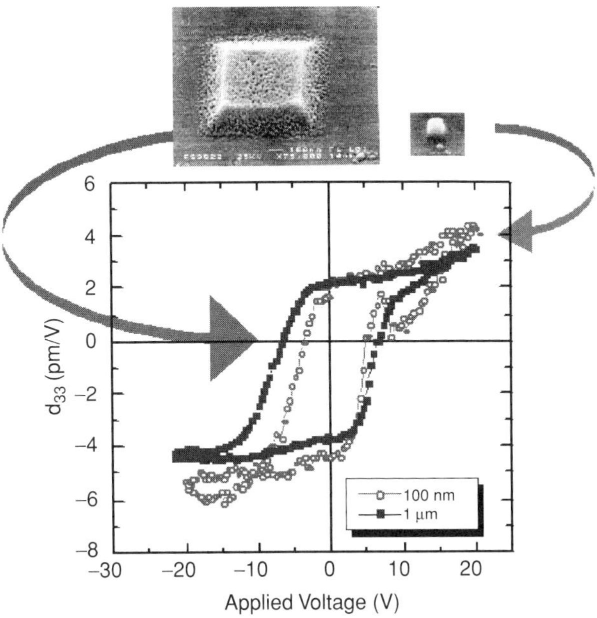


**Fig. 5.** Piezoelectric hysteresis curves of two PZT cells prepared by EBDW, one with 1 μm lateral size, the other with 100 nm lateral size. Insets: SEM micrographs of the two PZT cells.

scanning force microscope (SFM) and the sample is very complex [17–19], neither the absolute polarisation values nor the true coercive field can be deduced from these measurements.

The cells obtained in this way were used to study switching and size effects using PFM [14,20]. In PFM images, white and dark contrasts are associated with the two remanent polarisation states $\pm P_r$, and a grey contrast indicates the absence of out-of-plane piezoresponse. In-plane investigations, probing the in-plane components of $P_r$, are possible as well, *cf.* Chapt. 4.2. Figure 6 shows a PZT array consisting of cells with a lateral size of 250 nm. The entire array was first uniformly poled into a negative state (black colour) with a −20 V pulse (Fig. 6a), before then switching one selected cell into the positive state (white colour) applying a +20 V pulse to this cell (Fig. 6b). Due to the convolution with the tip shape, the shape of the cells appears distorted in Fig. 6. During switching, no cross-talk occurred, as is also obvious from Fig. 6b.

Detailed investigations of this kind proved the existence of imprint and size effects in PZT cells with lateral sizes between 1 μm and 75 nm [20]. It turned out that the free lateral surfaces of the cells "pin" a layer of about 35 nm thickness, preventing the switching of the near-surface volume of the cell. This layer may prevent the application of EBDW-prepared PZT cells with a lateral size below 70 nm, if this application requires switching. In addition, indications of a ~10 nm thick layer close to the bottom electrode were obtained, which does not switch. This layer may represent a depletion layer [21,22], or rather the high-density oxygen vacancy layer associated with domain pinning in the fatigue model of Scott and Dawber [23].

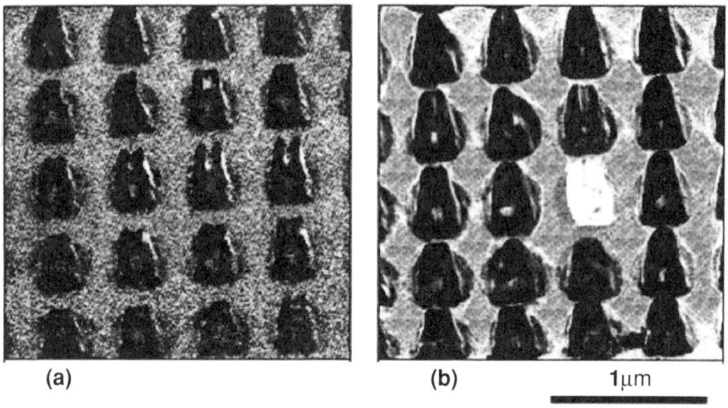


**Fig. 6.** PFM images of a PZT cell array prepared by EBDW, with cells of 250 nm lateral size., *viz.* (a) before, and (b) after switching a cell with a +20 V pulse. Note the switching of the entire cell volumes and the absence of cross-talk.

Overall, electron-beam direct writing proved a powerful method to prepare arrays of ferroelectric cells with lateral sizes down to 75 nm, although with quite an effort of expensive equipment and time. It is therefore rather suited for research purposes than for industrial mass production. For the latter, more economic means of preparation have to be developed.

## 2.3. Nanoimprint lithography

A method to prepare ferroelectric cell arrays that is potientially suitable for industrial mass production, is nanoimprint lithography (NIL). Traditional NIL has emerged as a rapid, low-cost technique for the preparation of nanostructures in polymer films like PMMA with structure sizes well below the limit of other lithography techniques, down to 20 nm feature size [24–26]. Similar to imprint into polymer films, we have proposed the use of NIL to obtain large arrays of ferroelectric cells. Using a variant of this technique, we succeeded in preparing arrays of ferroelectric cells with lateral sizes down to 250 nm so far, and we proved the ferroelectric nature of these cells using PFM [27].

The standard NIL technique implies the use of a thermoplastic resist which becomes a viscous liquid above its glass transition temperature and which thus can flow above this temperature. Thus it can be deformed easlly. NIL is usually performed in two steps. In the imprint step, a mold with the negative of the desired pattern on its surface is pressed into a thin resist cast on a substrate, followed by removal of the mold. This step transfers the pattern of the mold into the resist film, or, in other words, it creates a thickness contrast pattern in the resist. The second step is the pattern transfer in which an anisotropic etch process, such as reactive ion etching (RIE), is used to remove the residual resist in the compressed area. This step transfers the thickness contrast pattern into the entire resist. Preparing ferroelectric arrays begins with spin-coating a substrate with a precursor film, most similar to the first step of EBDW. This precursor film is prepared by a chemical solution deposition (CSD) technique. Subsequently, a thickness contrast is created in this precusor film by pressing a mold with a defined pattern into

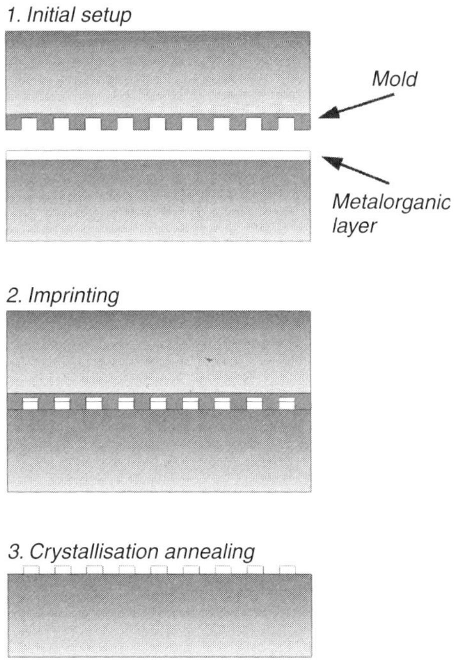


**Fig. 7.** Process flow chart of the preparation of ferroelectric cell arrays by nanoimprint lithography.

it. After removing the mold, the precursor is transformed into a ferroelectric by drying and annealing (Fig. 7).

Two chemical routes were explored for preparation of ferroelectric PZT cell arrays on conductive substrates, *viz.* Pt-coated silicon (Pt/Si) or Nb-doped $SrTiO_3$ (STO) serving as the bottom electrode: metalorganic deposition (MOD) and sol-gel deposition. In the MOD route, precursor solutions were spun onto the subtrates, forming thin films of thicknesses between 200 and 700 nm, depending on solution concentration and rotation speed. These precursor films were dried by annealing at 70 °C, before they were imprinted. After imprint, the mesas were converted into the oxide by heating at 300 °C, and finally crystallised into PZT by annealing at 650 °C for 1h. The sol-gel route was identical to the MOD route, except for the drying before the imprint step. An optimum imprint pressure of 1 kbar was applied using a laboratory press at room temperature. The molds used consisted either of macroporous silicon, prepared by electrochemical methods [28], or of $SiO_2$-covered silicon wafers patterned by photolithography and reactive ion etching. The smallest size of the negative features in the mold (holes, pores) was ca. 400 nm for porous silicon and 2 μm for the $SiO_2$/Si molds.

An important question is whether the shape of the imprinted pattern is maintained during crystallisation. For maintaining the shape, the size of the grains formed after annealing has to be well below the feature size of the pattern, at least by one order of magnitude. Figure 8a compares the grain size distributions of the ferroelectric films obtained on STO substrates using the MOD and sol-gel techniques. The maximum for

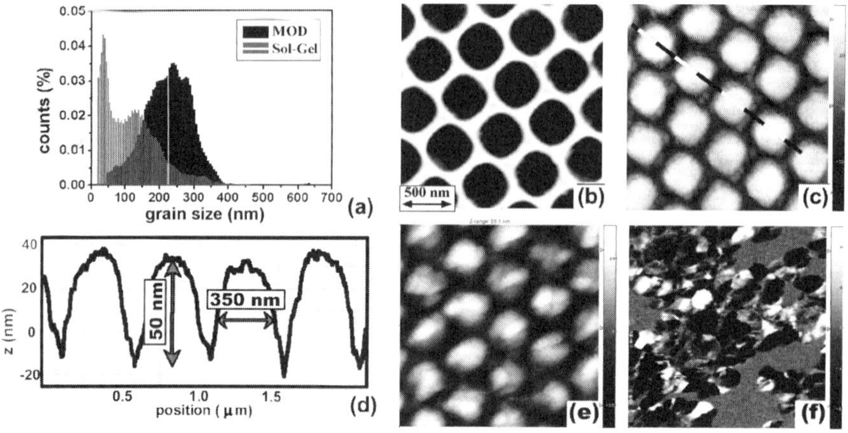


**Fig. 8.** (a) Grain size distribution of continuous ferroelectric films obtained on STO substrates using the MOD and sol-gel technique, respectively. (b) Macroporous silicon with an interpore distance of 500 nm, i.e. ca. 400 nm pore diameter. (c) Topographic image of the cells obtained by imprinting a sol-gel film using the mold shown in (b). (d) Cross-section line scan of (c) along the marked dashed line. (e) Topography image of the cells obtained using MOD and the mold shown in (b). (f) Piezoresponse image of the area corresponding to (e).

MOD is at 240 nm, while that for the sol-gel route is at 35 nm. This explains why the size of the smallest structures we could obtain using MOD, without losing the shape of the structure, was 1.25 µm. The sol-gel method is most suitable for achieving structures with a lateral size in the 100 nm range, while still preserving the shape of the mold. A typical Si mold of 500 nm pitch is shown in Fig. 8b. The diameter of the pores is 350–450 nm while the walls are 50–150 nm thick. The cells obtained using such a mold and the sol-gel route are shown in Fig. 8c. The shape of the mold is well preserved due to the small grain size after crystallisation. The cells have an apparent diameter of 350 nm (full width at half maximum) and an average height of 50 nm, as shown by the cross-section profile in Figure 6d. Using tip-deconvolution software packages we found the lateral size to be even smaller by some 50 to 80 nm, i.e. in the 250–300 nm range. The images in Figs. 8e and f are the topography and the corresponding PFM image of a sample prepared by MOD. The topographic image shows that the sizes of the grains formed is too large to preserve the shape of the mold after crystallisation. The PFM image reveals that the cells are not entirely crystallised. The flat grey areas representing the complete absence of piezoresponse show that there is no component of the spontaneous polarisation out of the substrate plane. An attempt to orient the polarisation out of the substrate plane failed. In contrast, nice hysteresis loops were obtained from the region showing strong (bright or dark) contrast. From all these findings it is clear that the sol-gel route is more suitable for the fabrication of ferroelectric cells, since the shape of the cells is well preserved down to a size of about 250 nm, and the cells are entirely crystallised (not shown here). The sol-gel solution, however, is sticky and often adheres to the mold. To overcome this problem, the mold should be coated with an inert anti-sticking layer to reduce the adhesion between gel and mold.

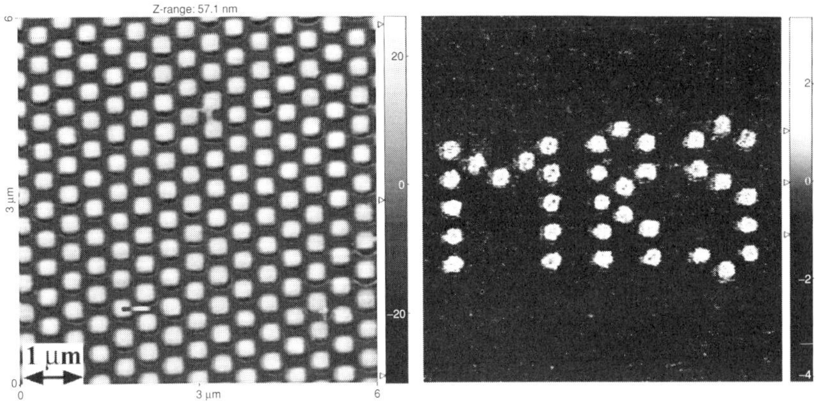


**Fig. 9.** Topography image of an array of ferroelectric PZT cells on STO substrate (left) and piezoresponse image of the same array (at the same magnification) after switching several cells using d.c. pulses applied to the tip (right).

The functionality of the ferroelectric cells obtained—using the sol-gel method for the film deposition prior to imprint—is demonstrated in Fig. 9. Selected cells were poled by applying pulses of 15 V to the scanning probe microscope tip for a time of 3 s. The lateral size of the switched areas is around 200 nm, a size that is most probably due to the large radius of the tip apex. This is also suggested by the fact that the switched areas do not take the shape of the ferroelectric cell, but they are rather circular. The switched areas were stable at the time scale of the experiment: For at least three days no backswitching and no crosstalk occurred. Several aspects of the preparation of ferroelectric cell arrays by NIL still need further consideration, see Ref. [27]. However, we believe that it should be possible to achieve large-area, low-cost ferroelectric cells below 100 nm in lateral size with good ferroelectric properties using nanoimprint lithography.

## 2.4. Next generation lithography

Conventional photolithography is based on the old principle of photography, in which a negative is used to multiply an image (the positive pattern) as many times as necessary. The negative is called mask and the pattern (positive image) is transferred into some special photosensitive polymeric material, the photoresist, that after patterning acts as a mask for either an etch or a lift-off process. The intrinsic limit of this exceptionally simple and productive method is connected to the wavelength of the light used to transfer the mask pattern into the photoresist. Modern photolithography techniques can create features of about 100 nm in mass production and about 70 nm in experimental setups.

To improve the resolution of the above process up to several tens of nanometers would require the use of an electromagnetic radiation with a smaller wavelength, such as either extreme ultraviolet light with wavelengths between 10 nm and 70 nm, or X-rays with wavelengths between 0.1 and 10 nm. The high costs of such lithography tools, that could exceed the annual budget of a University or a research center, has initiated the emergence of a research field in which the aim is to establish simple and inexpensive nanofabrication methods. Along with nanoimprint lithography described previously, two other methods based on stamps have been developed, *viz.* microcontact printing and micromolding in capillaries. Both have been recently developed by the Whitesides group at Harvard University [29,30]. These methods, also called soft lithography methods, are based on the fabrication of soft molds or stamps using a soft rubbery polymer (polydimethylsiloxane-PDMS). In a first step, a master mask is fabricated on a substrate of choice, e.g. silicon, using conventional photo- or e-beam lithography. Then stamps are fabricated by replication of the master hard mask by simply pouring PDMS and subsequently curing it. The stamp can then be used in two modes. One mode is microcontact printing, in which the stamp is covered by a special solution containing molecules known to form self-assembled monolayers (SAMs), for instance thiol molecules on gold. When the stamp is brought into contact with the sample surface, the desired pattern of SAMs is transferred. In the second mode, the micromolding technique, the stamp is placed on a hard surface and a liquid polymer flows by capillary forces into the free space between the mold and the surface. While microcontact printing has a relatively poor resolution of only few microns, micromolding can replicate structures smaller than 10 nm.

Although the soft lithography is not an appropriate method for microelectronics, several advantages over the classical photolithography can be enumerated. Soft lithography methods are extremely useful in the case of patterning on non-flat surfaces, are relatively tolerant against a dusty environment, are cheap and can be used in combination with a wide range of materials, ferroelectrics being among the most interesting ones.

There is only a very small number of reports in literature on patterning ferroelectrics by combining soft lithography and chemical solution deposition methods such as sol-gel or metaloganic decomposition. Microcontact printing has been used together with sol-gel processing to selectively deposit oxide thin films with micron-scale lateral resolution. Patterned structures, such as capacitors and waveguides, have been fabricated from $LiNbO_3$, $PbTiO_3$, and $BaTiO_3$ on different substrates, including Si, Al, Pt, sapphire, and TiN. The technique involved functionalisation of substrate surfaces by microcontact printing of octadecyltrichlorosilane (OTS) self-assembled monolayers (SAMs). Sol-gel precursors are then spin-coated on the SAMs-patterned surfaces and heat-treated to deposit amorphous oxide layers of a thickness between 20 nm and 300 nm. The oxide on the functionalised regions is removed with mild polishing, yielding patterned films with features as small as 5 μm [31].

Micromolding has found a niche in MEMS technology, particularly in patterning of micron-scale piezoelectric structures [32,33]. Ceramic thick film and usual photolithography technologies have been used to develop micromolding of PZT thick films with lateral features down to 10 μm and aspect ratios up to 7:1 using soft reusable polydimethylsiloxane (PDMS) molds in order to fabricate piezoelectric actuators and transducers [34].

Dip-pen nano-lithography, developed by Mirkin of Northwestern University, is a method based on "literarily" writing patterns with an AFM tip and using a special ink that usually is based on thiol molecules, as in the microcontact printing [35]. Although the method has not yet been applied to pattern ferroelectric nanosize structures, it is obvious that dip-pen lithography has a certain potential in the field of fabrication of nanoscale ferroelectrics. If an appropriate stable precursor "ink" would be developed to replace the thiol-based ink, features of several tens of nanometers across can be easily written. Nonetheless, the main disadvantage remains the serial processing and the specific instability of the precursor solution against hydrolysis.

# 3. Bottom-Up Approaches

All lithography-based patterning techniques described above are top-down approaches and represent the basis of nowadays microelectronics technology. They are very well suited to fabricate structures with extremely good spatial resolution and positioning precision, but are intrinsically limited to structures, the sizes of which are well above the lowest nanometer-scale sizes. Therefore, there is a considerable interest in other fabrication methods that are not based on carving thin films, but rather on building structures from the bottom using atoms and molecules. These bottom-up methods will in principle allow the inexpensive fabrication of structures with sizes of 10–20 nm in a parallel approach. The primary disadvantage of the bottom-up methods is the random positioning of the obtained nanostructures that will make a precise interconnection of them virtually impossible. Successful strategies and routes have been developed to synthesize nanoscale materials of numerous simple systems such as semiconductors (Si, CdS, InAs/GaAs) or metals. Complex systems such as ferroelectric oxides or any muticomponent oxides belonging to the class of functional materials have not yet been systematically addressed, despite the possibility of discovering new materials with unique properties.

Recently the first attempts to prepare ferroelectric nanostructures *via* bottom-up approaches have been published. The approaches can already be classified into two main routes, *viz.* physical and chemical, which are obviously based on different principles. Both routes will be briefly described in the following.

## 3.1. Self-pattering *via* physical routes

In the quest for smaller and smaller ferroelectric structures, the physical routes have proven simple and reliable, making use of simple physical processes. Well-known physical growth concepts, such as island growth, have already been successfully applied to grow nanosize dots of germanium on silicon, or of InAs on GaAs. An important role is played by the lattice constant mismatch between the substrate and the deposited film. Depositing an epitaxial thin film onto a single-crystal substrate with high lattice mismatch, the initial stages of the growth process are characterised by either the island (Volmer-Weber) or layer-than-island (Stranski-Krastanov) growth modes. In contrast to the layer-by-layer (Frank-van der Merwe) growth mode, which results in a smooth uniform film, the two former growth modes are suitable to grow crystalline nano-size dots. The effect is well known and has been carefully studied in the case of simple systems such as germanium on silicon [36], and has even found application in the growth of compound semiconductor lasers [37]. Only recently the same

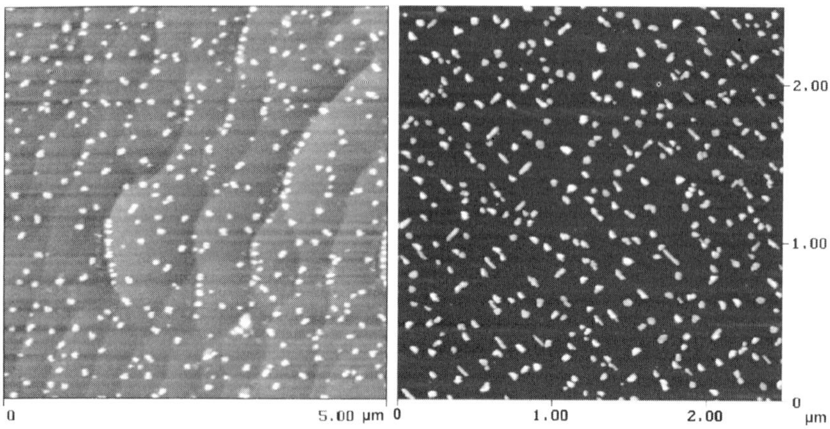


**Fig. 10.** SFM topography images of $BaTiO_3$ nanostructures deposited by pulsed laser deposition onto vicinal $SrTiO_3(100)$ single-crystal surfaces.

concept has been applied to the growth of complex oxide islands on different substrates, demonstrating ferroelectricity in PZT islands as small as 70 nm wide and 30 nm high [38, 39].

Pulsed laser deposition (PLD) and pulsed laser ablation are also potential methods to obtain nanoparticles deposited on a substrate or in the form of a dispersed powder, respectively. Pulsed laser ablation has been used to produce PZT nanoparticles with a relatively narrow size distribution. The experimental setup is rather complex and consists of a laser ablation chamber, a charger, and a furnace [40]. Single crystalline nanoparticles with an average diameter of about 7 nm and a standard deviation of 5.4% have been obtained. Visinoiu et al. have grown $BaTiO_3$ nanoparticles on $SrTiO_3$ substrates by a simple PLD process [41]. Using vicinal $SrTiO_3$ substrates, a certain degree of registration and self-organisation could be induced as shown in Fig. 10.

Another method used to fabricate nanostructures is based on the concept of structural and morphological instability of ultrathin films. This possibility has been found during an effort of growing single-crystal epitaxial films by chemical solution deposition (CSD). Lange, Speck, and coworkers have theoretically and experimentally established that an ultrathin amorphous layer of PZT (or another perovskite oxide) deposited onto a single-crystal substrate such as $SrTiO_3$, during high-temperature crystallisation breaks up into small islands due to a microstructural instability [42]. The driving force of this process is the minimisation of the free energy of the film-substrate system, by lowering the interface area (forming islands) and by forming low-energy surfaces *via* faceting [43].

Polycrystalline nanosize structures have been fabricated by depositing ultrathin lead titanate ($PbTiO_3$) films onto Pt-coated silicon substrates using CSD. As a result of the high-temperature crystallisation by rapid thermal annealing, the deposited amorphous film breaks up into randomly oriented nanosize crystals with sizes ranging from 10 to 100 nm [44]. Using piezoresponse SFM, Roelofs et al. have shown that grains with lateral sizes of about 20 nm do not display an out-of-plane piezoelectric signal,

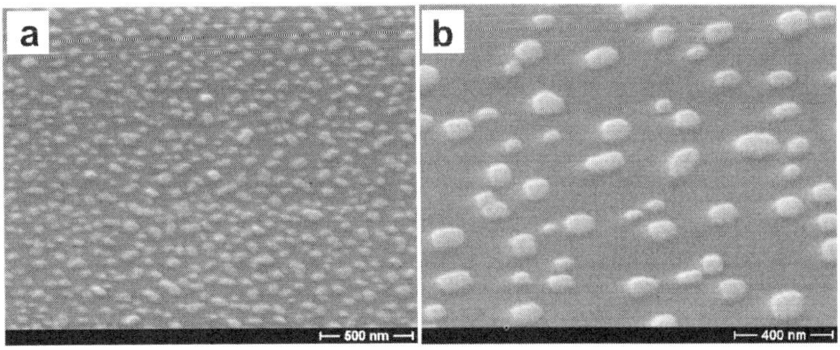


**Fig. 11.** Scanning electron micrographs of PZT nanocrystals crystallized at (a) 800 °C and b) 1100 °C.

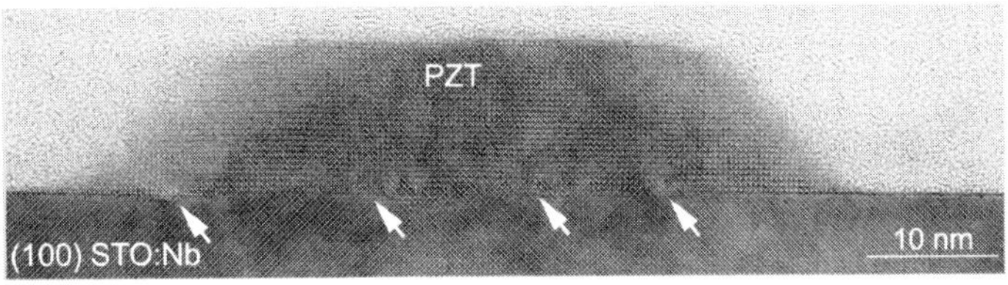


**Fig. 12.** High-resolution cross-section transmission electron micrograph of a small epitaxial PZT island grown on a $SrTiO_3$ single-crystal substrate. Arrows point to misfit dislocations.

suggesting that the size at which ferroelectricity vanishes in $PbTiO_3$ is about this size of 20 nm [45].

The main drawback of this experiment is the polycrystalline nature, i.e. the random orientation of the obtained grains. Without knowing the exact orientation of the measured crystallite it is extremely difficult to correctly interpret the piezoresponse signal [18], *cf.* Chapt. 4.2 below. Therefore it would be advantageous, at least in principle, to obtain epitaxial nanostructures rather than polycrystalline grains, so that the crystal orientation of all the epitaxial structures will be known. Szafraniak et al. have fabricated nanosize epitaxial PZT crystals using the microstructural instability method, Fig. 11 [46]. Structures with lateral dimensions of 40–90 nm have been obtained by conventional annealing of ultrathin amorphous oxide films at temperatures ranging from 800 °C to 1100 °C. The size, shape and distribution of the nanocrystals could be tuned to some extent by modifying the initial film thickness and the crystallisation temperature (*cf.* Figure). The epitaxial nature of the crystals is revealed both by X-ray diffraction analysis and high resolution TEM investigations (Fig. 12). The ferroelectric properties of the individual islands were determined by piezoresponse SFM. There were no direct correlations between the lateral size of the structures and the presence of the piezoelectric activity. Well-developed hysteresis loops were acquired on structures either with large lateral size or with high thickness, suggesting that the piezoelectric activity is connected rather to the volume of the structure than simply to the lateral size.

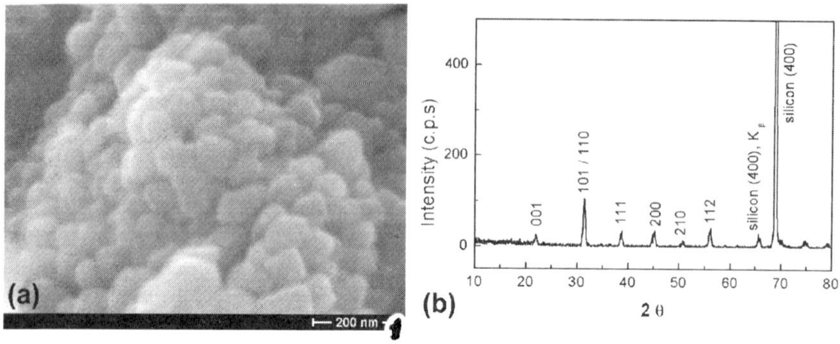


**Fig. 13.** (a) Microemulsion-derived $BaTiO_3$ nanopowder annealed at 450 °C for 1 h, and (b) a corresponding XRD analysis showing the formation of single-phase $BaTiO_3$ nanopowder.

### 3.2. Self-pattering *via* chemical routes

A number of interesting fabrication approaches appear to be based on chemical routes. Using simple chemical routes it is possible to fabricate nanosize crystals or nanoparticles in colloidal suspensions, which can later be spread onto any substrate surface. After solvent evaporation the nanoparticles can be crystallised in two- or even three-dimensional arrays. Although obtaining a regular array of particles is not a simple and easy task, self-assembly concepts based on the reduction of the interfacial energy at the fluid-solid or fluid-fluid interface—concepts that have proven valid for more simple systems such as CdS [47]—can be used for more complicated systems such as ferroelectric oxides. In any case, one has first to develop methods to fabricate nanoscale ferroelectric particles. One of the most promising routes is based on the microemulsion concept in which a water-in-oil emulsion is produced using a surfactant. Such an emulsion consists of nanometer-size water droplets uniformly dispersed in an oily solvent. This microemulsion can be used to hydrolyse a complex metalorganic precursor. The nano-droplets act as nano-reactors in which the hydrolysis of the precursors takes place and, if the optimum conditions are fulfilled, the final reaction product consists of mono-dispersed nanosize particles [48].

A cyclohexane/Brij 30 (Polyethylene glycol dodecyl ether)/$H_2O$ water-in-oil microemulsion was used to hydrolyse a metallic alkoxide precursor (e.g., barium titanium ethylhexano-isopropoxide for $BaTiO_3$ particles) [49]. By adding the microemulsion to an appropriate amount of bimetallic precursor, the latter reacts with the water droplets present in the microemulsion and forms oxide particles. After one hour of thermal treatment at 450 °C, fully crystallised, single-phase $BaTiO_3$ nanosize powders are obtained (Fig. 13). To avoid agglomeration of the nano particles, tetrabutylammonium hydroxide can be used as a peptising agent.

A slightly different route to prepare nanosize $BaTiO_3$ was implemented by O'Brian et al. [50]. A single bimetallic alkoxide precursor (barium titanium ethylhexano-isopropoxide) is injected into a mixture of diphenyl ether and oleic acid as a stabilising agent at 140 °C under inert atmosphere. After distillation of 2-propanol, the resulting solution is cooled to 100 °C, and hydrogen peroxide is injected. The solution is maintained at this temperature for 48 h to promote hydrolysis and crystallisation, in the given inverse micelle (microemulsion) conditions. Monodispersed $BaTiO_3$ particles of 4 to 8 nm diameter are obtained. The diameter could be tuned by modifying the ratio between the alkoxide precursor and the oleic acid, which also passivates the particle surface, preventing agglomeration and enabling the transfer of the particles into a nonpolar solvent.

Concerning this route, two remarks are worth mentioning. First, the crystalline phase occurs at temperatures as low as 100 °C, directly from the reaction. This is the only known route to obtain phase-pure perovskite oxides at such a low temperature. Second, at a more general level, the above route involving reactions in a so called inverse micelle, or colloidal-type solution (microemulsion route) is a generic route to synthesize nanoparticles. Due to its simplicity, its excellent output, and its low cost it might have an important impact in nanofabrication of ferroelectrics.

## 4. Ferroelectric Testing of Nanopatterned Electroceramics

Due to their small size, the characterisation of the cells or structures obtained cannot be achieved using conventional measurement techniques. For this purpose, scanning probe techniques, in particular scanning force microscopy (SFM), have proven to be most valuable. The principle of an SFM is quite simple [51]: A sharp tip mounted on a cantilever is brought close to a sample surface. The (attractive or repulsive) force between the tip and the surface is detected using the bending of the cantilever. A feedback loop adjusts the z-position of the cantilever using a piezoelectric element, so that the interaction force is kept constant. By scanning the tip over the surface and recording the z-position of the cantilever, a map of the sample topography can be obtained[1].

Of particular interest in the characterisation of electroceramic materials is the use of a conductive SFM tip as a movable top electrode, in order to locally measure electrical or electromechanical properties of the sample. In this case the tips are either coated with a conductive layer or made of a conductive material. The sample itself is provided with a bottom electrode on which the electroceramic material is deposited or nano-fabricated. This geometry allows the application of a voltage to an area of a few hundreds of square nanometers, permitting to probe individual cells of sub-100 nm lateral sizes.

### 4.1. Electronic measurements

Electrical d.c. measurements using the SFM tip as a top electrode are possible in a similar way as normal macroscopic measurements. However, the experiments are very sensitive to the local environment and the reported results are quite contradictory [52, 53]. Our experimental data (Fig. 14), e.g, confirm the space-charge limited type conduction mechanism in EBDW-patterned PZT structures of 100 nm lateral size.

Direct measurements of polarisation hysteresis loops from FIB-milled PZT structures was also achieved [54]. In this case, the charge signal from single submicron

[1] Here it is assumed that the interaction force depends only on the distance between the tip and the surface.

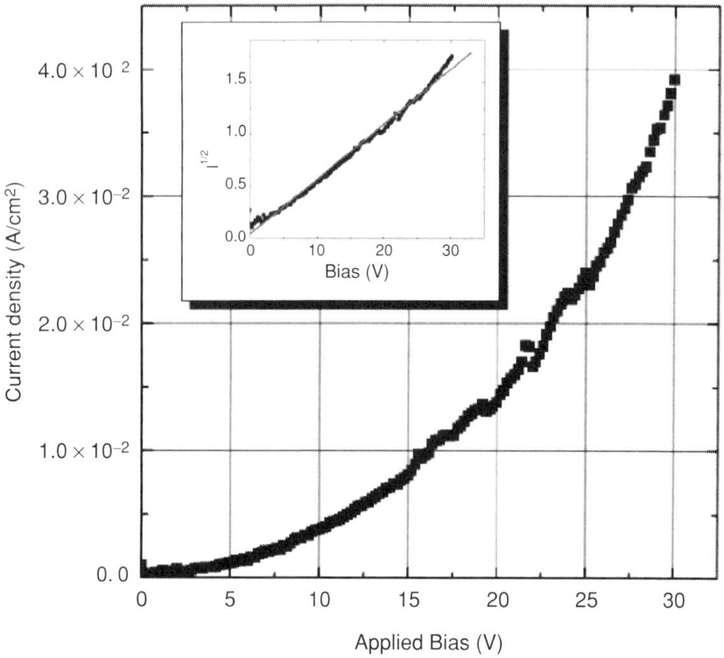


**Fig. 14.** I-V curve of a PZT cell of 100 nm lateral size, measured with a Pt-coated SFM tip consisting of diamond. The positive voltage was applied to the Pt-coated tip serving as the top electrode. The inset shows the same curve replotted into a $I^{1/2}$ *vs.* voltage plot, the straight line corresponding to the linear fit.

capacitors is lower than the shunt capacitance signal of the cantilever. A special correction was employed to get the true polarisation hysteresis loop. The conclusion, however, confirmed previous results [8, 9, 14, 15], according to which there is no size effect in PZT down to a lateral size of 100 nm.

### 4.2. Piezoresponse scanning force microscopy

The local polarisation state and the electromechanical properties of ferroelectric thin films and nanostructures can be probed *via* the converse piezoelectric effect using SFM combined with a lock-in technique. This technique is based on the detection of local vibrations of a ferroelectric sample induced by a testing a.c. signal applied between the conductive tip of the SFM and the bottom electrode of the sample [55, 56]. The mechanical oscillations of the sample underneath the tip are transmitted to the cantilever, provided that the contact between tip and sample is firm. This condition is fulfilled when the contact force is dominating over the electrostatic and adhesion forces, i.e. the SFM is in the contact mode of operation with stiff cantilevers. The oscillations of the cantilever are then detected from the global cantilever deflection using a standard lock-in technique (Fig. 15). Since the first harmonic of the oscillations represents the piezoelectric constant, these vibrations converted into an electrical a.c. signal are further referred to as the piezoresponse signal, or PRS.

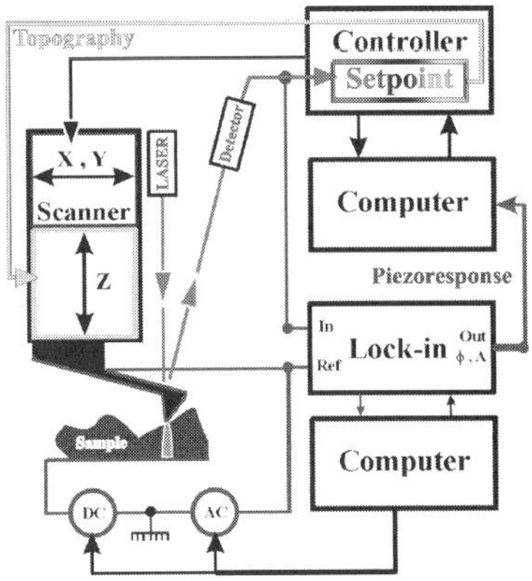


**Fig. 15.** Typical experimental setup for piezoresponse imaging.

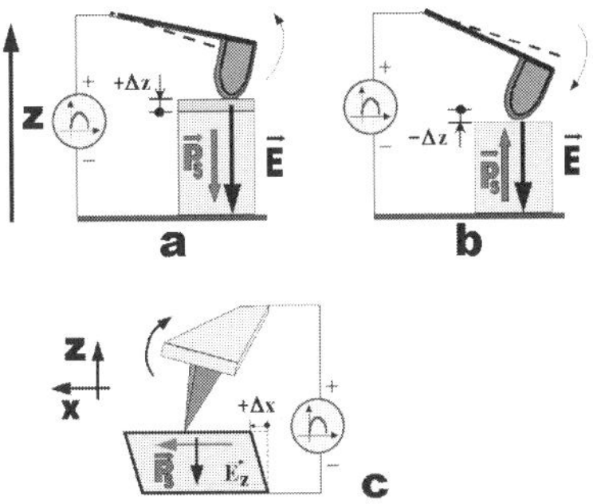


**Fig. 16.** Principle of piezoresponse force microscopy. a) Piezoelectric expansion when the electric field and the polarization have the same direction. b) Piezoelectric contraction of the material when the directions of the polarization and of the electric field are opposite. c) Lateral displacement of the surface due to the induced shear strain.

The most prominent feature in piezoresponse detection of the polarisation at the surface of an electroceramic material is the strong relation between the piezoelectric coefficient and the spontaneous polarisation. The sign of the piezoelectric coefficient, which determines whether the material expands or contracts, is given by the direction of the spontaneous polarisation with respect to that of the applied electric field. This allows the detection of the direction of the out-of-plane (OP) component of polarisation ($P_z$). Figures 16a and b illustrate the principle: If the electric field and the polarisation are in the same direction, the material will expand (positive piezoelectric coefficient), otherwise it will contract. The magnitude of the PRS is related to the magnitude of the OP-component of the polarisation. In the general case the relationship between the polarisation $P_z$ and the piezoelectric coefficient $d_{zz}$ along a certain direction $z$ is given by tensorial transformations [18], which sometimes results in a non-univocal relation. However, in several particular cases there is a favourable univocal relationship. One of these cases is that of a material of tetragonal crystal symmetry, for which the shear piezoelectric coefficient is not excessively high compared to the longitudinal piezoelectric coefficient [57].

The above depiction of the processes involved in the detection of the PRS can be described mathematically, in the most simplified way, as follows: The OP-PRS is proportional to the surface displacement $\Delta z$ via a calibration factor $\gamma_{AFM}$, which is given by the cantilever and the optical detection system of the SFM:

$$OPPRS = \gamma_{AFM}\Delta z.$$

Assuming a uniform electric field, the surface displacement is

$$\Delta z = \begin{cases} \Delta z^{+} = -d_{zz}V_{AC}, & P_z > 0 \\ \Delta z^{-} = d_{zz}V_{AC}, & P_z < 0 \end{cases}$$

where $V_{AC}$ is the amplitude of the applied voltage.

In a.c. measurements, a negative sign can be interpreted as a phase shift of 180° between the respective physical quantities. In other words, domains with opposite orientations of the polarisation will result in out-of-phase piezoresponse signals. The magnitude of $d_{zz}$ is related to the standard piezoelectric coefficients *via* tensor transformations representing rotations of the piezoelectric tensor. For the simple tetragonal symmetry the final relationship is

$$d_{zz}(\theta) = (d_{31} + d_{15})\sin^2\theta\cos\theta + d_{33}\cos^3\theta,$$

where $\theta$ is the angle between the direction of measurement z and the vector of the spontaneous polarisation. Since all piezoelectric coefficients are related to the magnitude of the spontaneous polarisation[2] [57] it can be concluded that the magnitude of OP-PRS gives an idea of the OP component of the polarisation $P_z = P_S\cos(\theta)$ if the above equation is univocal.

The detection of an in-plane (IP) component of the polarisation vector (parallel to the sample surface) using the converse piezoelectric effect is also possible [58]. The basic principle of this detection consists in the existence of piezoelectric shear deformations (Fig. 16c). If the polarisation vector is perpendicular to the electric field, there is no piezoelectric deformation along the field direction, but a shear strain appears in the ferroelectric, leading to displacements of the sample surface parallel to itself, along the polarisation direction. The IP displacements of the surface are transferred *via* friction to the SFM tip as lateral movements. The component of these movements perpendicular to the cantilever axis induces a torsion of the cantilever end, which can be detected using the capability of the SFM to measure the cantilever torsion.

As in the case of detection of the domains with OP polarisation, an a.c. field induces IP oscillations phase shifted by 180° in regions with opposite orientations of the polarisation. However, due to the complex nature of the friction interaction it is difficult to describe them quantitatively. It should be noted that a shear deformation can be present along both the x- and y- axes, but due to the cantilever geometry, domains having a polarisation along the y-axis can be recorded only by physically rotating the sample by 90°.

Figure 17 shows an example of piezoresponse imaging at the surface of a bulk (001)-oriented barium titanate single crystal. The two PRS images (Figs. 17b and c)

[2] An exception from this rule is barium titanate, for which $d_{33} = 80$ pm/V and $d_{15} = 400$ pm/V.

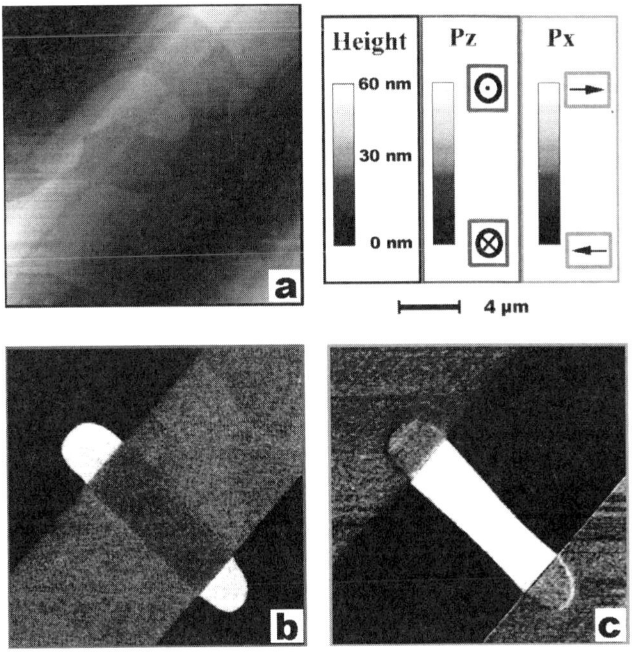


**Fig. 17.** Example of topography and piezoresponse images of the same sample region. a) Sample topography, b) OP-PRS and c) IP-PRS of a barium titanate single crystal exhibiting an *a-c* lamellar ferroelectric domain structure. (Sample courtesy to Prof. L.M. Eng, Dresden, Germany.)

were acquired simultaneously with the sample topography (Fig. 17a). The topographic data show the presence of facets making an angle of 0.5°, close to the theoretical value of the tetragonal distortion for barium titanate of 0.6°. Surface steps, about 10 nm in height, are also visible, which have no correlation with the actual ferroelectric domain structure. The images (b) and (c) in Fig. 17 display the OP-PRS and the IP-PRS, respectively, for the same region. The complementary character of the two images is obvoius: Regions with high $P_Z$-contrast (black or white color in image b) exhibit zero $P_X$-contrast (grey color in image c). In other words, a certain point of the surface can have a polarisation either perpendicular or parallel to the surface, in complete agreement with the crystallographic orientation of the sample. However, it should be kept in mind that in order to determine all three components of the vector of polarisation, additional information is required. In the case of a barium titanate single crystal it is sufficient to know the crystallographic orientation of the sample, but this may not be the case for materials with a lower symmetry, or for polycrystalline films.

By definition, ferroelectricity implies that the domains can be switched by an external electric field. In addition to ferroelectric domain imaging, PFM permits the control of ferroelectric domains, i.e. it allows to induce switching by applying a d.c. bias voltage between the tip and the bottom electrode. The most important proof of ferroelectricity is the presence of a piezoelectric hysteresis. The latter is generally recorded using a d.c. bias source connccted in series with the a.c. voltage source (*cf.* Fig. 15). Hysteresis loops are obtained by sweeping the bias voltage and recording the piezoresponse signal

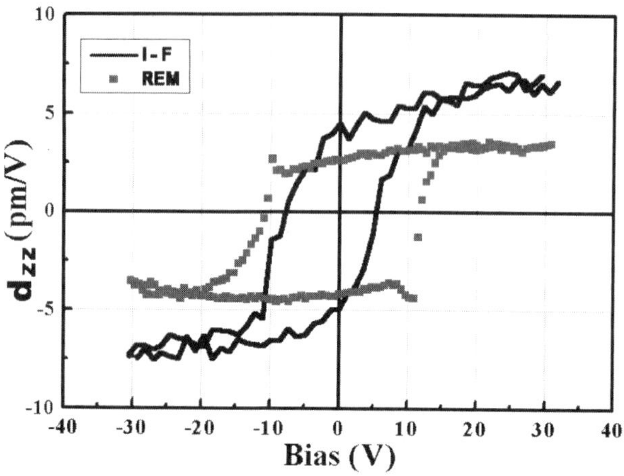


**Fig. 18.** In-field (I-F) and remanent (REM) piezoelectric hysteresis loops of a PZT grain.

while the SFM tip is kept fixed above the desired region of the sample. The result of this procedure is named *in-field* (I-F) hysteresis loop. In addition to this procedure, the d.c. bias may be applied in the form of pulses and the PRS recorded between the pulses, i.e. while no bias is applied to the sample [59]. In this way any electrostatic interaction is avoided. The resulting hysteresis loop is named *remanent* (REM) loop [62]. It provides additional information on the switching process, on retention and imprint. As an example, Fig. 18 shows an I-F and a REM hysteresis loop recorded from the same PZT (40/60) grain in a 400 nm thick film. Whereas the I-F loop (dark curve) shows a symmetric hysteresis, the REM loop (bright curve) demonstrates that the positive retention value is actually only about 70% of the negative one, and therefore the grain is imprinted. Additionally, if switching of polarisation is achieved using voltage pulses (instead of a continuous voltage), the coercive voltage required for switching is doubled. A more detailed discussion on the above topics is given in Cap. 2 of [63].

## 5. Summary and Outlook

Several non-conventional patterning techniques are available for micro- and nanopatterning of electroceramics. They are based either on top-down or on bottom-up approaches and have successfully been applied to obtain sub-micron ferroelectric nanostructures.

Focussed ion beam patterning is one of the most versatile methods. It has an important advantage—it can pattern features in a well-known and well characterized material (thin film). It has also some drawbacks, which, however, can be minimized by using the method carefully. Electron beam direct writing is an inexpensive method to prepare arrays of sub-100 nm ferroelectric structures. The high resolution and the versatility are unfortunately obscured by the extremely low writing speed which is due to the serial process flow. Nanoimprint lithography is a potential solution for large area patterning. It has been proven that it works very well in the micron size, but in the submicron range there are still some difficulties which might be overcome with some effort. Nanoimprint has a major advantage over classical lithography methods due to its possibility of patterning structures on non-flat surfaces. This will most probably have a certain impact in optical applications where the feature size is a micron or larger, and where the surface in use is not always flat.

Bottom-up approaches offer the possibility of fabricating ferroelectric structures with lateral sizes of 10–50 nm, well below the sizes accessible by any of the top-down methods. Unfortunately, registration of the structures is still a major problem of all bottom-up approaches, and quite an effort will be necessary to solve this problem. On the positive side, all bottom-up approaches have two considerable advantages, *viz.* low cost and large area, which make them potentially attractive for large scale fabrication.

The characterization of sub-micron and nanoscale ferroelectrics has been boosted by the development of piezoresponse SFM (PFM). PFM offers an elegant solution to observe and modify ferroelectric domains, to locally measure piezoelectric properties, and along with other scanning probe microscopy methods it can give insight into various phenomena occurring in ferroelectrics at the nanoscale. Due to its extremely simple setup, PFM has become a standard method to characterize ferroelectric materials.

The successful preparation of sub-micron and nanosize ferroelectric structures and the development of appropriate nanoscale measurement methods open a new emerging field, *viz.* that of *nanoscale ferroelectrics*. Besides new and exciting physics that waits to be revealed, new fields of application are likely to emerge. Nanoscale ferroelectrics may, e.g., have a certain impact in photonics, where tunable photonic band gap materials might open unlimited perspectives, or in MEMS and NEMS, where, for example, mesoscopic and nanosize ferroelectric/piezoelectric wires and tubes might open the possibility of creating active cantilevers or even on-chip scanning tunneling microscopes [60,61]. In any case, *nanoscale ferroelectrics* will be part of our near future.

## 6. Acknowledgements

This report is based on experimental work performed together with S. Bhattacharyya, S. Chattopadyay, M.-W. Chu, U. Doß, V. Dragoi, W. Erfurth, N. Kurowsky, A. Pignolet, J. Schilling, R. Scholz, I. Szafraniak, A. Visinoiu, and R.B. Wehrspohn, to whom the authors are grateful for their respective contributions. The authors thank J.F. Scott, R. Ramesh, and L.M. Eng for many fruitful discussions. Special thanks are due to U.M. Gösele for initiating investigations on nano-sized ferroelectrics as well as for numerous stimulating discussions and continuous support.

## References

[1] J.F. Scott, Ferroelectric Memories (Springer, Berlin, 2000).
[2] M.E. Lines and A.M. Glass, Principles and Applications of Ferroelectrics and Related Materials (Clarendon Press, Oxford, 1977).
[3] Y.G. Wang, W.L. Zhong, and P.L. Zhang, Phys. Rev. B **51**, 5311 (1995).
[4] I. Stolichnov, A. Tagantsev, E. Colla, S. Gentil, S. Hiboux, J. Baborowski, P. Muralt, and N. Setter, J. Appl. Phys. **88**, 2154 (2000).

[5] S. Bühlmann, B. Dwir, J. Baborowski, and P. Muralt, Appl. Phys. Lett. **80**, 3195 (2002).
[6] J.-F. Lin, J.P. Bird, L. Rotkina, and P.A. Benett, Appl. Phys. Lett. **82**, 802 (2003).
[7] C.S. Ganpule, A. Stanishevsky, Q. Su, S. Aggarwal, J. Melngaillis, E. Williams, and R. Ramesh, Appl. Phys. Lett. **75**, 409 (1999).
[8] C.S. Ganpule, A. Stanishevsky, S. Aggarwal, J. Melngailis, E. Williams, R. Ramesh, V. Joshi, and C.P. de Araujo, Appl. Phys. Lett. **75**, 3874 (1999).
[9] V. Nagarajan, A. Roytburd, A. Stanishevsky, S. Prasertchoung, T. Zhao, L. Chen, J. Melngailis, O. Auciello, and R. Ramesh, Nature Materials **2**, 43 (2003).
[10] A. Stanishevsky, B. Nagaraj, J. Melngailis, R. Ramesh, L. Khriachtchev, and E. McDaniel, J. Appl. Phys. **92**, 3275 (2002).
[11] H.G. Craighead and L.M. Schiavone, Appl. Phys. Lett. **48**, 1748 (1986).
[12] J. Lochau, S. Friedrichowski, D. Dumpich, E.F. Wassermann, M. Winter, and M.T. Reetz, J. Vac. Sci. Technol. B **16**, 77 (1998).
[13] S. Okamura, K. Mori, T. Tsukamoto, and T. Shiosaki, Integr. Ferroelectr. **18**, 311 (1997).
[14] M. Alexe, C. Harnagea, D. Hesse, and U. Gösele, Appl. Phys. Lett. **75**, 1793 (1999).
[15] M. Alexe, C. Harnagea, W. Erfurth, D. Hesse, and U. Gösele, Appl. Phys. A: Mater. Sci. and Processes **70**, 247 (2000).
[16] D. Damjanovic, Rep. Prog. Phys. **61**, 1267 (1998).
[17] C.H. Ahn, T. Tybell, L. Antognazza, K. Char, R.H. Hammond, M.R. Basley, O. Fischer, and J.-M. Triscone, Science **276**, 1100 (1997).
[18] C. Harnagea, A. Pignolet, M. Alexe, and D. Hesse, Integrated Ferrelectrics **44**, 113 (2002).
[19] C. Harnagea, M. Alexe, D. Hesse, and A. Pignolet, Appl. Phys. Lett. **83**, 338 (2003).
[20] M. Alexe, C. Harnagea, D. Hesse, and U. Gösele, Appl. Phys. Lett. **79**, 242 (2001).
[21] S.K. Day, J-J. Lee, and P. Alluri, Jap. J. Appl. Phys., Pt. 2 **34**, 3142 (1995).
[22] J.F. Scott, K. Watanabe, A.J. Hartmann, and R.N. Lamb, Ferroelectrics **225**, 83 (1999).
[23] J.F. Scott and M. Dawber, Appl. Phys. Lett. **76**, 3801 (2000).
[24] S.Y. Chou, P.R. Krauss, and P.J. Renstrom, Appl. Phys. Lett. **67**, 3114 (1995).
[25] S.Y. Chou, P.R. Krauss, and P.J. Renstrom, J. Vac. Sci. Technol. B **14**, 4129 (1996).
[26] S.Y. Chou, P.R. Krauss, and P.J. Renstrom, Science **272**, 84 (1996).
[27] C. Harnagea, M. Alexe, J. Schilling, R.B. Wehrspohn, D. Hesse, and U. Gösele, Mater. Res. Soc. Symp. Proc. **748**, U.4.2.1. (2003).
[28] R. B. Wehrspohn and J. Schilling, MRS Bull. **26**, 623 (2001).
[29] R.J. Jackman, J.L. Wilbur, and G.M. Whitesides, Science **269**, 664 (1995).
[30] E. Kim, Y. Xia, and G.M. Whitesides, Advanced Materials **8**, 245 (1996).
[31] P.G. Clem, N.L. Jeon, R.G. Nuzzo, and D.A. Payne, Mater. Res. Soc. Symp. Proc. **435**, 521 (1996).
[32] M. Sayer, M. Lukacs, T. Olding, G. Pang, L. Zou, and Y. Chen, Mater. Res. Soc. Symp. Proc. **541**, 599 (1999).
[33] L. Dellmann, S. Roth, C. Beuret, G.-A. Racine, H. Lorenz, M. Despont, P. Renaud, P. Vettiger, and N.F. De Rooij, Sensors and Actuators **A70**, 42 (1998).
[34] T. Rosqvist and S. Johansson, Sensors and **A97–98**, 512 (2002).
[35] R.D. Piner, J. Zhu, F. Xu, S. Hong, and C.A. Mirkin, Science **283**, 661 (1999).
[36] D. J. Egeleham and M. Cerullo, Phys. Rev. Lett. **64**, 1943 (1990).
[37] R. L. Selin, Ch. Ribbat, M. Grundmann, N.N. Ledentsov, and D. Bimberg, Appl. Phys. Lett. **78**, 1207 (2001).
[38] H. Fujisawa, K. Morimoto, M. Shimizu, H. Niu, K. Honda, and S. Ohtani, MRS Symp. Proc. **655**, C.C. 10.4.1. (2001).
[39] H. Fujisawa, M. Okaniwa, H. Nonomura, M. Shimizu, and H. Niu, Proc. 15th Int. Symp. Integrated Ferroelectrics (ISIF), March 9–12, 2003, Colorado Springs, Colorado/USA.

[40] K. S. Seol, S. Tomita, K. Takeuchi, T. Miyagawa, T. Katagiri, and Y. Ohki, Appl. Phys. Lett. **81**, 1893 (2002).
[41] A. Visinoiu, M. Alexe, and D. Hesse, MPI Halle. Unpublished results (2002).
[42] A Seifert, A. Vjta, J. S. Speck, and F. F. Lange, J. Mater. Res. 11, 1470 (1996).
[43] L. Zhao, A. T. Chen, F. F. Lange, and J. S. Speck, J. Mater. Res. 11, 1325 (1996).
[44] R. Waser, T. Schneller, S. Hoffmann-Eifert, and P. Ehrhart, Integrated Ferroelectrics **36**, 3 (2001).
[45] A. Roelofs, T. Schneller, K. Szot, and R. Waser, Nanotechnology **14**, 250 (2003).
[46] I. Szafraniak, C. Harnagea, R. Scholz, S. Bhattacharyya, D. Hesse, and M. Alexe, Appl. Phys. Lett. **83**, 2211 (2003).
[47] Y. Lin, H. Skaff, T. Emrik, A. D. Dinsmore, and T. P. Russell, Science **299**, 226 (2003).
[48] H. Herrig and R. Hempelmann, Mater. Lett. **27**, 287 (1996).
[49] S. Bhattacharyya, S. Chattopadhyay, and M. Alexe, Mater. Res. Soc. Symp. Proc. **740**, I.10.3.1. (2003).
[50] S. O'Brien, L. Brus, and C.B. Murray, J. Amer. Chem. Soc. **123**, 12085 (2001).
[51] G. Binnig, F.C. Quate, and Ch. Gerber, Phys. Rev. Lett. **56**, 930 (1986).
[52] H. Fujisawa, M. Shimizu, T. Horiuchi, T. Shiosaki, and K. Matsushige, Appl. Phys. Lett. **71**, 416 (1997).
[53] Z. Xie, E. Z. Luo, J. B. Xu, I. H. Wilson, H. B. Peng, L. H. Zhao, and B. R. Zhao, Appl. Phys. Lett. **76**, 1923 (2000).
[54] S. Tiedke, T. Schmitz, K. Prume, A. Roelofs, T. Schneller, U. Kall, R. Waser, C. S. Ganpule, V. Nagarajan, A. Stanishevsky, and R. Ramesh, Appl. Phys. Lett. **79**, 3678 (2001).
[55] H. Birk, J. Glatz-Reichenbach, L. Jie, E. Schreck, and K. Dransfeld, J. Vac. Sci. Technol. B **9**, 1162 (1991).
[56] K. Franke, J. Besold, W. Haessler, and C. Seegebarth, Surf. Sci. **302**, L283 (1994).
[57] A.F. Devonshire, Philos. Mag. **42**, 1065 (1951). See also A.F. Devonshire, Philos. Mag. **40**, 1040 (1949).
[58] M. Abplanalp, L.M. Eng, and P. Günter, Appl. Phys. A: Mater. Sci. and Processes **66**, S231 (1998).
[59] C. Harnagea, A. Pignolet, M. Alexe, D. Hesse, and U. Gösele, Appl.Phys. A **70**, 261 (2000).
[60] M. Alexe, Y. Luo, I. Szafraniak, R. Wehrspohn, and M. Steinhart, European Patent Application No. 03000969 (2002).
[61] Y. Luo, I. Szafraniak, V. Nagarajan, R.B. Wehrspohn, M. Steinhart, J.H. Wendorff, N.D. Zakharov, R. Ramesh, and M. Alexe, Appl. Phys. Lett. 83, 440–442 (2003).
[62] C. Harnagea, A. Pignolet, M. Alexe, D. Hesse, and U. Gösele, Appl. Phys. A: Mater. Sci. and Processes **70**, 261 (2000).
[63] M. Alexe and A. Gruverman, (Eds.), *Nanoscale characterisation of ferroelectric materials—Scanning probe microscopy approach*, Nanoscience and Technology Series, Springer, Berlin, Germany (2004).

# Chapter 15

## Low-Cost Patterning of Ceramic Thin Films

C. R. Martin and I. A. Aksay
*Department of Chemical Engineering, Princeton University*
*Princeton, NJ 08544-5263 USA*

The patterning of ceramic thin films is of great interest for use in MEMS and other applications as discussed in the chapters by Maeda *et al.* and by Muralt and Baborowski. However, the complex chemistries of certain materials make the use of traditional photolithography techniques prohibitive. In this chapter, a number of low-cost, high throughput techniques for the patterning of ceramic thin films derived from chemical solution precursors, such as sol-gels and ceramic slurries, are presented. Most of these methods are derived from soft lithographic methods using elastomer molds, a method that is categorized as the next generation lithography in the chapter by Alexe *et al.* Two categories of techniques are discussed: first, the focus is on methods that rely on the principles of confinement within the physical features of the mold to define the pattern on the substrate surface. Then, subtractive patterning techniques that rely on transferring a pattern to a spin-cast large-area continuous thin film are described. Most techniques have been demonstrated with fidelities on the order of 100 nm; however, their inability to precisely register and align the patterns as part of a hierarchical fabrication scheme has hindered their commercial implementation thus far. This chapter has been updated from the original manuscript [1] to reflect the most recent available literature and complements the chapter by Baborowski on pattern formation by micromaching techniques.

## 1. Introduction

The advent of the electronic age has been the primary driving force for the development of thin film technologies for semiconductor applications, including techniques for the definition of submicrometer-scale patterns. [2] A typical microfabrication process utilizes photolithography [3] with photo-curable polymers to define patterns that are transferred to thin films of metal and semiconductor materials comprising the integrated circuit (IC) by wet or dry chemical etching. Many of these techniques have been transferred to materials with specialized properties for new engineering applications, such as microelectromechanical systems (MEMS) capable of sensing, [4] actuation, information processing, [5] optical waveguides, surface acoustic wave devices, and holographic memories. [6] For example, Ozawa and Yao [7] demonstrated the use of a photolithography lift-off process to pattern $TiO_2$ thin films with line widths as small as 1 μm. Similarly, Nashimoto *et al.* [8] fabricated 5 μm wide $(Pb,La)(Zr,Ti)O_3$ waveguides by using wet chemical etching with HCl prior to heat treatment.

Unfortunately, the utility of applying conventional photolithography techniques to ceramics in general can be limited by the chemistry of the material. For instance, the integration of ferroelectric materials, such as lead zirconate titanate ($PbZr_xTi_{1-x}O_3$, PZT), in electronic devices such as capacitors, transducers, pyroelectric devices, thermosistors, and dielectrics is a field attracting considerable attention. [9] However, traditional subtractive patterning steps are of limited utility to materials such as PZT because many of the solvents and acids used to etch silicon, such as HF, are not compatible with PZT. Instead, Reactive Ion Etching (RIE) must be used which is very costly due to the requirements of specialized equipment with controlled environments and waste control systems to minimize lead contamination. [10]

During the past decade, a number of alternative techniques to traditional photolithography for patterning a variety of materials have been developed. [11] The motivation for this work stems from four reasons: (1) lower cost and ease of use than the capital intensive equipment required for photolithography, (2) the ability to apply patterning techniques to new materials, (3) the potential to pattern large areas simultaneously, and (4) the ability to pattern on nanometer length scales that are not yet commercially produced by lithography. This chapter focuses on techniques developed to address the first three motivations. Specifically, we highlight the development of patterning methodologies on the submicrometer length scale that can be applied to a wide range of ceramic thin film materials. Patterning on smaller nanometer length scales is not our specific goal due to the propensity for many ferroelectric materials to lose particular properties that make them of interest for MEMS and other applications in the first place. For example, lead titanate may lose its piezoelectric properties in grains smaller than 20 nm [12] although results contrary to this have also been reported. [13]

Consistent with the objective of low cost processes, the discussion will be limited to methods that do not require large expenditures of capital equipment beyond what would be found in a basic microfabrication laboratory equipped for photolithography. As a result, our focus will be on materials that can be deposited and patterned using liquid precursors, such as sol-gels [14, 15] or colloidal suspensions. We will also further restrict our discussion by neglecting serial patterning techniques, such as ink-jet printing, [16–18] as these are not viable techniques for scaling production to the commercial applications that require the ability to pattern large areas simultaneously.

This chapter is divided into two sections. First, patterning methods that are based on the principle of physical confinement of the precursor solution are discussed. The second section highlights a pair of patterning techniques that are based on the spin-casting of a large area thin film and then producing the final pattern by subtractive means. The discussion of each technique includes examples of ceramic materials that have been patterned by the method with emphasis on the length scale of the resolution limit.

Many of the techniques illustrated in this chapter are derivatives of the methods developed by the Whitesides group at Harvard University, [11] as soft lithography is more convenient and lower in cost than the traditional subtractive methods. This is a very active area of research in its application to ceramic thin films, as the majority of the available literature has been published within in the past six years.

Soft lithography requires the use of an elastomeric mold patterned with relief structures on its surface. First, a lithographic master with complex three dimensional patterns is fabricated using conventional photolithography, micromachining, or electron beam writing techniques. Using cast molding, a negative of the patterns of the master is produced in the elastomeric stamp, as shown in the center panel of Figure 1. [11]

The most common elastomeric material used in soft lithography is poly(dimethylsiloxane) (PDMS), although many other materials may be used. The low surface energy of PDMS ($\gamma_{\text{PDMS-air}} = 0.0216$ N/m, compared to $\gamma_{\text{Si/SiO}_2\text{-air}} \sim 0.072$ N/m) allows it to be easily released from the master after curing, as well as from the substrate and the dried material after patterning. PDMS is not hydroscopic and does not swell when in contact with water. It has not been found to react with or swell due to contact with many of the compounds and solvents present in chemical precursor solutions. The PDMS is thermally stable to 186 °C, allowing for the use of moderate heat to quickly cure the mold. The material is isotropic and homogeneous, allowing it to be used to pattern over large areas in future applications. Further, PDMS is permeable to gaseous molecules, a critical property for allowing the patterned chemical solution or colloidal suspension to dry within the confines of the mold. [11]

Conveniently, the materials for casting PDMS molds are available commercially; most research groups use Sylgard™ 184, a two-part kit manufactured by Dow Corning (Midland, MI). To prepare the PDMS for casting, a 10:1 ratio of the liquid silicon rubber base, containing primarily vinyl-terminated PDMS, is mixed with a curing agent that promotes hydrosilylation reactions to cross-link the network and produce a solid elastomer. [11] Feature sizes as small as 30 nm have been patterned on a PDMS mold. [19]

With the exception of the preparation of the lithographic master, soft lithography does not require the use of a clean room or expensive microfabrication equipment. Once a lithographic master is available, the master can be reused multiple times for templating PDMS molds. Thus, the various soft lithography techniques are very much available in terms of time and cost. In addition, the use of elastomers for the mold in soft lithography adds a degree of flexibility not found in a mask used in photolithography, as the molds can be deformed by mechanical compression, tension, or strain to conform to curved substrates or manipulate the desired patterned features sizes. [11]

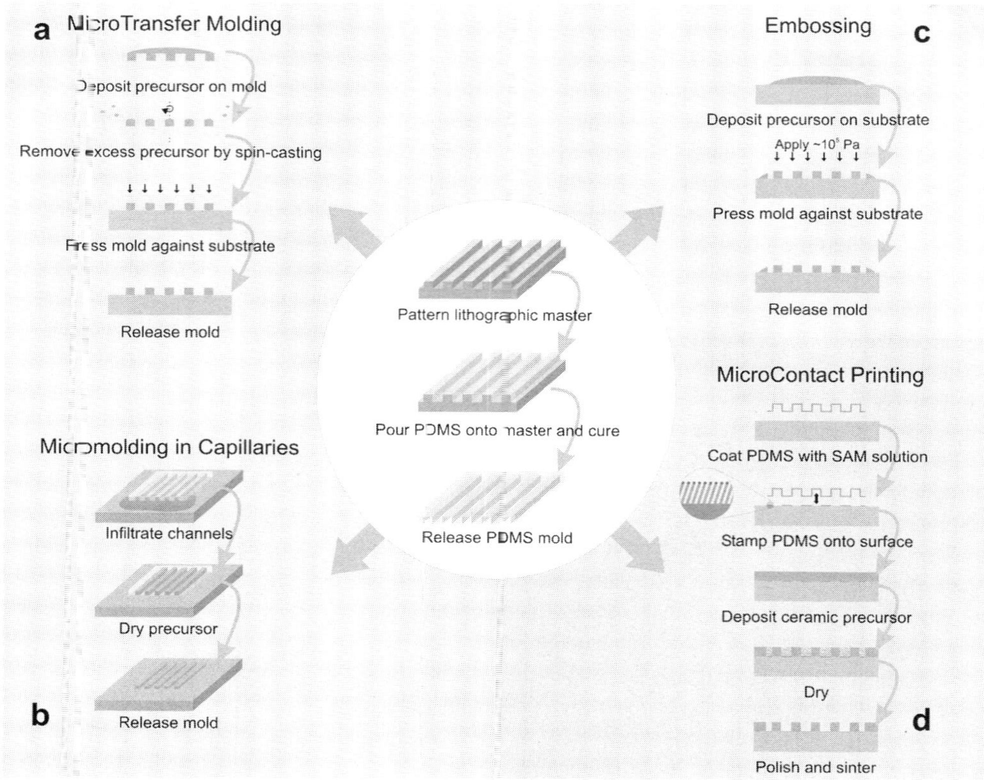


**Fig. 1** Schematic illustration of PDMS mold preparation and soft lithography techniques used to pattern ceramic materials. a) MicroTransfer Molding, b) Micromolding in Capillaries, and c) Embossing are confinement-based techniques while d) MicroContact Printing is based on the principles of surface modification.

There are potential disadvantages of soft lithography that need to be overcome before it is readily implemented in commercial practice. First, PDMS shrinks by approximately 1% during curing and can be swelled by nonpolar organic solvents such as toluene and hexane. The softness of the elastomeric material can cause features to deform or distort, such as sagging or buckling channels, if insufficient attention is paid to the aspect ratio of the microstructures. [11] This second issue can be partially mitigated by the use of composite stamps, particularly for producing free-standing features smaller than 100 nm. [20] But most critically, it is difficult to design a scheme to register the PDMS molds with the substrate during the fabrication of multiple layered structures. [21] It is this point in particular that led to the development of a class of hybrid photolithography/soft lithography casting schemes, such as the MicroChannel Molding (μCM) technique developed by the authors. [22]

## 2. Confinement-Based Techniques

The philosophy shared by the following confinement-based techniques is that the pattern is imparted to the thin film by using physical boundaries to define where the chemical precursor solution or colloidal suspension dries and rigidifies. Each of these methods is an additive patterning technique, whereby the material is deposited to the substrate only where desired in the final pattern. In general, this is advantageous to subtractive patterning techniques such as etching, as the methods tend to require fewer processing steps. In industrial practice, these methods may also result in less waste, reducing disposal and raw material costs.

The first two methods, MicroTransfer Molding (μTM) [23] and Micromolding in Capillaries (MIMIC), [24–26] are each soft lithography techniques that rely on the filling and duplication of the three dimensional topography of the mold followed by the release of the elastomer to transfer the pattern solely to the substrate. The third technique, embossing, relies on the displacement of material by the protruding features of the mold to effect the pattern to the thin film. Embossing can use either molds fabricated from hard materials, as well as elastomers. The resolution of the patterned material is determined by interfacial energies that determine the wetting of the precursor solution with the solid surfaces and the ability of the mold to detach from the patterned material after drying. Kinetic factors such as filling of the mold and gelation time also play a role in determining the fidelity of the final pattern. [11] Each of these techniques produces a negative replica of the pattern in the mold, or a positive replica of the original lithographic master. These techniques can be used to produce features that have variable feature heights unlike traditional photolithography. However, they also suffer some of the same disadvantages and limitations associated with soft lithography, such as the limited ability to register the patterns in hierarchical structures.

A fourth class of techniques based on the principles of casting is presented in this chapter. The problem of pattern registry is overcome by placing the relief pattern for confinement on the substrate to be patterned instead of the disposable mold. Well-established photolithography techniques, or even other soft lithography methods, can be used to pattern the relief features used to cast the thin film, but at the expense of additional processing steps required to remove the relief features at the end.

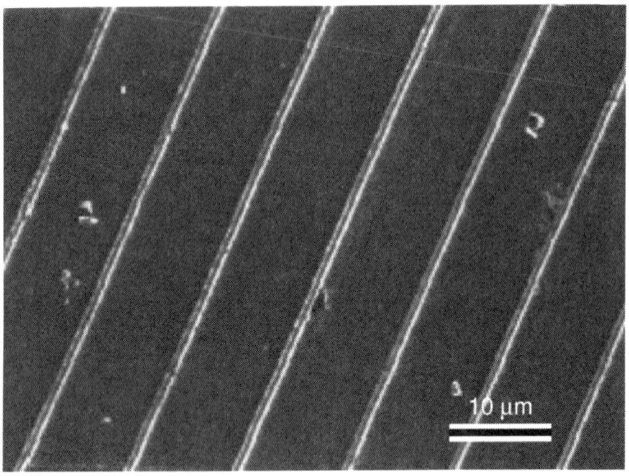


**Fig. 2.** 500 nm wide $ZrO_2$ lines patterned by μTM. [28]

## 2.1. MicroTransfer Molding

In μTM, [23] a prepolymer or chemical precursor solution is deposited into the recessed regions of the patterned side of the stamp, most commonly by spin-casting. The stamp is then brought into contact with the substrate. The lower interfacial energy of the PDMS relative to the substrate allows for the patterned material to remain on the surface as the stamp is peeled away. [11] The material is then ready for heat treatment. The μTM process is illustrated schematically in Figure 1(a). [27] μTM is a very rapid patterning method capable of patterning very large areas simultaneously.

An analysis of the process indicates that for proper pattern transfer poor adhesion of the "ink" with the stamp is required with a wetting angle of greater than 90 °. This allows for the solution to pull away from the walls of the recessed channels when the ink dries during the spin-coating step. Any remaining bonds between the ink and the PDMS must be broken during the stamping step from the imposed mechanical stress. [28,29] These constraints require the careful tuning of solution precursor properties to the interfacial energy of the elastomeric stamp and surface to obtain high quality films with complete pattern transfer to the substrate.

The Whitesides group [23] used commercial spin-on glass and $ZrO_2$ precursors to demonstrate the utility of μTM to pattern ceramic materials. Meanwhile, Moran and Lange [28] patterned monoclinic zirconia lines orientated epitaxially from a zirconyl 2-ethylhexanoate precursor solution with 500 nm line widths, as shown in the scanning electron microscope (SEM) image in Figure 2. Kim *et al.* [27] demonstrated the use of μTM with perovskite metal oxide structures, by patterning 50 μm wide $SrBi_2Ta_2O_9$ (SBT) lines on a (001) $SrTiO_3$ substrate. Using ceramic powders of alumina, boehmite, [30] zirconia, cerium gadolinium oxide, and tin oxide, [29] the Gauckler group used PDMS molds to cast suspensions and sinter the dried patterns to produce line widths as small as 3 μm. However, instead of releasing the patterned film against a substrate, they released the ceramic structures as free-standing microscale objects. Similarly, Su *et al.* [31–32] fabricated free-standing alumina microgears using centrifugal casting of alumina suspensions into PDMS molds.

No one has yet demonstrated the use of μTM in the fabrication of a hierarchical structure. In addition to the pattern alignment issues associated with soft lithography, μTM adds the complexity of trying to align the pattern with the presence of a drying, time-sensitive material on the surface of the mold. This operational handicap may prevent the ability to produce well-controlled features with the degree of reproducibility required for commercial applications.

## 2.2. Micromolding in Capillaries

MIMIC has received perhaps the most attention of the primary soft lithography molding techniques for use in patterning ceramic materials. [24–26] It is a single step process that has been used successfully to pattern a wide variety of materials, including photo- and thermally-curable prepolymers, ceramics, metals (in conjunction with electroless deposition), colloidal particles, and biologically functional macromolecules. [11, 26] The PDMS mold is placed with the patterned surface down on the substrate to form a network of microcapillary channels. The elastomer must be cut beforehand to expose the entrance of the channels for the liquid to enter. A pool of solution is placed at the entrance of the channels. Capillary action spontaneously draws the liquid into the channels to fill them. The permeability of the PDMS to water and organic vapors permits the use of MIMIC in applications with sol-gels and colloidal suspensions that need to dry and cure. [11] After gelation and drying, the mold is released from the patterned substrate. This procedure is illustrated in Figure 1(b).

This is a process driven by the thermodynamics of minimizing the total interfacial energy of the system as the solid-vapor and solid-liquid interfacial areas change. To continue to fill the channels, the capillary pressure at the meniscus must be greater than the hydrodynamic pressure drop along the already-filled distance of the channel. As a result, the rate of capillary filling is inversely proportional to the distance that has already been filled, as expressed by

$$\frac{dz}{dt} = \frac{R\gamma_{LV}\cos\theta}{4\eta z} = \frac{R(\gamma_{SV} - \gamma_{SL})}{4\eta z}, \tag{2.1}$$

where R is the hydraulic radius of the channel, z is the penetration distance of the fluid in the channel, $\eta$ is the viscosity of the solution, $\theta$ is the contact angle of the liquid on the PDMS surface, and $\gamma_{SV}$, $\gamma_{SL}$, and $\gamma_{LV}$ correspond to the PDMS-vapor, PDMS-liquid, and liquid-vapor interfacial free energies, respectively. [24, 33] Typical capillary filling distances are on the order of 1 cm. [11] Because capillary pressures for channels of radius from 1 to 10 μm are on the same order as atmospheric pressure, the filling distances can be enhanced by the use of a vacuum pump at one end of the channels. [34] The rate of capillary filling is proportional to the hydraulic radius of the channel, indicating the higher capillary pressures of smaller capillaries is more than offset by the increased resistance from the channel surfaces. Further, it is important to control the viscosity of the filling fluid because the rate is inversely proportional to the viscosity of the precursor solution. The sol-gels that we use have viscosities on the order of 0.01 Pa·s.

While the size and shape of the features to be patterned are very flexible, it is also important to note that the method requires a network of connected channels. The

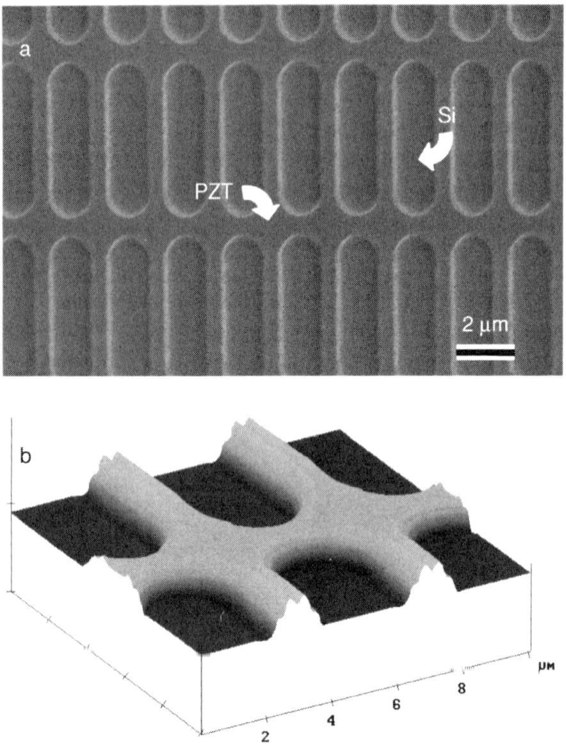


**Fig. 3.** a) SEM image of a grid pattern of PZT lines patterned from 1 μm wide channels. b) AFM topography scan of 2 μm wide PZT lines that shows the "double-peak" topographical distortion that forms during drying and sintering.

method cannot be used to pattern an array of isolated dots or features. The procedure requires the removal of excess dried precursor solution that gelled outside the confines of the mold, in this case from the liquid from the original pool that did not infiltrate the channels. Unlike μTM, patterning is not a very rapid process as the filling of the capillary channels as it can take over a day when using a sol-gel.

The most serious limitation of the process when applying the process to colloidal suspensions and sol-gels is that the solution being patterned rarely fills the entire channel in the vertical direction after the volumetric shrinkage that results from the drying of the solvent. In the case of the PZT synthesized by a chelation-modified sol-gel chemistry, [35], [36] the final sintered material comprises only 18% of the original volume. [37] This can result in deviations of the topography from the shape of the original mold, as demonstrated by the atomic force microscopy (AFM) scans of sintered PZT pattern shown in Figure 3(b). These topographical defects are most pronounced when using molds with features such as corners whose radius of curvature is much smaller than the width of the mold. The presence of sharp corners causes accelerated drying and slower dewetting which combine to produce "double-peak" topographies in the patterned thin films. [37]

The application of MIMIC to ceramics was first demonstrated by the Whitesides group using a $SiO_2$ sol-gel precursor. [26] Xia *et al.* have applied the method to the patterning of structural oxide ceramics such as $ZrO_2$ [11, 38] and tin dioxide ($SnO_2$). [38] The first ferroelectric ceramics patterned by MIMIC used PZT derived by the chelation synthesis route by researchers at the University of Washington [39] and the Aksay group. [36, 37] A grid of 1 μm wide PZT lines patterned on a silicon wafer is shown in Figure 3. The University of Washington group also reported the patterning of piezoelectric strontium niobate ($Sr_2Nb_2O_7$). [39] The Aksay group also augmented the capillary forces with electric fields to infiltrate the microcapillary channels with mesoscopic silica system with hexagonally-arranged cylindrical micelles; conveniently, this electric field was found to have the effect of aligning the nanometer-scale tubules in an axial alignment with the applied field. [40, 41] Yang *et al.* [42] used MIMIC to pattern a suspension of 200 nm polystyrene microspheres. The colloids arranged themselves in a FCC lattice within the confines of the channels. After drying with the mold in place, a precursor mesoscopic silica precursor solution was infiltrated into the channels. The polystyrene template was removed by pyrolysis after the sol-gel dried. The result is the simultaneous patterning of cubic mesoporous silica on the three discrete length scales of 10 nm, 100 nm, and 1 μm.

Most recently, researchers at the Chinese Academy of Sciences used MIMIC with sol-gel chemistry to pattern phosphor thin films, [43] perovskite-type organic/inorganic hybrid luminescent films, [44] and silica films doped with organic dyes. [45] Heule and Gauckler [46, 47] patterned $SnO_2$ from a slurry for use in a gas sensor, one of the first demonstrations of incorporating soft lithography with a ceramic material for use in functional device. Microscopy images and a schematic diagram of their device are shown in Figure 4. MIMIC has been used to pattern defect-free regions of line widths as small as 400 to 500 nm with a surface coverage of 1 $cm^2$. [36, 38] The viscosity of most chemical solution precursors limits the ability to fill capillary channels smaller than this length scale over appreciable distances, placing a lower bound on the possible resolution that can be achieved by MIMIC.

## 2.3. Embossing

Embossing is a very active area of research, particularly among those interested in fabricating complicated three-dimensional structures for MEMS. [48] A continuous film layer is deposited onto the substrate and then patterned by forceful application of a mold. It is similar in principle to injection molding except pressure is applied directly to the mold, not the fluid. It has gained tremendous use for patterning polymers, due to the ability to control the rheological properties using temperature. This feature has permitted the ability to pattern polymer thin films on the nanometer length scale, a process commonly referred to as nanoimprinting. [49, 50]

The use of embossing to patterning ceramic films derived from sol-gels or colloidal slurries adds the complexity of not being able precisely control the fluid rheological and wetting properties during the process. Care must be used to select the proper system of substrate, mold, and solution to produce crack-free patterns with the desired fidelity.

When using a liquid precursor such as a sol-gel, PDMS stamps are a common choice for the mold. A film of the precursor solution is deposited on the substrate. The PDMS

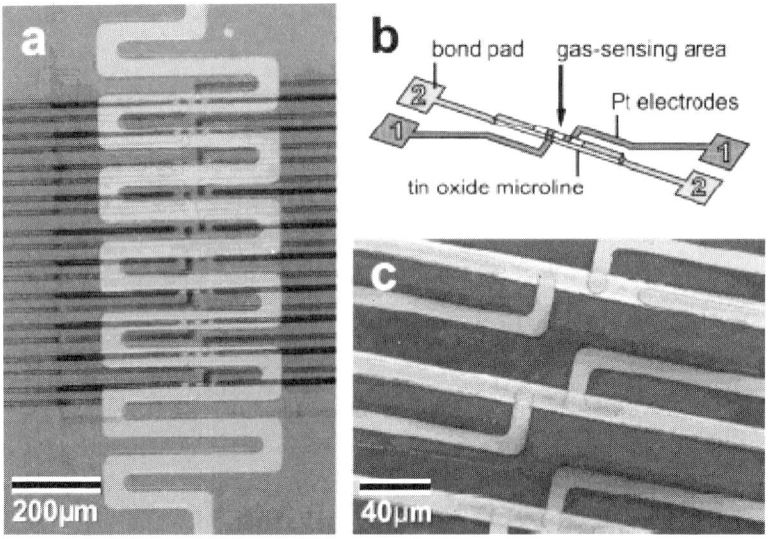


**Fig. 4.** Gas sensor fabricated by Heule and Gauckler [46] using micromolding in capillaries. a) Top view as viewed through an optical microscope in which the dark horizontal lines are the patterned tin oxide features. b) Schematic of the device. c) SEM image of the device showing the platinum contact lines contacting the tin oxide features.

mold is pressed against the substrate with a pressure of approximately $10^5$ Pa. [42,51] For a high quality pattern to develop, it is imperative for the solution to dewet from the regions where no solution is desired to allow the raised features of the mold to contact the substrate. The dewetting is driven by the applied pressure and difference in surface tensions of the solution and elastomer. [51] Frequently, the addition of a polar solvent is used to enhance the dewetting of the film from the substrate, a variation of the technique that is called Solvent-Assisted Micromolding (SAMIM). [11,51] When using PDMS stamps, it is important not to use too much force as this could cause the features of the stamp to buckle or deform. For the same reason, the features on the elastomer should be limited to an aspect ratio between 0.5 and 2. [52]

When the stamp is in contact with the substrate, the solution is allowed to gel and dry. Once cured, the mold can be released from the substrate, and the patterned material is ready for heat treatment for further drying and sintering. The process is illustrated schematically in Figure 1(c). [51] As for µTM and MIMIC, the production of a high resolution pattern in the thin film requires the selection of a stamp that has significantly lower adhesion energy with the material than the substrate has with the material. One limitation of this process is that some dried material may remain on the substrate around the edges of stamp that may need to be removed.

Liquid embossing with PDMS molds was first used by the Whitesides group [51] to selectively deposit silica glass doped with boron oxide, titanium oxide, or aluminum oxide from sol-gel solutions on silicon wafers. These patterned waveguides are shown in the SEM images in Figure 5. Yang *et al.* [42] patterned mesostructured $SiO_2$, $Nb_2O_5$, $TiO_2$, $ZrO_2$, $WO_3$, $AlSiO_{3.5}$, and $SiTiO_4$ with features on the 100 nm length scale.

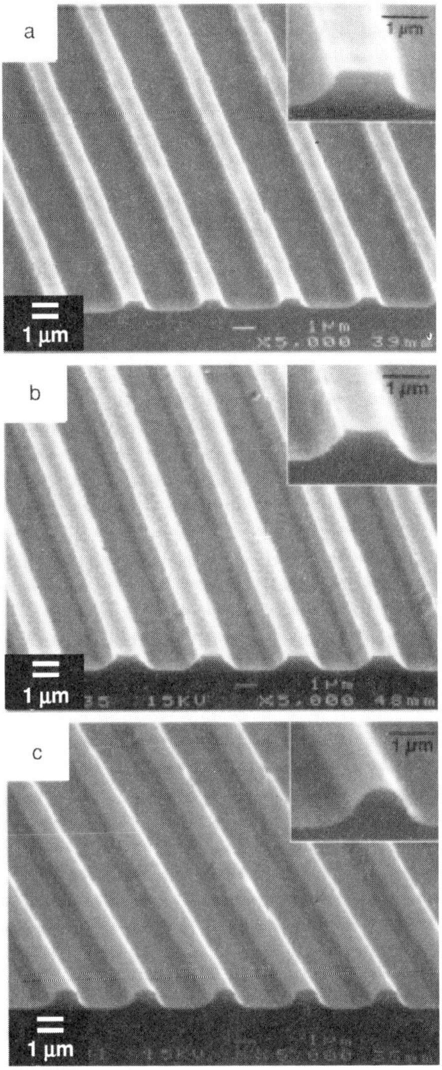


**Fig. 5.** Waveguides patterned by embossing on a silicon substrate by Marzolin *et al.* [51]: a) aluminosilicate (Al/Si = 0.09); borosilicate (B/Si = 0.1) annealed for 10 min at 800 °C (b) and 900 °C (c).

More recently, Harnagea *et al.* [53] patterned 300 nm sized arrays of PZT cells using $SiO_2$-Si molds. Bulthaup *et al.* [48] demonstrated the use of liquid embossing to fabricate inorganic hierarchical structures such as transistors, photodetectors, and resistors with features as small as 120 nm. Commercial spin-on glass, semiconducting CdSe nanocrystals, and Au nanocrystals dispersed in $\alpha$-terpineol were each embossed to fabricate a thin film transistor (TFT) with a field effect mobility of 0.1 $cm^2 \cdot V^{-1} \cdot s^{-1}$ and an on/off ratio of $10^3$. These results compare favorably with all-printed inorganic TFTs.

Due to the difficulty in achieving complete dewetting of the fluid where the relief structures press into the thin film, embossing has found much greater use in applications where complete removal of the fluid is not required. For example, Schueller *et al.* [54] fabricated a two-dimensional photonic crystal structure for lasing applications by curing a 2 μm thick film of organically-modified silicate compressed between the substrate and a PDMS mold with 50 nm high circular protrusions arranged in a hexagonal arrangement. The smallest lateral dimension of the features was approximately 150 nm and the measured optical threshold was consistent with resonators fabricated by conventional photolithography techniques in thermally-grown silicon oxide. Matsuda *et al.* [55] patterned grooves of pitch 1.6 μm and depth 86 nm in gel films of $CH_3SiO_{3/2}$-$SiO_2$ for optical data storage applications.

When using ceramic slurries, harder masks that are less prone to deformation are favored over elastomers due to the requirement of increased pressures. Petronis *et al.* [56] used $SiO_2$ masks with a hexagonally-arranged pattern of 200 μm circles spaced 100 μm to produce titania green bodies by both embossing and injection-molding. After heat treatment, a ceramic scaffold was produced for hepatocyte cell cultures. Su *et al.* [31,57,58] embossed the surface of PZT viscous polymer processed (VPP) tapes with 10 to 150 μm features and 3 to 10 aspect ratios using polymethyl methacrylate (PMMA) molds. The embossing required pressures of $10^7$ Pa and the green body was removed from the mold by chemical dissolution of the PMMA or laser ablation.

While embossing may be a fast and practical method for producing patterned features, the use of significant pressures to impart the pattern may prohibit the use of embossing in certain applications with brittle substrates. The process has not yet been demonstrated in a functional device requiring hierarchical processing steps. Embossing may be limited to the steps at the beginning of the procedure when the substrate surface is relatively planar; the applied pressures used in embossing may potentially distort or alter any features that may have been previously fabricated on a surface.

## 2.4. Casting

Despite the large interest in soft lithography techniques, they have not yet met with wide commercial success. The primary reason is that registry of the patterns produced with soft lithography to fabricate hierarchical structures remains problematic. Some progress has been made by attaching the PDMS mold to a glass slide and aligning the patterns with a light optical microscope. [21] New techniques that combine the principles of soft lithography with traditional photolithography equipment and materials may hold the greatest promise for commercial use. In the following casting techniques, the substrate surface is patterned with the relief features that form the mold for patterning the ceramic precursor solutions.

Schönholzer *et al.* [59] used a casting technique called photoresist molding to pattern alumina and zirconia thin films derived from colloidal suspensions. Nearly defect-free regions of 1 $cm^2$ with features with line widths of five microns and aspect ratios less than one have been produced. Traditional photolithography is used to pattern a polymer photoresist on a flat substrate. The combination of the substrate and patterned photoresist forms the "mold." A ceramic suspension with high solids loading is poured on top of the mold and a polymer blade is swept across the surface to remove the excess liquid. The

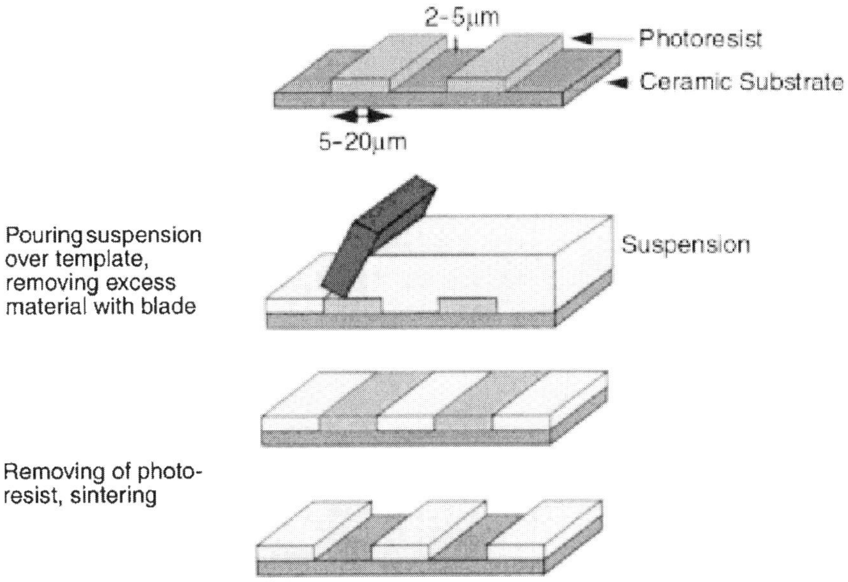


**Fig. 6.** Schematic diagram of photoresist molding developed by Schönholzer *et al.* [59].

process is illustrated in Figure 6. The thin ceramic layer dries nearly instaneously and leaves behind residual supernatant particles on the photoresist regions. These particles can be removed by contacting the filled mold with an adhesive tape, gently pressing into the sample, and then peeling off the tape. The photoresist is removed by dipping the substrate in a beaker of acetone prior to the heat treatment of the patterned ceramic thin film.

This technique has not yet been adapted for use with materials derived from sol-gels or other chemical solution precursors. One reason may be that many sol-gels contain organic solvents that dissolve commercial photoresist. To mitigate this problem, the photoresist mold would have to be replaced with a different polymer. This polymer could potentially be patterned by any of the soft lithography techniques introduced earlier to eliminate the reliance on using photolithography for definition of the pattern. Further, no studies of the patterned material have been performed to evaluate whether the shear stresses imposed during wiping step produce any anisotropic properties in the resulting material.

Using the same idea of preparing a pattern for casting on the substrate surface, we developed a hybrid approach called MicroChannel Molding (μCM) that combines photolithography with MIMIC. [22] The motivation stems from being able to leverage the key advantage of MIMIC, the ability to pattern on submicrometer length scales, while solving the problems of pattern registry and topographical deformation. In μCM, a continuous thin film of sacrificial material is deposited on the substrate surface and patterned using traditional photolithography. A flat, unpatterned PDMS membrane is placed across the top of the sample to form the network of capillary channels for infiltration. The substrate surface forms the bottom, the patterned thin film forms the sidewalls, and the PDMS membrane forms the top of the channel. The pool of liquid precursor is placed at the entrance to channels, allowing capillary forces to draw the fluid into the network. As in MIMIC, this approach is limited to patterns with interconnected capillary channels; isolated features can not be patterned.

The topographical profile of the resulting patterns is much more representative of the shape of the original channels. This is because the permeable PDMS membrane is located only at the top surface of the channel, producing a one dimensional evaporation profile. Without the amplified drying rates at the corners of the noncircular capillary channels, the drying material does not lock in the shape of the corner significantly earlier than the rest of the thin film, eliminating the formation of the double-peaked topographical profile that develop in thin films patterned by MIMIC.

One disadvantage of μCM is that thin film used for "casting" or molding must be compatible with the sol-gel or suspension to be patterned. For example, when patterning our PZT thin films, the isopropanol in the sol-gel partially dissolves most commercial photoresists. Further, the dried PZT gel is removed when using acetone to remove the photoresist so it must be heat treated to convert it to desired crystal structure first. But sintering takes place at 600°C, well above the pyrolysis temperature of most polymers. To resolve this issue, patterned thin films of Cr that have been etched with the commercial etchant Cr-7 (Cyantek Corporation, Fremont, CA) are used to define the channels for μCM to pattern PZT.

## 3. Surface Modification Techniques

The surface modification methods described here require the deposition of a uniform thin film of the material to be patterned across the substrate. These techniques are subtractive because material is removed from the regions where it is not desired in the final pattern. This is analogous to the first step in patterning in traditional semiconductor fabrication, whereby the photosensitive polymer film is spin-cast from a liquid suspension. Similarly, the thin film of ceramic precursors can be deposited by spin-coating or dip-coating from a solution, such as suspension of solid particles or a sol-gel, followed by a heat treatment for densification and crystallization.

The success of photolithography is due to the photosensitive nature of the polymer films used which permits the definition of patterns due to the selective exposure to light. Using the same principles, some researchers [60–69] have chemically added this functionality to certain sol gels and chemical solution precursors. However, this technique is not accessible to all chemistries and does not have much value to ceramic thin films derived from colloidal suspensions. A more general surface modification approach called MicroContact Printing (μCP) using self-assembled monolayers [76,77] can be used instead. The section concludes with a letterpress printing technique that relies on surface tension to confine the liquid precursor solution to the protruding tips of an inked stamp for pattern transfer.

### 3.1. Selective Exposure Techniques

The following paragraphs highlight some of the methods by which a deposited thin film can be patterned. Selective exposure methods use an external energy source to convert selected regions of a deposited thin film to non-soluble forms prior to leaching. These processes are analogous to the patterned deposition of an organic thin film positive photoresist layer in semiconductor manufacturing. [2] First, the photosensitive polymer is spin-cast onto the wafer. Next, the film is exposed to ultraviolet (UV) light through a mask. In a positive photoresist, the regions in the film that are exposed to the ultraviolet radiation are cross-linked and rendered insoluble when the wafer is submerged in a developer solution. In the same manner, UV radiation is used to chemically convert regions of the deposited inorganic precursor thin film to a cross-linked form that is insoluble to a leaching solution that removes the undesired regions. The difference between traditional photolithography and selective exposure is that instead of using the patterned photoresist as a mask to transfer the image to another layer of material by wet chemical etch or lift-off techniques, the pattern is directly defined in the inorganic material of interest.

After exposure of the film to UV light, the unexposed regions are removed by dipping the sample in a solvent bath. The patterned ceramic thin film is then converted to its final desired crystalline form by heat treatment. Similar to the other methods described in this chapter, shrinkage due to the evaporation of residual solvent or organic compounds, as well due to densification, occurs, resulting in changes to the lateral and vertical dimensions of the patterns.

Preliminary work relied on the effect of using the UV light to accelerate particular sol-gel reactions and reducing the carbon content of the films. [60] Shinmou *et al.* [61] aimed to broaden the technique by introducing acetylacetone into the gelled films containing Zr-n-butoxides. When exposed to UV light with a wavelength of 300 nm, the p-p* transition in the chelate ring of the acetylacetone dissociates the chelate bonds in the gel films, reducing the solubility of the films to acidic leaching solutions. After heat treatment, crystalline zirconium oxide ($ZrO_2$) films with feature sizes of 200 μm were produced. Similarly, Yogo *et al.* [62] patterned films by adding 1-phenyl-1,3-butanedione, a photosensitive chemical designed to absorb ultraviolet (UV) light at 330 nm, to their precursor solution to pattern lithium niobate ($LiNbO_3$) films on sapphire substrates. Kikuta *et al.* patterned titania [63] and tin oxide [64] thin films from precursor solutions doped with UV-sensitive diethanolamine or N-phenyldiethanolamine followed by an ethanol wash. The Whitesides group [65] synthesized a glass photoresist and used photolithography to pattern 1.5 μm wide lines, as shown in Figure 7. As in photolithography, the feature sizes than can be achieved are limited by the resolution of the mask, equipment optics, and the wavelength of the light source. Using a focused laser beam operating in the UV range, Nakao *et al.* [66]patterned feature sizes of 0.5 μm.

The primary advantage of this method is that previous experience with photolithography can be used to transfer the desired patterns. Many of the equipment required, such as the UV photomasks, are readily available. Careful control of the irradiation intensity and exposure time are required to achieve the desired resolution; for instance, overexposure can result in roughening of the edge of the patterned lines. [62] A severe disadvantage of the process is that the vertical thickness of the film is limited by the nature of the spin-coating process. The fidelity of the patterns in thicker films may be limited due the need for the sufficient penetration of the energy beam, though Kärkkäinen *et al.* [67,68] have patterned siloxane-based hybrid glasses for microoptics applications with thicknesses over 100 μm.

**Fig. 7.** 1.5 μm $SiO_2$ lines patterned by photoexposure by Marzolin *et al.* [65].

The primary drawback of this technique is that most materials require the alteration of the chemical precursor solution recipe for the addition of compounds sensitive to the wavelength of the UV beam to effect the desired reaction. This greatly limits the choices of materials that can be patterned by this technique. Further, this technique is only applicable to sol-gel based chemistries and can not be used to pattern films derived from colloidal suspensions unless the particles are suspended in a photocurable medium. For instance, Lu and coworkers [69] fabricated 64 nm thick micropatterns with lateral dimensions as small as 4 μm using mercaptoacetic acid-capped CdS nanoparticles.

A greater selection of materials can be patterned by changing the choice of external energy source to lasers, [60–71] ion beams, [72] or electron beams. [73–75] These methods selectively pyrolyze the desired regions in the surface using the external energy supplied by the photons, ions, or electrons, eliminating the need to alter the chemistry of the precursor solution. However, these options are beyond the scope of this chapter because they require substantial capital investment and are not capable of patterning large areas. These techniques rely upon the serial scanning of the beam across the thin film; while this feature may be desirable due to the ability to custom-write patterns into the thin film oxide, the rate of throughput in an actual production process would be prohibitively slow compared to mask-based processes.

### 3.2. MicroContact Printing

While selective exposure techniques are based upon transferring a pattern into a previously deposited uniform thin film, μCP [76, 77] relies on the modification to the substrate surface prior to the deposition of the film to effect the final pattern. In the μCP process shown in Figure 1(d), [10] the PDMS stamp is used to transfer a self-assembled monolayer (SAM) pattern to the substrate. Self-assembly relies on the use of weak interactions to spontaneously organize molecules into a thermodynamically stable structure. SAMs are prepared from ligand molecules of the form $Y(CH_2)_nX$. The anchoring group, Y, is designed to bind to the surface of the substrate, while the head group, X, is tailored to control the surface properties of the monolayer, such as the degree of hydrophilicity. The number of methylene units in the alkyl backbone is used to control the thickness of the SAM. The fidelity of μCP relies on the ability of the molecules to quickly order into a SAM with an autophobicity that prevents the spreading of the monolayer across the surface to non-deposited regions. [11]

The patterned SAM layer is then used as a template for the selective deposition of the metal oxide thin film. A chemical precursor solution is spin-coated on top of the substrate with a patterned SAM with hydrophobic end groups, followed by a heat treatment of approximately 300–400 °C to pyrolyze the organic components and to densify the thin film. The regions of the film deposited on top of the SAM show poor adhesion to the substrate surface and crack while the material deposited on the unmodified substrate surface remains firmly bonded. The crazed material is "lifted-off" by mild mechanical abrasion using a cotton felt dampened with isopropanol. [10, 78] Optical microscope images of the film prior to and after this polishing step are shown in Figure 8. [10, 81] A higher temperature heat treatment is then used to convert the remaining precursor to the desired crystal structure. [10,78] The delamination process determines the fidelity of the transferred pattern and appears to be governed by the roughness of the grain boundaries of the underlying substrate. [79] Most researchers choose to deposit the thin film via spin-casting, although selective dip-coating could also be used on surfaces that have been micropatterned into regions of different hydrophilicity. [80]

Jeon *et al.* [78] first reported the use of μCP to pattern a SAM from octadecyltrichlorosilane (OTS, $CH_3(CH_2)_{17}SiCl_3$) in hexanol as a template for depositing lithium niobate ($LiNbO_3$) and lead lanthanum titanate ($Pb_{1-3x/2}La_xTiO_3$) on sapphire, silicon, and indium tin oxide (ITO) substrates. Later work by the same research groups at the University of Illinois at Urbana-Champaign demonstrated the use of the technique for patterning tantalum oxide ($Ta_2O_5$) on silicon, aluminized silicon, and platinized silicon wafers. [79] They also fabricated ferroelectric memory cells by patterning 200 μm diameter PZT posts on platinized silicon and then using μCP again to pattern platinum on top of the posts. [81] They achieved line widths as small as 4 μm with 100 nm resolution and vertical thickness of 80 to 120 nm with their resulting patterned thin film oxides. [79] Their work is well summarized in a review article. [10]

Hwang *et al.* [82] used chemical bath deposition of CdS particles on Si substrates prepared with OTS patterns by μCP. After ultrasonic agitation to lift off the particles deposited on top of the OTS regions, patterns with boundary resolutions of 100 nm were fabricated. Shin *et al.* [83] used μCP to pattern an arrangement of hydrophobic SAMs and coated the remainder of the surface with hydrophilic SAMs prior to deposition of iron oxide thin films. Bartz *et al.* [84] patterned alkane thiols on a gold surface and then coated the unpatterned regions with a thiol terminated with a styrene monomer. The monomer was polymerized prior to the precipitation of a titania film on the surface. A solution of chloroform and trimethylsilyl chloride was used to cleave the polystyrene, leaving behind the patterned titania thin film on top of the alkane SAM. Yan *et al.* [85] used a variation called Hot MicroContact Printing (HμCP) to transfer a hydrophobic long-chain alkyltrichlorosilane to a less-reactive silicon surface that is heated on a hot plate. Atomic layer epitaxy is used to grow ZnO films on the Si regions not covered by the SAM to produce feature sizes as small as 1 μm. Similarly, Bechiniger *et al.* [86] used the contrast between hydrophobic and hydrophilic regions to pattern an array of 150 nm tungsten oxide dots from an alkoxide precursor solution. Their PDMS stamp

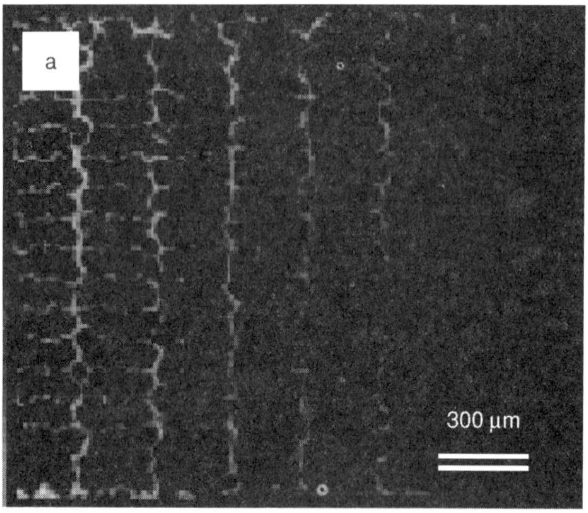


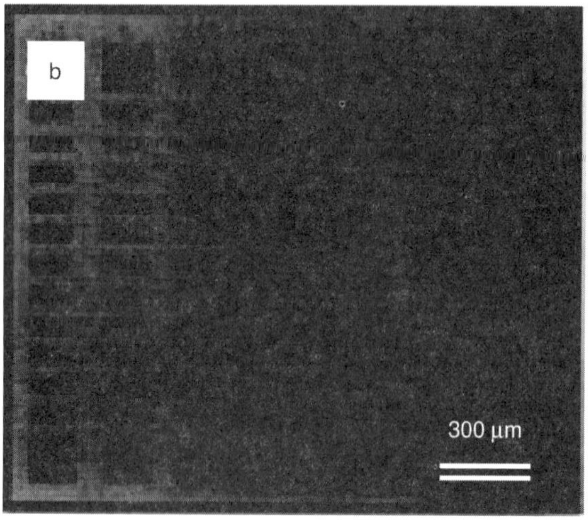


**Fig. 8.** Optical micrographs of a patterned tantalum oxide film a) before and b) after the polishing step. [10, 81]

was cast from a self-assembled, hexagonally-arranged, close-packed monolayer of 600 nm diameter polystyrene colloidal spheres.

Another variation of μCP is to print a pattern of acid on top of a continuous thin film to effect the pattern. The Whitesides group [65] inked their elastomeric stamp with hexadecanesulfonic acid in t-butyl methyl ether and 2-propanol and printed a pattern on top of a spin-cast glass photoresist. After baking the sample at 130 °C for 20 min, the glass was developed in anisole for 20 s. The regions exposed to the acid were cross-linked and rendered insoluble to the developer. This technique was used to pattern lines with edge irregularities of 100 nm.

A number of researchers have produced patterned ceramic thin films, such as $TiO_2$, [87,88] ZnO, [89,90] $SnO_2$, [91] and $SiO_2$ [92] on top of SAMs that were patterned using exposure to UV light through a mask. Typically, the hydrophobic functional group at the end of the chain is replaced by a hydrophilic hydroxyl. The ceramic material deposited from aqueous solutions preferentially adheres to the hydrophilic regions. Pyrolysis of the SAM during heat treatment steps does not lead to cracking or delamination of the ceramic thin film on the substrate. [93] However, the pyrolysis may leave behind impurities embedded into the oxide layer of the substrate that may impact the performance of the hierarchical device. This issue is not completely mitigated by using μCP to define a pattern of SAMs either, because molecules such as OTS may leave behind carbonaceous contamination after the heat treatment steps and render these regions of the surface unavailable for depositing other materials. The only hierarchical structures that have been fabricated on thin films patterned by μCP are devices whose layers are vertically stacked such as the isolated ferroelectric capacitors by Jeon *et al.* [81]

### 3.3. Letterpress Printing

Letterpress printing is a related technique to μCP whereby the final material to be deposited serves as the "ink" and is directly printed by the mold or "printing plate". The process was first demonstrated with an ink of palladium colloids suspended in toluene printed by PDMS stamps by the Whitesides group. [94] Wang and colleagues [95] patterned dots of polymers using PDMS stamps. Due to their reliance on the presence of solvent to aid the transfer of the polymer from the stamp surface to the substrate, they referred to their process as Micro-Fluid-Contact Printing (μFCP). Miller *et al.* [96] patterned more complex patterns using both rigid and flexible stamps by maintaining the temperature of the ink and target substrate above the glass transition temperature of the polymer.

As this technique is relatively new, only one demonstration of the technique to ceramic systems has thus been published. Wang *et al.* [97] applied the same μFCP technique to a composite ink dispersion of $TiO_2$ colloids dispersed in a hydrolyzed poly(styrene-*alt*-maleic anhydride) copolymer (HSMA). Two-dimensional arrays of 6.8 μm diameter dots were patterned. Currently, the technique is limited to producing discrete circular "dishes" whose thickness at the rims is an order of magnitude greater than at the interior. The dynamics of dewetting of the ink from the stamp to the substrate and the evaporation of the carrier solvent need to be better controlled for this to become a viable technique for ceramic materials.

## 4. General Limitations

### 4.1. Materials Limitations

With the exception of letterpress printing, each of the methods presented in this chapter has been successfully demonstrated with chemical solution precursors such as sol-gel. Theoretically, letterpress printing could be used with a sol-gel, but it would require solving a number of engineering challenges. Specifically, precise control of the rheological properties to effect the desired pattern transfer from the stamp to the surface is needed. The soft lithography method of μTM suggests that elastomer stamps can be good carriers for sol-gel because the low interfacial energy allows the release of the material when contacted to a substrate with a more beneficial binding energy. For the technique to be a viable alternative, further study on the pressing conditions with both sol-gels and colloidal suspensions is required to eliminate the major topographical distortions that the technique produces. [97]

Each of the techniques can also be applied to powder-based chemistries, such as colloidal suspensions. Only one instance of patterning of thin films derived from ceramic slurries by selective exposure has been reported [67]. To use selective exposure, the colloidal particles need to be suspended in a photosensitive solvent such as photocurable monomer whose adhesion properties are affected by the application of UV light. The particles themselves can not be directly patterned as they are already comprised of crystalline solid particles whose adhesion properties will not change in response to the external energy source.

## 4.2. Resolution Limits

Most of the confinement-based techniques are capable of resolutions on the sub-micrometer length scale. With the possible exception of MIMIC, there is no reason why these fidelities can not be further improved with the use of finer molds. MIMIC is inherently limited by the requirement of filling the network of capillary channels with relatively viscous chemical solution precursors to produce the pattern. The casting techniques discussed in this chapter have not yet been pushed to their resolution limits. Since μCM relies on the same capillary filling principles as MIMIC, there is no reason why it can not be used to define patterns with the same 400 nm feature sizes. Because the work on letterpress printing is relatively new and unproven with ceramic materials, it is not clear what the defining resolution limit of this method is. Because of the technique's similarity to μTM in terms of using differences in interfacial energy between the stamp and substrate, letterpress printing should be able to pattern features smaller than one micron once better control of the solution chemistry and printing conditions is established.

## 4.3. Cracking

The subtractive patterning techniques of selective exposure and μCP have one key limiting feature that may prohibit their use in many applications. Because both techniques rely on spin-casting continuous thin films before production of the final pattern, the films are limited in their thickness. By the Griffith criterion, thin films will crack if they exceed a critical thickness $t_c$

$$t \geq t_c = \frac{4\omega \cdot E}{Z(1-\nu)\sigma^2} \tag{4.1}$$

where ω, E, and υ are the specific surface energy, elastic modulus, and Poisson's ratio of the thin film material, Z is a dimensionless geometrical factor, and σ is a constant biaxial stress applied to the film. [14,98] For many sol-gels, this critical thickness is on the order of 0.1 μm. The fabrication of thicker films requires the repeated spin-casting and heat treatment of multiple layers; the requirement of heat treatment to densify each

**Table 1.** Summary of patterning techniques for ceramic thin films

| Technique | Sol-Gel | Colloidal Dispersions | Resolution |
|---|---|---|---|
| μTM | 19, 27–28 | 29–32 | 500 nm (28) |
| MIMIC | 26, 36–40, 42–45 | 46–47 | 400 nm (38) |
| Embossing | 42, 48, 51, 53–55 | 29, 48, 56–58 | 120 nm (48) |
| Casting | 22 | 59 | 5 μm (59) |
| Selective Exposure | 61–68 | 69 | 1.5 μm (65) |
| μCP | 10, 65, 78–79, 81, 83–86 | 82 | 4 μm (79) |
| Letterpress Printing | Not yet demonstrated | 97 | 6.8 μm (97) |

thin film layer may prohibit the application of the subtractive patterning methods to thicker films.

The issue of cracking is not mitigated by the use of confinement-based techniques. Most of the techniques require a post-patterning heat treatment to densify the thin film and convert the material to the desired crystalline phase. The issue of shrinkage and cracking is still present, but the patterned material is able to release stress in both a lateral direction and the vertical direction. These methods allow the patterning films thicker than permitted in large area continuous thin films as long as the lateral dimension is sufficiently small for the release of sufficient strain energy. For example, Beh *et al.* [38] reported patterning $ZrO_2$ lines as wide as 14 μm from molds with 2 μm high channels without cracking, while we have patterned crack-free PZT lines over 20 μm wide from 1.85 μm high channels using MIMIC. [37]

## 5. Conclusion

A summary of the techniques reviewed in this chapter is presented in Table 1 with a list of the smallest reported resolution achieved to date. Ultimately, the success of each of the techniques described in this chapter will be measured by whether or not it is incorporated into the fabrication scheme for a working device. To date, these techniques have been used for very few applications. Perhaps the most note-worthy example of the incorporation of one of these methods in microfabrication is the use of MIMIC to pattern tin oxide in a gas sensor by Heule and Gauckler. [46] The gas sensor was demonstrated to be able to detect hydrogen and carbon monoxide for concentrations on the order of 100 ppm. The electrode and insulating layers for the device were patterned by traditional photolithography, an indication of how much more convenient and reproducible pattern registry is than by soft lithography. This issue may very well prove to be the Achilles' heel for many of these submicrometer patterning methods as researchers attempt to incorporate them into hierarchical fabrication schemes.

## 6. Acknowledgements

The main portion of the authors' work is supported by NASA grant NAG-2-1475. Support by the NASA University Research, Engineering, and Technology Institute on BioInspired Materials (BIMat) under award NCC-1-02037 is also acknowledged.

## References

[1] C.R. Martin and I.A. Aksay, *J. Electroceramics* **12**, 53 (2004).
[2] S.A. Campbell, *The Science and Engineering of Microelectronic Fabrication* (Oxford University Press, New York, 1996).
[3] T. Ito and S. Okazaki, *Nature* **406**, 1027 (2000).
[4] C. Hageleitner, A. Hierlemann, D. Lange, A. Kummer, N. Kerness, O. Brand, and H. Baltes, *Nature* **414**, 293 (2001).
[5] D.L. Polla and L.F. Francis, *Annu. Rev. Mater. Sci.* **28**, 563 (1998).
[6] S. Ono and S. Hirano, *J. Am. Ceram. Soc.* **80**, 2533 (1997).
[7] N. Ozawa and T. Yao, *Solid State Ionics* **151**, 79 (2002).
[8] K. Nashimoto, K. Haga, M. Watanabe, S. Nakamura, and E. Osakabe, *Appl. Phys. Lett.* **75**, 1054 (1999).
[9] J.F. Scott and C.A. Paz de Araujo, *Science* **246**, 1400 (1989).
[10] D.A. Payne and P.G. Clem, *J. Electroceram.* **3**, 163 (1999).
[11] Y. Xia and G.M. Whitesides, *Annu. Rev. Mater. Sci.* **28**, 153 (1998).
[12] A. Roelofs, T. Schneller, K. Szot, and R. Waser, *Appl. Phys. Lett.* **81**, 5231 (2002).
[13] S. Bühlmann, B. Dwir, J. Baborowski, and P. Muralt, *Appl. Phys. Lett* **80**, 3195 (2002).
[14] F.F. Lange, *Science* **273**, 903 (1996).
[15] R.W. Schwartz, *Chem. Mater.* **9**, 2325 (1997).
[16] P.F. Blazdell, J.R.G. Evans, M.J. Edirisinghe, P. Shaw, and M.J. Binstead, *J. Mater. Sci. Lett.* **14**, 1562 (1995).
[17] M. Mott, J.-H. Song, and J.R.G. Evans, *J. Am. Ceram. Soc.* **82**, 1653 (1999).
[18] P. Calvert, *Chem. Mater.* **13**, 3299 (2001).
[19] X.-M. Zhao, Y. Xia, and G.M. Whitesides, *J. Mater. Chem.* **7**, 1069 (1997).
[20] T.W. Odom, J.C. Love, D.B. Wolfe, K.E. Paul, and G.M. Whitesides, *Langmuir* **18**, 5314 (2002).
[21] A. Folch and M.A. Schmidt, *J. Microelectromech. S.* **8**, 85 (1999).
[22] C.R. Martin and I.A. Aksay, submitted.
[23] X.M. Zhao, Y.N. Xia, and G.M. Whitesides, *Adv. Mater.* **8**, 837 (1996).
[24] E. Kim, Y. Xia, and G.M. Whitesides, *Nature* **376**, 581 (1995).
[25] Y. Xia, E. Kim, and G.M. Whitesides, *Chem. Mater.* **8**, 1558 (1996).
[26] E. Kim, Y. Xia, and G.M. Whitesides, *J. Am. Chem. Soc.* **118**, 5722 (1996).
[27] J.H. Kim, F.F. Lange, and C.-I. Choen, *J. Mater. Res.* **14**, 1194 (1999).
[28] P.M. Moran and F.F. Lange, *Appl. Phys. Lett.* **74**, 1332 (1999).
[29] U.P. Schönholzer, N. Stutzmann, T.A. Tervoort, P. Smith, and L.J. Gauckler, *J. Am. Ceram. Soc.* **85**, 1885 (2002).
[30] U.P. Schönholzer and L.J. Gauckler, *Adv. Mater.* **11**, 630 (1999).
[31] B. Su, D. Zhang, and T.W. Button, *J. Mater. Sci.* **37**, 3123 (2002).
[32] D. Zhang, B. Su, and T.W. Button, *Adv. Eng. Mater.* **5**, 924 (2003).
[33] E. Kim and G.M. Whitesides, *J. Phys. Chem. B* **101**, 855 (1997).
[34] N.L. Jeon, I.S. Choi, B. Xu, and G.M. Whitesides, *Adv. Mater.* **11**, 946 (1999).
[35] G. Yi, Z. Wu, and M. Sayer, *J. Appl. Phys.* **64**, 2717 (1988).
[36] J.S. Vartuli, M. Özenbaş, C.-M. Chun, M. Trau, and I.A. Aksay, *J. Mater. Res.*, **18**, 1259 (2003).
[37] C.R. Martin and I.A. Aksay, *J. Phys. Chem. B.*, **107**, 4261 (2003).
[38] W.S. Beh, Y. Xia, and D. Qin, *J. Mater. Res.* **14**, 3995 (1999).
[39] S. Seraji, Y. Wu, N.E. Jewell-Larson, M.J. Forbess, S.J. Limmer, T.P. Chou, and G. Cao, *Adv. Mater.* **12**, 1421 (2000).

[40] M. Trau, N. Yao, E. Kim, Y. Xia, G.M. Whitesides, and I.A. Aksay, *Nature* **390**, 674 (1997).
[41] A.Y. Ku, doctoral dissertation, Princeton University, 2004.
[42] P. Yang, T. Deng, D. Zhao, P. Feng, D. Pine, B.F. Chmelka, G.M. Whitesides, and G.D. Stucky, *Science* **282**, 2244 (1998).
[43] M. Yu, J. Lin, Z. Wang, J. Fu, S. Wang, H.J. Zhang, and Y.C. Han, *Chem. Mater.* **14**, 2224 (2002); M. Yu, J. Lin, and Y.C. Han, *Chem. Phys. Lett.* **371**, 178 (2003); M. Yu, J. Lin, J. Fu, H.J. Zhang, and Y.C. Han, *J. Mater. Chem.* **13**, 1413 (2003); M.L. Pang, J. Lin, Z.Y. Cheng, J. Fu, R.B. Xing, and S.B. Wang, *Mater. Sci. Eng.* **B100**, 124 (2003); X.M. Han, J. Lin, R.B. Xing, J. Fu, S.B. Wang, and Y.C. Han, *J. Phys.: Condens. Matter* **15**, 2115 (2003); X.M. Han, J. Lin, J. Fu, R.B. Xing, M. Yu, Y.H. Zhou, M.L. Pang, *Solid State Sci.* **6**, 349 (2004).
[44] Z.Y. Cheng, Z. Wang, R.B. Xing, Y.C. Han, and J. Lin, *Chem. Phys. Lett.* **376**, 481 (2003).
[45] X.M. Han, J. Lin, R.B. Xing, J. Fu, and S.B. Wang, *Mater. Lett.* **57**, 1355 (2003).
[46] M. Heule and L.J. Gauckler, *Adv. Mater.* **13**, 1790 (2001).
[47] M. Heule and L.J. Gauckler, *Sens. Actuators B* **93**, 100 (2003).
[48] C.A. Bulthaup, E.J. Wilhelm, B.N. Hubert, B.A. Ridley, and J.M. Jacobson, *Appl. Phys. Lett.* **79**, 1525 (2001).
[49] S.Y. Chou, P.R. Krauss, and P.J. Renstrom, *Science* **272**, 85 (1996).
[50] S.Y. Chou, P.R. Krauss, W. Zhang, L. Guo, and L. Zhuang, *J. Vac. Sci. Technol. B* **15**, 2897 (1997).
[51] C. Marzolin, S.P. Smith, M. Prentiss, and G.M. Whitesides, *Adv. Mater.* **10**, 571 (1998).
[52] E. Delamarche, H. Schmid, H.A. Biebuyck, and B. Michel, *Adv. Mater.* **9**, 741 (1997).
[53] C. Harnagea, M. Alexe, J. Schilling, J. Choi, R.B. Wehrspohn, D. Hesse, and U. Gösele, *Appl. Phys. Lett.* **83**, 1827 (2003).
[54] O.J.A. Schueller, G.M. Whitesides, J.A. Rogers, M. Meier, and A. Dodabalapur, *Appl. Optics* **38**, 5799 (1999).
[55] A. Matsuda, Y. Matsuno, M. Tatsumisago, and T. Minami, *J. Am. Ceram. Soc.* **81**, 2849 (1998).
[56] S. Petronis, K.-L. Eckert, J. Gold, and E. Wintermantel, *J. Mater. Sci.-Mater. Med.* **12**, 523 (2001).
[57] B. Su, T.W. Button, A. Schneider, L. Singleton, and P. Prewett, *Microsyst. Technol.* **8**, 359 (2002).
[58] A. Schneider, B. Su, T.W. Button, L. Singleton, O. Wilhelmi, S.E. Huq, P.D. Prewett, and R.A. Lawes, *Microsyst. Technol.* **8**, 88 (2002).
[59] U.P. Schönholzer, R. Hummel, and L.J. Gauckler, *Adv. Mater.* **12**, 1261 (2000).
[60] T. Ohishi, S. Maekawa, and A. Katoh, *J. Non-Cryst. Solids* **147/148**, 493 (1992).
[61] K. Shinmou, N. Tohge, and T. Minami, *Jpn. J. Appl. Phys.* **33**, L1181 (1994).
[62] T. Yogo, Y. Takeichi, K. Kikuta, and S.-I. Hirano, *J. Am. Ceram. Soc.* **78**, 1649 (1995).
[63] K. Kikuta, K. Takagi, and S. Hirano, *J. Am. Ceram. Soc.* **82**, 1569 (1999).
[64] K. Kikuta, K. Suzumori, K. Takagi, and S. Hirano, *J. Am. Ceram. Soc.* **82**, 2263 (1999).
[65] C. Marzolin, A. Terfort, J. Tien, and G.M. Whitesides, *Thin Solid Films* **315**, 9 (1998).
[66] Y. Nakao, T. Nakamura, K. Hoshiba, K. Sameshima, A. Kamisawa, K. Abe, N. Soyama, and K. Ogi, *Jpn. J. Appl. Phys.* **32**, 4141 (1993).
[67] A.H. Kärkkäinen, J.T. Rantala, A. Maaninen, G.E. Jabbour, and M.R. Descour, *Adv. Mater.* **14**, 535 (2002).
[68] A.H. Kärkkäinen, J.T. Rantala, J.M. Tamkin, and M.R. Descour, *J. Lightwave Technol.* **21**, 614 (2003).
[69] C. Lu, N. Wu, F. Wei, X. Zhao, X. Jiao, J. Xu, C. Luo, and W. Cao, *Adv. Funct. Mater.* **13**, 548 (2003).

[70] J.V. Mantese, A.B. Catalan, A.M. Mance, A.H. Hamdi, A.L. Micheli, J.A. Sell, and M.S. Meyer, *Appl. Phys. Lett.* **53**, 1335 (1988).
[71] T. Toshiyuki, N. Ichinose, S. Kawanishi, M. Nishii, T. Sasuga, I. Hashida, and K. Mizuno, *Chem. Mater.* **9**, 2674 (1997).
[72] L.S. Hung and L.R. Zheng, *Appl. Phys. Lett.* **60**, 2210 (1992).
[73] S. Okamura, Y. Yagi, S. Ando, T. Tsukamoto, and K. Mori, *Jpn. J. Appl. Phys.* **33**, 6579 (1996).
[74] A. Kakimi, S. Okamura, Y. Yagi, K. Mori, and T. Tsukamoto, *Jpn. J. Appl. Phys.* **33**, 5301 (1994).
[75] K. Mori and S. Okamura, *Jpn. J. Appl. Phys.* **31**, L1143 (1992).
[76] A. Kumar and G.M. Whitesides, *Appl. Phys. Lett.* **63**, 2002 (1993).
[77] A. Kumar, H.A. Biebuyck, and G.M. Whitesides, *Langmuir* **10**, 1498 (1994).
[78] N.L. Jeon, P.G. Clem, R.G. Nuzzo, and D.A. Payne, *J. Mater. Res.* **10**, 2996 (1995).
[79] P.G. Clem, N.-L. Jeon, R.G. Nuzzo, and D.A. Payne, *J. Am. Ceram. Soc.* **80**, 2821 (1997).
[80] A.A. Darhuber, S.M. Troian, J.M. Davis, S.M. Miller, S. Wagner, *J. Appl. Phys.* **88**, 5119 (2000).
[81] N.L. Jeon, P. Clem, D.Y. Jung, W. Lin, G.S. Girolami, D.A. Payne, and R.G. Nuzzo, *Adv. Mater.* **9**, 891 (1997).
[82] Y.K. Hwang, S.Y. Woo, J.H. Lee, D.-Y. Jung, and Y.-U. Kwon, *Chem. Mater.* **12**, 2059 (2000).
[83] H. Shin, J.U. Jeon, Y.E. Pak, H. Im, and E.S. Kim, *J. Mater. Res.* **16**, 564 (2001).
[84] M. Bartz, A. Terfort, W. Knoll, and W. Tremel, *Chem. Eur. J.* **6**, 4149 (2000).
[85] M. Yan, Y. Koide, J.R. Babcock, P.R. Markworth, J.A. Belot, T.J. Marks, and R.P.H. Chang, *Appl. Phys. Lett.* **79**, 1709 (2001).
[86] C. Bechinger, H. Muffler, C. Schäfle, O. Sundberg, and P. Leiderer, *Thin Solid Films* **366**, 135 (2000).
[87] R.J. Collins, H. Shin, M.R. DeGuire, A.H. Heuer, and C.N. Sukenik, *Appl. Phys. Lett.* **69**, 860 (1996).
[88] K. Koumoto, S. Seo, T. Sugiyama, W.S. Seo, and W.J. Dressick, *Chem. Mater.* **11**, 2305 (1999).
[89] N. Saito, H. Haneda, W.-S. Seo, and K. Koumoto, *Langmuir* **17**, 1461 (2001).
[90] N. Saito, H. Haneda, T. Sekiguchi, N. Ohashi, I. Sakaguchi, and K. Koumoto, *Adv. Mater.* **14**, 418 (2002).
[91] N. Shirahata, Y. Masuda, T. Yonezawa, and K. Koumoto, *Langmuir* **18**, 10379 (2002).
[92] Y. Masuda, W.S. Seo, K. Koumoto, *Jpn. J. Appl. Phys.* **39**, 4596 (2000).
[93] H. Shin, Y. Wang, U. Sampathkumaran, M.R. DeGuire, A.H. Heuer, and C.N. Sukenik, *J. Mater. Res.* **14**, 2116 (1999).
[94] P.C. Hidber, W. Helbig, E. Kim, and G.M. Whitesides, *Langmuir* **12**, 1375 (1996).
[95] M. Wang, H.-G. Braun, T. Kratzmuller, and E. Meyer, *Adv. Mater.* **13**, 1312 (2001).
[96] S.M. Miller, S.M. Troian, and S. Wagner, *J. Vac. Sci. Technol. B* **20**, 2320 (2002).
[97] M. Wang, H.-G. Braun, and E. Meyer, *Chem. Mater.* **14**, 4812 (2002).
[98] J.W. Hutchinson and Z. Suo, *Adv. Appl. Mech.* **29**, 63 (1992).

# Index

Accelerometer 57–58, 201, 327, 328, 353
Acoustic resonator 200
Acoustic sensor 37–48, 200, 350–352
Acoustic coupler 51
Actuator (micro-) 12–15, 19–35, 38, 46, 49, 55–57, 73–74, 93, 134, 159, 162, 163, 164, 165, 169 170, 193, 199, 203, 204, 206, 353, 372
Adaptable system 133, 134, 152, 153
Aerosol deposition 26
AlN 13–17, 115–131, 199, 202, 205–206, 325–360
Angular rate sensor 28
Antenna 149–153, 235, 240–241

$BaTiO_3$ 98, 172 , 272, 290, 293, 372, 374, 376, 377
$(Ba,Sr)TiO_3$ (BST) 39, 138, 139, 140, 141, 142, 148, 217–234, 238, 239, 286, 292–305, 311–317
BAW 115–131, 200
$Bi_4Ti_3O_{12}$ 365
Bolometer 82, 84, 105, 107–108, 165–167
Bulk acoustic resonator 115–131, 200

Calorimetric sensor 10
$Cd_2Nb_2O_7$ 299–300
Chemoresistive sensor 8–10
Chemical sensor 3–18, 62
Clamped plate structure 24, 40–42
cMUT 39, 347
Commutation quality factor 239, 242, 244, 245, 266–268, 292, 298, 299, 300, 313, 317–318
Composite 98–100, 178, 185–187, 191, 193, 203, 236, 237, 250–252, 263–268, 273–274, 294–299, 316–318
Cracking 108, 405, 406–407
CSD film 13, 21, 25, 64, 93, 96, 98, 99, 101, 106, 107, 178, 185, 191, 192, 193, 207, 209, 313, 325, 328, 329, 355, 356, 363, 368, 369, 370, 371, 372, 374
CVD film 9, 32, 93, 104, 171, 172, 178, 217, 218, 222, 223, 232, 363

Defects 87, 98, 139, 140, 237, 245, 248–250, 254, 262, 263, 268, 291, 298, 308, 312, 317, 364, 394, 395
Delay line 133, 134, 135, 136, 138, 149–153
Depletion 137, 262, 263, 274–277, 304, 308, 318, 367
Dewetting 394, 396, 398, 405
Dielectric constant 81, 87, 88, 94, 96, 97, 98, 99, 109, 136–142, 150, 185, 205, 235–324
  effect of bias field on 42, 146
  effect of sintering temperature on 54, 190
  high frequency measurement technique of 280–289
Dielectric loss 81, 87, 88, 89, 93, 109, 125, 126, 133–157, 204, 205, 206, 218, 219, 221, 227, 231, 232, 235–324
  high frequency measurement technique of 280–289
Diffusion barrier 53, 108, 177, 179–180, 217, 230, 231, 233, 326
Direct patterning 184, 185, 186
Direct writing 177, 188–190, 360, 365–368, 382
Disk drive head 22, 26–27
Displacement sensor 75–75
DLP 161–165
DMD 158–163
Doppler effect 38
Dry etching 54, 100, 163–164, 230, 325–360, 365, 368, 369

E-beam direct writing 360, 365–368, 382
Electrode 93, 94, 95, 101, 115, 121, 129, 143, 146, 180, 186–187, 209, 211, 217, 223, 224–227, 269–277, 280–289, 314, 318, 325–360, 363, 369, 377,
  Cr-Ni 104
  Cr-Au 103
  Cu 218, 230–234
  high temperature super-conductor 136, 302, 308
  reference electrode 3, 12, 187
  silver-palladium 52, 67
  gold electrode 140, 150
  used as etching mask 54
  transparent 104
  $IrO_2$ 329, 336, 337, 338, 339–340, 353
  $RuO_2$ 329, 338, 340–342, 353
  Ir 329, 331, 336–340, 353

Electrolytic based sensor 10–12
Electromechanical coupling 37–48, 115, 120, 123, 124, 129, 204, 205, 206, 327
Electrooptic 170, 172
Electrophoretic deposition 177, 185–187
Emission power 44

Ferroelectric material/film 6, 133–156, 217–234, 235–324
Ferroelectric memory 336, 338, 339, 340, 353, 362
Filter 115–132, 133–136, 147–149, 165, 203, 235, 239, 243, 244
Focal plane array 81, 92, 104–107
Focused ion beam (FIB) patterning 361, 362–365, 382, 377

Gas sensor 3–16, 65, 81, 106, 165, 395, 396
Gelcasting 190
Grating light valve 159, 160, 164–165, 169

Infrared detector 81–114, 172, 166–167
Ink jet printing 189, 194
Ink jet 24–27, 38
insertion loss 125, 126–129, 147, 228, 229, 235, 242, 352
Interfacial reaction 139, 140, 146, 177, 178, 209, 217, 219, 224, 226, 227, 231, 232, 332
Interfacial energy 391, 392, 393, 406
IR absorption 103–104
IR detector 81–114, 172, 166–167

$KNbO_3$ 300
$K(Nb,Ta)O_3$ (KTN) 301–302
$KTaO_3$ 138, 247, 249, 258, 260, 261, 290–292, 300

Ladder type filter 117, 119, 127, 128
Lamb-wave 27–28, 44
$La_{1/2}Sr_{1/2}CoO_3$ (LSCO) 363
Leakage 139, 141, 142, 219, 222, 224, 225, 227, 231, 233, 364
Letterpress printing 405
Light modulator 162, 167, 172
$LiNbO_3$ 117, 372, 401, 403
Liquid crystal 134, 135–136
$LiTaO_3$ 92–94, 105, 117
Lithography
  Dip-pen nano-lithography 372–373, 403
  E-beam lithography 354, 372, 402
  micro stereolithography 189–190
  nano-imprint lithography 361, 368–372, 382, 383, 395
  soft lithography 371–372, 387–410
  synchrotron radiation lithography 31

Loss (dielectric) 44, 81, 87, 88, 89, 93, 109, 125, 126, 133–157, 204, 205, 206, 218, 219, 221, 227, 231, 232, 235–324

Magnetic tunable device 133, 134, 141, 147
Magnetostriction 68–70
Magnetostrictive metal 69–79
Membrane structure 24–31, 37–48, 101–103, 119, 166, 169, 171, 203, 325, 327–330, 338–339, 343–356
MOD film 369–370
Morphology 9, 14
Morphotropic phase boundary 183, 199, 206, 207–208, 209, 210, 212, 213
Micro actuator 12–15, 19–35, 38, 46, 49, 55–57, 73–74, 93, 134, 159, 162, 163, 164, 165, 169–170, 193, 193, 199, 203, 204, 206, 353, 372
Micro gears 392
Micro hot-plate 6–8, 354–355
Micro cantilever beam based sensor 12–14, 21–24, 203, 204, 325–332, 342, 347–353, 383
Micro-contact printing 371–372, 390, 400, 403, 405
  micro fluid contact printing 405
Micro flluidic system 24–26, 66–67
Micro manipulation 31
Micro mirror 20–28, 158–160, 161–162, 162–164, 167, 171
Micro mixer 30
Micro molding 371–372, 390, 391, 393, 396
  micro channel molding 391, 399
  micro molding in capillaries 391, 393, 396
  micro transfer molding 391–399
Micro optics 401
Microstructure 103, 121–124, 138, 217, 218, 219, 220, 223–224, 264, 295, 298, 315–317, 326, 374, 375, 391
Micro motor 23, 30–32, 41, 58–60, 201, 328
Micro pen direct writing 189
Micro positioning 27–28, 157, 158, 161, 163, 164, 168, 169, 170, 172
Micro photonic device 157–175, 383
Micro pump 25–26, 29, 46, 53–56, 64, 73–74, 201
Micro valve 26, 74–75
Micromachining 24, 29–30, 40, 100, 102, 104, 106, 107, 163, 171, 325–360, 362–365
Microwave applications 115–131, 133–157, 217–235, 235–245
Microwave scanning microscopy 286–287, 312
MOEMS 157–175
Multi-modal sensor 65–66
MW applications 115–131, 133–157, 217–234, 235–245

$Na_{1/2}$ $K_{1/2}NbO_3$ 145
Nano patterning 361–386, 388, 395

Optical fiber 158, 160, 171
Optical switch 75, 160, 168, 171
Orientation (crystallographical) 94–98, 121–124, 126–128, 138, 199, 202, 205, 206, 207, 208–213, 219, 272, 302, 328, 333, 355, 375, 392

Patterning 40, 52, 191–193, 325–339, 343, 353, 354, 361–386, 387–410
$PbMg_{1/3}Nb_{2/3}O_3$ (PMN) 209, 210, 212, 331, 333, 334
$PbTiO_3$ 92–97, 99, 208, 209, 210, 212, 326, 329, 346, 372, 374, 375
$PbYb_{1/2}Nb_{1/2}O_3$ 208, 209
Permittivity 81, 87, 88, 94, 96, 97, 98, 99, 109, 136–142, 150, 185, 205, 235–324
Phase shifter 133, 134, 136, 149–153, 217, 218, 227–230, 233, 244
Phased array 133, 134, 136, 151, 153, 241
Piezoelectric AFM 21–23, 361, 366, 367, 368, 370, 378–382, 383
Piezoelectric buzzer 55
Piezoelectricity 50–51, 200, 201
Piezoelectric material/film 12–14, 19–36, 37–48, 49–67, 115–131, 177–197, 199–215, 235–359, 361–385
Piezoelectric sensor 5, 12–14, 37–49, 204
Piezoelectric properties
  general 13, 22, 50–51, 97–98, 121
  of thick films 52, 53, 54–55, 58, 190–191, 199–217
  of thin films 21, 40–43, 87, 94, 117, 120, 124, 127, 199–217, 366–367, 374, 375, 378–382, 383, 388
Plasma etching 54, 328, 330–342
PLD films 140, 145, 171, 172, 178, 192, 303, 305, 306, 309–312
PLZT 31, 107
pMUT 37–48, 347–350
Polychromator 160, 165–166, 169
Potentiometric sensor 8–10
Powder blasting 192–195
Power handling 134, 146–147, 164
Presence sensor 38, 46
Pressure sensor 75, 325, 350–352
Prostatic hand 60
Pyroelectric array 81, 90, 96, 101–106, 166
Pyroelectric figure of merit 87–88, 91, 92–93
Pyroelectricity 82–88, 97–98
Pyroelectric material/film 81–113
Pyroelectric sensor 67, 81–113, 166
Pyrolysis 185, 188, 209, 355, 395, 400, 405
PZT 6, 13–17, 20–32, 37–48, 50, 53, 55–67, 92, 93, 98, 99, 101, 103, 105, 106, 107, 163–164, 169, 172, 177–197, 199, 205, 207–212, 315, 325–360, 363–382, 388, 395, 397, 398, 400, 403, 407

Q-factor (acoustic) 38, 44, 125, 126
Q-factor (dielectric) 142, 143, 144, 147, 148, 149, 235, 239, 284, 285–287, 315
Quality factor (commutation-) 239, 242, 244, 245, 266–268, 292, 298, 299, 300, 313, 317–318
Quasi-Debye loss 246, 254, 258–263, 266, 277, 291, 292, 294, 310, 317

Reactive ion etching (RIE) 24, 325, 356, 328, 330, 333–345, 368, 402
Reconfigurable system 133, 134, 152, 153, 235–245
Resonant sensor 5, 12–14, 21–23, 38, 60–64
Resonator 115–133, 135, 138, 147–149, 205, 218, 235, 239, 280, 283–289, 291, 292
RF MEMS 115–131, 133–157, 217–234, 235–324, 347
Robocasting 189

SAW device 12, 64–65, 115, 117, 128, 203, 205, 206
Screen printing 25, 26, 32, 51–54, 177, 180, 184–187, 194
Screen printed film 25, 32, 49–79, 180, 184–187, 190, 191, 243–244
Self assembled mono layers (SAM) 403–405
Self-patterning 9, 373–377
SiC 8, 14
Sintering aid 183–184
Slip sensor 60
$SnO_2$ 6, 7, 9
SOI 325, 326, 328, 329, 344–348, 351, 356
Sol-gel film 13, 21, 25, 55, 64, 93, 96, 98, 99, 101, 106, 107, 207, 209, 313, 325, 328, 329, 355, 356, 363, 368, 369, 370, 371, 372, 374, 387–406
  in composite technology 185
  thick 187–188
Spray coating 25–27, 185
Sputtered film 9, 13, 25, 28, 32, 70, 101, 121, 178, 192, 217, 218, 219, 225, 227, 230, 231, 232, 302, 303, 308, 309, 310, 328, 329, 333, 347, 353, 355,
$SrBi_2Ta_2O_9$ (SBT) 363, 365, 366, 392
$Sr_2Nb_2O_7$ 395
$SrTiO_3$ (STO) 138, 139, 140, 148, 209, 210, 212, 290–294, 300, 302–318, 354, 369, 374, 375, 392

Strain-tunable optical device 169–170
Stress (mechanical) 12–14, 41, 55, 68, 71, 94–98, 108, 180–183, 206, 208, 210, 306, 325, 326, 327, 328, 343, 351, 355, 356, 392, 399, 406, 407
Substrate 81, 93, 94–97, 100, 101, 119, 121, 123, 125, 137, 139, 140, 166, 178–198, 203, 205, 208–212, 219, 225–232, 277–279, 282–285, 301–315, 325–329, 340, 341, 344, 346, 364, 372, 373, 374, 376, 387–406
  Si 71–72, 137, 145
  STO 93, 95
  MgO 93, 95, 106, 139, 172, 279, 306
  SiC 14
  damage 27
  SOI 42, 321, 329
  alumina 51–53, 139, 178, 315
  stainless steel 62
  $LaAlO_3$ 139, 148, 302, 306, 312, 315
  $DyScO_3$ 305
Surface acoustic wave device 12, 64–65, 115, 117, 128, 203, 205, 206
Surface modification technique 403–405

Temperature detector 81, 86
Thermal detector 81–113
Thermal imaging 81–113
Thick film 21, 22, 32, 33, 49–79, 94, 177–197, 243–244, 331, 333, 334
$TiO_2$ 94, 140, 179, 180, 181, 184, 209, 219, 239, 326, 329, 332, 339, 346, 355, 388, 396, 405
Tunability 133–157, 165, 168, 218, 221, 222, 228, 231, 235–324
Tunable ferroelectric 133–157, 217–235, 235–324
Tunable filter 134, 135, 147–149, 235, 239–241

Ultrasonic transducer 37–48, 325, 328, 347, 356
Ultrasonic medical imaging 37–48
Ultrasonic sensor 65
Ultrasonic motor 23, 30–32, 41, 58–60, 201, 328

Varactor 133–157, 228, 229, 235, 239–245, 289, 304, 347
Voltage controlled oscillator 149, 243

Wet etching 53, 54, 57, 64, 68, 73, 100, 101, 192, 194, 326, 328, 332–333, 338, 347, 388, 401
Wetting 391, 392, 395

ZnO 13–17, 22, 28, 115, 119, 120, 121, 128, 199, 202, 205–206, 327–328, 347